ADVANCES IN CHEMICAL PHYSICS

VOLUME XXII

Advances in CHEMICAL PHYSICS

EDITED BY

I. PRIGOGINE

University of Brussels,
Brussels, Belgium

AND

STUART A. RICE

Department of Chemistry
and
The James Franck Institute
The University of Chicago
Chicago, Illinois

VOLUME XXII

WILEY-INTERSCIENCE

A DIVISION OF JOHN WILEY AND SONS, INC.

NEW YORK · LONDON · SYDNEY · TORONTO

Library of Congress Catalog Card Number: 58-9935

ISBN 0-471-69926-8

Printed in the United States of America.

10 9 8 7 6 5 4 3 2 1

INTRODUCTION

In the last decades, chemical physics has attracted an ever-increasing amount of interest. The variety of problems, such as those of chemical kinetics, molecular physics, molecular spectroscopy, transport processes, thermodynamics, the study of the state of matter, and the variety of experimental methods used, makes the great development of this field understandable. But the consequence of this breadth of subject matter has been the scattering of the relevant literature in a great number of publications.

Despite this variety and the implicit difficulty of exactly defining the topic of chemical physics, there are a certain number of basic problems that concern the properties of individual molecules and atoms as well as the behavior of statistical ensembles of molecules and atoms. This new series is devoted to this group of problems which are characteristic of modern chemical physics.

As a consequence of the enormous growth in the amount of information to be transmitted, the original papers, as published in the leading scientific journals, have of necessity been made as short as is compatible with a minimum of scientific clarity. They have, therefore, become increasingly difficult to follow for anyone who is not an expert in this specific field. In order to alleviate this situation, numerous publications have recently appeared which are devoted to review articles and which contain a more or less critical survey of the literature in a specific field.

An alternative way to improve the situation, however, is to ask an expert to write a comprehensive article in which he explains his view on a subject freely and without limitation of space. The emphasis in this case would be on the personal ideas of the author. This is the approach that has been attempted in this new series. We hope that as a consequence of this approach, the series may become especially stimulating for new research.

Finally, we hope that the style of this series will develop into something more personal and less academic than what has become the standard scientific style. Such a hope, however, is not likely to be completely realized until a certain degree of maturity has been attained—a process which normally requires a few years.

At present, we intend to publish one volume a year, and occasionally several volumes, but this schedule may be revised in the future.

In order to proceed to a more effective coverage of the different aspects of chemical physics, it has seemed appropriate to form an editorial board. I want to express to them my thanks for their cooperation.

I. PRIGOGINE

CONTRIBUTORS TO VOLUME XXII

GÉRARD G. EMCH, The Theoretical Physics Institute, Nijmegen, The Netherlands

KARL F. FREED, Department of Chemistry and The James Franck Institute, The University of Chicago, Chicago, Illinois

N. JACOBI, Department of Chemistry, University of Southern California, Los Angeles, California

ELLIOTT W. MONTROLL, Department of Physics and Chemistry, University of Rochester, Rochester, New York

C. S. PARMENTER, Department of Chemistry, Indiana University, Bloomington, Indiana

O. SCHNEPP, Department of Chemistry, University of Southern California, Los Angeles, California

ROGER M. WARTELL, Department of Physics and Chemistry, University of Rochester, Rochester, New York (Present address: Department of Biochemistry, The University of Wisconsin, Madison, Wisconsin)

CONTENTS

ADVANCES IN CHEMICAL PHYSICS

VOLUME XXII

FUNCTIONAL INTEGRALS AND POLYMER STATISTICS

KARL F. FREED*

The Department of Chemistry and The James Franck Institute, The University of Chicago, Chicago, Illinois

CONTENTS

* Alfred P. Sloan Foundation Fellow.

I. INTRODUCTION

Because of free, or almost free, rotation about the single bonds in a polymer molecule, a single long polymer chain can have an enormous number of different conformations.[1–5] These conformations are often referred to as polymer configurations. Given the potentials describing the hindered (or free) rotation about these bonds and/or the steric interactions between nearby bonds, it is a simple matter to express formally the (unnormalized) probability of finding a particular chain configuration. If we consider an ensemble of chains having the same average energy, the statistical weight is, of course, merely the Boltzmann factor $\exp(-\beta V)$. Here $\beta = 1/kT$ and V is the total potential energy evaluated for the chosen molecular configuration. As usual, in order to obtain the normalized probability for a given chain configuration, as well as the thermodynamic properties of the chain molecules, it is necessary to evaluate the partition function, the integral of $\exp(-\beta V)$ over all possible polymer configurations. This partition function can be evaluated in only a few idealized cases. Even in the case of some of the simplest physically acceptable models of polymer configurations, we are quickly faced with highly complicated mathematical problems.

The difficulty of a complete description of the configuration statistics of a single polymer chain arises from the very large number of internal degrees of freedom possessed by a polymer molecule. Thus the description of a single polymer already is a many-body problem. In fact, the mathematical problems encountered in a discussion of the properties of polymers have a number of features in common with many-bodylike problems in other areas of chemistry and physics.

In real experiments, we are almost never able to perform measurements upon a single polymer chain (excluding for the moment polymerized materials in bulk). Thus, even if the polymer concentration in the system is sufficiently low that chains are isolated from one another, all measurable properties are molecular properties which are averaged over very many different configurations. These then also are averages over a large number of polymer molecules, all of which exist in identical environments.* Since the experimental observables relate to averages over many polymer molecules, which may be in many different configurations, we describe these properties by means of statistical mechanics. The Boltzmann factor then corresponds to the use of a canonical ensemble.

As in the case of the statistical mechanics of a fluid, these Boltzmann factors contain more information than is necessary in order to characterize the experimental properties of polymer systems. We therefore focus attention upon reduced distribution functions in order to make contact with the macroscopic observable properties of polymers. In the usual many-body problems encountered in statistical mechanics, the reduced distribution functions are the solutions to coupled sets of integro-differential equations.[6,7] On the other hand, because a polymer is composed of several atoms (or groups of atoms) that are sequentially joined together by chemical bonds, these reduced distributions for polymers will obey difference equations. Therefore, by employing the limit in which a polymer molecule is characterized by a continuous chain, these reduced probability distributions can be made to obey differential, instead of difference, equations. This limit of a continuous chain then enables the use of mathematical analogies between polymers and other many-body systems. The use of this limit naturally leads to the use of the technique of functional integration.

Although the use of functional integral methods is not widespread among polymer chemists and theorists, these techniques have been applied to a number of polymer problems. This review is intended to be an introduction to the use of these methods in the description of polymer systems. There are already some reviews of the techniques of functional integration;[8–12] however, not one of these is within the context of polymer problems. Actually a discussion of polymer problems is a very heuristic way of introducing and motivating the use of functional integrals. In the present review we assume no prior knowledge of, or familiarity with, functional integrals.

Discussions of the properties of polymers are now becoming quite numerous. For some of the background material concerning polymers we can refer to the books by Flory,[1,4] Volkenstein,[2] and Birshtein and Ptitsyn.[3] The new book by Yamakawa[5] critically summarizes all the theoretical descriptions of polymers in dilute solution. We review only a sufficient portion of this material to introduce the notation and concepts

* Ideally, all the polymer chains are chemically equivalent; however, in practice there is usually some molecular weight distribution in any experimental sample.

that are necessary for subsequent developments. Although an effort is made to keep the review self-contained, we refer to these monographs for more of the experimental and theoretical details.

The techniques of functional integration have been applied only to selected problems in the theory of polymer systems. We discuss some of these problems, choosing those that best enable the elucidation of the mathematical methods. The rest of the literature on polymer problems should then be accessible to the reader.

One of the first polymer problems to be treated by these methods was the so-called excluded volume problem,[13,14] which arises from an attempt to account for the fact that two different constituent atoms (or group of atoms) of a polymer chain cannot occupy the same region of space. The simplest approach to the analogous excluded volume problem in the case of simple fluids is van der Waals' theory of fluids. In van der Waals' theory an equation of state is easily written and the thermodynamic properties of the fluid are obtained from it. By contrast the polymer excluded volume problem is of the same order of mathematical difficulty and is very similar in nature to the general many-body problem.[13–15] Yamakawa reviews the large number of theoretical approaches to the excluded volume problem and the resultant controversies over the properties of polymers with excluded volume.[5]

The problems of the description of the observable properties corresponding to realistic models of polymer chains often appear to be insurmountable. They are, however, minute in comparison to those encountered in the characterization of the properties of polymers in bulk, e.g., rubber elasticity, crystallization upon stretching.[1,2,16] The statistical mechanical description of polymers in bulk would be difficult enough if the system could be considered to be just a liquid of the constituent atoms or groups thereof (the monomers). But even this approximation is untenable, since the connectivity of polymer chains and networks is of prime importance in determining the properties of polymers in bulk.[1,2,16]

For assaulting difficult problems of this nature, the following approaches are usually available. The particular choice depends upon the problem at hand and the investigator's interest, needs, etc.

1. Simple kinetic theory type arguments, based upon simple physical models, often provide a physically reasonable result in a quick and efficient manner.

2. The usual approaches to statistical mechanics may be employed in those cases for which the kinetic theory type arguments fail. Also, statistical mechanics is used to provide a molecular basis for the phenomenological coefficients that enter into the successful kinetic theories. Once this

task is accomplished, statistical mechanics is used to extend the range of validity of these kinetic theories and to approach new problems.

3. When these two lines of attack meet with difficulty, there is often the temptation to look to more general, formal, "powerful" techniques with the hope that they will shed some light upon these problems.

It is already quite clear that this review is concerned with the third type of investigation of polymer systems. As noted, this is partly because functional integration is the natural language with which to describe continuous polymer chains. The use of continuous models of polymer chains is acceptable when the physical properties of interest are not dependent upon molecular properties over very short distances, say, a few monomer units. The use of functional integration also enables the comparison of polymer problems with mathematically analogous problems arising in, for instance, quantum mechanics. These analogies are often helpful in devising "new" approximation schemes, although extreme care must be taken not to use these mathematical analogies blindly.[17–19]

The use of very formal mathematical methods is, of course, fraught with a number of perils:

1. We may just be transcribing out ignorance into yet another elegant formulation.
2. Often formal derivations are provided for already well-known results.
3. Finally, nontrivial problems might be solved which were not amenable to simpler techniques.

The result is usually an admixture of all three. However, the usefulness of a theory should be measured by the physical insight it provides as well as by the problems that it is capable of solving to the accuracy that is required.

Functional integral techniques have already been useful in extending our physical pictures in many areas of chemistry and physics. For instance, the Feynman path integral formulation of quantum mechanics presents a very interesting (to some, at least) insight into the wave and interference aspects of quantum mechanical systems.[12] In some cases, also, these functional integral techniques have led to new methods of solution or approximation. In particular, in the case of polymer systems, the techniques of integration in function space have already been employed to discuss the configurational (and dynamical) properties of stiff (wormlike) chains,[20–22] to develop a self-consistent field theory for the excluded volume problem (for a single chain and for solutions at intermediate concentrations),[13–15,23] to describe chain entanglements and elasticity,[22,24–26] and finally to investigate the properties of polymer networks,[27–29] etc.[30–31]

In this review, we concentrate on the static problems involving polymer systems, as the dynamic (e.g., the transport) properties have been less extensively studied.[22] Nevertheless, since an understanding of static properties is usually simpler to obtain than that of the dynamic properties for a given system, a discussion of only the former properties introduces a number of the techniques and concepts that are useful in the treatment of the latter properties. Within the category of static problems, the following qualitatively different cases can be classified:

1. There are a number of mathematically "simple" problems involving single polymer chains which are basically "exactly soluble" by the standard methods.[1–4] By the term "exactly soluble" we imply that a closed form analytical solution is available which may be evaluated, possibly with the use of computers, to any desired numerical accuracy. This category of problems includes the configurational statistics of polymers in "ideal" dilute solutions[1–5] (i.e., at the Θ-point). (See Section II for a brief discussion of the Θ-point.)
2. There are several problems that involve single polymer chains for which an "exact solution" is virtually out of the question. The polymer excluded volume problem is one such case. It is then necessary to search for suitable approximation schemes for these problems.
3. The problems associated with polymers in bulk or in concentrated solutions form a class unto themselves.[1,16] A discussion of such problems immediately raises questions about the validity of ordinary statistical mechanics for systems with a large number of internal constraints.[24–29] We do not consider the interesting topological problems associated with theories of polymer entanglements. In order to pursue any statistical mechanical theory of polymers in bulk, it is imperative that the question of internal constraints be fully understood.

In order to introduce the notation and some of the necessary concepts, as well as to motivate the introduction of the functional integral techniques, first some "exact" results from the configurational statistics of individual polymer chains are introduced. Functional integral techniques are then applied to these "simpler" problems before discussing the more "difficult" problems of polymer excluded volume and the description of polymers in bulk.

In discussing these problems, it is, of course, important and natural to indicate the parallels in fields other than polymer chemistry and physics. Thus it is natural to discuss briefly any connections or similarities with Feynman path integrals,[12] with descriptions of Brownian motion and other random processes,[8–10] and with the functional integral formulation of many body or quantum field theories.[9,10,29,32] Only the most rudimentary notions from these subjects are discussed, so extensive familiarity is not assumed.

There have already been a number of reviews of functional integration and its application. Some, or most, of these reviews are devoted to the "standard" material,[8–12] leaving some uneasiness as to whether anything new can be accomplished. In the present review, this "standard" material is used as an introduction. Reference is given to earlier reviews for some of the mathematical details. Thus this review has a threefold purpose:

1. It reviews recent applications of functional integration methods to problems involving polymers.
2. It provides an introduction to the theories of polymer systems.
3. It reviews functional integral techniques in general.

II. BRIEF SUMMARY OF THE CONFIGURATIONAL STATISTICS OF POLYMER CHAINS

A. Background

The object of any statistical mechanical theory of polymer systems is ultimately to relate the measurable physical properties of the system to the properties of the constituent monomers and their mutual interactions. It is imperative that the initial statistical mechanical theories of these physical properties of polymer systems not depend on the exact details of a particular polymer. Instead, these theories should reflect those generic properties of polymer systems that are a result of the chainlike structures of polymer molecules. Once the properties of simple, yet general, models of polymers are well understood, it is natural to focus attention upon the particular aspects of a polymer of interest. The initial use of simple models of polymers is not solely dictated by an attempt to obtain those general features of polymer systems. The mathematical simplicity of the model is required so that we avoid the use of uncontrollable mathematical approximations which necessarily arise with the use of more complicated models. When the model is sufficiently simple, yet physically nontrivial, we are able to test different approximation schemes to find those that are useful. Presumably these methods of approximation would also be useful for more complicated models. This emphasis upon mathematical simplicity has its analog in the theory of fluids.[6,7] First hard-core interactions can be used to test the physical principles associated with various methods of approximation. Once physically sound approximation schemes have been obtained with this model, they may be applied with more realistic potentials, e.g., the Lennard-Jones potential, which require subsequent numerical approximations.[6,7] Thus we wish to separate approximations of a physical origin from those of purely a numerical nature. This separation is difficult if the most general, and hence mathematically complicated model is initially chosen for the system. In this section, therefore, we discuss the basis for a mathematically simple, yet physically general model of a single polymer chain. This material is already well known, so only the important points are summarized. More details can be found in the books on polymers.[1–5]

Polymers, of course, do not exist in the gas phase. Therefore, if we wish to consider the configurational statistics of single polymer chains, these polymers must be found in very dilute solutions. The system then consists of a polymer chain in an infinite medium of solvent. Since all measurements reflect some average over a large number of polymers and/or a time average for the individual chains, average properties of polymer solutions are evaluated, as usual, by considering an ensemble average over all possible equivalent replicas of the system. (In Section VIII questions of ergodicity for polymer systems are discussed.) The presence of the solvent in addition to the polymer molecule may appear at first to be an enormous complication. However, it turns out that in some instances the solvent leads to essential simplification.[1]

For a single chain in solution, the following different interactions (or forces) must be considered:[1]

1. The polymer-solvent interaction.

2. The osmotic forces, which alone would tend to make the polymer have uniform concentration throughout the medium.

3. Finally, there are polymer-polymer interactions. These interactions are usually divided into two categories:

a. The interactions between neighboring or nearby monomers *along the chain* are usually called short-range interactions. This terminology indicates that these interactions are of short range *along the chain*.

b. There are also interactions between monomers which are far removed from each other along the chain. These are called long-range interactions. The long-range interactions are typically short-range in space. They are of the same nature as intermolecular forces and are negligible when the distance between the interacting particles exceeds some finite distance.

The separation of polymer-polymer interactions into these two types is a division into qualitatively different kinds of interactions. Type *a* interactions involve only a few monomers "at a time" and therefore represent a few-body problem. The configurational statistics of polymers which have only short-range interactions can usually be treated exactly by using, say, the mathematical methods of the one-dimensional Ising model, or, equivalently, of s-fold Markov process for s *small.*[1–5] On the other hand, b interactions can, in principle, involve all the monomers. Thus a single polymer with long-range interactions is already a full many-body problem.

In summary, for a single polymer chain in solution there are polymer-solvent interactions 1 and 2, "simple" polymer-polymer interactions ($3a$), and long-range polymer interactions ($3b$). It turns out that there is a particular temperature, called the Θ temperature, at which the polymer-solvent interactions 1 and 2 exactly cancel the long-range polymer-polymer interactions $3b$. Thus at the Θ-point only the short-range polymer-polymer interactions $3a$ remain. The Θ-point therefore bears some analogy to the Boyle point for real gases. For nonideal gases, it is only at the Boyle point T_B that $PV = nkT_B$, and the gas appears to obey the ideal gas law. The Boyle point occurs because of a net balancing between the contributions of the attractive and repulsive interactions to the free energy. Flory has provided arguments demonstrating the occurrence of the Θ-point, so they need not be repeated here.[1]

At the Θ-point, therefore, the solvent introduces an essential simplification. Very dilute polymer solutions at the Θ-point are called ideal polymer solutions. In these ideal solutions, it is assumed that there is sufficient dilution that mutual interactions and interpenetration of different polymer chains can be neglected. Thus in considering the configurational statistics of ideal polymer solutions, the interactions 1, 2, and $3b$ are ignored; in this class we need to discuss only single polymer chains with only the short-range interactions $3a$.

A basic experimental problem with polymers is that it is not always possible to find different Θ-solvents for a given polymer over a range of temperatures of interest. Thus, in practice, it is often necessary to perform some experiments away from the Θ-point. Ultimately, it is therefore necessary to consider the statistical mechanics of nonideal polymer solutions (i.e., those with all interactions 1, 2, 3). Of course, we always begin with the simpler ideal case.

The study of the configurational statistics of polymers in ideal solutions has already reached a high level of sophistication. This arises in part from the fact that many problems of this kind are soluble in closed form and the resultant expressions are readily evaluated by numerical techniques. The maturity of this field is a tribute to the matrix methods of the one-dimensional Ising model. As this subject is thoroughly presented in the literature,[1–5] it is not necessary to repeat the material here, except insofar as it is convenient to introduce the notation and concepts that are necessary in any discussion of polymer systems. Before proceeding with a brief discussion of the configurational statistics of ideal polymer chains, we should also indicate an additional motivation for considering this subject in a review concerning functional integral techniques in the statistical mechanics of polymer systems.

B. Flexible and Stiff Polymer Chains

Because of the large number of internal-rotational degrees of freedom about the single bonds forming the polymer chain backbone, a polymer may take up virtually an infinite number of configurations, necessitating the use of statistical mechanics. There are numerous models of varying levels of complexity and reality for ideal polymers, i.e., polymers in infinitely dilute ideal solutions. The simplest model, the random flight chain, considers the polymer to be composed of segments which are joined by bonds of fixed length. For the random flight chain the bond angles are completely random variables and there is free rotation about these bonds. In the freely hinged model of ideal polymers the bond lengths are again fixed as are the bond angles; however, there is free rotation about the bonds. More complicated models include hindered rotation about the bonds which form the chain backbone, and possibly they also include the steric hindrances which result in the interdependence of the internal rotational degrees of freedom. After studying the properties of these models, it becomes apparent that they all have certain fundamental features in common.[1–5] Before indicating the similar aspects of these models, it is necessary to digress and mention some of the objectives of the statistical mechanics of ideal polymer solutions.

As noted in the introduction, the thermodynamic properties of a chain can be derived from the chain's partition function. For ideal polymers one of the quantities of great experimental interest is the molecular weight (distribution) of the polymer. The dimensions of the polymer chain must be related to the polymer's weight as well as to the structural nature of the chain. This structural information, in turn, relates to the monomer-monomer interactions. Thus one of the goals of the statistical mechanics of ideal polymers is to relate the short-range polymer-polymer interactions to the dimensions of the polymer chain and its molecular weight. (The dimensions of polymer chains are deduced from light scattering and osmotic pressure measurements.[5] See Section VD and VIG.) Rather than considering the probability distributions for individual chain conformations, it is convenient to consider the probability distribution $P(\mathbf{R})$ for the end-to-end vector of the chain. (Other distributions are introduced later.) This is analogous to the focus upon one- and two-particle reduced distribution functions in ordinary equilibrium statistical mechanics.[6,7]

The foregoing models of ideal polymers represent models which increasingly approximate the expected properties of vinyl polymers, e.g., the simplest polyethylene. Their common features correspond to similarities in the end-to-end vector distribution. For all of these models, a polymer chain of a given total number of segments either has or has not a sufficiently large number of degrees of freedom that the nature of $P(\mathbf{R})$ is determined by the central limit theorem. If the chain has enough degrees of freedom, the central limit theorem requires that $P(\mathbf{R})$ be gaussian. In that case it is solely determined by the mean $\langle \mathbf{R} \rangle \equiv \int d\mathbf{R}\mathbf{R}P(\mathbf{R})$ and the variance $\langle (\mathbf{R} - \langle \mathbf{R} \rangle)^2 \rangle$. In the absence of an external field space is isotropic, and $\langle \mathbf{R} \rangle$ must vanish, so $P(\mathbf{R})$ is determined by only $\langle \mathbf{R}^2 \rangle$. If the central limit theorem determines $P(\mathbf{R})$, the chain is called flexible. If not, the chain is termed stiff, and $P(\mathbf{R})$ is not a simple gaussian, in general. This immediately provides a general characterization of all models of ideal polymers which emphasizes the general aspects of the chainlike structure of polymers: the flexible and stiff polymer chains. From the arguments of the preceding paragraph, it should be clear that $\langle \mathbf{R}^2 \rangle$ contains a great deal of information concerning the structure and molecular weight of ideal flexible polymers.

The techniques of functional integration are naturally introduced into the study of the statistical mechanics of polymer systems because they are ideally suited to the description of simple generic models of flexible and stiff ideal polymer chains. Thus we discuss ideal polymer chains to introduce the simple generic models which then naturally lead to the use of functional integrals.

Having thoroughly discussed the basic approach and philosophy of this section, we proceed to the mathematics. We follow parts of Yamakawa's Chapter II, providing only the basic definitions and results. We refer to the book for details, references, and related materials.[5]

C. Review of Configurational Statistics of Flexible Polymers

The chain is considered to be composed of $n + 1$ elements (the monomers) which are joined successively. They are numbered, $0, 1, 2, \ldots, n$, from one end of the chain to the other. Let the coordinates $\{\mathbf{r}_k\}$ be the positions of the elements, or segments, with respect to some arbitrary origin in space. When there is no external force, space is isotropic, and the choice of the origin of coordinates is arbitrary. We can conveniently take $\mathbf{r}_0 \equiv \mathbf{0}$ as the origin. The potential energy is written as

$$U(\{\mathbf{r}_k\}) = \sum_{j=1}^{n} u_j(\mathbf{r}_{j-1}, \mathbf{r}_j) + W(\{\mathbf{r}_k\}) \tag{2.1}$$

In this equation u_j formally, or explicitly, accounts for the connectivity of the chain. W contains all other interactions. All of the statistical properties of the chain are contained in the distribution

$$G(\{\mathbf{r}_k\}) = \exp\left[-\beta U(\{\mathbf{r}_k\})\right] \tag{2.2}$$

where $\beta = 1/kT$. From (2.2), we can in principle calculate the partition function

$$Z = \int d\{\mathbf{r}_k\} \exp\left[-\beta U(\{\mathbf{r}_k\})\right] \equiv \int d\{\mathbf{r}_k\} G(\{\mathbf{r}_k\}) \tag{2.3}$$

where

$$d\{\mathbf{r}_k\} = \prod_{j=1}^{n} d\mathbf{r}_j \qquad \text{when } \mathbf{r}_0 \equiv \mathbf{0} \tag{2.3a}$$

The distribution function for the entire chain configuration is

$$P(\{\mathbf{r}_k\}) = Z^{-1} G(\{\mathbf{r}_k\}) \tag{2.4}$$

As usual, $P(\{\mathbf{r}_k\})\, d\{\mathbf{r}_k\}$ is the probability that the segments lie between $\{\mathbf{r}_k\}$ and $\{\mathbf{r}_k + d\mathbf{r}_k\}$ (all $k \neq 0$), so $P(\{\mathbf{r}_k\})$ is the probability distribution. It is, however, useful to avoid the repetition of the preceding sentence, and we refer to $P(\{\mathbf{r}_k\})$ as the probability of having $\{\mathbf{r}_k\}$ with the understanding that differentials and intervals are to be taken.

Since the essential nature of the polymer resides in the fact that the segments are sequentially *bonded*, it is often convenient to introduce the bond vectors

$$\mathbf{R}_j = \mathbf{r}_j - \mathbf{r}_{j-1} \tag{2.5}$$

The chain connectivity is then expressed in terms of the bond "probabilities"

$$\tau(\mathbf{R}_j) \equiv \exp\left[-\beta u_j(\mathbf{R}_j)\right] \tag{2.6}$$

where $u_j(\mathbf{R}_j) \equiv u_j(\mathbf{r}_j, \mathbf{r}_{j-1})$. The zero of energy is chosen to make $\tau(\mathbf{R}_j)$ normalized

$$\int d\mathbf{R}_j \tau(\mathbf{R}_j) = 1 \tag{2.7}$$

so that $\tau_j(\mathbf{R}_j)$ is the "probability" that the jth bond has length R_j. $G(\{\mathbf{r}_k\})$ as defined by (2.2) is then

$$G(\{\mathbf{R}_k\}) = \prod_{j=1}^{n} \tau_j(\mathbf{R}_j) \exp\left[-\beta W(\{\mathbf{R}_k\})\right] \tag{2.8}$$

As is usual in statistical mechanics, (2.8) contains more information than is really needed. Experimental quantities are obtainable from reduced distributions, integrals of (2.8) over a subset of the bond vectors or of (2.2) over a subset of the segment positions. Consider the end-to-end distribution function

$$G(\mathbf{R}; n) = \int G(\{\mathbf{R}_k\})\, \delta(\mathbf{r}_n - \mathbf{R})\, d\{\mathbf{R}_k\} \tag{2.9}$$

which, apart from a normalization, is the "probability" that the end segment $\mathbf{r}_n$ is at $\mathbf{R}$. Using (2.5) we have

$$\mathbf{r}_n = \sum_{j=1}^{n} \mathbf{R}_j \tag{2.10}$$

Equation (2.10) expresses the end-to-end vector $\mathbf{r}_n$ as the sum of bond vectors, and

$$G(\mathbf{R}; n) = \int G(\{\mathbf{R}_k\})\, \delta\left(\sum_{j=1}^{n} \mathbf{R}_j - \mathbf{R}\right) d\{\mathbf{R}_k\} \tag{2.11}$$

The partition function is

$$Z = \int d\mathbf{R}\, G(\mathbf{R}; n) \tag{2.12a}$$

and the end-to-end probability distribution is

$$P(\mathbf{R}; n) = Z^{-1} G(\mathbf{R}; n) \tag{2.12b}$$

The introduction of the Dirac delta function in (2.9) and (2.11) merely selects out of all possible chain configurations $\{\mathbf{R}_k\}$ those that have $\mathbf{r}_n = \mathbf{R} = \sum_{j=1}^{n} \mathbf{R}_j$. Note that $P(\mathbf{R}; n)$ is properly normalized such that

$$\int d\mathbf{R}\, P(\mathbf{R}; n) = 1 \tag{2.12c}$$

As will be shown, this use of delta functions is basically Markov's method for random variables.

Since we are interested in discussing generic, rather than specific, properties of polymers, we consider the case for which all the bonds are the same. Then $\tau_j(\mathbf{R}_j) = \tau(\mathbf{R}_j)$ all j, and

$$G(\{\mathbf{R}_k\}) = \exp\left[-\beta W(\{\mathbf{R}_k\})\right] \prod_{j=1}^{n} \tau(\mathbf{R}_j) \tag{2.13}$$

The simplest case of (2.13) is of course when $W \equiv 0$, in which case by (2.7), $Z \equiv 1$ and

$$P(\mathbf{R}; n) = G(\mathbf{R}; n) = \int d\{\mathbf{R}_k\} \left[\prod_{j=1}^{n} \tau(\mathbf{R}_j)\right] \delta\left(\sum_{j=1}^{n} \mathbf{R}_j - \mathbf{R}\right) \tag{2.14}$$

The chain with $W \equiv 0$ is called the random flight chain because the evaluation of (2.14) is formally identical to the calculation of the probability distribution for a random walk of n steps where the length of each step is governed by the probability distribution $\tau(\mathbf{R}_j)$.[1–5,33]

Thus (2.14) connects the study of chain statistics to that of random walks, and hence to Brownian motion and diffusion. Since the diffusion equation is similar to the Schrödinger equation, we should expect to find analogies between polymer chains and quantum-mechanical systems.

Given that bond-bond (i.e., monomer-monomer) interactions are neglected, the simplest model of a real polymer chain is one in which all the bonds have the length l, i.e.,

$$\tau(\mathbf{R}_j) = \frac{1}{4\pi l^2}\,\delta(|\mathbf{R}_j| - l) \tag{2.15}$$

Thus $P(\mathbf{R}; n)$ represents a random walk of n steps, each of length l. Equation (2.14) can be evaluated for a given $\tau(\mathbf{R}_j)$ by Markov's method[5,33] as follows. Introducing the Fourier representation of the delta function

$$\delta\left(\sum_{j=1}^{n}\mathbf{R}_j - \mathbf{R}\right) = \int \frac{d^3k}{(2\pi)^3} \exp\left[i\left(\sum_{j=1}^{n}\mathbf{R}_j - \mathbf{R}\right)\cdot\mathbf{k}\right] \tag{2.16}$$

into (2.14) gives a product of n identical factors:

$$P(\mathbf{R}; n) = \int \frac{d^3k}{(2\pi)^3} \exp(-i\mathbf{k}\cdot\mathbf{R})\left[\int d\mathbf{R}_j \tau(\mathbf{R}_j) \exp(i\mathbf{R}_j\cdot k)\right]^n \tag{2.17}$$

$$\equiv \int \frac{d^3k}{(2\pi)^3} \exp(-i\mathbf{k}\cdot\mathbf{R}) K(\mathbf{k}; n) \tag{2.17a}$$

where

$$K(\mathbf{k}; n) = \int d\mathbf{R} P(\mathbf{R}; n) \exp(i\mathbf{k}\cdot\mathbf{R}) \tag{2.18}$$

is the characteristic function. For the case of (2.15), $K(\mathbf{k}; n)$ can be explicitly evaluated as

$$K(\mathbf{k}; n) = \left[\frac{\sin(kl)}{kl}\right]^n \tag{2.19}$$

where $k = |\mathbf{k}|$. However, $P(\mathbf{R}; n)$ can be obtained from (2.17a) only in the form of an infinite series. As is well known,[33] the moments of $P(\mathbf{R}; n)$ can be obtained directly from $K(\mathbf{k}; n)$, e.g., the pth moment

$$\langle\mathbf{R}^p\rangle = \int d\mathbf{R} P(\mathbf{R}; n)\mathbf{R}^p \tag{2.20}$$

is proportional to the coefficient of $\mathbf{k}^p$ in the power series expansion of $K(\mathbf{k}; n)$ in terms of $\mathbf{k}$. Explicitly,

$$\langle\mathbf{R}^p\rangle = 0 \qquad \text{if } p \text{ is odd}$$

since $K(\mathbf{k}; n)$ depends only on k, and, for instance,

$$\langle\mathbf{R}^2\rangle = nl^2 = Ll \tag{2.21}$$

where $L = nl$ is the contour length of the chain.

For long enough polymer chains, $P(\mathbf{R}; n)$ must be determined solely by the central limit theorem and be gaussian. This is because $\mathbf{R} = \sum_{j=1}^{n} \mathbf{R}_j$ in this limit is just a sum of a large number of independent random variables. When n is large, (2.19) is sharply peaked about $\mathbf{k} = 0$. For other choices of $\tau(\mathbf{R}_j)$, similar behavior is expected. From (2.17) we then have

$$\begin{aligned}\lim_{n\to\text{large}} & P(\mathbf{R}; n) \\ &= P_a(\mathbf{R}) \\ &= \lim_{n\to\text{large}} \int \frac{d^3k}{(2\pi)^3} \exp(-i\mathbf{k}\cdot\mathbf{R}) \\ &\quad \times \exp\left(n \log \left\{\int d\mathbf{R}_j \tau(\mathbf{R}_j)[1 + i\mathbf{k}\cdot\mathbf{R}_j - \tfrac{1}{2}(\mathbf{k}\cdot\mathbf{R}_j)^2 + \cdots]\right\}\right) \end{aligned} \tag{2.22}$$

In (2.22) the power series expansion reflects the expectation that the Fourier transform of $\tau(\mathbf{R}_j)$ will be sharply peaked about $\mathbf{k} = \mathbf{0}$. Since $\tau(\mathbf{R}_j)$ is spherically symmetric, the term linear in $\mathbf{k}$ in (2.22) vanishes. Define l by

$$\tfrac{1}{3}\delta_{\alpha\beta} l^2 = \int d\mathbf{R}_j \tau(\mathbf{R}_j) \mathbf{R}_j^{\alpha} \mathbf{R}_j^{\beta} \qquad (\alpha, \beta = x, y, z)$$

and expand the log in (2.22) to order k^2 to give

$$\begin{aligned} P_a(\mathbf{R}) &= \int \frac{d^3k}{(2\pi)^3} \exp(-i\mathbf{k}\cdot\mathbf{R}) \exp\left(-\frac{nl^2k^2}{6}\right) \\ &= \left(\frac{3}{2\pi l^2 n}\right)^{3/2} \exp\left(-\frac{3\mathbf{R}^2}{2nl^2}\right) \end{aligned} \tag{2.23}$$

It can be shown that (2.23) is a valid asymptotic limit [for the case (2.15)] when $n \gg 1$ (e.g., $n \gtrsim 10$) and $\mathbf{R} \ll nl$. This can be verified by retaining higher terms in the expansion of the log in (2.22) and by approximating $\int d\mathbf{R}_j \tau(\mathbf{R}_j) \mathbf{R}_j^4 \approx \mathcal{O}(l^4)$, etc. Note that (2.23) is very small, but nonzero, for $\mathbf{R} > nl$, whereas for (2.15), (2.17) is identically zero for those lengths that are greater than maximum chain extension.

The detailed derivation is not of great importance for our purposes (see Yamakawa[5]); however, the fact that all random flight polymers give the gaussian distribution (2.23) for long enough chains and for distances much less than full extension (i.e., $n \gg 1$, $\mathbf{R} \ll 1$) is central to the subsequent development. As the particular "physical" choice of constant bond lengths (2.15) leads for general n and $\mathbf{R}$ to a rather complicated distribution $P(\mathbf{R}; n)$, it is tempting to introduce a different $\tau(\mathbf{R}_j)$ from (2.15), which reproduces the properties (2.21) and (2.23), but which is mathematically simpler to handle. When treating cases in which there are polymers containing monomer-monomer interactions in bulk or in nonideal solutions, the simplified form of $P(\mathbf{R}; n)$ can be used to check the general aspects of a theory.

The distribution that satisfies (2.21) and gives (2.23) identically for all n is the chain with gaussian links:[33]

$$\tau(\mathbf{R}_j) = \left(\frac{3}{2\pi l^2}\right)^{3/2} \exp\left(-\frac{3\mathbf{R}_j^2}{2l^2}\right) \tag{2.24}$$

This can be verified easily by substitution into (2.17). [Approximate distributions have also been obtained in closed form which are also valid for $\mathbf{R}$ approaching maximum extension, but since they are much more cumbersome than (2.23), they are not considered further here.]

Now (2.24) "physically" appears to be a poor approximation to the freely hinged chain (2.15) whose bonds are of constant length. However, the freely hinged chain is a poor approximation to many real polymer chains in which bond angles are fixed and there may be barriers to internal rotation and other steric hindrances which result in the interdependence of bond angles. Thus the next level of sophistication in dealing with ideal polymer solutions consists of including the fixed bond angles. After that we should include hindered rotations, and finally steric interactions between *nearby* monomers (*along the chain*). In all of these cases, the end-to-end vector

$$\mathbf{R} = \mathbf{r}_n = \sum_{j=1}^{n} \mathbf{R}_j$$

is a sum of a large number of random variables, $\{\mathbf{R}_k\}$ when n is large. In general, there may be some interdependence between $\mathbf{R}_j$ and say $\mathbf{R}_{j-s}$, $\mathbf{R}_{j-s-1}, \ldots, \mathbf{R}_{j+s}$, *for* $s \ll n$. (We are dealing only with short-range interactions now.) Now either n is large enough so that the probability distribution $P(\mathbf{R}; n)$ is governed by the central limit theorem and is then gaussian, or n is not large enough. Those are the only two possibilities, the flexible and stiff polymers, respectively. In the absence of external fields, the asymptotic distribution for the end-to-end vectors of *all flexible chains* must then be

$$P_a(\mathbf{R}; n) = \left(\frac{3}{2\pi\langle\mathbf{R}^2\rangle_n}\right)^{3/2} \exp\left[-\frac{3\mathbf{R}^2}{2\langle\mathbf{R}^2\rangle_n}\right] \tag{2.25}$$

where $\langle\mathbf{R}^2\rangle_n \equiv \int d\mathbf{R} P_a(\mathbf{R}; n)\mathbf{R}^2$. Now if two flexible chains are attached end-to-end, the resultant chain must also be flexible, i.e., for $n = n_1 + n_2$,

$$P_a(\mathbf{R}; n) = \int d\mathbf{R}' P_a(\mathbf{R} - \mathbf{R}'; n_2) P_a(\mathbf{R}'; n_1) \tag{2.26}$$

This requires that

$$\langle \mathbf{R}^2 \rangle_n = \langle \mathbf{R}^2 \rangle_{n_1} + \langle \mathbf{R}^2 \rangle_{n_2} \tag{2.27}$$

The mean square end-to-end distance must then be an additive function of chain length, i.e.,

$$\frac{\langle \mathbf{R}^2 \rangle_n}{n} = \text{constant} \equiv l^2 \tag{2.28}$$

for a flexible polymer chain. Alternatively, all chains for which (2.25) and (2.28) are not valid are stiff chains.

One important aspect which is not necessarily apparent from this use of the central limit theorem is that in the general case of hindered rotations, or the interdependence of rotations about single bonds, the "step lengths" l are functions of temperature.

There are several alternate ways to express (2.28). If L represents the maximum contour length of the chain (a quantity which is independent of temperature but just determined by geometrical considerations),

$$\langle \mathbf{R}^2 \rangle = lL \tag{2.29}$$

defines l (as a function of temperature) given the mean square-end-to-end distance $\langle \mathbf{R}^2 \rangle$ for a flexible polymer.

The simplicity of (2.25) is to be contrasted with the complexity of the exact $P(\mathbf{R}; n)$ for realistic models of flexible chains for all $\mathbf{R}$ (and for small n). When dealing with the complicated problems of nonideal polymer solutions, etc., it is therefore customary to replace the real polymer chain by the so-called Kuhn effective random flight chain. An effective chain is one with N (in general different from n) links of size Δs such that $N\,\Delta s = L$ and $\langle \mathbf{R}^2 \rangle$ is as given by (2.29). This substitution of a real chain by its equivalent chain is often a necessity so that we may separate errors in principle from errors arising from a poor mathematical approximation to the exact $P(\mathbf{R}; n)$ when dealing with problems which are not exactly soluble. This equivalent chain therefore provides us with reasonable approximations to the properties of real polymer chains, provided the physical properties of interest do not depend heavily upon those chain configurations with $R \gtrsim L$ or upon chain properties over small distances for which the real chain is stiff.

Since the equivalent chain is not a real physical polymer chain, and since the equivalent links do not represent real chemical bonds, we are free to choose $\tau(\mathbf{R}_j)$ for the equivalent link in any manner we choose. One possibility is the freely hinged chain of (2.15), but that leads to fairly complicated mathematical expressions $P(\mathbf{R}_{ji}; |j - i|)$ for the distribution of distances between the ith and jth "equivalent" segments—this distribution is often of interest. On the other hand, a chain with equivalent gaussian links identically reproduces (2.25) for any pair of effective segments which are separated by $n = |i - j|$ equivalent links. Thus the equivalent gaussian-link chain given by

$$\tau(\mathbf{R}_j) = \left(\frac{3}{2\pi l\,\Delta s}\right)^{3/2} \exp\left[-\frac{3\mathbf{R}_j^{\,2}}{2l\,\Delta s}\right] \tag{2.30}$$

with N links such that $N\,\Delta s = L$ identically gives (2.25) and (2.29). [In general, we need not require that Δs be the same for all bonds; there can be different values $\{\Delta s_j\}$, provided that $\sum_{j=1}^{N} \Delta s_j = L$.] Since L is independent of temperature, N and Δs can also be taken to have that property. The equivalent chain given by (2.30) is then a generic statement about the common characteristics of flexible chainlike polymers for properties that depend primarily on $N\,\Delta s \gg R \gg \Delta s$. In the case of stiff chains, it is of course necessary to introduce alternate "mathematically simple" models and this is done later in this section.

From (2.25) to (2.28) it is apparent that the separation of chains into flexible and stiff classes can be made on the basis of the dependence of $\langle \mathbf{R}^2 \rangle_n$ on n, the number of links in the real polymer chain.[1–5] The first nontrivial case beyond the random flight polymer chain is the freely rotating chain. In this chain the bond angles are fixed at say θ and the bond lengths are fixed at l, as shown in Fig. 1. The only degree of freedom left is the rotation about the single bonds, which is taken to be free (unhindered). If $\mathbf{R}_{j-1}$ and $\mathbf{R}_j$ are taken to be in the plane of the paper, $\mathbf{R}_{j+1}$ may lie on a cone which is specified by $|\mathbf{R}_{j+1}| = l$ and the polar and azimuthal angles with respect to an origin at the jth segment and a polar axis along $-\mathbf{R}_j$. The other angle ϕ is arbitrary and this implies that

$$\langle \mathbf{R}_j \cdot \mathbf{R}_{j+1} \rangle = l^2 \cos(\pi - \theta) = -l^2 \cos\theta \tag{2.31}$$

where the average is over all values of ϕ.

Using (2.10) with $\mathbf{R} \equiv \mathbf{r}_n$ gives

$$\langle \mathbf{R}^2 \rangle = \sum_{j=1}^{n} \langle \mathbf{R}_j^{\,2} \rangle + 2\sum_{j=1}^{n} \sum_{s=1}^{n-j} \langle \mathbf{R}_j \cdot \mathbf{R}_{j+s} \rangle \tag{2.32}$$

The second term in (2.32) can be evaluated easily by successively projecting $\mathbf{R}_j$ on a unit vector in the direction of $\mathbf{R}_{j+1}$, then $\mathbf{R}_{j+1}$ on a unit vector along $\mathbf{R}_{j+2}$, etc.,

$$\mathbf{R}_j = \frac{(\mathbf{R}_j \cdot \mathbf{R}_{j+1})\mathbf{R}_{j+1}}{|\mathbf{R}_{j+1}|^2} = \frac{(\mathbf{R}_j \cdot \mathbf{R}_{j+1})}{l^2} \frac{(\mathbf{R}_{j+1} \cdot \mathbf{R}_{j+2})\mathbf{R}_{j+2}}{|\mathbf{R}_{j+2}|^2} = \cdots \tag{2.33}$$

so that

$$\langle \mathbf{R}_j \cdot \mathbf{R}_{j+s} \rangle = l^2[-\cos\theta]^s \tag{2.34}$$

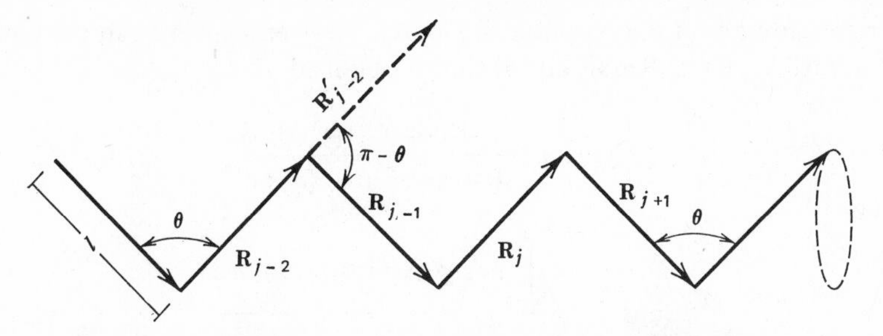

Fig. 1. The freely rotating chain Bond lengths are of fixed length l and bond angles are all $\pi - \theta$. There is free rotation about the skeletal bonds.

The geometric series obtained by substituting (2.34) into (2.32) can be summed to give

$$\langle \mathbf{R}^2 \rangle = nl^2 \left[\frac{1 - \cos\theta}{1 + \cos\theta} + \frac{2\cos\theta}{n} \frac{1 - (-\cos\theta)^n}{(1 + \cos\theta)^2} \right]. \tag{2.35}$$

the well-known result of Eyring.[1–5] For large enough n and $|\cos\theta| < 1$, the second term in (2.35) is negligible and

$$\langle \mathbf{R}^2 \rangle_n \to nl^2 \frac{1 - \cos\theta}{1 + \cos\theta} \tag{2.36}$$

Since (2.36) satisfies (2.28), the long freely rotating chain qualifies as a flexible chain. When the second term in (2.35) is nonnegligible, the freely rotating chain has stiffness.

In order to define the effective gaussian link chain corresponding to (2.36), the maximum contour length is required. The greatest elongation of the chain occurs when all of the bonds lie in a plane, and

$$L = nl \sin\left(\frac{\theta}{2}\right) \tag{2.37a}$$

The effective length l_{eff} to be used in (2.29) and (2.25) is

$$l_{\text{eff}} \equiv \frac{\langle \mathbf{R}^2 \rangle_n}{L} = \frac{2l \sin(\theta/2)}{(1 + \cos\theta)} \tag{2.37b}$$

which is temperature independent because there are no interaction energies involved.

If account is taken of the fact that in real polymers there is hindered rotation about each of the single bonds, then l_{eff} will be temperature dependent. The results are standard and summarized in a number of places,[1–5] and we only quote them here. Consider the chain with hindered rotations which are described by the symmetric (even) potential $w(\phi_j)$

for rotation about the jth bond. When n is large enough, i.e., in the limit of a flexible chain, Benoit and Kuhn[1–5] obtained

$$\langle \mathbf{R}^2 \rangle_n = nl^2 \frac{1 - \cos\theta}{1 + \cos\theta} \frac{1 + \langle \cos\phi \rangle}{1 - \langle \cos\phi \rangle} \tag{2.38}$$

where

$$\langle \cos\phi \rangle = \frac{\int_{-\pi}^{\pi} \cos\phi \exp\left[-\beta w(\phi)\right] d\phi}{\int_{-\pi}^{\pi} \exp\left[-\beta w(\phi)\right] d\phi} \tag{2.39}$$

Equation (2.38) and hence l_{eff} is temperature dependent. Of course, when $w \equiv 0$, (2.38) reduces to (2.36).

The model of a polymer chain with fixed bond angles and hindered internal rotation is still not entirely realistic for simple polymers like polyethylene in which there are steric interactions which result in the interdependence of rotations about neighboring bonds, i.e., interactions of the form $w_2(\phi_j, \phi_{j-1})$. In the popular rotational isomeric model, the internal rotation hindrances are taken to limit the angles $\{\phi_j\}$ to only a small number of discrete values $\phi^{(1)}, \phi^{(2)}, \ldots, \phi^{(p)}$ with energies $w(\phi^{(p)})$ independent of j.[2–5] The interaction w_2 can then take on the values $w_2(\phi^{(\alpha)}, \phi^{(\beta)})$ independent of j, and the calculation of the partition function becomes mathematically identical to the calculation of the partition function for a one-dimensional Ising model of spin $S(2S + 1 = p)$ with the most general type of nearest neighbor interactions—*an exactly soluble problem.* Matrix techniques have been developed to evaluate averages like $\langle \mathbf{R}^2 \rangle$, $\langle \mathbf{R}^4 \rangle, \ldots$, and these are aptly summarized in Flory's book,[4] so they need not be repeated here. The basic point for this development, is that, again, for a long enough chain (n large enough), the chain is flexible, and $\langle \mathbf{R}^2 \rangle$ is proportional to n. In this limit, the distribution of end-to-end vectors is given by (2.25), with (2.29) defining a l_{eff}, which is temperature dependent.

Thus the use of an equivalent gaussian link chain, which appeared artificial in the case of the freely hinged chain, is quite acceptable for the freely rotating chain, the chain with hindered internal rotation, or the rotational isomeric model. In these cases there is no pretense of reality of the equivalent links, so they need not be given fixed lengths l_{eff}. This equivalent chain is therefore a summary of the chainlike character of flexible polymer chains, and its simplicity enables the investigation of those properties of polymer systems which are solely dependent on the flexible chainlike nature of the polymer molecules. It is then only natural to attempt to formulate simple models which are characteristic of stiff chains, so that these models may then be used as the "building blocks" for problems which require approximations. One such model is the Kratky-Porod wormlike chain.

D. The Wormlike Chain

If a chain has some degrees of stiffness, there must be some residual correlations between the initial and final links of the chain which persist as a result of the short-range interactions. In fact, there is even some persistence of correlations in flexible chains. For the simple case of the freely rotating chain, $\mathbf{R}_1/|\mathbf{R}_1| = \mathbf{R}_1/l$ is a unit vector along the first bond. The average projection of the end-to-end vector $\mathbf{R}$ along the chain's initial direction is then

$$\left\langle \frac{\mathbf{R} \cdot \mathbf{R}_1}{l} \right\rangle = \frac{1}{l} \sum_{j=1}^{n} \langle \mathbf{R}_j \cdot \mathbf{R}_1 \rangle \tag{2.40a}$$

$$= \frac{1}{l} \sum_{j=1}^{n} l^2(-\cos\theta)^{j-1} \tag{2.40b}$$

by the projection method leading to (2.34). Summing the geometric series again gives

$$\left\langle \frac{\mathbf{R} \cdot \mathbf{R}_1}{l} \right\rangle = l \frac{1 - (-\cos\theta)^n}{1 + \cos\theta} \tag{2.41a}$$

$$\xrightarrow[|\cos\theta| < 1]{n \to \infty} \frac{l}{1 + \cos\theta} \equiv \frac{1}{2\lambda} \tag{2.41b}$$

where $1/2\lambda$ is appropriately called the persistence length. The model of the wormlike chain is that model continuous chain which is obtained by taking the limit $l \to 0$, $n \to \infty$, $nl \to L$, $\theta \to \pi$, $l/(1 + \cos\theta) \to 1/2\lambda$ of the discrete freely rotating chain.[5,34] Denoting this limit as $\lim_{\text{worm}}$, we find

$$\lim_{\text{worm}} \left\langle \frac{\mathbf{R} \cdot \mathbf{R}_1}{l} \right\rangle = \frac{1}{2\lambda} \left[1 - \lim_{\text{worm}} (-\cos\theta)^n \right] \tag{2.42}$$

Since

$$(-\cos\theta)^n = \exp\{n \log[1 - \cos\theta - 1]\}$$

we see that

$$\lim_{\text{worm}} (-\cos\theta)^n = \lim_{\text{worm}} \exp\left\{ -nl \frac{(1 + \cos\theta)}{l} - \frac{nl}{2} \frac{(1 + \cos\theta)^2}{l} + \cdots \right\}$$

$$= \exp(-2\lambda L) \tag{2.42a}$$

Finally,

$$\lim_{\text{worm}} \left\langle \frac{\mathbf{R} \cdot \mathbf{R}_1}{l} \right\rangle = \frac{1}{2\lambda} [1 - \exp(-2\lambda L)] \tag{2.43}$$

Similarly, applying $\lim\limits_{\text{worm}}$ to the expression for $\langle \mathbf{R}^2 \rangle$ in (2.35) gives

$$\lim_{\text{worm}} \langle \mathbf{R}^2 \rangle = \frac{L}{\lambda} - \frac{1}{\lambda^2} [1 - \exp(-2\lambda L)] \tag{2.44}$$

This, of course, does not give $\langle \mathbf{R}^2 \rangle \propto L$. It is then easily verified that $\langle \mathbf{R}^4 \rangle_{\text{worm}} \neq \frac{5}{3} \langle \mathbf{R}^2 \rangle^2_{\text{worm}}$, so that the end-to-end distribution is not gaussian.

When the chain length L is much greater than the persistence length, i.e., $\lambda L \to \infty$, the effects of stiffness become negligible. In this limit, it is found that

$$\begin{aligned} \lim_{\lambda L \to \infty} \langle \mathbf{R}^2 \rangle_{\text{worm}} &= \frac{L}{\lambda} \\ \lim_{\lambda L \to \infty} \langle \mathbf{R}^4 \rangle_{\text{worm}} &= \frac{5}{3} \left(\frac{L}{\lambda} \right)^2 \end{aligned} \tag{2.45}$$

The wormlike chain then reduces to a random flight chain. In the opposite limit of small λL the chain is very stiff, i.e.,

$$\begin{aligned} \lim_{\lambda L \to 0} \langle \mathbf{R}^2 \rangle_{\text{worm}} &= L^2 \\ \lim_{\lambda L \to 0} \langle \mathbf{R}^2 \rangle_{\text{worm}} &= L^4 \end{aligned} \tag{2.46}$$

and the wormlike chain reduces to the rigid rod.

The wormlike chain is described by a continuous space curve (not as a number of sequentially bonded segments) which is a simple representation of the generic chainlike structure of a polymer molecule. The space curve is denoted by $\mathbf{r}(s)$, $0 \leqslant s \leqslant L$, and represents the position of a polymer segment in space as a function of the contour distance (a proper time!) from the beginning of the chain (taken arbitrarily as either end) to that point. In the limit (2.45) the wormlike chain is the same as a flexible equivalent model chain. Therefore in this limit a continuous space curve representation is automatically a continuous version of the equivalent chain representation. In other words, the space curve representation has the short-range (along the chain) interaction built into it and is thus an acceptable model of generic chainlike structure. The only requirements that were imposed upon the representation of a flexible polymer chain by an equivalent chain are expressed in (2.25) and (2.29). Thus, if the equivalent chain is composed of N random flight links, each of contour length Δs such that $L = N\,\Delta s$, there is no difficulty, in principle, in passing to the limit of a continuous equivalent chain, i.e., $N \to \infty$, $\Delta s \to 0$, $N\,\Delta s = L$. In this limit there is also very little difference whether we use the gaussian equivalent links (2.30) or freely hinged equivalent links. After all for small Δs, (2.30) is "almost a delta function."

The representation of a flexible polymer chain by an equivalent random flight model chain which is subsequently taken to be a continuous space curve leads directly to the introduction of a functional integral representation of polymer chains.

III. INTRODUCTION TO FUNCTIONAL INTEGRATION: FLEXIBLE POLYMER CHAINS

A. The Wiener Integral

Although the Wiener integral formulation for the distribution functions of flexible polymer chains rests upon general considerations of random walks and Brownian motion, it is easily introduced, heuristically, through the concept of an equivalent chain. In this section, only those flexible polymer chains are considered which are composed of equivalent gaussian links. Here L is the maximum contour length of the real chain at full extension, and $\langle \mathbf{R}^2 \rangle$ for the equivalent chain is taken to be that for the real chain. Thus we have

$$l \equiv \frac{\langle \mathbf{R}^2 \rangle}{L} \tag{3.1}$$

as a given function of temperature. If the equivalent chain is taken to have n gaussian links, each of average (rms) length Δs, the bond probabilities are

$$\tau(\mathbf{R}_j) = \tau(\mathbf{r}_j - \mathbf{r}_{j-1}) = \left(\frac{3}{2\pi l\,\Delta s}\right)^{3/2} \exp\left(\frac{-3\mathbf{R}_j^{\,2}}{2l\,\Delta s}\right) \tag{3.2}$$

where

$$n\,\Delta s \equiv L \tag{3.3}$$

Upon substitution of (3.2) and (3.3) into (2.17), it is easily verified that $P(\mathbf{R}; n)$ for the equivalent chain is given by (2.25) with (3.1). Using the normalization (2.7) for $\tau(\mathbf{R}_j)$, and setting $W = 0$ in (2.8), leads to the probability distribution function $P(\{\mathbf{r}_k\})$ for the entire chain configuration, which is

$$P(\{\mathbf{r}_k\}) = \prod_{j=1}^{n} \left\{ \left(\frac{3}{2\pi l\,\Delta s}\right)^{3/2} \exp\left[-\frac{3(\mathbf{r}_j - \mathbf{r}_{j-1})^2}{2l\,\Delta s}\right] \right\} \tag{3.4a}$$

$$= \mathcal{N} \exp\left[-\sum_{i=1}^{n} \frac{3(\mathbf{r}_j - \mathbf{r}_{j-1})^2}{2l\,\Delta s}\right], \tag{3.4b}$$

where

$$\mathcal{N} = \left(\frac{3}{2\pi l\,\Delta s}\right)^{3n/2} \tag{3.4c}$$

is just the normalization factor. $\mathcal{N}$ is fixed by the condition that $P(\{\mathbf{r}_k\})$ be a probability distribution, i.e., the probability that the chain has any configuration is unity:

$$\int d\{\mathbf{r}_k\}P(\{\mathbf{r}_k\}) \equiv 1. \tag{3.5}$$

Since (3.4b) gives the "probability" of a particular chain configuration $\mathbf{r}_0 \equiv \mathbf{0}, \mathbf{r}_1, \ldots, \mathbf{r}_n$, it is clear that this chain configuration $\{\mathbf{r}_k\}$ can be taken to be the discrete representation of the continuous curve $\mathbf{r}(s)$. Therefore let

$$\mathbf{r}_j = \mathbf{r}(j\,\Delta s) \equiv \mathbf{r}(s_j) \tag{3.6}$$

denote the position of the jth segment with respect to the origin ($s_0 = 0$) which is taken as $\mathbf{r}_0 = \mathbf{0}$ in the absence of external forces.

We now take the limit $\Delta s \to 0, n \to \infty, n\,\Delta s = L$ to obtain a representation of a continuous equivalent random flight chain. In this limit, denoted by $\lim_{FI}$, the expression in the exponential in (3.4b) is

$$\lim_{FI} \sum_{j=1}^{n} \left[\frac{\mathbf{r}(s_j) - \mathbf{r}(s_j - \Delta s)}{\Delta s}\right]^2 \Delta s \equiv \lim_{FI} \sum_{j=1}^{n} \left[\left.\frac{\partial \mathbf{r}(s)}{\partial s}\right|_{s=s_j}\right]^2 \Delta s$$

$$\equiv \int_0^L ds \left[\frac{\partial \mathbf{r}(s)}{\partial s}\right]^2 = \int_0^L ds[\dot{\mathbf{r}}(s)]^2 \tag{3.7}$$

Equation (3.7) follows from the definition of derivatives and integrals in elementary calculus. In this limit, the probability $P(\{\mathbf{r}_k\})$ becomes

$$P(\{\mathbf{r}_k\}) \prod_{j=1}^{n} d\mathbf{r}_j \to P[\mathbf{r}(s)]\,\delta\mathbf{r}(s) \tag{3.8}$$

and $P[\mathbf{r}(s)]\,\delta\mathbf{r}(s)$ is the probability that the chain configuration lies between the continuous space curves $\mathbf{r}(s)$ and $\mathbf{r}(s) + \delta\mathbf{r}(s)$. There is no need to worry about the fact that

$$\lim_{FI} \mathcal{N} \to (\infty)^{\infty} \tag{3.4d}$$

because the normalization is there just to preserve the probabilistic nature of $P[\mathbf{r}(s)]$. Usually the normalization is absorbed into a single differential for curve $\mathbf{r}(s)$,

$$\mathcal{N}\,\delta\mathbf{r}(s) \equiv \mathscr{D}[\mathbf{r}(s)] \tag{3.9}$$

Finally,

$$P[\mathbf{r}(s)]\,\delta\mathbf{r}(s) = \mathscr{D}[\mathbf{r}(s)] \exp\left[-\frac{3}{2l}\int_0^L [\dot{\mathbf{r}}(s)]^2\,ds\right] \tag{3.10}$$

gives the probability of the chain configuration $\mathbf{r}(s)$ and is the well-known Wiener measure.

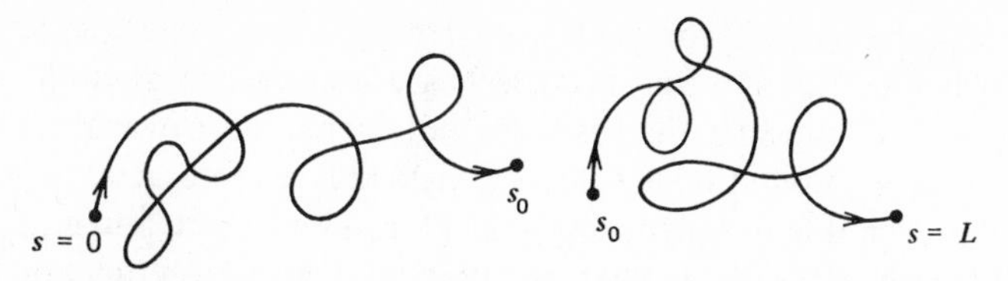

Fig. 2. A discontinuous polymer chain which is given zero weight by the Wiener measure (3.10). The arrows denote the arbitrarily chosen direction from chain beginning ($s = 0$) to chain end ($s = L$).

Equation (3.4d) has led mathematicians frequently to claim that the representation of the Wiener measure in (3.10) is "undefined." Their complaint is reminiscent of the disrepute in which Dirac delta functions were held by mathematicians for a number of years. There are mathematically acceptable formulations, or notational transcriptions, of these functional integrals. These formulations may make for good mathematics, but they are physically unnecessary.[35] When in doubt, we just remember that the functional integrals are defined in terms of the limit of an iterated integral.

The transition to a continuous curves does not physically lead to nonsense. If $\mathbf{r}(s)$ were a discontinuous curve, as in Fig. 2 where there is a discontinuity in $\mathbf{r}(s)$ at s_0, $|\dot{\mathbf{r}}(s_0)| = \infty$, and $\exp\,[-3/2l \int_0^L [\dot{\mathbf{r}}(s)]^2\,ds] \equiv 0$ for that curve. Thus only *continuous curves* have nonzero weight or measure. (Sometimes taking the limit of a continuous curve leads to difficulties if there are physical properties which depend on the total number of degrees of freedom. These properties are finite as long as Δs is finite. These divergences are easily spotted and remedied.) In this continuous limit (3.5) becomes

$$\int_{\mathbf{r}(0)=0} P[\mathbf{r}(s)]\,\delta\mathbf{r}(s) = 1$$
$$= \int_{\mathbf{r}(0)=0} \mathscr{D}[\mathbf{r}(s)] \exp\left\{-\frac{3}{2l}\int_0^L [\dot{\mathbf{r}}(s)]^2\,ds\right\} \tag{3.11}$$

where the integration is over all continuous curves $\mathbf{r}(s)$, $0 \leqslant s \leqslant L$, such that $\mathbf{r}(0) \equiv \mathbf{0}$. Thus the probability that the chain will take up some configuration is unity and the "infinite" normalization (3.4c, 3.4d) is necessary to give this value. The functional integral (3.11) is defined only as the limit of the corresponding integral for a discrete chain. This point is discussed in more detail later.

B. The Relation to Diffusion and Brownian Motion

We have taken the continuous limit of a gaussian equivalent chain for a flexible polymer chain. This procedure leads to a nice physical picture of continuous chains and compact formal expressions. The question may remain whether this limiting process has accomplished anything useful. As an example, consider the case of a polymer in an external field. This field may be a gravitational field, the field due to interactions with the solvent, or even due to other polymers. Thus, given some potential energy per unit length of the chain $W(\mathbf{r})$, the thermal distribution function for the discrete chain is

$$G(\{\mathbf{r}_k\})\,d\{\mathbf{r}_k\} = \mathcal{N}\,d\{\mathbf{r}_k\}\exp\left[-\frac{3}{2l\,\Delta s}\sum_{j=1}^{n}(\mathbf{r}_j-\mathbf{r}_{j-1})^2 - \beta\,\Delta s\sum_{j=1}^{n}W\left(\frac{\mathbf{r}_j+\mathbf{r}_{j-1}}{2}\right)\right] \tag{3.12}$$

In (3.12) the center of force is taken at the center of each bond. This approximation becomes exact as $\Delta s \to 0$. The presence of an external force implies that space is no longer isotropic. Thus, let $\mathbf{r}_0 \equiv \mathbf{R}'$, and

$$d\{\mathbf{r}_k\} = \prod_{j=1}^{n} d\mathbf{r}_j \tag{3.13}$$

The partition function is obtained from (3.12) by taking

$$Z = \int d\mathbf{r}_0 \int d\{\mathbf{r}_k\}G(\{\mathbf{r}_k\}) \tag{3.14}$$

and the probability distribution for the chain is

$$P(\{\mathbf{r}_k\}) = Z^{-1}G(\{\mathbf{r}_k\}) \tag{3.15}$$

Letting

$$V \equiv \beta W \tag{3.16}$$

in the limit of a continuous chain, we obtain

$$\beta\sum_{j=1}^{n}W\left(\frac{\mathbf{r}_j+\mathbf{r}_{j-1}}{2}\right)\Delta s \to \int_0^L ds V[\mathbf{r}(s)] \tag{3.17}$$

using the association (3.6) for a given configuration $\{\mathbf{r}_k\}$ for the discrete chain and the definition of a definite integral as the limit of a sum! Thus we have

$$G[\mathbf{r}(s)]\,\delta\mathbf{r}(s) = \mathcal{D}[\mathbf{r}(s)]\exp\left(-\int_0^L ds\left\{\frac{3}{2l}\dot{\mathbf{r}}^2(s) + V[\mathbf{r}(s)]\right\}\right) \tag{3.18}$$

As noted in Section II, $G[\mathbf{r}(s)]$ contains all the information pertaining to the configurational statistics of the polymer chain. However, this is more information than we can handle or need. A number of reduced distributions are therefore of interest. One of the simplest is the distribution of the end vectors $G(\mathbf{R}, \mathbf{R}'; L)$, which is obtained from (3.12) by selecting out those chains that have $\mathbf{r}_0 = \mathbf{R}'$ and $\mathbf{r}_n = \mathbf{R}$. In other words, in the discrete case we use

$$G(\mathbf{R}, \mathbf{R}'; L) \equiv \int d\mathbf{r}_0 \int d\{\mathbf{r}_k\}\, \delta(\mathbf{r}_0 - \mathbf{R}')\, \delta(\mathbf{r}_n - \mathbf{R}) G(\{\mathbf{r}_k\}) \tag{3.19a}$$

which in the continuous limit becomes

$$G(\mathbf{R}, \mathbf{R}'; L) = \int \delta\mathbf{r}(s) \int d\mathbf{r}_0\, \delta[\mathbf{r}(0) - \mathbf{R}']\, \delta[\mathbf{r}(L) - \mathbf{R}] G[\mathbf{r}(s)] \tag{3.19b}$$

Explicitly,

$$\begin{aligned} G(\mathbf{R}, \mathbf{R}'; L) &= \int_{\mathbf{r}(0)=\mathbf{R}'}^{\mathbf{r}(L)=\mathbf{R}} \mathscr{D}[\mathbf{r}(s)] \exp\left\{-\int_0^L ds \left(\frac{3}{2l}\dot{\mathbf{r}}^2(s) + V[\mathbf{r}(s)]\right)\right\} \\ &= \lim_{FI} \int d\mathbf{r}_0\, \delta(\mathbf{r}_0 - \mathbf{R}') \int d\{\mathbf{r}_k\}\, \delta(\mathbf{r}_n - \mathbf{R}) \mathscr{N} \\ &\quad \times \exp\left\{-\frac{3}{2l\,\Delta s}\sum_{j=1}^{n}[\mathbf{r}_j - \mathbf{r}_{j-1}]^2 - \Delta s \sum_{j=1}^{n} V\left[\frac{\mathbf{r}_j + \mathbf{r}_{j-1}}{2}\right]\right\} \end{aligned}$$

(3.20

gives the definition of the functional integral for $G(\mathbf{R}, \mathbf{R}'; L)$ *as a limit of an iterated integral*! Just as the operation of differentiation and integration are *defined* as particular limits of differences and sums, so are functional integrals like (3.20) and (3.11) *defined* only as particular limits. More rigorously and generally, we might have divided the equivalent gaussian chain into a number of links each of differing contour lengths Δs_j, where $\sum_{j=1}^{n} \Delta s_j = L$. The expressions in (3.2), (3.4), (3.7), (3.12), and (3.20) are easily rewritten in that case. The functional integrals (3.11) and (3.20) are then defined as the limits $n \to \infty$, max $(\Delta s_j) \to 0$, $\sum_{j=1}^{n} \Delta s_j = L$ of the discrete expression when these limits exist and are independent of the manner in which the limits are taken. Thus the introduction of functional integrals as particular limiting processes is wholly analogous to the introduction of the operations of differentiation and integration in elementary calculus—they are defined only as particular limits. Having performed the limits explicitly, we can tabulate the results and use them directly. Functional integrals thus represent *merely another kind of limiting process*. Just as we learn in elementary calculus that integrals may be "evaluated" by the inverse of differentiation, the particular type (3.20) of (Wiener) functional integrals may be "evaluated" by other more familiar means.

We now demonstrate that $G(\mathbf{R}, \mathbf{R}'; L)$ satisfies[8-12] the diffusion equation for a particle in the external potential $V(\mathbf{R})$. In particular, it is well known that G satisfies

$$\left[\frac{\partial}{\partial L} - \frac{l}{6}\nabla_{\mathbf{R}}^{2} + V(\mathbf{R})\right] G(\mathbf{R}, \mathbf{R}'; L) = 0, \qquad L \neq 0 \tag{3.21}$$

subject to the boundary condition that

$$\lim_{L\to 0} G(\mathbf{R}, \mathbf{R}'; L) = \delta(\mathbf{R} - \mathbf{R}') \tag{3.22}$$

That (3.21) and (3.22) are obtained should be very clear. We started with a random flight model of a polymer where the contour length represents a timelike variable. For long enough "times," random walks, or Brownian motion, can be considered to be diffusion processes. Here the "diffusion constant" is defined as the mean square displacement per unit "time." Using (3.2), we find

$$\begin{aligned} D_{\alpha\beta} &\equiv \frac{1}{\Delta s}\langle \mathbf{R}^{\alpha}\mathbf{R}^{\beta}\rangle_{\Delta s} \\ &= \frac{1}{\Delta s}\int d\mathbf{R}_j \mathbf{R}_j^{\alpha}\mathbf{R}_j^{\beta}\tau(\mathbf{R}_j) \\ &= \frac{l}{6}\,\delta_{\alpha\beta}, \qquad \alpha, \beta = x, y, z \end{aligned} \tag{3.23}$$

The following derivation of (3.21) and (3.22), which follows Feynman and Hibbs,[12] is presented because it uses operations that are required many times. The generalization of this derivation is also required for functional integrals that are more complicated than (3.20). It is easiest to introduce these simpler aspects first. First the boundary condition (3.22) is established. For small ϵ,

$$\lim_{\epsilon\to 0^+} G(\mathbf{R}, \mathbf{R}'; \epsilon) = \lim_{\epsilon\to 0^+}\left(\frac{3}{2\pi l\epsilon}\right)^{3/2} \exp\left[-\frac{3(\mathbf{R}-\mathbf{R}')^2}{2l\epsilon} - \epsilon V\left(\frac{\mathbf{R}+\mathbf{R}'}{2}\right)\right] \tag{3.24}$$

Provided the potential is well behaved, $\epsilon V \to 0$ as $\epsilon \to 0$, so

$$\begin{aligned} \lim_{\epsilon\to 0^+} G(\mathbf{R}, \mathbf{R}'; \epsilon) &= \lim_{\epsilon\to 0^+}\left(\frac{3}{2\pi l\epsilon}\right)^{3/2} \exp\left[\frac{-3(R-R')^2}{2l\epsilon}\right] \\ &\equiv \delta(\mathbf{R}-\mathbf{R}') \end{aligned} \tag{3.25}$$

is just one particular *definition* of a delta function.

In order to find a differential equation satisfied by $G(\mathbf{R}, \mathbf{R}'; L)$, we consider $G(\mathbf{R}, \mathbf{R}'; L + \epsilon)$ for infinitesimal ϵ, so only terms of $\mathcal{O}(\epsilon)$ need be kept.

$$G(\mathbf{R}, \mathbf{R}'; L + \epsilon) \equiv \int_{\mathbf{r}(0)=\mathbf{R}'}^{\mathbf{r}(L+\epsilon)=\mathbf{R}} \mathscr{D}[\mathbf{r}(s)] \exp\left(-\int_0^{L+\epsilon} ds\left\{\frac{3}{2l}\dot{\mathbf{r}}^2(s) + V[\mathbf{r}(s)]\right\}\right) \tag{3.26}$$

Now the definite integral in the exponent of (3.26) can always be written, for every continuous curve $\mathbf{r}(s)$, as a sum of integral for two intervals,

$$\int_0^{L+\epsilon} ds \cdots = \int_0^L ds \cdots + \int_L^{L+\epsilon} ds \cdots \tag{3.27}$$

The functional integral is defined only as the limit of the finite multiple integral, but we may write $\mathscr{D}[\mathbf{r}(s)]$ as defined on some interval $0 < s < L'$ via

$$\mathscr{D}[\mathbf{r}(s)] \equiv \mathscr{N} \prod_{j=1}^{n-1} d\mathbf{r}(s_j) \equiv \frac{\mathscr{N}\left[\prod_{s=0}^{L'} d\mathbf{r}(s)\right]}{d\mathbf{r}(0)\, d\mathbf{r}(L')} \tag{3.28}$$

In the first equality we imply that $n\,\Delta s = L'$ (for equal intervals, etc., if unequal). In the last part of (3.28) we imply that the initial and final points $s = 0$ and $s = L'$ are to be omitted from the product. Since $\mathscr{D}[\mathbf{r}(s)]$ in (3.28) is a product over all s in the interval $0 < s < L'$, we can always write it as a product of the terms

$$\mathscr{D}_1[\mathbf{r}(s)] = \frac{\mathscr{N}_1\left[\prod_{s=0}^{L''} d\mathbf{r}(s)\right]}{d\mathbf{r}(0)\, d\mathbf{r}(L'')} \tag{3.29a}$$

and

$$\mathscr{D}_2[\mathbf{r}(s)] = \frac{\mathscr{N}_2\left[\prod_{s=L''}^{L'} d\mathbf{r}(s)\right]}{d\mathbf{r}(L'')\, d\mathbf{r}(L')} \tag{3.29b}$$

$\mathscr{D}[\mathbf{r}(s)]$ is then a product of $\mathscr{D}_j[\mathbf{r}(s)]$ for each of the intervals $0 < s < L''$ and $L'' < s < L'$ and of course the differential $d\mathbf{r}(L'')$ for the point $s = L''$,

$$\mathscr{D}[\mathbf{r}(s)] = \mathscr{D}_1[\mathbf{r}(s)]\, d\mathbf{r}(L'')\mathscr{D}_2[\mathbf{r}(s)] \tag{3.30}$$

Although (3.26) requires only an infinitesimal ϵ, [with $L = L''$, $L + \epsilon = L'$ in (3.29) and (3.30)] it is instructive to first allow ϵ to be arbitrary. In this case we take each path $\mathbf{r}(s)$ for $0 \leqslant s \leqslant L + \epsilon$, which runs from $\mathbf{r}(0) = \mathbf{R}'$ to $\mathbf{r}(L + \epsilon) = \mathbf{R}$, and in (3.26) we insert a factor of unity in the form

$$1 = \int d\mathbf{R}''\, \delta[\mathbf{R}'' - \mathbf{r}(L')] \tag{3.31}$$

Interchanging the orders of the $\mathbf{R}''$ and the functional integrals gives

$$G(\mathbf{R}, \mathbf{R}'; L + \epsilon) \equiv \int d\mathbf{R}'' \int_{\mathbf{r}(0)=\mathbf{R}'}^{\mathbf{r}(L+\epsilon)=\mathbf{R}} \mathscr{D}[\mathbf{r}(s)]\, \delta[\mathbf{R}'' - \mathbf{r}(L'')] \times \exp\left(-\int_0^{L+\epsilon} \left\{\frac{3}{2l}\dot{\mathbf{r}}^2(s) + V[\mathbf{r}(s)]\right\}\right) \quad (3.32)$$

Substituting (3.27) and (3.30) into (3.32) with $L = L''$, $L + \epsilon = L'$ gives

$$G(\mathbf{R}, \mathbf{R}'; L + \epsilon) = \int d\mathbf{R}'' \int d\mathbf{r}(L) \left\{\int_{\mathbf{r}(L)}^{\mathbf{r}(L+\epsilon)=\mathbf{R}} \mathscr{D}_2[\mathbf{r}(s)] \times \exp\left[-\int_L^{L+\epsilon} \left(\frac{3}{2l}\dot{\mathbf{r}}^2 + V\right)\right]\right\} \delta[\mathbf{R}'' - \mathbf{r}(L)] \times \left\{\int_{\mathbf{r}(0)=\mathbf{R}'}^{\mathbf{r}(L)} \mathscr{D}_1[\mathbf{r}(s)] \exp\left[-\int_0^L \left(\frac{3}{2l}\dot{\mathbf{r}}^2 + V\right)\right]\right\} \quad (3.33)$$

where the lower (upper) limit $\mathbf{r}(L)$ on the $\mathscr{D}_2(\mathscr{D}_1)$ integration implies that the endpoints of those integrations are still integration variables. The integral $\int d\mathbf{r}(L)\, \delta[\mathbf{R}'' - \mathbf{r}(L)]$ fixes these limits at $\mathbf{R}''$, and then the terms in braces represent $G(\mathbf{R}, \mathbf{R}''; L + \epsilon, L)$ and $G(\mathbf{R}'', \mathbf{R}'; L, 0)$, respectively. Thus (3.33) becomes

$$G(\mathbf{R}, \mathbf{R}'; L + \epsilon, 0) = \int d\mathbf{R}'' G(\mathbf{R}, \mathbf{R}''; L + \epsilon, L) G(\mathbf{R}'', \mathbf{R}'; L, 0) \quad (3.34)$$

where the initial and final contour lengths have been included. The steps (3.31) to (3.33) simply result from the fact that all curves $\mathbf{r}(s)$, $0 \leqslant s \leqslant L + \epsilon$ such that $\mathbf{r}(0) = \mathbf{R}'$ and $\mathbf{r}(L + \epsilon) = \mathbf{R}$ can be separated into two parts, one beginning and one ending at $\mathbf{r}(L) = \mathbf{R}''$ for some $\mathbf{R}''$. The functional integral over all $\mathbf{r}(s)$ subject to the endpoints $\mathbf{R}$, $\mathbf{R}'$ is then the $\int d\mathbf{R}''$ of the functional integral over all functions $\mathbf{r}(s)$ which pass through $\mathbf{R}''$ and the endpoints $\mathbf{R}$ and $\mathbf{R}'$. Equation (3.34) is the general Smoluchowski-Chapman-Kolmogorov equation for a Markov process. Thus $G(\mathbf{R}, \mathbf{R}'; L + \epsilon, 0)$, the (unnormalized) "probability" for going from $\mathbf{R}'$, 0 to $\mathbf{R}$, $L + \epsilon$ is equal to the product of the "probability" of going from $\mathbf{R}'$, 0 to some intermediate $\mathbf{R}''$, L and the "probability" of going from $\mathbf{R}''$, L to $\mathbf{R}$, $L + \epsilon$, "summed" over all intermediate $\mathbf{R}''$.

Having established the Markov nature of G as defined by (3.20), we now return to the case of ϵ as an infinitesimal. As in (3.24), before taking the limit we can write

$$G(\mathbf{R}, \mathbf{R}''; L + \epsilon, L) \simeq \left(\frac{3}{2\pi l\epsilon}\right)^{3/2} \exp\left[\frac{-3(\mathbf{R} - \mathbf{R}'')^2}{2l\epsilon} - \epsilon V\left(\frac{\mathbf{R} + \mathbf{R}''}{2}\right)\right] \quad (3.35)$$

Substituting (3.35) into (3.34) and changing variables to $\mathbf{R}'' = \mathbf{R} + \boldsymbol{\eta}$, so $\int d\mathbf{R}'' = \int d\boldsymbol{\eta}$ gives

$$G(\mathbf{R}, \mathbf{R}'; L + \epsilon, 0)$$
$$= \int d\boldsymbol{\eta} \left(\frac{3}{2\pi l\epsilon}\right)^{3/2} \exp\left[\frac{-3\eta^2}{2l\epsilon} + V\left(\mathbf{R} + \frac{\boldsymbol{\eta}}{2}\right)\right] G(\mathbf{R} + \boldsymbol{\eta}, \mathbf{R}'; L, 0) \quad (3.36)$$

Since the term $\mathcal{N} \exp[-3\eta^2/2l\epsilon]$ has an essential singularity as $\epsilon \to 0^+$ (i.e., the delta function), it cannot be expanded in a power series in ϵ. Since ϵ is taken to be an infinitesimal this "almost delta function" implies that only infinitesimal $\boldsymbol{\eta}$ can contribute to the integral. Formally only $|\boldsymbol{\eta}| \sim \mathcal{O}(\epsilon^{1/2})$ can give nonnegligible contributions. Hence expanding in powers of ϵ gives

$$\epsilon V\left(\mathbf{R} + \frac{\boldsymbol{\eta}}{2}\right) = \epsilon V(\mathbf{R}) + \frac{\epsilon\boldsymbol{\eta}}{2} \cdot \nabla_{\mathbf{R}} V(\mathbf{R}) + \mathcal{O}(\epsilon\eta^2) \quad (3.37)$$

The expansion of the nonsingular part of the exponential of (3.37) leads to

$$\exp\left[-\epsilon V\left(\mathbf{R} + \frac{\boldsymbol{\eta}}{2}\right)\right] = 1 - \epsilon V(\mathbf{R}) + \mathcal{O}(\epsilon\boldsymbol{\eta}, \epsilon\eta^2, \epsilon^2) \quad (3.38)$$

while

$$G(\mathbf{R}, \mathbf{R}'; L + \epsilon, 0) = G(\mathbf{R}, \mathbf{R}'; L, 0) + \epsilon \frac{\partial}{\partial L} G(\mathbf{R}, \mathbf{R}'; L, 0) + \mathcal{O}(\epsilon^2) \quad (3.39a)$$

and

$$G(\mathbf{R} + \boldsymbol{\eta}, \mathbf{R}'; L, 0) = G(\mathbf{R}, \mathbf{R}'; L, 0) + \boldsymbol{\eta} \cdot \nabla_{\mathbf{R}} G(\mathbf{R}, \mathbf{R}'; L, 0)$$
$$+ \tfrac{1}{2}\boldsymbol{\eta}\boldsymbol{\eta} : \nabla_{\mathbf{R}}\nabla_{\mathbf{R}} G(\mathbf{R}, \mathbf{R}'; L, 0) + \mathcal{O}(\eta^3, \eta^4) \quad (3.39b)$$

In (3.39b) we use the definition

$$\mathbf{AB} : \mathbf{CD} \equiv \sum_{ij} A_i B_j C_j D_i = (\mathbf{A} \cdot \mathbf{D})(\mathbf{B} \cdot \mathbf{C})$$

Substituting (3.38) and (3.39) into (3.36) and performing the trivial $\boldsymbol{\eta}$ integrations

$$\int d\boldsymbol{\eta} \left(\frac{3}{2\pi l\epsilon}\right)^{3/2} \begin{pmatrix} 1 \\ \boldsymbol{\eta} \\ \eta_i\eta_j \\ \eta^3 \\ \eta^4 \end{pmatrix} \exp\left(\frac{-3\eta^2}{2l\epsilon}\right) = \begin{pmatrix} 1 \\ 0 \\ \dfrac{\delta_{ij} l\epsilon}{3} \\ 0 \\ \epsilon^2 \end{pmatrix} \quad (3.40)$$

gives

$$G(\mathbf{R}, \mathbf{R}'; L, 0) + \epsilon \frac{\partial}{\partial L} G(\mathbf{R}, \mathbf{R}'; L, 0) + \mathcal{O}(\epsilon^2)$$

$$= G(\mathbf{R}, \mathbf{R}'; L, 0) + \epsilon\left[-V(\mathbf{R}) + \frac{l}{6}\nabla_{\mathbf{R}}^2\right]G(\mathbf{R}, \mathbf{R}'; L, 0) + \mathcal{O}(\epsilon^2) \quad (3.41)$$

where $\mathbf{1}:\nabla_{\mathbf{R}}\nabla_{\mathbf{R}} = \nabla_{\mathbf{R}}^2$ has been used. Dividing by ϵ and taking the $\lim \epsilon \to 0$, finally yields (3.21).

Since negative chain lengths $L = -|a|$ are meaningless, we can define a new function

$$\begin{aligned} \tilde{G}(\mathbf{R}, \mathbf{R}'; L) &= G(\mathbf{R}, \mathbf{R}'; L), \qquad L \geqslant 0 \\ &= 0 \qquad\qquad\qquad\qquad L < 0 \end{aligned} \quad (3.42)$$

$\tilde{G}$ has a discontinuity of $\delta(\mathbf{R} - \mathbf{R}')$ across $L = 0$, and therefore it satisfies the equation

$$\left[\frac{\partial}{\partial L} - \frac{l}{6}\nabla_{\mathbf{R}}^2 + V(\mathbf{R})\right]\tilde{G}(\mathbf{R}, \mathbf{R}'; L) = \delta(L)\,\delta(\mathbf{R} - \mathbf{R}') \quad (3.43)$$

$\tilde{G}$ is then the Green's function for the diffusion equation with the "diffusion" constant $D = l/6$ [see (3.23)]. The "diffusing particle" is initially at $\mathbf{R}'$ at "time" $L = 0$. [Wiener integrals were initially used to describe Brownian motion!]

This example of a flexible polymer chain in the presence of an external field serves to demonstrate how the use of the limit of a continuous chain leads to distributions that satisfy differential equations. This limit therefore introduces an analogy between the configurations of the continuous equivalent chain and the path of a particle which is undergoing Brownian or diffusive motion.[8–12] It is also useful to relate this continuous chain description to Feynman's path integral formulation of quantum mechanics.[12] This relationship permits the exploitation of mathematical analogies which might lead to suitable approximation schemes. Reference is made to the book by Feynman and Hibbs for certain mathematical details which can be transcribed between quantum mechanical and polymer problems.

IV. FEYNMAN PATH INTEGRALS IN QUANTUM MECHANICS

The diffusion equation in an external field (3.21) or (3.43) is the same as the nonrelativistic Schrödinger equation for a particle in an external field, apart from the appearance of the imaginary unit i which appears in the Schrödinger equation. Specifically, this Schrödinger equation for the space-time propagator $K(\mathbf{R}, \mathbf{R}'; t, t')$, which gives the probability amplitude that the particle initially at $\mathbf{R}'$ at t' is at $\mathbf{R}$ at time t, is given by

$$\left[i\hbar \frac{\partial}{\partial t} + \frac{\hbar^2}{2m} \nabla_{\mathbf{R}}{}^2 - V(\mathbf{R})\right] K(\mathbf{R}, \mathbf{R}'; t, t') = \delta(t - t')\, \delta(\mathbf{R} - \mathbf{R}') \tag{4.1}$$

In (4.1) we impose the boundary condition $K = 0$ if $t < t'$. Because of the similarity between (4.1) and (3.43), it is not surprising that K can be expressed as a functional integral. We could, at this point, quote the result, and then a derivation analogous to (3.26) to (3.43) would prove that indeed the result is the solution to (4.1). It is, however, instructive to derive the result as opposed to Feynman's approach of introducing it as a postulate.

A numerical analyst might solve (4.1) by breaking the time interval $t - t'$ into a number of small intervals ϵ_j such that $\sum_{j=1}^{n} \epsilon_j = t - t'$ (the intervals need not be all the same size). Then the equation is solved to the desired accuracy for each interval. For the $\{\epsilon_j\}$ small enough, (4.1) can be solved on each interval to first order in ϵ. The result is

$$K(\mathbf{r}_j, \mathbf{r}_{j-1}; t_j, t_{j-1}) = \left(\frac{m}{2\pi i \hbar \epsilon_j}\right)^{3/2} \times \exp\left[\frac{i}{\hbar} \frac{m}{2} \frac{(\mathbf{r}_j - \mathbf{r}_{j-1})^2}{\epsilon_j} - \frac{i\epsilon_j}{\hbar} V\left(\frac{\mathbf{r}_j + \mathbf{r}_{j+1}}{2}\right)\right] \tag{4.2}$$

as is easily verified. In solving (4.1) via (4.2) for the first interval ϵ_1, we obtain the probability amplitude that a particle at $\mathbf{R}'$ at t' be at some generic point $\mathbf{r}_1$ at $t_1 = t + \epsilon_1$. The solution for the second interval gives the probability amplitude of going from $\mathbf{r}_1$ at t_1 to some $\mathbf{r}_2$ at t_2, etc. Clearly, the probability amplitude for going from $\mathbf{R}'t' \to \mathbf{R}t$ is not dependent on the intermediate points $\mathbf{r}_1, \mathbf{r}_2, \ldots, \mathbf{r}_{n-1}$, so the overall probability amplitude is the product of the probability amplitudes for each of the intervals "summed" (i.e., integrated) over all intermediate points. Thus in this finite difference approximation,

$$K(\mathbf{R}, \mathbf{R}'; t, t') \cong \frac{\prod_{j=1}^{n} \left[\int d\mathbf{r}_j K(\mathbf{r}_j, \mathbf{r}_{j-1}; t_j, t_{j-1})\right]}{d\mathbf{r}_n} \tag{4.3}$$

Analogous to (3.6), the points $\{\mathbf{r}_k\}$ are the discrete representations of the continuous curve $\mathbf{r}(t)$. Here $\mathbf{r}(t)$ is a possible trajectory followed by the particle. Thus the product of the K's in (4.3) gives the probability amplitude density that the particle follow the trajectory $\{\mathbf{r}_k\}$. As the integration is over all possible intermediate points $\{\mathbf{r}_k \mid k = 1, \ldots, n - 1\}$, the multiple integration corresponds to an integration over all possible particle trajectories (for the discrete subdivision of time) from the initial point $\mathbf{R}'$, t' to the final point $\mathbf{R}$, t.

The theoretician tells us that the solution (4.3) is an exact solution to (4.1) provided that the ϵ_j are taken to be arbitrarily small, or according to the mathematician $\max \epsilon_j \to 0$, $n \to \infty$, $\sum_{j=1}^{n} \epsilon_j = t - t'$. In this limit (4.3) is then a functional integral which we write as

$$K(\mathbf{R}, \mathbf{R}'; t, t') = \int_{\mathbf{r}(t')=\mathbf{R}'}^{\mathbf{r}(t)=\mathbf{R}} \mathscr{D}[\mathbf{r}(\tau)] \exp\left[\frac{i}{\hbar}\int_{t'}^{t} d\tau \mathscr{L}[\dot{\mathbf{r}}(\tau), \mathbf{r}(\tau)]\right] \tag{4.4}$$

where

$$\mathscr{L}[\dot{\mathbf{r}}(\tau), \mathbf{r}(\tau)] = \frac{m}{2}\dot{\mathbf{r}}^2(\tau) - V[\mathbf{r}(\tau)] \tag{4.5}$$

is the classical Lagrangian, and $\mathscr{D}[\mathbf{r}(\tau)]$ denotes that the normalization is included along with $\delta\mathbf{r}(t)$. [The mathematician is still unhappy because all of the exponentials in (4.2) are highly oscillatory, but his complaint is easily dismissed by taking $\mathscr{Im}\, \hbar = -|\gamma|$ and then letting $|\gamma| \to 0^+$ at the end of the calculation.]

Equations (4.4) and (4.5) present an interesting physical picture of nonrelativistic quantum mechanics. A similar approach can be employed in the treatment of thermal properties in quantum statistical mechanics.[8–12] In quantum statistical mechanics, the Slater sum, or thermal Green's function, is written in Dirac notation as

$$\rho(\mathbf{rr}'; \beta) = \langle \mathbf{r}| \exp(-\beta H) |\mathbf{r}'\rangle \tag{4.6}$$

where H is the Hamiltonian, and the canonical partition function is

$$Z = \int d\mathbf{r} \rho(\mathbf{rr}; \beta) \tag{4.6a}$$

Since

$$\lim_{\beta \to 0} \rho(\mathbf{rr}'; \beta) = \langle \mathbf{r}| \mathbf{1} |\mathbf{r}'\rangle = \delta(\mathbf{r} - \mathbf{r}') \tag{4.6b}$$

if $\rho(\ldots; \beta)$ is defined to be zero for $\beta < 0$, ρ satisfies the Bloch equation

$$\left(\frac{\partial}{\partial\beta} + H\right)\rho = \delta(\beta)\mathbf{1} \tag{4.7a}$$

Explicitly, for a single particle in an external field $V(\mathbf{r})$, we have

$$\left[\frac{\partial}{\partial\beta} - \frac{\hbar^2}{2m}\nabla_{\mathbf{R}}^{\ 2} + V(\mathbf{r})\right]\rho(\mathbf{rr}'; \beta) = \delta(\beta)\,\delta(\mathbf{r} - \mathbf{r}') \tag{4.7b}$$

But (4.7b) is also a diffusion equation which admits the Wiener integral representation

$$\rho(\mathbf{r}, \mathbf{r}'; \beta) = \int_{\mathbf{r}(0)=\mathbf{r}'}^{\mathbf{r}(\beta)=\mathbf{r}} \mathscr{D}[\mathbf{r}(\tau)] \exp\left(-\int_0^\beta d\tau \left\{\frac{m\dot{\mathbf{r}}^2(\tau)}{2\hbar^2} + V[\mathbf{r}(\tau)]\right\}\right) \quad (4.7c)$$

Since our purpose is to give a better insight into the representation of the configurational of polymer chains in terms of functional integrals and to enable the use of any results obtained from analogous quantum mechanical problems, we need not pursue the discussion of Feynman's path integral formulation of quantum mechanics any further. Therefore we return now to a discussion of polymer systems, digressing to Feynman path integrals when the analogy is fruitful or when references are needed. (We also note in passing that functional integrals have also been employed to provide a formal representation of the propagator for the Liouville equation of classical nonequilibrium statistical mechanics.[36])

V. INTRODUCTION TO FUNCTIONAL INTEGRATION: STIFF POLYMER CHAINS

A. Introduction

Thus far we have the functional integral representation for the distribution functions involved in the configurational statistics of flexible polymers. By considering the polymer in the presence of an external field, we are able to relate the configurations of a polymer to the paths of a particle when this particle is undergoing Brownian or diffusive motion, or when it is evolving according to the laws of quantum mechanics. This establishes connections with other familiar concepts.

In this section we consider functional integral representations for the distribution functions for stiff polymer chains. Aside from providing a class of useful models of stiff polymer chains, these results illustrate the following:

1. They provide functional integrals which can be evaluated directly.
2. They give some examples in which the derivation of the equation of motion analogous to (3.21) becomes interesting.
3. They relate stiff chains to cases of Brownian motion where the particle's position, velocity, etc., are considered. Connection is also established between stiff polymers and the quantum mechanics of systems which classically would have velocity, acceleration, etc., dependent forces.

The excluded volume problem is considered in the next section, and then we discuss problems involving polymers in bulk.

We have already seen how the introduction of a continuous equivalent chain leads to the use of (Wiener) functional integrals. The Kratky-Porod wormlike chain, which is a popular model for stiff chains, is explicitly obtained in Section II as the limit of a continuous chain. Therefore the techniques of functional integration must also be naturally suited to the study of the configurational statistics of stiff "ideal" polymers. Saitô, Takahashi, and Yunoki (STY) have used such an approach to provide a functional integral representation for the wormlike chain.[20] Unfortunately, the end-to-end vector distribution for this model has not been obtained in a simple, useful closed form. The first few moments $\langle \mathbf{R}^2 \rangle$, $\langle \mathbf{R}^4 \rangle$, and $\langle \mathbf{R}^6 \rangle$, however, have been evaluated. As stressed in Sections I and II, it is imperative to have these simple models when considering problems which are not exactly soluble. Harris and Hearst have introduced a model similar to that of STY without the use of functional integrals.[37] Harris and Hearst use an approximation in which the polymer chains are no longer locally inextensible; they are only globally inextensible. If an approximation of this spirit is introduced into the functional integral formulation, the resultant integrals are exactly soluble and lead to the requisite simple model. This model implies polymer properties which are qualitatively the same as those obtained for the wormlike chain.

The stiff chain is taken to be described by the continuous spare curve $\mathbf{r}(s)$, $0 \leqslant s \leqslant L$, with respect to an arbitrary origin of coordinates, and s again represents the contour length along the chain. From elementary differential geometry, the vector

$$\mathbf{u}(s) = \frac{\partial \mathbf{r}(s)}{\partial s} \equiv \dot{\mathbf{r}}(s) \tag{5.1}$$

must be a unit vector.[38] This follows because the element of arc length ds along the curve $\mathbf{r}(s)$ is defined by $\{\mathbf{r}(s) = [x(s), y(s), z(s)]\}$,

$$ds = [dx^2 + dy^2 + dz^2]^{1/2} = \left[\left(\frac{\partial x}{\partial s}\right)^2 + \left(\frac{\partial y}{\partial s}\right)^2 + \left(\frac{\partial z}{\partial s}\right)^2\right]^{1/2} ds \tag{5.2}$$

so that

$$|\mathbf{u}(s)|^2 \equiv |\dot{\mathbf{r}}(s)|^2 \equiv 1 \tag{5.2a}$$

Thus $\mathbf{u}(s)$ is a unit vector which is tangent to the curve $\mathbf{r}(s)$.

The chain configuration could equally well be described by the tangent curve $\mathbf{u}(s)$ and the condition that the end-to-end vector have some chosen value $\mathbf{R}$ [for the case $\mathbf{r}(0) \equiv \mathbf{0}$]. Given $\mathbf{u}(s)$ and $\mathbf{R}$ we can always obtain $\mathbf{r}(s)$ from

$$\mathbf{r}(s) = \int_0^s ds' \mathbf{u}(s') \equiv \int_0^s ds' \frac{\partial \mathbf{r}(s')}{\partial s'} \tag{5.3}$$

with

$$\mathbf{R} = \int_0^L ds \mathbf{u}(s) = \mathbf{r}(L) - \mathbf{r}(0) \tag{5.4}$$

STY relate averages, such as $\langle \mathbf{R}^2 \rangle$ and $\langle \mathbf{R}^4 \rangle$, to averages over the auxiliary distribution of initial ($s = 0$) and final ($s = L$) tangent vectors.[20] (Even this auxiliary distribution is expressed only as an infinite series expansion.)

In the approach of STY, the condition (5.2a) is imposed for all s ($0 \leqslant s \leqslant L$).[20,38] However, following Harris and Hearst, we might attempt to relax the condition (5.2a) in the Wiener integral formulation of stiff chains of STY. Such an approximation rests upon much more fundamental grounds than the associated approximation in the treatment of polymer dynamics. The Wiener integral formulation of flexible polymer chains *exactly* reproduces the asymptotic gaussian properties of these flexible chains. The use of the condition (5.2a) in this case *would be wrong mathematically and in fact would spoil this agreement.*

This new model of stiff polymer chains leads to some general considerations of Markov processes in multidimensional spaces and to some generalized Wiener integral descriptions of these Markov processes. Furthermore, such generalized discussions of Wiener integrals also lead to a large number of other possible models of stiff polymer chains. These models can be considered to be the continuous limit analogs of rotational-isomeric type models just as the simple Wiener integral represents the continuous limit of the random flight chain.

B. Stiffness

As STY discuss, for chains with stiffness, we associate a potential energy with the bending of the chain. The potential energy per unit length stored in a bent rod is[20,37,38]

$$v = \frac{\frac{1}{2}\epsilon}{\mathscr{R}_c^{\,2}} \tag{5.5}$$

where ϵ is the bending force constant and $\mathscr{R}_c$ is the radius of curvature. This radius is defined by

$$\mathscr{R}_c^{\,-1} = \left| \frac{\partial \mathbf{u}(s)}{\partial s} \right| \equiv |\ddot{\mathbf{r}}(s)| \equiv \left| \frac{\partial^2 \mathbf{r}(s)}{\partial s^2} \right| \tag{5.6}$$

For a chain of contour length L the total potential energy is

$$V_2 = \tfrac{1}{2}\epsilon \int_0^L \left| \frac{\partial \mathbf{u}(s)}{\partial s} \right|^2 ds = \tfrac{1}{2}\epsilon \int_0^L |\ddot{\mathbf{r}}(s)|^2 \, ds \tag{5.7}$$

As in (2.2) and (2.8), this potential energy gives rise to the Boltzmann factor $\exp(-\beta V_2)$. If gaussian equivalent links are used for the discrete chain, (2.8) and (5.7) give

$$G(\{\mathbf{r}_k\}) = \prod_{j=1}^{n} \left(\frac{3}{2\pi l \, \Delta s} \right)^{3/2} \exp \left\{ - \left(\frac{3 \dot{\mathbf{r}}_j^{\,2}}{2 l \, \Delta s} \right) - \left(\frac{\beta \epsilon [\dot{\mathbf{r}}_j - \dot{\mathbf{r}}_{j-1}]^2}{2 \, \Delta s} \right) \right. \tag{5.8}$$

where we define

$$\mathbf{u}_j = \dot{\mathbf{r}}_j \equiv \frac{\mathbf{r}_j - \mathbf{r}_{j-1}}{\Delta s} = \frac{\mathbf{R}_j}{\Delta s} \tag{5.9}$$

(We could allow a set of different Δs_j, but in the continuous limit this would just provide a more rigorous definition of the functional integral.) Note that (5.8) contains terms like $\mathbf{R}_j \cdot \mathbf{R}_{i-1}$ which implies correlations between adjacent bond vectors. The potential V_2 therefore can give rise to stiffness. Equation (5.8) also has $\dot{\mathbf{r}}_0$ from the $j = 0$ term in the product. Since we can choose $\mathbf{r}_0 \equiv \mathbf{0}$, by (5.9) $\dot{\mathbf{r}}_0$ involves a quantity $\mathbf{r}_{-1}$ which is undefined. Thus it is necessary to specify the initial value $\dot{\mathbf{r}}_0 = \mathbf{u}_0$ in order that (5.8) have a definite meaning. Since it is necessary to specify $\mathbf{u}_0$, it is also convenient to consider the specification of the other endpoint $\mathbf{u}_n$. This is wholly analogous to the specification of $\mathbf{r}_0$ and/or $\mathbf{r}_n$ in the Wiener integral representation for flexible polymer chains.

In the limit of a continuous chain, (5.8) becomes

$$G[\mathbf{r}(s)] = \mathcal{N} \exp\left\{-\frac{3}{2l}\int_0^L \dot{\mathbf{r}}^2(s)\, ds - \frac{\beta\epsilon}{2}\int_0^L \ddot{\mathbf{r}}^2(s)\, ds\right\} \tag{5.10}$$

along with the requisite endpoint constraints. Equation (5.10) then represents the (unnormalized) "probability" for the configuration $\mathbf{r}(s)$. Apart from a change in normalization, (5.10) implies that

$$G'[\mathbf{u}(s)] = \mathcal{N}' \exp\left\{-\frac{3}{2l}\int_0^L \mathbf{u}^2(s)\, ds - \frac{\beta\epsilon}{2}\int_0^L \dot{\mathbf{u}}^2(s)\, ds\right\} \tag{5.11}$$

along with the conditions on the endpoints, also provides the (unnormalized) "probability" for the configuration $\mathbf{u}(s)$. However, according to the relations (5.3) and (5.4), (5.10) and (5.11) differ only in their overall normalization. [Our choice of $\mathbf{r}(0) = \mathbf{0}$ is, of course, valid for (5.10) and (5.11), because of the isotropy and translational invariance of space.] Thus we may focus attention upon (5.11).

C. An Analogy with Brownian Motion

Before we consider the reduced endpoint distributions, it is convenient to note the analogy between (5.10) and (5.11) and situations involving Brownian motion.[6–10,39] If we consider cases of Brownian motion for which the inertial term $m\ddot{\mathbf{r}}(t)$ (t is the time) in the equations of motion can be neglected, then the Wiener measure

$$P[\mathbf{r}(\tau)] = \mathscr{D}[\mathbf{r}(\tau)] \exp\left\{-\frac{1}{4D}\int_0^t d\tau \dot{\mathbf{r}}^2(\tau)\right\} \tag{5.12}$$

provides the "probability" that the particle follow the trajectory $\mathbf{r}(\tau)$. The reduced distribution

$$P(\mathbf{r}t \mid \mathbf{r}'0) = \int_{\mathbf{r}(0)=\mathbf{r}'}^{\mathbf{r}(t)=\mathbf{r}} \mathscr{D}[\mathbf{r}(\tau)] \exp\left\{-\frac{1}{4D}\int_0^t d\tau \dot{\mathbf{r}}^2(\tau)\right\} \tag{5.13}$$

gives the conditional probability density that a particle initially at $\mathbf{r}'$ at time $\tau = 0$ be found at time $\tau = t$ at $\mathbf{r}$. By comparison of (5.13) with (3.20) for $V = 0$, (3.21) implies that $P(\mathbf{r}t \mid \mathbf{r}'0)$ satisfies the usual diffusion equation

$$\left[\frac{\partial}{\partial t} - D\nabla_{\mathbf{r}}^2\right] P(\mathbf{r}t \mid \mathbf{r}'0) = \delta(\mathbf{r} - \mathbf{r}')\,\delta(t) \tag{5.14}$$

if $P(\mathbf{r}t \mid \mathbf{r}'0)$ is taken to vanish for $t < 0$. Equation (5.14) results from the well-known fact that diffusion is the long time limit of Brownian motion.

Since (5.10) contains $\ddot{\mathbf{r}}(s)$, it is formally analogous to cases of Brownian motion for which the inertial term is not negligible. Because the equation of motion (e.g., the Langevin equation) contains $\ddot{\mathbf{r}}(t)$, it is necessary to specify the particle's initial position $\mathbf{r}(0)$ and velocity $\mathbf{v}(0)$. Because of the random force acting upon this Brownian particle, its final position $\mathbf{r}(t)$ and velocity $\mathbf{v}(t)$ are not uniquely specified. Thus there is a conditional probability distribution $P(\mathbf{rv}t \mid \mathbf{r}'\mathbf{v}'0)$ that a particle initially having position and velocity $\mathbf{r}'$ and $\mathbf{v}'$, respectively, be at $\mathbf{r}$ and $\mathbf{v}$ at time t. This analogy between (5.10) and (5.11) and cases of Brownian motion may aid in clarifying the necessity for the specification of the endpoint(s) for the tangent vector curve $\mathbf{u}(s)$ in the polymer case.

D. The Reduced Distributions

Although we wish to evaluate the end vector distribution $G(\mathbf{R0}; L0)$ for the stiff chain, we are forced to consider first the distribution function $G(\mathbf{R0}, \mathbf{uu}'; L0)$ for chains having $\mathbf{r}(L) = \mathbf{R}$ and $\mathbf{r}(0) = \mathbf{0}$ *and* $\mathbf{u}(0) = \mathbf{U}'$ and $\mathbf{u}(L) = \mathbf{U}$! This distribution is obtained by integrating (5.11) over all curves $\mathbf{u}(s)$, subject to the condition (5.4),

$$G(\mathbf{R0}, \mathbf{UU}'; L0) = \int_{\mathbf{u}(0)=\mathbf{U}'}^{\mathbf{u}(L)=\mathbf{U}} \mathscr{D}[\mathbf{u}(s)]\, \delta\left[\mathbf{R} - \int_0^L \mathbf{u}(s)\, ds\right] \times \exp\left\{-\int_0^L ds\left[\frac{\beta\epsilon}{2}\dot{\mathbf{u}}(s)^2 + \frac{3}{2l}\mathbf{u}(s)^2\right]\right\} \tag{5.15}$$

The Dirac delta function in (5.15) merely selects out those curves $\mathbf{u}(s)$ that satisfy the constraint (5.4). The desired end-to-end vector distribution $G(\mathbf{R0}; L0)$ is then to be obtained from (5.15) by some (as yet unspecified) average over $\mathbf{U}$ and $\mathbf{U}'$,

$$G(\mathbf{R0}; L0) = \langle\langle G(\mathbf{R0}, \mathbf{UU}'; L0)\rangle\rangle \tag{5.16}$$

If we take $\int d\mathbf{R}$ of (5.15), the useful auxiliary distribution $G(\mathbf{UU}'; L0)$ is obtained,

$$G(\mathbf{UU}'; L0) \equiv \int_{\mathbf{u}(0)=\mathbf{U}'}^{\mathbf{u}(L)\mathbf{U}} \mathscr{D}[\mathbf{u}(s)] \exp\left\{-\int_0^L ds\left[\frac{\beta\epsilon}{2}\dot{\mathbf{u}}^2(s) + \frac{3}{2l}\mathbf{u}^2(s)\right]\right\} \quad (5.17)$$

[In obtaining (5.17) from (5.15), the orders of the $\mathbf{R}$ and the functional integration have been interchanged.]

The normalization $\mathscr{D}[\mathbf{u}(s)]$ is chosen so that $G(\mathbf{UU}'; L0)$ has the usual normalization required to have

$$\lim_{L\to 0} G(\mathbf{UU}'; L0) \to \delta(\mathbf{U} - \mathbf{U}') \quad (5.18a)$$

Therefore, since

$$\lim_{L\to 0} \delta\left[\mathbf{R} - \int_0^L ds\mathbf{u}(s)\, ds\right] \equiv \delta(\mathbf{R}) \quad (5.18b)$$

we immediately find that

$$\lim_{L\to 0} G(\mathbf{R}0, \mathbf{UU}'; L0) = \delta(\mathbf{R})\, \delta(\mathbf{U} - \mathbf{U}') \quad (5.18c)$$

Essentially, "Markov's method" corresponds to the use of the Fourier representation of the Dirac delta function in (5.15),

$$\delta\left[\mathbf{R} - \int_0^L ds\mathbf{u}(s)\right] = \int \frac{d^3k}{(2\pi)^3} \exp\left\{-i\mathbf{k}\cdot\left[\mathbf{R} - \int_0^L \mathbf{u}(s)\, ds\right]\right\} \quad (5.19)$$

This enables (5.15) to be rewritten as

$$G(\mathbf{R}0, \mathbf{UU}'; L0) = \int \frac{d^3k}{(2\pi)^3} \exp(-i\mathbf{k}\cdot\mathbf{R}) I(\mathbf{k}, \mathbf{UU}'; L) \quad (5.20)$$

where

$$I(\mathbf{k}, \mathbf{UU}'; L) \equiv \int_{\mathbf{u}(0)=\mathbf{U}'}^{\mathbf{u}(L)=\mathbf{U}} \mathscr{D}[\mathbf{u}(s)]$$

$$\times \exp\left\{-\int_0^L ds\left[\frac{\beta\epsilon}{2}\dot{\mathbf{u}}^2(s) + \frac{3}{2l}\mathbf{u}^2(s) - i\mathbf{k}\cdot\mathbf{u}(s)\right]\right\} \quad (5.21)$$

Inverting the transform (5.20) implies that

$$I(\mathbf{k}, \mathbf{UU}'; L) = \int d\mathbf{R} \exp(i\mathbf{k}\cdot\mathbf{R}) G(\mathbf{R}0, \mathbf{UU}'; L0) \quad (5.22)$$

and I is the characteristic function of $G(\mathbf{R}0, \mathbf{UU}'; L0)$ with respect to $\mathbf{R}$. As is well known, many of the moments of G with respect to $\mathbf{R}$ can easily be obtained from I (see below). Furthermore, I is related by[5,20]

$$I(\theta) = Z^{-1}\int_0^L ds \int_0^L ds' \langle\langle I(\mathbf{k}, \mathbf{UU}'; s - s')\rangle\rangle \quad (5.23)$$

to the angular distribution of the intensity, $I(\theta)$, of light scattered by the polymer. In (5.23), θ is the scattering angle

$$|\mathbf{k}| = \frac{2\pi}{\lambda} \sin\left(\frac{\theta}{2}\right) \tag{5.23a}$$

λ is the wavelength of the light, and Z is the partition function

$$Z = \int d\mathbf{R} G(\mathbf{R0}; L0)$$

The averaging in (5.23) is taken to be that appropriate to (5.16).

Equations (5.15) and (5.21) are familiar Wiener integrals which are easily evaluated in closed form, *provided* (5.2*a*) *is not invoked.* Before considering this case, it is instructive to recover the results of STY.

E. The STY Approach[20]

Applying the STY constraint (5.2a) to (5.17) and (5.21), we find that the term containing $\mathbf{u}(s)^2$ gives

$$\int_0^L ds \mathbf{u}(s)^2 \equiv L \tag{5.24}$$

This uninteresting factor of $\exp(-3L/2l)$ can be absorbed into the normalization. Thus, if the subscript c labels quantities which are constrained according to (5.2a), then (5.17) becomes

$$G_c(\mathbf{UU}'; L0) = \int_{\mathbf{u}(0)=\mathbf{U}'}^{\mathbf{u}(L)=\mathbf{U}} \mathscr{D}[\mathbf{u}(s)] \exp\left[-\frac{\beta\epsilon}{2}\int_0^L \dot{\mathbf{u}}^2(s)\, ds\right] \tag{5.25}$$

where

$$|\mathbf{u}(s)| = 1, \qquad \text{all } s$$

Equation (5.25) is a simple Wiener integral which is subject to (5.2a). Therefore, from (5.25) with (3.20)–(3.22), G_c satisfies the diffusion equation

$$\left[\frac{\partial}{\partial L} - \frac{1}{2\beta\epsilon}\nabla_{\mathbf{U}}^2\right] G_c(\mathbf{UU}'; L0) = \delta(L)\,\delta(\mathbf{U} - \mathbf{U}') \tag{5.26}$$

where $\mathbf{U}$ and $\mathbf{U}'$ are unit vectors. Equation (5.26) is understood to be taken in polar coordinates with, e.g., $\mathbf{U} = (1, \theta, \phi)$, etc., for $\mathbf{U}'$, $\nabla_{\mathbf{U}}^2$, and $\delta(\mathbf{U} - \mathbf{U}')$ in terms of θ, ϕ, θ', and ϕ'. Equation (5.26) can be solved- as an infinite series in the eigenfunctions, $Y_{lm}(\theta, \phi)$, the spherical harmonics of the angles of $\mathbf{U}$. Although (5.25) and (5.26) do not describe the desired $G_c(\mathbf{R0}; L0)$, STY show how $\langle \mathbf{R}^2 \rangle$ and $\langle \mathbf{R}^4 \rangle$ can be obtained as moments of $G_c(\mathbf{UU}'; L0)$. These results are identical to those obtained for the wormlike chain. Thus $G_c(\mathbf{R0}; L0)$ is not explicitly needed in order to obtain these averages. Note that STY obtain $\langle \mathbf{R}^2 \rangle$, $\langle \mathbf{R}^4 \rangle$, etc., by interpreting the averages in (5.16), etc., to be

$$\langle\langle F \rangle\rangle \equiv \int d\Omega_{\mathbf{U}} \int d\Omega_{\mathbf{U}'} F \tag{5.27}$$

where

$$d\Omega_{\mathbf{U}} = \frac{1}{4\pi} \sin\theta \, d\theta \, d\phi$$

Equation (5.27) simply implies that all initial orientations of the chain which are described by the unit vector $\mathbf{U}'$ are equally probable. This result is physically clear.

STY use the obvious analog of (5.20) to relate $G_c(\mathbf{R0}, \mathbf{UU}'; L0)$ to $I_c(\mathbf{k}, \mathbf{UU}'; L0)$, which gives the angular distribution of scattered light [see (5.23)] when the average (5.27) is employed. Explicitly, I_c is given by

$$I_c(\mathbf{k}, \mathbf{UU}'; L0) = \int_{\mathbf{u}(0)=\mathbf{U}'}^{\mathbf{u}(L)=\mathbf{U}} \mathscr{D}[\mathbf{u}(s)] \times \exp\left\{-\int_0^L ds \left[\frac{\beta\epsilon}{2} \dot{\mathbf{u}}(s)^2 - i\mathbf{k}\cdot\mathbf{u}(s)\right]\right\} \tag{5.28a}$$

subject to (5.2a), and therefore

$$\left[\frac{\partial}{\partial L} - \frac{1}{2\beta\epsilon}\nabla_{\mathbf{U}}^2 - i\mathbf{k}\cdot\mathbf{U}\right] I_c(\mathbf{k}; \mathbf{UU}'; L) = \delta(L)\,\delta(\mathbf{U}-\mathbf{U}') \tag{5.28b}$$

is also to be interpreted in polar coordinates. But the simpler $G_c(\mathbf{UU}'; L0)$ of (5.26) is already expressed as an infinite series, and the $\mathbf{k}$ dependence in (5.28b) is not expected to simplify matters.

Therefore, in an attempt to obtain simple analytic expressions for the distribution functions of stiff polymer chains, condition (5.2a) is relaxed. The relaxation of this condition is in the original spirit of the use of Wiener integrals. If this condition were imposed for flexible polymer chains, the Wiener measure would be $\mathscr{D}[\mathbf{r}(s)] \exp(-3L/2l)$ and would give equal weight (measure) to all continuous configurations of the polymer. Thus the use of (5.2a) would not yield the correct gaussian distribution for flexible chains.

The relaxation of condition (5.2a) might appear to be "unphysical" since it allows the curve $\mathbf{u}(s)$ to be arbitrary given the measure (5.11). However, we note that the equivalent chain itself is not a physical chain. It is but a mathematical construct which properly reproduces the generic properties of flexible polymer chains. Hence the utility of the present approximation resides in its ability to reflect those common features of stiff polymer chains.

F. The Simpler Model[21]

As noted previously (5.21) is a simple Wiener integral which can be evaluated by using the diffusion equation

$$\left[\frac{\partial}{\partial L} - \frac{1}{2\beta\epsilon}\nabla_{\mathbf{U}}^2 + \frac{3}{2l}\mathbf{U}^2 - i\mathbf{k}\cdot\mathbf{U}\right] I(\mathbf{k}, \mathbf{UU}'; L) = \delta(L)\,\delta(\mathbf{U}-\mathbf{U}') \tag{5.29}$$

where $\mathbf{U}$ and $\mathbf{U}'$ are not constrained to be unit vectors. However, it should be expected that (5.29) will only approximate the physical behavior of stiff chains if somehow $\mathbf{U}$ and $\mathbf{U}'$ are on the average "close" to unit vectors. Equivalently, by introducing (5.20) and (5.22), we can write

$$\left[\frac{\partial}{\partial L} - \frac{1}{2\beta\epsilon}\nabla_{\mathbf{U}}^2 + \frac{3}{2l}\mathbf{U}^2 + \mathbf{U}\cdot\nabla_{\mathbf{R}}\right] G(\mathbf{R}0, \mathbf{UU}'; L0) = \delta(L)\,\delta(\mathbf{R})\,\delta(\mathbf{U}-\mathbf{U}') \tag{5.30}$$

Equation (5.30) is analogous to a Fokker-Planck equation with a harmonic-velocity-dependent external field instead of a viscous damping force,[6,7] and it is therefore not surprising that it can be solved exactly in closed form. Comparing either (5.29) with (4.7b) or (5.21) with (4.7c), we see that (5.2a) and (5.21) resemble the displaced harmonic oscillator of quantum statistical mechanics.

Therefore, introducing in (5.21) the change in variables

$$\mathbf{v}(s) = \mathbf{u}(s) - \frac{il\mathbf{k}}{3}, \qquad \mathbf{v}(0) = \mathbf{U}' - \frac{il\mathbf{k}}{3}, \qquad \mathbf{v}(L) = \mathbf{U} - \frac{il\mathbf{k}}{3}$$

$$\dot{\mathbf{v}}(s) = \dot{\mathbf{u}}(s), \qquad \mathscr{D}[\mathbf{v}(s)] = \mathscr{D}[\mathbf{u}(s)]$$

leads to

$$\begin{aligned} I(\mathbf{k}, \mathbf{UU}'; L0) &= \exp\left(-\frac{\mathbf{k}^2 Ll}{6}\right)\int_{\mathbf{v}(0)=\mathbf{U}'-il\mathbf{k}/3}^{\mathbf{v}(L)=\mathbf{U}-il\mathbf{k}/3} \mathscr{D}[\mathbf{v}(s)] \\ &\quad\times \exp\left\{-\int_0^L ds\left[\frac{3}{2l}\mathbf{v}^2(s) + \frac{\beta\epsilon}{2}\dot{\mathbf{v}}^2(s)\right]\right\} \\ &\equiv \exp\left(-\frac{\mathbf{k}^2 Ll}{6}\right) G\left(\mathbf{U} - \frac{il\mathbf{k}}{3}, \mathbf{U}' - \frac{il\mathbf{k}}{3}; L0\right) \end{aligned} \tag{5.31}$$

G in (5.31) is the "harmonic oscillator" Slater sum of (5.17), which is well known:[12]

$$\begin{aligned} G(\mathbf{UU}'; L0) &= \left(\frac{b}{\pi \sinh a}\right)^{3/2} \\ &\quad\times \exp\left\{-\frac{b}{\sinh a}\left[(\mathbf{U}^2 + \mathbf{U}'^2)\cosh a - 2\mathbf{U}\cdot\mathbf{U}'\right]\right\} \end{aligned} \tag{5.32}$$

Equation (5.17) is evaluated directly in Appendix A. This calculation involves the solution of the functional integral (5.17) by the expansion of the random variable $\mathbf{u}(s)$ in terms of a Fourier series. This technique is also useful in treatments of problems involving polymers in bulk.

Substituting (5.32) into (5.31) leads to

$$I(\mathbf{k}, \mathbf{UU}'; L) = G(\mathbf{UU}'; L0) \times \exp\left\{-\frac{k^2Ll}{6}\left[1 - \frac{2}{a}\tanh\left(\frac{a}{2}\right)\right] + i\frac{L}{a}\tanh\left(\frac{a}{2}\right)\mathbf{k}\cdot(\mathbf{U}+\mathbf{U}')\right\} \tag{5.33}$$

where

$$a = L\left[\frac{3}{\beta\epsilon l}\right]^{1/2}, \qquad b = \left[\frac{3\beta\epsilon}{4l}\right]^{1/2} \tag{5.33a}$$

Equation (5.33) is a simple closed-form expression as sought. The Fourier integral (5.20) is then a trivial gaussian integral, which implies that $G(\mathbf{R0}; \mathbf{UU}'; L0)$ is a multivariate gaussian distribution in $\mathbf{R}$, $\mathbf{U}$, $\mathbf{U}'$. The simplicity of (5.32)–(5.33) is, of course, of no value to the problem unless it adequately describes the physical properties of stiff polymer chains. We therefore consider the expectation values of $\mathbf{R}^2$, $\mathbf{R}^4$, and $\mathbf{R}\cdot\mathbf{U}'/|\mathbf{U}'|$, the latter being related to the persistence of the initial direction because $\mathbf{U}'/|\mathbf{U}'|$ is a unit vector along the initial bond. By using the Fourier representation (5.20), integrating by parts, dropping surface terms as $\mathbf{R}\to\infty$ in the usual way,[6,7] all of these expectation values can be related to the "harmonic oscillator" G of (5.32) via

$$\left\langle\frac{\mathbf{R}\cdot\mathbf{U}'}{|\mathbf{U}'|}\right\rangle = \frac{L}{a}\tanh\left(\frac{a}{2}\right)\left\langle\!\left\langle\left[|\mathbf{U}'| + \frac{\mathbf{U}\cdot\mathbf{U}'}{|\mathbf{U}'|}\right]G(\mathbf{UU}'; L0)\right\rangle\!\right\rangle \tag{5.34}$$

and

$$\langle\mathbf{R}^2\rangle = Ll\left[1 - \frac{2}{a}\tanh\left(\frac{a}{2}\right)\right]\langle\langle G(\mathbf{UU}'; L0)\rangle\rangle + \left[\frac{L}{a}\tanh\left(\frac{a}{2}\right)\right]^2\langle\langle(\mathbf{U}+\mathbf{U}')^2 G(\mathbf{UU}'; L0)\rangle\rangle \tag{5.35}$$

There is a slightly more involved expression for $\langle\mathbf{R}^4\rangle$ which, as will be shown, is not necessary.

Note that the angular brackets in (5.34) and (5.35) are still to be taken as the average over $\mathbf{U}$ and $\mathbf{U}'$ of (5.16) and (5.23), which is now to be specified. If $G(\mathbf{R0}, \mathbf{UU}'; L0)$ is interpreted, apart from an overall normalization, to be the probability density that a chain of length L have the end-to-end vector $\mathbf{R}$ and $\mathbf{U}(\mathbf{U}')$ along the final (initial) directions, then the average of some quantity $f(\mathbf{R}, \mathbf{U}, \mathbf{U}')$ would be calculated analogously to (5.27), except that an average must be taken over the magnitudes of $\mathbf{U}$ and $\mathbf{U}'$:

$$\langle\langle f(\mathbf{R}, \mathbf{U}, \mathbf{U}')\rangle\rangle \doteq \frac{\int d\mathbf{R} \int d\mathbf{U} \int d\mathbf{U}' f(\mathbf{R}, \mathbf{U}, \mathbf{U}') G(\mathbf{R0}, \mathbf{UU}'; L0)}{\int d\mathbf{R} \int d\mathbf{U} \int d\mathbf{U}' G(\mathbf{R0}, \mathbf{UU}'; L0)} \tag{5.36}$$

The normalization has also been included in (5.36). However, using (5.36), we would obtain $\langle \mathbf{R}^2 \rangle = Ll$, $\langle \mathbf{R}^4 \rangle = \frac{5}{3}(Ll)^2$, i.e., $G(\mathbf{R0}; L0)$ is that for a completely flexible gaussian chain, *which is totally independent of the bending force constant* ϵ *in* (5.10), (5.15), *etc.* Thus the averaging process in (5.36) is not the correct one in this case. The $\int d\mathbf{U}'$in (5.36) implies that all initial values of $\mathbf{U}'$ are equally probable. But in order to describe stiff chains, $\mathbf{U}'$ must somehow be close to a unit vector on the average. Therefore all initial values of $\mathbf{U}'$ are not equally probable, and there exists a probability distribution for initial values $p(\mathbf{U}')$. The analogy between (5.30) and a Fokker-Planck type equation for Brownian motion provides an example in which the kind of averaging (5.36) is also not appropriate.[6,7] For instance, in calculating the equilibrium velocity autocorrelation function $\langle \mathbf{v}(t) \cdot \mathbf{v}(0) \rangle$, (analogous to $\langle\langle \mathbf{U} \cdot \mathbf{U}'G \rangle\rangle$ in the polymer case) there is also a probability distribution for the initial velocity $\mathbf{v}(0)$, namely the Maxwell-Boltzmann velocity distribution.[6,7]

Thus by analogy and from the more general discussion that follows, the averages must be taken as

$$\langle\langle f(\mathbf{R}, \mathbf{U}, \mathbf{U}')\rangle\rangle = \frac{\int d\mathbf{R} \int d\mathbf{U} \int d\mathbf{U}' f(\mathbf{R}, \mathbf{U}, \mathbf{U}') G(\mathbf{R0}, \mathbf{UU}'; L0) p(\mathbf{U}')}{\int d\mathbf{R} \int d\mathbf{U} \int d\mathbf{U}' G(\mathbf{R0}, \mathbf{UU}'; L0) p(\mathbf{U}')} \tag{5.37}$$

The arguments of Collins and Wragg[40] show that $G(\mathbf{R0}; \mathbf{UU}'; L)$ is (apart from a constant factor) the *number* of chain configurations for a chain of total contour length L which have the end-to-end vector $\mathbf{R}$ and $\mathbf{U}'(\mathbf{U})$ initially (finally). G is not proportional to a joint probability density! The probability of having $\mathbf{U}'$ initially must therefore be proportional to the number of chains which have $\mathbf{U}'$ initially and have any $\mathbf{U}$ and $\mathbf{R}$, i.e.,

$$p(\mathbf{U}') \propto \int d\mathbf{R} \int d\mathbf{U} G(\mathbf{R0}, \mathbf{UU}'; L0) \tag{5.38}$$

Thus

$$p(\mathbf{U}') = \left(\frac{b \tanh a}{\pi}\right)^{3/2} \exp\left[-b\mathbf{U}'^2 \tanh a\right] = p(\mathbf{U}'; L) \tag{5.39}$$

is the properly normalized distribution. We note that if in (5.15) ff. we took $\mathbf{r}(0) = \mathbf{R}'$ and $\mathbf{r}(L) = \mathbf{R}$, we would get $G(\mathbf{RR}', \mathbf{UU}'; L)$. This G depends only on $\mathbf{R} - \mathbf{R}'$. Using the preceding argument, we find that the probability of initially having $\mathbf{R}'$ is independent of $\mathbf{R}$, as it must be, because space is isotropic in the absence of external forces.

The fact that (5.39) is dependent on L is at first somewhat disturbing. However, this arises because $\mathbf{u}(s)$ is required to satisfy the constraint (5.4) to give the chosen end-to-end vector $\mathbf{R}$. For short chains, $\mathbf{u}(s)$ must adjust its initial value $\mathbf{U}'$ severely to accomplish this condition, but for long enough chains, the curve $\mathbf{u}(s)$ can "adiabatically" adjust to this requirement. Indeed, for long enough chains

$$\lim_{L\to\infty} \langle \mathbf{U}'^2 \rangle_{\mathbf{U}'} \equiv \lim_{L\to\infty} \int d\mathbf{U}' p(\mathbf{U}'; L)\mathbf{U}'^2$$
$$\equiv \lim_{L\to\infty} \frac{3}{2b \tanh a} = \frac{3}{2b} \tag{5.40}$$

because from (5.33a) a is proportional to L. Thus (5.40) implies that on the (mean square) average, for a very long chain, $\mathbf{U}'$ is a unit vector if b is chosen to be 3/2. That there is the freedom to choose b in this way should be clear from (5.15); since this equation has three parameters L, l, and $\beta\epsilon$, whereas the wormlike chain is described in terms of the two parameters L and λ. We can, of course, keep all three parameters to provide a more general model.

Calculating averages via (5.39) and (5.40) gives the following:

1.
$$\langle \mathbf{U}^2 \rangle = \frac{3}{2b} [\tanh a + \operatorname{csch} a] \tag{5.41a}$$

In the limit of large a (large L), for $b = 3/2$ this is equal to

$$\langle \mathbf{U}^2 \rangle \to 1 + 2 \exp(-4a) + \mathcal{O}[\exp(-8a)] \tag{5.41b}$$

which is "close" to a unit vector on the average.

2.
$$\left\langle \frac{\mathbf{R} \cdot \mathbf{U}'}{|\mathbf{U}'|} \right\rangle = \left(\frac{8}{\pi b}\right)^{1/2} \frac{L}{2a} [\tanh a]^{1/2} \tag{5.42a}$$

If we set $b = 3/2$ and define $a = \lambda L$, for $a \gg 1$ we find

$$\left\langle \frac{\mathbf{R} \cdot \mathbf{U}'}{|\mathbf{U}'|} \right\rangle \to \left[\frac{4}{3}\left(\frac{3}{\pi}\right)^{1/2}\right] \frac{1}{2\lambda} [1 - \exp(-2\lambda L) + \tfrac{1}{2} \exp(-4\lambda L) + \cdots] \tag{5.42b}$$

which is close to the wormlike result

$$\langle \mathbf{R} \cdot \hat{\mathbf{U}}' \rangle = \frac{1}{2\lambda} [1 - \exp(-2\lambda L)] \tag{5.42c}$$

Note that the choice of b and the definition of λ imply that

$$l = \frac{1}{\lambda} \tag{5.43}$$

Similarly,

3.
$$\langle \mathbf{R}^2 \rangle = \frac{3L^2}{2ab}\left[1 - \frac{1}{2a} \tanh a\right] \tag{5.44a}$$

and for $a \gg 1$, $b = 3/2$,

$$\langle \mathbf{R}^2 \rangle \to \frac{L}{\lambda} - \frac{1}{2\lambda^2}\{1 - 2\exp(-2\lambda L) + \mathcal{O}[\exp(-4\lambda L)]\} \tag{5.44b}$$

which is to be compared with the wormlike chain for which

$$\langle \mathbf{R}^2 \rangle = \frac{L}{\lambda} - \frac{1}{2\lambda^2} [1 - \exp(-2\lambda L)] \tag{5.44c}$$

4. Some straightforward, but tedious, algebra can be used to obtain the end-to-end vector distribution

$$P(\mathbf{R}; L) \equiv \frac{\int d\mathbf{U} \int d\mathbf{U}' G(\mathbf{R}\mathbf{0}, \mathbf{U}\mathbf{U}'; L0) p(\mathbf{U}'; L)}{\int d\mathbf{R} \int d\mathbf{U} \int d\mathbf{U}' G(\mathbf{R}\mathbf{0}, \mathbf{U}\mathbf{U}'; L0) p(\mathbf{U}'; L)}$$

$$= \left(\frac{3}{2\pi\langle \mathbf{R}^2 \rangle}\right)^{3/2} \exp\left[-\frac{3\mathbf{R}^2}{2\langle \mathbf{R}^2 \rangle}\right] \tag{5.45}$$

where $\langle \mathbf{R}^2 \rangle$ is given by (5.44a). Thus the present model of chains with a small degree of stiffness has the deficiency of giving a gaussian end-to-end vector distribution. The latter, however, does reflect stiffness because $\langle \mathbf{R}^2 \rangle / L$ is still a function of L. $G(\mathbf{R0}; \mathbf{UU}'; L0)$ also includes the added persistence of the initial direction, thereby making the foregoing model more appropriate to stiff chains than the use of (5.45) alone with $\langle \mathbf{R}^2 \rangle$ as given by, say, the wormlike chain result (5.44a).

In the limit of very stiff chains, we could recover some of the asymptotic properties of rodlike chains by taking the limit

$$a \to 0, \qquad b \to \infty, \qquad ab \to \tfrac{3}{4} \tag{5.46a}$$

with the results

1.
$$\langle \mathbf{U}^2 \rangle \to 1 \tag{5.46b}$$

2.
$$\left\langle \frac{\mathbf{R} \cdot \mathbf{U}'}{|\mathbf{U}'|} \right\rangle \to \left[\frac{8}{3\pi}\right]^{1/2} L \tag{5.46c}$$

and

3.
$$\langle \mathbf{R}^2 \rangle \to L^2 \tag{5.46d}$$

But $P(\mathbf{R}; L)$ is still gaussian as in (5.45) with the rodlike mean square $\langle \mathbf{R}^2 \rangle$ of (5.46d). The limit (5.46a) implies that the average bond length $l \to 2L$ and that $\beta\epsilon \to \infty$. This is of course consistent with rodlike behavior. However, in this limit there is probably no real need for the use of Wiener integral formulations in the first place.

The fact that the model (5.15) yields a gaussian $P(\mathbf{R}; L)$ might, at first, be unexpected. However, this property rests upon much more general considerations and would be present in a whole class of *algebraically tractable* models of which (5.15) is the simplest case. Therefore we now consider the possible generalizations of the model (5.15), and the associated Markov processes.

G. Generalized Models[21]

The fact that (5.45) is gaussian might seem to arise from considering only the curvature (bending) in (5.7) and (5.15). It would then be of interest to include a torsional energy.[22,38] By ignoring the cross-terms between the derivatives (since they do not yield any complications), (5.7) would be rewritten as, say,

$$V_3 = \tfrac{1}{2}\epsilon \int_0^L |\ddot{\mathbf{r}}(s)|^2 \, ds + \tfrac{1}{2}\gamma \int_0^L |\dddot{\mathbf{r}}(s)|^2 \, ds \tag{5.47}$$

Equation (5.15) then becomes

$$G(\mathbf{R0}, \ldots ; L0) \doteq \int_{\mathbf{r}(0)=\mathbf{0}}^{\mathbf{r}(L)=\mathbf{R}} \mathscr{D}[\mathbf{r}(s)] \times \exp\left\{-\frac{3}{2l}\int_0^L \dot{\mathbf{r}}(s)^2 \, ds - \frac{\beta\epsilon}{2}\int_0^L ds \ddot{\mathbf{r}}(s)^2 - \frac{\gamma\beta}{2}\int_0^L ds \dddot{\mathbf{r}}(s)^2\right\} \tag{5.48}$$

Now (5.48) is a special case of a more general type of potential. Therefore we consider the general case first, and then we return to the particular case of (5.48).

Defining

$$\mathbf{r}^{(n)}(s) \equiv \frac{d^n \mathbf{r}(s)}{ds^n} \tag{5.49}$$

we find that the most general potential V_N, which maintains the simplicity of only having $\mathbf{r}^2$-type terms (and therefore possibly meeting the requirements of being exactly soluble in a simple form), is

$$V_N = \sum_{n=2}^{N} \frac{g_n}{\beta} \int_0^L ds[\mathbf{r}^{(n)}(s)]^2 \tag{5.50}$$

Writing

$$g_1 = \frac{3\beta}{2l} \tag{5.50a}$$

gives the Green's function for the end-to-end vector as

$$G(\mathbf{R0}, \ldots ; L0) \doteq \int_{\mathbf{r}(0)=\mathbf{o}}^{\mathbf{r}(L)=\mathbf{R}} \mathscr{D}[\mathbf{r}(s)] \exp\left\{-\int_0^L ds \sum_{n=1}^{N} g_n[\mathbf{r}^{(n)}(s)]^2\right\} \tag{5.51}$$

where, again, it is necessary to specify endpoint values for $\dot{\mathbf{r}}(0)$, $\dot{\mathbf{r}}(L)$, etc. (see below). Assuming the $\{g_n\}$ are chosen such that G exists, we can show that the general case (5.51), and hence (5.48) and (5.15), have the gaussian character (5.45) where $\langle \mathbf{R}^2 \rangle / L$ is some general function of L which is determined by the $\{g_n\}$. This assertion can be demonstrated by direct evaluation of (5.51). Since the details are involved, they are not reproduced here.[21] The techniques employed in this derivation are similar to those used in Appendix A. However, as noted in the case of Section VB, the introduction of $\mathbf{u}(s)$ and its initial (final) values $\mathbf{U}'(\mathbf{U})$ enables us to deal with quantities that depend upon the persistence of the initial direction. Similar observations hold for (5.51) where higher derivatives of $\mathbf{r}(s)$ are present.

The simplest way to avoid obtaining a gaussian $G(\mathbf{R}, L)$ is to have powers of $\mathbf{r}^{(n)}(s)$ other than quadratic, i.e., to introduce $\int_0^L ds[\mathbf{r}^{(n)}(s)]^m$, $m \neq 2$ or more generally scalar functions of $\mathbf{r}^{(n)}(s)$ like $\int_0^L dsf[\mathbf{r}^{(n)}(s), s]$, into (5.50) and (5.51). The resulting functional integral, however, generally cannot be evaluated exactly in a simple closed form, so this does not meet the requirements for a zeroth-order model of stiff chains. However, Edwards has discussed the need for this general type of Wiener integral in the solution of certain problems, and these cases are briefly discussed. For instance, if we had [ignoring (5.2a)]

$$G(\mathbf{R0}, \mathbf{UU}'; L0) = \int_{\mathbf{u}(0)=\mathbf{U}'}^{\mathbf{u}(L)=\mathbf{U}} \mathscr{D}[\mathbf{u}(s)]\, \delta\left[\mathbf{R} - \int_0^L \mathbf{u}(s)\, ds\right] \times \exp\left\{-\int_0^L dsf[\mathbf{u}(s), s] - \frac{\beta\gamma}{2}\int_0^L \dot{\mathbf{u}}(s)^2\, ds\right\} \tag{5.52}$$

instead of (5.15), with f arbitrary, (5.52) would still be a Wiener integral.[8–12] G would obey the differential equation

$$\left[\frac{\partial}{\partial L} - \frac{1}{2\beta\gamma}\nabla_{\mathbf{U}}^2 + f(\mathbf{U}, L) + \mathbf{U}\cdot\nabla_{\mathbf{R}}\right]G(\mathbf{R}0, \mathbf{U}\mathbf{U}'; L0)$$
$$= \delta(L)\,\delta(\mathbf{R})\,\delta(\mathbf{U} - \mathbf{U}') \quad (5.53)$$

as compared to (5.30). Those limited f for which (5.53) is exactly soluble in simple closed form would then qualify as possible simple models for stiff polymer chains. We could also have included terms in $\int_0^L ds\dot{\mathbf{u}}(s)\cdot g[\mathbf{u}(s)]$ into the exponent of (5.52) as these terms are well known in the case of Fokker-Planck type equations.[39]

Returning now to the special case of (5.48), or more generally, consider

$$G(\mathbf{R}0, \ldots; L0) \doteq \int_{\mathbf{r}(0)=\mathbf{0}}^{\mathbf{r}(L)=\mathbf{R}} \mathscr{D}[\mathbf{r}(s)] \exp\left\{-\int_0^L ds(f[\dot{\mathbf{r}}(s), \ddot{\mathbf{r}}(s), s] + \gamma\dddot{\mathbf{r}}(s)^2\right\} \quad (5.54)$$

which is still a Wiener integral (see below). [Again terms in $\int_0^L \dddot{\mathbf{r}}(s)\cdot g[\ddot{\mathbf{r}}(s)]\,ds$ could also be included.] If we write (5.54) in finite difference form, i.e., in terms of individual bonds as in (5.8), it is clearly necessary to specify the initial (final) values of $\dot{\mathbf{r}}(s)$, $\mathbf{U}'(\mathbf{U})$, as well as the initial (final) values of $\ddot{\mathbf{r}}(s)$, $\mathbf{A}'(\mathbf{A})$. Alternatively, had (5.54) been obtained from some type of Brownian motion, it would imply the necessity of specifying the initial "velocities," $\mathbf{U}'$, and "accelerations," $\mathbf{A}'$, because of the $\dddot{\mathbf{r}}(s)$ in the equations of motion. Because of the random nature of the external force acting on this fictitious Brownian particle, for a given initial $\mathbf{U}'$ and $\mathbf{A}'$, there is a probability distribution of final $\mathbf{U}$ and $\mathbf{A}$. Therefore by analogy with the development in Sections VB and VD, it is eventually necessary to consider the auxiliary quantity $G(\mathbf{R}0, \mathbf{U}\mathbf{U}', \mathbf{A}\mathbf{A}'; L0)$. We could follow the approach of Edwards by considering generalized "Lagrangians" which contain $\ddot{\mathbf{r}}$, $\dddot{\mathbf{r}}$, etc., in order to obtain the equations of motion for this auxiliary G;[9] however, instead we follow the "standard" method for converting Wiener integrals to differential equations.[12]

First use (5.1), (5.4), and (5.19) to write

$$G(\mathbf{R}0, \mathbf{U}\mathbf{U}', \ldots; L0) \doteq \int_{\mathbf{u}(0)=\mathbf{U}'}^{\mathbf{u}(L)=\mathbf{U}} \mathscr{D}[\mathbf{u}(s)]\,\delta\left[\mathbf{R} - \int_0^L \mathbf{u}(s)\,ds\right]$$
$$\times \exp\left(-\int_0^L ds\{f[\mathbf{u}(s), \dot{\mathbf{u}}(s), s] + \gamma\ddot{\mathbf{u}}(s)^2\}\right)$$
$$= \int \frac{d^3k}{(2\pi)^3} \exp(i\mathbf{k}\cdot\mathbf{R}) I(\mathbf{k}, \mathbf{U}\mathbf{U}' \ldots; L) \quad (5.55)$$

where

$$I(\mathbf{k}, \mathbf{UU}', \ldots; L)$$

$$\doteq \int_{\mathbf{u}(0)=\mathbf{U}'}^{\mathbf{u}(L)=\mathbf{U}} \mathscr{D}[\mathbf{u}(s)] \exp\left\{-\int_0^L ds[f(\mathbf{u}, \dot{\mathbf{u}}, s) + \gamma\ddot{\mathbf{u}}(s)^2 + i\mathbf{k}\cdot\mathbf{u}(s)]\right\} \quad (5.56)$$

Now defining

$$\mathbf{a}(s) = \dot{\mathbf{u}}(s) \quad (5.57a)$$

with boundary conditions

$$\mathbf{a}(0) = \mathbf{A}', \qquad \mathbf{a}(L) = \mathbf{A} \quad (5.57b)$$

permits (5.57) to be inserted into (5.55) and (5.56). This is to be done by introducing Dirac delta functions requiring (5.57a) for all s. Explicitly,

$$I(\mathbf{k}, \mathbf{UU}', \mathbf{AA}'; L) \equiv \int_{\mathbf{u}(0)=\mathbf{U}'}^{\mathbf{u}(L)=\mathbf{U}} \mathscr{D}[\mathbf{u}(s)] \int_{\mathbf{a}(0)=\mathbf{A}'}^{\mathbf{a}(L)=\mathbf{A}} \mathscr{D}'[\mathbf{a}(s)] \prod_s \delta[\dot{\mathbf{u}}(s) - \mathbf{a}(s)]$$

$$\times \exp\left\{-\int_0^L ds[f(\mathbf{u}, \mathbf{a}, s) + \gamma\dot{\mathbf{a}}^2 + i\mathbf{k}\cdot\mathbf{u}]\right\} \quad (5.58)$$

where

$$\mathscr{D}'[\mathbf{a}(s)] \prod_s \delta[\dot{\mathbf{u}}(s) - \mathbf{a}(s)]$$

$$\equiv \lim_{FI} \prod_{sj}{}' d\mathbf{a}_j\, \delta\left[\mathbf{u}_j - \mathbf{u}_{j-1} - \frac{\Delta s_j}{2}(\mathbf{a}_j + \mathbf{a}_{j-1})\right] \quad (5.59a)$$

$$\equiv \lim_{FI} \prod_{sj}{}' d\mathbf{a}_j \int \frac{d^3q_j}{(2\pi)^3} \exp\left\{i\left[\mathbf{u}_j - \mathbf{u}_{j-1} - \frac{\Delta s_j(\mathbf{a}_j + \mathbf{a}_{j-1})}{2}\right]\cdot\mathbf{q}_j\right\} \quad (5.59b)$$

$$= \delta_F[\mathbf{a}(s)] \int \delta'_F[\mathbf{q}(s)] \exp\left\{i\int_0^L ds\mathbf{q}(s)\cdot[\dot{\mathbf{u}}(s) - \mathbf{a}(s)]\right\} \quad (5.59c)$$

gives the desired constraint. The $\lim_{FI}$ in (5.59) implies the usual limit of a functional integral as in (3.7). The primes in (5.59a) and (5.59b) imply that the $s = 0$ term is to be omitted, since for $s_j = 0$, $\mathbf{u}_{j-1}$ and $\mathbf{a}_{j-1}$ are undefined. In particular, $\delta_F[\mathbf{a}(s)] \equiv \prod_s d\mathbf{a}(s)$, and (5.59c) is just $\delta_F[\mathbf{a}(s)]$ times the usual functional delta function, giving (5.57a).

Substituting (5.59) into (5.58) leads to

$$I(\mathbf{k}, \mathbf{UU}'; \mathbf{AA}'; L)$$

$$= \int_{\mathbf{u}(0)=\mathbf{U}'}^{\mathbf{u}(L)=\mathbf{U}} \mathscr{D}[\mathbf{u}(s)] \int_{\mathbf{a}(0)=\mathbf{A}'}^{\mathbf{a}(L)=\mathbf{A}} \delta_F[\mathbf{a}(s)] \int \delta'_F[\mathbf{q}(s)]$$

$$\times \exp\left\{-\int_0^L ds[f(\mathbf{u}, \mathbf{a}, s) + \gamma\dot{\mathbf{a}}^2 + i\mathbf{k}\cdot\mathbf{u} + i\mathbf{q}\cdot\mathbf{a} - i\mathbf{q}\cdot\dot{\mathbf{u}}]\right\} \quad (5.60)$$

which describes a "Markov process" in the sense that we can write a Smoluchowski-Chapman-Kolmogorov equation[6,7,10,12,33]

$$I(\mathbf{k}, \mathbf{UU}', \mathbf{AA}'; L + \epsilon) = \int d\mathbf{U}'' \int d\mathbf{A}'' I(\mathbf{k}, \mathbf{UU}'', \mathbf{AA}''; \epsilon) I(\mathbf{k}, \mathbf{U}''\mathbf{U}', \mathbf{A}''\mathbf{A}'; L) \quad (5.61)$$

This follows from the fact that

$$\int_0^{L+\epsilon} ds \rightarrow \int_0^L ds + \int_L^{L+\epsilon} ds,$$

that the measures

$$\mathscr{D}[\mathbf{u}(s)], \qquad \delta_F[\mathbf{a}(s)], \qquad \delta'_F[\mathbf{q}(s)]$$

are all separable into products for the two intervals $0 < s < L$ and $L < s < L + \epsilon$, and that the point L is just as in (3.27)–(3.30). If ϵ is an infinitesimal, (5.61) and (5.60) imply that

$$\begin{aligned} I(\mathbf{k}, \mathbf{UU}', \mathbf{AA}'; L + \epsilon) &= \int d\mathbf{U}'' \int d\mathbf{A}'' \left(\frac{\gamma}{\pi\epsilon}\right)^{3/2} \int \frac{d^3q}{(2\pi)^3} \\ &\times \exp\left\{-\epsilon\left(f\left[\frac{\mathbf{U}+\mathbf{U}''}{2}, \frac{\mathbf{A}+\mathbf{A}''}{2}, L\right] + \gamma\left(\frac{\mathbf{A}-\mathbf{A}''}{\epsilon}\right)^2 + \frac{i\mathbf{k}\cdot(\mathbf{U}+\mathbf{U}'')}{2} \right.\right. \\ &\left.\left. + \frac{i\mathbf{q}\cdot(\mathbf{A}+\mathbf{A}'')}{2} - \frac{i\mathbf{q}\cdot(\mathbf{U}-\mathbf{U}'')}{\epsilon}\right)\right\} I(\mathbf{k}, \mathbf{U}''\mathbf{U}', \mathbf{A}''\mathbf{A}'; L) \end{aligned} \quad (5.62)$$

Letting

$$\boldsymbol{\omega} = \mathbf{A} - \mathbf{A}'', \qquad \boldsymbol{\eta} = \mathbf{U} - \mathbf{U}'', \qquad d\mathbf{U}''\, d\mathbf{A}'' = d\boldsymbol{\omega}\, d\boldsymbol{\eta} \quad (5.63)$$

and expanding the exponential terms of $\mathcal{O}(\epsilon)$ in (5.62) except for the singular $\exp(-\gamma\boldsymbol{\omega}^2/\epsilon)$ term, we find

$$\begin{aligned} I(\mathbf{k}, \mathbf{UU}', \mathbf{AA}'; L + \epsilon) &= \int d\boldsymbol{\omega} \int \frac{d^3q}{(2\pi)^3} \int d\boldsymbol{\eta} \left(\frac{\gamma}{\pi\epsilon}\right)^{3/2} \exp\left(-\frac{\gamma\boldsymbol{\omega}^2}{\epsilon} + i\mathbf{q}\cdot\boldsymbol{\eta}\right) \\ &\times \{1 - \epsilon[f(\mathbf{U}, \mathbf{A}, L) + i\mathbf{k}\cdot\mathbf{U} + i\mathbf{q}\cdot\mathbf{A}] + \mathcal{O}(\epsilon^2)\} \\ &\times \exp[-\boldsymbol{\eta}\cdot\nabla_{\mathbf{U}}] I(\mathbf{k}, \mathbf{UU}', \mathbf{A} - \boldsymbol{\omega}\mathbf{A}'; L) \end{aligned} \quad (5.64)$$

where we have used the formal definition

$$g(\mathbf{U} - \boldsymbol{\eta}) = \exp[-\boldsymbol{\eta}\cdot\nabla_{\mathbf{U}}] g(\mathbf{U}) \quad (5.65)$$

The $\mathbf{q}$, $\boldsymbol{\eta}$ integrals are of the form

$$\int \frac{d^3q}{(2\pi)^3} \int d\boldsymbol{\eta} \exp [i\mathbf{q}\cdot\boldsymbol{\eta}] \exp [-\boldsymbol{\eta}\cdot\boldsymbol{\nabla}_{\mathrm{U}}] = \int d\boldsymbol{\eta}\, \delta(\boldsymbol{\eta}) \exp [-\boldsymbol{\eta}\cdot\boldsymbol{\nabla}_{\mathrm{U}}] = 1 \tag{5.66a}$$

and

$$\int \frac{d^3q}{(2\pi)^3} \int d\boldsymbol{\eta}\mathbf{q}\cdot\mathbf{A} \exp [i\mathbf{q}\cdot\boldsymbol{\eta}] \exp [-\boldsymbol{\eta}\cdot\boldsymbol{\nabla}_{\mathrm{U}}]$$

$$= \int \frac{d^3q}{(2\pi)^3} \int d\boldsymbol{\eta} \left(\frac{1}{i}\mathbf{A}\cdot\boldsymbol{\nabla}_{\eta} \exp [i\mathbf{q}\cdot\boldsymbol{\eta}]\right) \exp [-\boldsymbol{\eta}\cdot\boldsymbol{\nabla}_{\mathrm{U}}]$$

$$= \frac{1}{i}\mathbf{A}\cdot\boldsymbol{\nabla}_{\mathrm{U}} \tag{5.66b}$$

where (5.66b) follows upon integration by parts and then use of (5.66a). Substituting (5.66) into (5.64), expanding the left-hand side in powers of ϵ and the right-hand side (apart from $\exp [-\gamma\boldsymbol{\omega}^2/\epsilon]$) in powers of $\boldsymbol{\omega}$ in the standard manner, in the limit $\epsilon \to 0$, we have

$$\left[\frac{\partial}{\partial L} - \frac{1}{4\gamma}\nabla_{\mathrm{A}}{}^2 + f(\mathbf{U}, \mathbf{A}, L) + i\mathbf{k}\cdot\mathbf{U} + \mathbf{A}\cdot\boldsymbol{\nabla}_{\mathrm{U}}\right] I(\mathbf{k}, \mathbf{UU}', \mathbf{AA}'; L)$$

$$= \delta(L)\,\delta(\mathbf{U}-\mathbf{U}')\,\delta(\mathbf{A}-\mathbf{A}') \tag{5.67}$$

Upon introducing the Fourier transform, we get

$$\left[\frac{\partial}{\partial L} - \frac{1}{4\gamma}\nabla_{\mathrm{A}}^2 + f(\mathbf{U}, \mathbf{A}, L) + \mathbf{U}\cdot\nabla_{\mathrm{R}} + \mathbf{A}\cdot\nabla_{\mathrm{U}}\right] G(\mathbf{R}0, \mathbf{UU}', \mathbf{AA}'; L0)$$

$$= \delta(L)\,\delta(\mathbf{R})\,\delta(\mathbf{U}-\mathbf{U}')\,\delta(\mathbf{A}-\mathbf{A}') \tag{5.68}$$

Any averaging procedure is then to be taken as analogous to (5.37) and (5.38). In the general case with higher derivatives $\mathbf{r}^{(n)}$, $n > 3$, it may be simpler to follow Edwards' technique than to introduce the additional functional delta functions analogous to (5.59c) for (5.57a).

H. Discussion

We have discussed a model of stiff polymer chains and have shown that it gives results very similar to those descriptive of the wormlike chain in the limits of small and large stiffness. Presumably, for intermediate ranges of stiffness, the presence of three basic parameters in this model should enable it to reflect the character of real chains. One of the important virtues of the present model is the fact that the relevant distribution functions may be obtained in a simple closed-form analytic expression. It therefore represents a possible zeroth-order model for discussions of nonideal, e.g., excluded volume, systems composed of stiff polymer chains.

The model discussed, which is simply related to the Saitô, Takahashi, and Yunoki prescription for the wormlike chain, and its possible generalizations lead us to consider general Wiener integrals which represent multidimensional Markov processes. In these general Wiener integrals we have terms in $\ddot{\mathbf{r}}(s)$ and possibly higher derivatives. They are therefore analogous to cases of Brownian motion, maintaining the inertial term, or including forces dependent on velocity, acceleration, etc. Such a system is therefore not a simple Markov process, for in the polymer case the presence of $\ddot{\mathbf{r}}(s)$, $\dddot{\mathbf{r}}(s)$, etc., implies correlations between neighboring bonds, next nearest bonds, etc. However, a non-Markovian process can often be considered as the projection of a higher dimensional Markov process. In the models discussed, the introduction of $\mathbf{U}$, $\mathbf{A}$, etc., in fact leads to a Markov process of higher dimension [cf. (5.61)]. These models are then the continuous limit analogs of the popular rotational isomeric model in which we obtain a simple Markov process for independent bond rotations, a twofold Markov process when there are interactions between neighboring — rotations, etc.[2–5]

Appendix A. Direct Evaluation of the "Harmonic Oscillator" Green's Function (5.15)

All of the functional integrals encountered so far have been "evaluated" by relating them to the solution of inhomogeneous differential equations which are then solved by standard methods. There are only a few cases for which Wiener-type functional integrals can directly be evaluated from the definition of the functional integral. Often these integrals can then be evaluated by a number of different techniques. As noted earlier, the Green's function (5.15) can be shown to yield (5.29) in this manner. This result has been presented a number of times, using a variety of techniques. Our purpose in now illustrating of the direct evaluation of (5.15) is to provide various approximation schemes for functional integrals that cannot be exactly evaluated. The different methods of exact solution then form the basis for "new" kinds of approximations.

Equation (5.15) can be evaluated by first considering the integral in finite difference form, i.e., in terms of $\mathbf{u}_j$, Δs_j, etc. The multiple integral is then a multivariate gaussian integral which is readily evaluated. If the exponential part of the integrand is taken to be $\sum_{i,j} A_{ij}\mathbf{u}_i\mathbf{u}_j$, this integration leads to a bivariate gaussian in $\mathbf{U}$ and $\mathbf{U}'$ which is multiplied by $[\det \mathbf{A}]^{-3/2}$. In passing to the continuous limit it is shown that this determinant can be obtained as the solution of a differential equation. The results are summarized by Montroll[8] and others and need not be repeated here.

If we change notation, $\mathbf{u}(s) \to \mathbf{r}(s)$ and some of the constants, (5.15) becomes

$$G(\mathbf{RR}'; L) = \int_{\mathbf{r}(0)=\mathbf{R}'}^{\mathbf{r}(L)=\mathbf{R}} \mathscr{D}[\mathbf{r}(s)] \exp\left\{-\int_0^L ds\left[\frac{3}{2l}\dot{\mathbf{r}}^2(s) + k\mathbf{r}(s)^2\right]\right\} \quad \text{(A.1)}$$

which describes the end-vector distribution for a chain in a harmonic potential. Alternatively, if $l/6 \to D$ and $s \to \tau$, $L \to t$ (A.1) describes diffusion in a harmonic force field.

Equation (A.1) is a gaussian Wiener integral, since it contains powers of $\mathbf{r}(s)$ and $\dot{\mathbf{r}}(s)$, which are no higher than the second. For this case, Feynman shows how the solution, apart from a normalization that depends upon L, may be obtained.[12] This normalization can readily be obtained from the limiting condition (3.22). We briefly review this method, referring to Feynman for the details.[12] The path integral representation of the quantum mechanical propagator is given in (4.4). In the general case when the classical Lagrangian is also an explicit function of time $\mathscr{L}(\mathbf{r}, \dot{\mathbf{r}}, t)$, (4.4) becomes

$$K(\mathbf{RR}'; tt') = \int_{\mathbf{r}(t')=\mathbf{R}'}^{\mathbf{r}(t)=\mathbf{R}} \mathscr{D}[\mathbf{r}(\tau)] \exp\left[\frac{i}{\hbar}\int_{t'}^{t} d\tau \mathscr{L}(\mathbf{r}, \dot{\mathbf{r}}, \tau)\right] \quad \text{(A.2)}$$

When the particle is heavy, or has high energy, the action integral $S(t, t') = \int_{t'}^{t} d\tau \mathscr{L}$ is very large compared to $\hbar$. As the integrand ranges over all paths $\mathbf{r}(\tau)$, the exponential is highly oscillatory. Only those paths in the neighborhood of the path(s) that make the exponent stationary can contribute to the functional integral. The path(s) that make this exponent stationary thus are determined by the $\mathbf{r}_0(\tau)$ which are solutions to

$$\delta \int_{t'}^{t} d\tau \mathscr{L}[\mathbf{r}(\tau), \dot{\mathbf{r}}(\tau), \tau] = 0 \quad \text{(A.3)}$$

where the variation in (A.3) represents an arbitrary variation of the path $\mathbf{r}(\tau)$ subject to $\mathbf{r}(t) = \mathbf{R}$, $\mathbf{r}(t') = \mathbf{R}'$. But (A.3) is just Hamilton's principle, so $\mathbf{r}_0(\tau)$ is the classical path! Thus we write

$$\mathbf{r}(\tau) = \mathbf{r}_0(\tau) + \boldsymbol{\xi}(\tau) \quad \text{(A.4)}$$

where $\xi(t) = \xi(t') = 0$, and

$$\mathscr{D}[\mathbf{r}(\tau)] = \mathscr{D}[\boldsymbol{\xi}(\tau)] \quad \text{(A.5)}$$

Defining

$$\begin{aligned} S_{\text{cl}}(t, t') &= \int_{t'}^{t} d\tau \mathscr{L}[\mathbf{r}_0(\tau), \dot{\mathbf{r}}_0(\tau), \tau] \\ S(t, t') &= S_{\text{cl}}(t, t') + \int_{t'}^{t} d\tau \mathscr{L}'(\boldsymbol{\xi}, \dot{\boldsymbol{\xi}}, \tau) \end{aligned} \quad \text{(A.6)}$$

we see that (A.3) is identically

$$K(\mathbf{RR}'; tt') = \exp\left[\frac{i}{\hbar} S_{\mathrm{cl}}(t, t')\right] \int_{\boldsymbol{\xi}(t')=0}^{\boldsymbol{\xi}(t)=0} \mathscr{D}[\boldsymbol{\xi}(\tau)] \exp\left[\frac{i}{\hbar} \int_{t'}^{t} \mathscr{L}'(\boldsymbol{\xi}, \dot{\boldsymbol{\xi}}, \tau)\right] \tag{A.7}$$

Equation (A.7) is quite general (assuming a single classical path). As noted previously, when $\mathscr{L}$ is quadratic in $\mathbf{r}$ and $\dot{\mathbf{r}}$, the remaining functional integral in (A.7) is solely a function of t and t'; $S_{\mathrm{cl}}(t, t')$ contains all the $\mathbf{R}$ and $\mathbf{R}'$ dependence. This path integral need not be directly evaluated, since its value can be obtained from the condition that $K \to \delta(\mathbf{R} - \mathbf{R}')$ as $t' \to t$. This use of the classical path becomes more interesting, however, when $\mathscr{L}$ is more complicated than just a quadratic function. In this case, the remaining path integral in (A.7) still depends on the endpoints $\mathbf{R}$ and $\mathbf{R}'$. An approximation to K can be obtained by retaining the leading quadratic part of the expansion of $\mathscr{L}$ in terms of $\boldsymbol{\xi}(\tau)$ and then evaluating this gaussian functional integral. If this approximate K is then employed to obtain an approximate wavefunction, using stationary phase integration to evaluate the integrals, the result is the usual WKB semiclassical approximation.[41] Applying similar approximations to the representation of the thermal density matrix (4.7) we may find of the quantum corrections to the classical partition function.[10,12]

A third method of evaluation of (A.1) involves a particular "change in variables" that is often useful in the approximation of functional integrals. This technique is quite useful for the integrals that arise in discussions of polymers in bulk, so the method is repeated here. This approach has also been summarized by Brush.[11]

Since $\mathbf{r}(s)$ is a nice, well-behaved function [otherwise it would not contribute to (A.1)], it can always be expanded in terms of a complete set of functions $\phi_\alpha(s)$ on the interval $0 \leq s \leq L$,

$$\mathbf{r}(s) = \sum_\alpha \mathbf{r}_\alpha \phi_\alpha(s) \tag{A.8}$$

It is convenient to choose the functions $\phi_\alpha(s)$ to be orthonormal,

$$\int_0^L \phi_\alpha^*(s)\phi_\beta(s) \equiv \delta_{\alpha\beta} \tag{A.9}$$

Equation (A.8) represents a linear transformation between the variables of integration of (A.1) $\mathbf{r}(s)$ (*for each* s) and the new variables $\{\mathbf{r}_\alpha\}$. The Jacobian J for any such linear transformation must be constant,[12] and this constant will be considered shortly. As the generalized Fourier coefficients $\{\mathbf{r}_\alpha\}$ each range over all of three-dimensional space, they generate all possible functions $\mathbf{r}(s)$. Apart from this Jacobian, we must therefore have

$$\delta[\mathbf{r}(s)] \to \prod_{\alpha} d\mathbf{r}_{\alpha} \tag{A.10}$$

since to each continuous $\mathbf{r}(s)$ there corresponds only one $\{\mathbf{r}_{\alpha}\}$. Substitute (A.8) and (A.10) into (A.1), and the integration ranges for $\mathbf{r}_{\alpha}$ are $-\infty \leq x_{\alpha} \leq \infty$, etc. Obviously, we get

$$\int_0^L \mathbf{r}(s)^2 \, ds = \sum_{\alpha} \mathbf{r}_{\alpha}^{\,2} \tag{A.11}$$

by the orthonormality (A.9). On the other hand, the $\dot{\mathbf{r}}^2$ term gives

$$\int_0^L \dot{\mathbf{r}}^2(s) \, ds = \sum_{\alpha,\beta} \mathbf{r}_{\alpha} \cdot \mathbf{r}_{\beta} \int_0^L \frac{\partial \phi_{\alpha}^*(s)}{\partial s} \frac{\partial \phi_{\beta}(s)}{\partial s} \, ds \tag{A.12}$$

It would be convenient if the $\int_0^L ds$ in (A.12) were proportional to $\delta_{\alpha\beta}$, so that the exponential factors in (A.1) would then be a product of gaussians. Integrating (A.12) by parts would give

$$\int_0^L \frac{\partial \phi_{\alpha}^*}{\partial s} \frac{\partial \phi_{\beta}}{\partial s} \, ds = \phi_{\alpha}^*(s) \frac{\partial \phi_{\beta}}{\partial s} \bigg|_0^L - \int_0^L \phi_{\alpha}^*(s) \frac{\partial^2 \phi_{\beta}(s)}{\partial s^2} \, ds \tag{A.13}$$

which could be made proportional to $\delta_{\alpha\beta}$ if $\phi_{\alpha}^*(\partial \phi_{\beta}/\partial s)$ is zero at $s = 0$ and L and $\partial^2 \phi_{\beta}/\partial s^2$ is proportional to ϕ_{β}. The second term in (A.13) then would vanish because of the orthogonality (A.12). But this means that $\{\phi_{\alpha}\}$ can be taken as sines and/or cosines, making (A.8) an ordinary Fourier series. Since $\mathbf{r}(0) \neq \mathbf{r}(L)$, the cosines are chosen and

$$\mathbf{r}(s) = \frac{\mathbf{r}_0}{L^{1/2}} + \left(\frac{2}{L}\right)^{1/2} \sum_{n=1}^{\infty} \mathbf{r}_n \cos\left(\frac{n\pi s}{L}\right) \tag{A.8a}$$

Now it is easily shown that

$$\int_0^L ds \dot{\mathbf{r}}^2(s) = \sum_{n=1}^{\infty} \left(\frac{n\pi}{L}\right)^2 \mathbf{r}_n^{\,2} \tag{A.12a}$$

The endpoint conditions

$$\mathbf{R}' = \mathbf{r}(0) = \frac{\mathbf{r}_0}{L^{1/2}} + \left(\frac{2}{L}\right)^{1/2} \sum_{n=1}^{\infty} \mathbf{r}_n \tag{A.14a}$$

$$\mathbf{R} = \mathbf{r}(L) = \frac{\mathbf{r}_0}{L^{1/2}} + \left(\frac{2}{L}\right)^{1/2} \sum_{n=1}^{\infty} (-1)^n \mathbf{r}_n \tag{A.14b}$$

are conveniently introduced into (A.1) by the use of Dirac delta functions as used in (3.19). Using

$$\delta[\mathbf{R}' - \mathbf{r}(0)] = \int \frac{d^3\mu}{(2\pi)^3} \exp\{i\boldsymbol{\mu}\cdot[\mathbf{R}' - \mathbf{r}(0)]\} \tag{A.15a}$$

$$\delta[\mathbf{R} - \mathbf{r}(L)] = \int \frac{d^3\nu}{(2\pi)^3} \exp\{i\boldsymbol{\nu}\cdot[\mathbf{R} - \mathbf{r}(L)]\} \tag{A.15b}$$

with $\mathbf{r}(0)$ and $\mathbf{r}(L)$ given by (A.14), we see that all of the $\int d\mathbf{r}_\alpha$ are independent gaussian integrals which are evaluated to give

$$\begin{aligned} G(\mathbf{RR}'; L) = J\mathcal{N}\left\{\prod_{n=1}^{\infty}\left[\frac{3}{2l}\left(\frac{n\pi}{l}\right)^2 + k\right]\right\}^{-3/2} \int \frac{d^3\mu}{(2\pi)^3} \int \frac{d^3\nu}{(2\pi)^3} \\ \times \exp[i\boldsymbol{\mu}\cdot\mathbf{R}' + i\boldsymbol{\nu}\cdot\mathbf{R}] \\ \times \exp\left(-\frac{(\boldsymbol{\mu}^2 + \boldsymbol{\nu}^2)}{2}\left\{\frac{1}{2kL} + \frac{1}{L}\sum_{n=1}^{\infty}\left[k + \frac{3}{2l}\left(\frac{n\pi}{l}\right)^2\right]^{-1}\right\}\right. \\ \left. - \mathbf{u}\cdot\boldsymbol{\nu}\left\{\frac{1}{2kL} + \frac{1}{L}\sum_{n=1}^{\infty}(-1)^n\left[k + \frac{3}{2l}\left(\frac{n\pi}{l}\right)^2\right]^{-1}\right\}\right) \end{aligned} \tag{A.16}$$

The trivial gaussian integrals in (A.16) can be evaluated easily. Then using the relations[42]

$$\coth \pi x = (\pi x)^{-1} + \left(\frac{2x}{\pi}\right)\sum_{n=1}^{\infty}(x^2 + n^2)^{-1}$$

and

$$2\pi[e^{\pi x} - e^{-\pi x}]^{-1} = x^{-1} + 2x\sum_{n=1}^{\infty}(-1)^n(x^2 + n^2)^{-1}$$

gives, apart from the normalization, the result quoted in (5.29) with $\mathbf{U}, \mathbf{U}' \leftrightarrow \mathbf{R}, \mathbf{R}'$ and in this case

$$a = L\left(\frac{2lk}{3}\right)^{1/2}, \qquad b = \left(\frac{3k}{2l}\right)^{1/2}$$

Rather than using the limit $L \to 0$ to fix the overall normalization, we can consider its direct evaluation. This approach is useful when attempts are made to obtain rigorous bounds on G or Z. The normalization can be treated by an analog of the Hamiltonian formulation for path integrals,[43] but it is simpler to proceed directly. The normalization $\mathscr{D}[\mathbf{r}(s)]$ can always be written as

$$\mathscr{D}[\mathbf{r}(s)] = \lim_{FI} \prod_{j=1}^{n-1} \left[d\mathbf{r}_j \left(\frac{3}{2\pi l \, \Delta s_j} \right)^{3/2} \right]$$

$$\equiv \lim_{FI} \left(\left[\prod_{j=1}^{n} d\mathbf{r}_j \right] \left\{ \int d\mathbf{r}(0) \, \delta[\mathbf{r}(0)] \, d\mathbf{r}(L) \prod_{j=1}^{n} \int d\mathbf{r}_j \times \exp \left[-\frac{3(\mathbf{r}_j - \mathbf{r}_{j-1})^2}{2l \, \Delta s_j} \right] \right\}^{-1} \right) \tag{A.17}$$

Using the definition

$$\delta[\mathbf{r}(s)] = \frac{\prod_{s=0}^{L} d\mathbf{r}(s)}{d\mathbf{r}(0) \, d\mathbf{r}(L)} \tag{A.18}$$

and the notation

$$\int d\mathbf{R}' \int_{\mathbf{r}(0)=0}^{\mathbf{r}(L)=\mathbf{R}'} \delta[\mathbf{r}(s)] \equiv \int_{\mathbf{r}(0)=0} \bar{\delta}[\mathbf{r}(s)] \tag{A.19}$$

changes (A.17) to

$$\mathscr{D}[\mathbf{r}(s)] = \delta\mathbf{r}(s) \left\{ \int_{\mathbf{r}(0)=0} \bar{\delta}[\mathbf{r}(s)] \exp \left[-\int_0^L ds \frac{3}{2l} \dot{\mathbf{r}}^2(s) \right] \right\}^{-1} \tag{A.20}$$

Thus (A.1) can be rewritten as

$$G(\mathbf{R}, \mathbf{R}'; L) = \frac{\displaystyle\int_{\mathbf{r}(0)=\mathbf{R}'}^{\mathbf{r}(L)=\mathbf{R}} \delta[\mathbf{r}(s)] \exp \left\{ -\int_0^L ds[(3/2l)\dot{\mathbf{r}}^2(s) + k\mathbf{r}^2(s)] \right\}}{\displaystyle\int_{\mathbf{r}(0)=0} \bar{\delta}[\mathbf{r}(s)] \exp \left[-\int_0^L ds(3/2l)\dot{\mathbf{r}}^2(s) \right]} \tag{A.21}$$

The "infinite" normalization (3.4d) is not formally apparent in (A.21). Upon use of the transformation (A.8) in both the numerator and denominator of (A.21), the Jacobian that makes (A.10) an equality cancels. Upon performing the gaussian integrals that arise in the denominator of (A.21) and using the relation[42]

$$\frac{\sinh x}{x} = \prod_{n=1}^{\infty} \left(1 + \frac{x^2}{\pi^2 n^2} \right)$$

the final normalization can be shown to be that quoted.

VI. THE EXCLUDED VOLUME PROBLEM: THE SELF-CONSISTENT FIELD

It has already been noted that the configurational statistics of a single polymer chain with the long-range interactions is equivalent to the full many-body problem. Therefore, this excluded volume problem is not, in general, exactly soluble in a useful form. The theory of the excluded volume effect in polymers has been of considerable interest since it must be an essential part of any quantitative molecular basis for the understanding of the behavior of nonideal dilute polymer solutions (e.g., polymers in so-called good solvents).[5] In Chapter III of his book, Yamakawa thoroughly reviews all of the attempts to solve the polymer excluded volume problem and all the resultant controversies.[5] The reader is referred to this work for the details. We summarize some of the aspects of this work only to provide necessary background.

Consider a polymer chain with n equivalent links, each of rms length l, for which there is some short-range repulsive interaction

$$W_{ij} = W(\mathbf{r}_i - \mathbf{r}_j) \tag{6.1}$$

between all pairs of segments. Since the polymer chain exists in solution, W of (6.1) is the potential of mean force and includes the effects of the polymer-solvent interactions. It is assumed that this potential of mean force is pairwise additive as in (6.1). The introduction of three body mean forces, although probably realistic in many cases, would only serve to make the problem even less tractable. For the sake of mathematical simplicity the potential (6.1) is often taken as a delta function "pseudo-potential,"[5]

$$\frac{\beta}{2} W(\mathbf{r}_i - \mathbf{r}_j) = v\, \delta(\mathbf{r}_i - \mathbf{r}_j) \tag{6.2}$$

In the limit of low density, v is the negative of the second virial coefficient for the effective monomer-monomer interactions $w(\mathbf{r}_i - \mathbf{r}_j)$,

$$v \rightarrow \int d\mathbf{r}\{1 - \exp[-\beta w(r)]\} \tag{6.3}$$

Thus v represents the effective volume of one monomer which is excluded to all the others; hence the excluded volume problem.

In general, it is of interest to consider all of the measurable properties of polymers with excluded volume. However, a few basic quantities have been at the focus of attention. One such property is the asymptotic nature of the mean square end-to-end distance in the limit that $n \rightarrow \infty$. It is only in this limit of large n that the highly non-Markovian nature of the polymer with excluded volume becomes apparent. This asymptotic dependence often is expressed as

$$\langle \mathbf{R}^2 \rangle \rightarrow n^{\gamma} l_{\text{eff}} \tag{6.4}$$

as $n \rightarrow \infty$, where the "exponent" γ,

$$1 < \gamma < 2 \tag{6.4a}$$

is to be determined. We note that $\gamma = 1$ corresponds to the definition of flexible polymer chains which have no excluded volume effects, while $\gamma = 2$ is the rigid rod limit (2.46) which implies maximal segment-segment repulsions. It is therefore clear that (6.4a) must hold. As we shall see, γ appears to be solely a function of the dimensionality of space.

First we consider the distribution functions for the chain with excluded volume in order to demonstrate the many-body aspects of the problem. Then a brief review is given of the several approaches to the problem. Finally, the self-consistent field (SCF) approach is introduced. We shall consider a modified version of Edward's SCF theory.[13–15]

A. The Long-Range Interactions

Again we consider the simplest case of gaussian equivalent links. The interaction (6.1), or (2.1), implies that the thermal Green's function for the entire chain is

$$G(\{r_k\}) = \left[\prod_{j=1}^{n} p(r_j - r_{j-1}; \Delta s_j)\right] \exp\left[-\frac{\beta}{2}\sum_{i\neq j} W(\mathbf{r}_i - \mathbf{r}_j)\right] \quad (6.5)$$

where for generality the equivalent links have been allowed to be of differing contour lengths. Thus the bond probabilities are

$$p(\mathbf{r}_j - \mathbf{r}_{j-1}; \Delta s_j) = \left(\frac{3}{2\pi l\,\Delta s_j}\right)^{3/2} \exp\left[-\frac{3(\mathbf{r}_j - \mathbf{r}_{j-1})^2}{2l\,\Delta s_j}\right] \quad (6.6)$$

The continuous limit is again taken as

$$\lim_{FI} \equiv \lim_{n\to\infty,\max \Delta s_j \to 0}, \qquad \sum_{j=1}^{n} \Delta s_j = L \quad (6.7)$$

In order to pass to this limit, it is noted that W_{ij}/l^2 is the interaction energy between two equivalent bonds per unit length squared. The interaction energy for two equivalent bonds of "length" Δs_i and Δs_j is then approximately $W_{ij}\,\Delta s_i\,\Delta s_j/l^2$. Again, as in (3.12), this approximation becomes exact in the limit (6.7). Thus for the discrete chain the end-to-end vector Green's function is

$$G_d(\mathbf{R}; L) = \int d\mathbf{r}_0 \left[\prod_{j=1}^{n} \int d\mathbf{r}_j p(\mathbf{r}_j - \mathbf{r}_{j-1}; \Delta s_j)\right] \delta(\mathbf{r}_0)\,\delta(\mathbf{R} - \mathbf{r}_n)$$
$$\times \exp\left(-\tfrac{1}{2}\sum_{i,j} V_{ij}\,\Delta s_i\,\Delta s_j\right) \quad (6.8)$$

where

$$V_{ij} \equiv (1 - \delta_{ij})\left(\frac{\beta}{l^2}\right) W_{ij} \quad (6.9)$$

Then the end-to-end vector probability distribution is

$$P_d(\mathbf{R}; L) = Z_d^{-1}(L)\, G_d(\mathbf{R}; L) \tag{6.10}$$

with the partition function

$$Z_d(L) = \int d\mathbf{R} G_d(\mathbf{R}; L) \tag{6.11}$$

Upon passage to the continuous limit (6.7) G_d becomes the Green's function for the continuous chain,[13]

$$G(\mathbf{R0}; L0) = \int_{\mathbf{r}(0)=\mathbf{0}}^{\mathbf{r}(L)=\mathbf{R}} \mathscr{D}[\mathbf{r}(s)] \exp\left\{-\frac{3}{2l}\int_0^L ds\dot{\mathbf{r}}^2(s) - \tfrac{1}{2}\int_0^L ds \int_0^L ds' V[\mathbf{r}(s) - \mathbf{r}(s')]\right\} \tag{6.12}$$

We note that (6.12) exhibits translational invariance, due to the isotropy of space, as it must. Thus, if in (6.12) the transformation

$$\mathbf{r}(s) \to \mathscr{R}(s) = \mathbf{r}(s) + \boldsymbol{\rho}, \qquad \mathscr{R}(0) = \boldsymbol{\rho}, \qquad \mathscr{R}(L) = \mathbf{R} + \boldsymbol{\rho}$$

is performed, then $G(\mathbf{R0}; L0)$ would just become $G(\mathbf{R} + \boldsymbol{\rho}, \boldsymbol{\rho}; L0)$ with the same end-to-end vector $\mathbf{R}$.

The relationship between the Wiener integral (3.20) and the simple diffusion equation (3.21) suggests that it might be instructive to convert (6.12) to a differential equation. As also noted by Whittington,[44] for the case of a discrete chain (6.12) can be expressed only in terms of the solution of a hierarchy of integro-differential equations. The derivation in the continuous case is presented in Appendix B for convenience, although the result is quoted here. Define the three-point "Green's" function as

$$G_3(\mathbf{RR'0}; LL'0) = \int_{\mathbf{r}(0)=\mathbf{0}}^{\mathbf{r}(L)=\mathbf{R}} \mathscr{D}[\mathbf{r}(s)]\, \delta[\mathbf{r}(L') - \mathbf{R}'] \times \exp\left\{-\frac{3}{2l}\int_0^L ds\dot{\mathbf{r}}^2(s) - \tfrac{1}{2}\int_0^L ds \int_0^L ds' V[\mathbf{r}(s) - \mathbf{r}(s')]\right\} \tag{6.13}$$

Equation (6.13) gives the number of chains of contour length L which have segments at $\mathbf{0}0$, $\mathbf{R}'L'$, and $\mathbf{R}L$. The probability distribution that a polymer chain of contour length L and end-to-end vector $\mathbf{R}$ have its segment at contour length L' (measured from the initial segment at $s = 0$) located at the space point $\mathbf{R}'$ is then

$$P(\mathbf{R}'L' \mid \mathbf{R}L) = \frac{G_3(\mathbf{RR'0}; LL'0)}{G(\mathbf{R0}; L0)} \tag{6.14}$$

If we do not require that both ends of the polymer be fixed in space, the probability distribution that a chain of length L and *any* end-to-end vector have a segment at $\mathbf{R}'$, L' is

$$P(\mathbf{R}'L' \mid L) = \frac{\int d\mathbf{R} G_3(\mathbf{R}\mathbf{R}'\mathbf{0};\, LL'0)}{\int d\mathbf{R} G(\mathbf{R}\mathbf{0};\, L0)} \tag{6.15}$$

Thus the hierarchy begins with

$$\left[\frac{\partial}{\partial L} - \frac{l}{6}\nabla_{\mathbf{R}}{}^2\right] G(\mathbf{R}\mathbf{0};\, L0) + \int d\mathbf{R}' V(\mathbf{R} - \mathbf{R}') \int_0^L ds G_3(\mathbf{R}\mathbf{R}'\mathbf{0};\, Ls0) = \delta(\mathbf{R})\, \delta(L) \tag{6.16}$$

and has an analogous equation for G_3 in terms of a four point G_4, etc. Equation (6.16) is then explicitly of the form of the hierarchies of equilibrium and nonequilibrium statistical mechanics,[6,7] reinforcing the analogy between γ in (6.4a) and critical exponents.[45]

For a continuous chain, the hierarchy *never* terminates. On the other hand, in the case of a discrete chain, it terminates at $G_n \equiv G(\{\mathbf{r}_k\})$ of (2.8). However, in (6.4) we imply the limit $n \to \infty$, so the chain might just as well be continuous.

B. Survey of the Different Approaches to the Problem

At the Θ-point, the long-range interactions are cancelled by the polymer-solvent interactions. Thus near the Θ-point the excluded volume v is small, and perturbation theory may be used to express $G(\mathbf{R};\, L)$, $\langle \mathbf{R}^2 \rangle$, etc., in a power series in v. Perturbation theory can be developed by expanding (6.12) in a power series in v or by using the standard cluster theory on (6.5). Results[5] have been obtained through third-order, giving

$$\frac{\langle \mathbf{R}^2 \rangle}{nl^2} = 1 + \tfrac{4}{3}z + \left(\tfrac{16}{3} - \tfrac{28}{27}\pi\right)z^2 + 6.459z^3 + \cdots \tag{6.17}$$

where

$$z = \left(\frac{3}{2\pi l^2}\right)^{3/2} v n^{1/2} \tag{6.17a}$$

However, the series is very slowly convergent. Furthermore, since (6.17) represents a function which is asymptotically n^{γ}, $\gamma \neq 1.5$, 1, in a power series in $n^{1/2}$, there must be terms of the form $n \log n$ in this series. No such terms have yet been found, even though they must exist.

As noted in the introduction, in dealing with difficult many-body problems, like the excluded volume problem, there are usually a few rather crude, but physically reasonable, kinetic-theory type arguments which often give the correct results apart from a constant factor. In the excluded volume problem, there are some smoothed density approaches where, e.g., in (6.16), the field due to the excluded volume from the polymer density

$$\rho(\mathbf{R}' \mid \mathbf{R}L) = \int_0^L dL' P(\mathbf{R}'L' \mid \mathbf{R}L) \tag{6.18}$$

is approximated by some smooth distribution about the center of mass of the polymer.[5] Some of the distributions that have been employed are spherically symmetric gaussian and constant over an ellipsoid. All of these kinetic-theory type arguments lead to $6/5 \leqslant \gamma \leqslant 4/3$, thereby slightly narrowing the range of (6.4a).

As in the case of the Ising model, self-avoiding random walks on a lattice have been investigated for short n leading to values for $\langle \mathbf{R}^2 \rangle_n$, etc.[46] Then extrapolations for $n \to$ large are used to obtain γ. The exponents γ appear only to depend on the dimensionality of the lattice and not on the lattice type. They give

$$\begin{aligned} \gamma &\cong 1.20 \text{ in 3 dimensions} \\ \gamma &\cong 1.50 \text{ in 2 dimensions} \end{aligned} \tag{6.19}$$

[Note that in one dimension, the only self-avoiding random walk is a rigid rod for which $\gamma = 2$, whereas in an infinite number of dimensions, a random walk would never intersect itself so $\gamma \to 1$.] Although the lengths of the walks used in these calculations are not necessarily long enough to insure asymptotic behavior, the results obtained are indicative of the expected behavior. Monte Carlo calculations of the properties of self-avoiding random walks appear to agree with (6.19)

There have been some approaches to the problem based upon equation of motion methods. However, by far the most sophisticated approximations have involved the SCF theories. There have also been some diagrammatic approaches to the polymer excluded volume problem, but these can be shown to be intimately related to the SCF and perturbational methods.

In the category of SCF theories we wish to classify three theories: Reiss' SCF formulation employs Kirkwood's variational principle for the free energy.[47,48] Edwards has presented both a probabilistic approach[13] and a formal approach using functional integrals.[13,14] Yeh and Isihara[48] claim that Reiss missed a term in obtaining the variational equations, and therefore such equations are, in general, insoluble. They then introduce approximations in order to obtain the explicit formal solutions. It has been shown that the introduction of a limiting process enables us to write the correct variational equations of Yeh and Isihara in the form given by Reiss, provided the summation indices are properly interpreted.[49] The final result is then one-half of that originally given by Reiss [Reiss dropped this factor in passing from his equation (25) to (26)]. Edwards' highly misunderstood probabilistic SCF theory has been adequately reviewed by De Gennes[31] and therefore required little comment here. Edwards' functional integral approach to the problem stands out as a separate method even though the final results are identical to those from his simpler approach. However, this more formal method appears to be of greater generality as witnessed, in part, by its use in other, more complicated problems. This functional integral approach is generalized here to provide a hierarchy of SCF theories. The simplest such theory is that of Edwards, while the next one is almost identical to the corrected form of Reiss' variational approach.

Edwards has also noted the strong mathematical analogies between the functional integral representation of the polymer excluded volume problem and questions associated with electronic structure in disordered systems.[17–19] Thus a detailed discussion of this formal approach is of more than just academic interest. In the polymer case, approximation may be guided by probabilistic arguments, whereas in the disordered system analogous mathematical approximations rest upon less intuitive grounds.

C. Introduction of the SCF

We could attempt to introduce an SCF approximation directly into (6.16). Such a discussion would be instructive, but only heuristic. The formal derivation is presented and generalized in Section VID. The assumption of the existence of a suitable self-consistent field implies that somehow we destroy the isotropy of space. The anisotropy associated with the introduction of an SCF is introduced either by specifying that the initial segment is at some fixed point in space (conveniently chosen as the origin) or by specifying the end-to-end vector $\mathbf{R}$ in addition. In the first case, the assumption that $\mathbf{r}(0) \equiv \mathbf{0}$ leads to a polymer distribution which is spherically symmetric about the origin. The field representing the excluded volume then of course has the same symmetry. We want to introduce some approximation that will permit us to calculate both the distribution and the field in a completely self-consistent manner. In the second approach, the specification of $\mathbf{r}(0) \equiv \mathbf{0}$ and $\mathbf{r}(L) \equiv \mathbf{R}$ leads to a field of $D_{\infty h}$ symmetry about these two end (focal) points.

In addition to destroying the isotropy of space, the introduction of the SCF field implies a Markovian (albeit self-consistent) approximation to the inherently non-Markovian process described by (6.16). Thus, if V_{SCF}

is the SCF, the Markovian approximation to (6.12) is[13]

$$G_{\mathrm{SCF}}(\mathbf{R0};\,L0[\mathbf{R}L])$$
$$= \int_{\mathbf{r}(0)=\mathbf{0}}^{\mathbf{r}(L)=\mathbf{R}} \mathscr{D}[\mathbf{r}(s)] \exp\left\{-\frac{3}{2l}\int_0^L ds\,\dot{\mathbf{r}}(s)^2 - \int_0^L ds\,V_{\mathrm{SCF}}[\mathbf{r}(s)]\right\}$$
$$= \lim_{FI} \int d\mathbf{r}_0 \left\{\prod_{j=1}^{n} \int d\mathbf{r}_j\,p(\mathbf{r}_j - \mathbf{r}_{j-1};\,\Delta s_j)\exp(-V_j\,\Delta s_j)\right\}\delta(\mathbf{r}_0)\,\delta(\mathbf{r}_n - \mathbf{R}) \tag{6.20}$$

where the $[\mathbf{R}L]$ notation implies that $\mathbf{R}$ and L have been fixed. $[V_0(\mathbf{0}) \equiv 0.]$ Equation (6.20) implies that G_{SCF} obeys the diffusion equation

$$\left[\frac{\partial}{\partial s} - \frac{l}{6}\nabla_{\mathbf{R}'}{}^2 + V_{\mathrm{SCF}}(\mathbf{R}'[\mathbf{R}L])\right]G_{\mathrm{SCF}}(\mathbf{R}'\mathbf{R}'';\,ss'[\mathbf{R}L]) = \delta(\mathbf{R}' - \mathbf{R}'')\,\delta(s - s') \tag{6.21}$$

where V_{SCF} is some functional of G_{SCF}. Thus (6.21) is a closed equation for G_{SCF} and not a member of a hierarchy of equations. The fact that (6.21) exhibits G_{SCF} as the Green's function for a diffusion equation verifies that it describes a Markov process in an external field.[10]

In order to obtain the Markovian approximations to G_3 in (6.16) we remember that, apart from a normalization, $G_3(\mathbf{RR}'\mathbf{0};\,Ls0)$ represents the number of chain configurations that pass through $\mathbf{0}0$, $\mathbf{R}'s$, and $\mathbf{R}L$.[21,40] Hence the desired Markovian approximation to G_3 is the number of chains running from $\mathbf{0}0$ to $\mathbf{R}'s$ times the number running from $\mathbf{R}'s$ to RL or

$$G_{3\mathrm{SCF}}(\mathbf{RR}'\mathbf{0};\,Ls0[\mathbf{R}L]) = G_{\mathrm{SCF}}(\mathbf{RR}';\,Ls[\mathbf{R}L])G_{\mathrm{SCF}}(\mathbf{R}'\mathbf{0};\,s0[\mathbf{R}L]) \tag{6.22}$$

The price we pay for the Markovian approximation is the introduction of SCF Green's functions involving an intermediate point along the chain in addition to an endpoint. Note that (6.22) does not ignore chain connectivity or excluded volume from any part of the chain. Closure of the SCF equations then requires an equation of motion for the general propagator $G_{\mathrm{SCF}}(\mathbf{R}'\mathbf{R}'';\,ss'[\mathbf{R}L])$. As shown in Section VID, if V_{SCF} is chosen to give the "best" approximation to $G(\mathbf{R0};\,L0)$ to enable the direct calculation of $\langle \mathbf{R}^2\rangle$ via

$$\langle \mathbf{R}^2\rangle_L = \frac{\int d\mathbf{R}\,\mathbf{R}^2 G(\mathbf{R0};\,L0)}{\int d\mathbf{R}\,G(\mathbf{R0};\,L0)} \tag{6.23}$$

then the equation of motion obtained is the diffusion equation (6.21) with

$$V_{\mathrm{SCF}}(\mathbf{R}'[\mathbf{R}L]) = \int d\mathbf{r} \int_0^L ds V(\mathbf{R}' - \mathbf{r}) G_{\mathrm{SCF}}(\mathbf{Rr}; Ls[\mathbf{R}L]) \times \frac{G_{\mathrm{SCF}}(\mathbf{r0}; s0[\mathbf{R}L])}{G_{\mathrm{SCF}}(\mathbf{R0}; L0[\mathbf{R}L])} \tag{6.24}$$

$$\equiv \int d\mathbf{r} \int_0^L ds V(\mathbf{R}' - \mathbf{r}) P_{\mathrm{SCF}}(\mathbf{r}; s \mid \mathbf{R}L) \tag{6.25}$$

This result follows from the SCF approximation to the segment distribution for chains with $\mathbf{R}L$ fixed.

Alternatively, if V_{SCF} is chosen to give the "best" approximation to $Z(L)$, we obtain

$$V_{\mathrm{SCF}}(\mathbf{R}'[L]) = \int d\mathbf{r} \int_0^L ds V(\mathbf{R}' - \mathbf{r}) P_{\mathrm{SCF}}(\mathbf{r}, s \mid L) \tag{6.26}$$

where the SCF segment distribution is the Markovian approximation

$$P_{\mathrm{SCF}}(\mathbf{r}; s \mid L) = Z_{\mathrm{SCF}}^{-1}(L) \int d\mathbf{R} G_{\mathrm{SCF}}(\mathbf{Rr}; Ls[L]) G_{\mathrm{SCF}}(\mathbf{r0}; s0[L]) \tag{6.27}$$

in obvious notation. If we substitute (6.26) and (6.27) in (6.21) with $[L]$ instead of $[\mathbf{R}L]$, the result is that of Edwards' probabilistic approach.[13,31] It should be noted that Edwards' functional integral approach formally would give (6.23)–(6.25), i.e., it assumes fixed $\mathbf{R}$, but it could easily be modified to give (6.27)–(6.28). The fields (6.24), or (6.26) when substituted into (6.21) generate SCFs which are dependent only on the chain segment position with respect to the spatial origins—the fixed endpoints—but not on the segment positions along the chain. Physically, this result arises because the specification of the chain endpoints sets up a distribution of chain segments $\int_0^L ds P(\mathbf{r}; s \mid \cdots)$ throughout space. This distribution then leads to an excluded volume field which can only be given by (6.24) or (6.27) in the SCF approximation.[13,14,31]

The Edwards SCF therefore results from introducing the Markovian approximation to the lowest member of the hierarchy (6.16). This approximation can be introduced into higher members of the hierarchy also. If this is done in the equation for G_3 in terms of G_4, we obtain a result that is almost identical to the corrected Reiss result. Specifically, if we use Edwards' formal approach (see next section for details) to find the "best" approximation to $G_3(\mathbf{RR}'\mathbf{0}; Ls0)$ and therefore to $P(\mathbf{R}'; s[\mathbf{R}L])$ via (2.18) the result is

$$\left[\frac{\partial}{\partial s'} - \frac{l}{6}\nabla_{\mathbf{r}}^2 + V_{\mathrm{SCF}}(\mathbf{r}[\mathbf{R}L;\mathbf{R}'s])\right]G_{\mathrm{SCF}}(\mathbf{rr}'; s's''[\mathbf{R}L;\mathbf{R}'s]) = \delta(\mathbf{r}-\mathbf{r}')\,\delta(s'-s'') \quad (6.28)$$

with

$$V_{\mathrm{SCF}}(\mathbf{r}[\mathbf{R}L;\mathbf{R}'s]) = \int d\mathbf{r}'' \int ds''' V(\mathbf{r}-\mathbf{r}'')P_{\mathrm{SCF}}(\mathbf{r}''; s''' \mid \mathbf{R}L, \mathbf{R}'s) \quad (6.29)$$

and

$$P_{\mathrm{SCF}}(\mathbf{r}''; s''' \mid \mathbf{R}L, \mathbf{R}'s) = \begin{cases} \dfrac{G_{\mathrm{SCF}}(\mathbf{RR}'; Ls[\ ])G_{\mathrm{SCF}}(\mathbf{R}'\mathbf{r}''; ss'''[\ ])G_{\mathrm{SCF}}(\mathbf{r}''\mathbf{0}, s'''0[\ ])}{G_{\mathrm{SCF}}(\mathbf{RR}'; Ls[\ ])G_{\mathrm{SCF}}(\mathbf{R}'\mathbf{0}; s0[\ ])}, & s > s''' \\ \dfrac{G_{\mathrm{SCF}}(\mathbf{Rr}''; Ls'''[\ ])G_{\mathrm{SCF}}(\mathbf{r}''\mathbf{R}'; s'''s[\ ])G_{\mathrm{SCF}}(\mathbf{R}'\mathbf{0}; s0[\ ])}{G_{\mathrm{SCF}}(\mathbf{RR}'; Ls[\ ])G_{\mathrm{SCF}}(\mathbf{R}'\mathbf{0}; s0[\ ])}, & s''' > s \end{cases} \quad (6.30)$$

Here $[\ \ ] = [\mathbf{R}L, \mathbf{R}'s]$. Alternatively, if the method is used to find the "best" $\int G_3(\mathbf{RR}'0; Ls0]\, d\mathbf{R}$, thereby giving $P(\mathbf{R}'; s \mid L)$, the result is (6.28)–(6.30) expressed in terms of $V_{\mathrm{SCF}}(\mathbf{R}[L, \mathbf{R}'s])$, $P_{\mathrm{SCF}}[\mathbf{r}; s' \mid L, \mathbf{R}'s)$, $G_{\mathrm{SCF}}(\cdots [L, \mathbf{R}'s])$ with $\int d\mathbf{R}$ in the numerator and denominator of (6.30).

The connection with Reiss' method is best established by taking $\mathbf{r} = \mathbf{R}'$, $s = s'$, $\mathbf{r}' = \mathbf{0}$, $s'' = 0$ in (6.28)–(6.30). In this case, we consider the equation (say $\mathbf{R}$ fixed)

$$\left[\frac{\partial}{\partial s} - \frac{l}{6}\nabla_{\mathbf{R}'}^2 + V_{\mathrm{SCF}}(\mathbf{R}'[\mathbf{R}L, \mathbf{R}'s])\right]G_{\mathrm{SCF}}(\mathbf{R}'\mathbf{0}; s0[\mathbf{R}L, \mathbf{R}'s]) = \delta(\mathbf{R}')\,\delta(s) \quad (6.31)$$

in which $G_{\mathrm{SCF}}(\mathbf{R}'\mathbf{0}; s0[\mathbf{R}L; \mathbf{R}'0])$ describes the number of chain segments of contour length s and the end-to-end vector $\mathbf{R}'$ which are present in chains of length L and end-to-end vector $\mathbf{R}$. Thus the excluded volume field at $\mathbf{R}'s$ is generated by the polymer density due to all the other segments $P(\mathbf{r}''; s''' \mid \mathbf{R}L, \mathbf{R}'s)$, where evidently $\mathbf{R}L$, $\mathbf{R}'s$ have been specified.

The corrected SCF obtained from the Kirkwood variational principle (Reiss' approach)[47,48] is then one-half of (6.29).[49] The difference between the two fields can be expressed in simple physical terms. Equation (6.29) is a usual Hartree-type field in which the field at $\mathbf{r} = \mathbf{R}'(s = s')$ is a sum of contributions from all other segments. Thus, in using (6.29) to evaluate the free energy, the pairwise interactions are counted twice. In order to correct for this overcount, the first-order correction to the free energy is the negative of the average pairwise interaction free energy. In the Reiss approach $V_{SCF}(\mathbf{R}'[\mathbf{R}L, \mathbf{R}'s])$, or $V_{SCF}(\mathbf{R}'[L, \mathbf{R}'s])$, is only one-half the field due to all other segments. Therefore in the calculation of the free energy with this potential there is no double counting, and the first-order correction to the free energy vanishes identically. The Edwards and Reiss formulations come from different self-consistent schemes; hence there is no necessity that they be identical. However, they both are related to the well-known SCF theories in other branches of chemistry and physics.

D. Formal Derivation of the SCF Equations

The derivation here is in the same spirit as Edwards' approach in terms of functional integrals.[15] Since Edwards wanted to obtain explicit solutions to the SCF equations to determine γ of (6.4), he found it convenient to introduce other mathematical approximations first and then introduce the "SCF" approximation. The formal physical content is, however, best exhibited by initially employing the "SCF" approximation to obtain the SCF equations (6.21) and (6.24), (6.26), (6.29), etc. In our approach, the resulting equations then still remain to be solved self-consistently.

In order to avoid getting lost in the interesting mathematical details of the derivation, the major steps are sketched and then the details are supplied. Any SCF approximation must be of the form (6.21), i.e., a diffusion equation with some "external potential" which represents the average volume exclusion field due to the other segments in $0 \leqslant s \leqslant L$. It turns out that the exact G of (6.12), or equivalently (6.16), can be expressed in terms of an auxiliary Green's function $G([\phi])$ which satisfies

$$\left[\frac{\partial}{\partial L} - \frac{l}{6}\nabla_{\mathbf{R}}^2 + i\phi(\mathbf{R})\right] G(\mathbf{R}\mathbf{0}; L0[\phi]) = \delta(L)\,\delta(\mathbf{R}) \tag{6.25}$$

Equation (6.25) describes the Green's function for "diffusion" in the (real) external field $i\phi(\mathbf{R})$, where $\phi(\mathbf{R})$ is a *random field.* [The imaginary unit i appears in (6.25) only for mathematical convenience.] Specifically, the correspondence between G and $G([\phi])$ is written as

$$G(\mathbf{R}\mathbf{0}; L0) \equiv \langle G(\mathbf{R}\mathbf{0}; L0[\phi])\rangle_\phi \tag{6.26}$$

where G is the average of $G([\phi])$ over the gaussian random field $\phi(\mathbf{R})$ with zero mean and variance $V(\mathbf{R} - \mathbf{R}')$:

$$\langle \phi(\mathbf{R})\rangle_\phi \equiv 0 \tag{6.27a}$$

$$\langle \phi(\mathbf{R})\phi(\mathbf{R}')\rangle_\phi = V(\mathbf{R} - \mathbf{R}') \tag{6.27b}$$

Before discussing gaussian random fields in more detail and explicitly proving (6.25)–(6.27), we examine the implications of these equations for the existence of a SCF approximation.

Since (6.25) expresses $G([\phi])$ as the Green's function for some external field $i\phi(\mathbf{R})$, the average (6.26) over all ϕ implies that if an SCF approximation V_{SCF} is meaningful, some particular ϕ, call it $\phi_0(\mathbf{R})$, must dominate the average (6.26). Thus a reasonable approximation is obtained by just taking $\phi \to \phi_0$, so

$$G \simeq G([\phi_0]) \equiv G_{\text{SCF}} \tag{6.28}$$

By comparing (6.21) and (6.25), we find that the SCF potential V_{SCF} must be this dominant potential in (6.26),

$$V_{\text{SCF}} \equiv i\phi_0(\mathbf{R}) \tag{6.29}$$

The validity of such an SCF approximation can in principle be checked by examining the contribution from the fluctuations about $\phi_0(\mathbf{R})$ in (6.26), but in practice this is mathematically difficult. The natural method for obtaining the dominant contribution to averages [including (6.26)] is by Laplace's method of steepest descent. This is the basic spirit of Edwards' formal method.[13,14] He first approximates the solution to (6.25) by WKB techniques and then obtains the dominant contribution to (6.26) by steepest descent. The generalization of this method to obtain the Reiss-like field (6.29) is provided by finding the dominant contribution ϕ_1 to G_3, where

$$G_3(\mathbf{RR}'\mathbf{0};\, Ls0) \equiv \langle G(\mathbf{RR}';\, Ls[\phi])G(\mathbf{R}'\mathbf{0};\, s0[\phi])\rangle_\phi \tag{6.30}$$

The $G([\phi])$ in (6.30) is the same as that in (6.25), and the average over ϕ is also that of (6.26). Note that under the ϕ-averaging, we rigorously have a Markov process. Thus the procedure for obtaining generalized Hartree-type SCF's should now be obvious (apart from the details which now follow).

Although Gaussian random functions are commonly encountered, their representation in terms of functional integrals is often unnecessary; the averages (6.27) are usually sufficient for the development. A familiar example occurs in the description of Brownian motion by means of the Langevin equation:[6,7]

$$\dot{\mathbf{v}}(t) = -\zeta\mathbf{u}(t) + \mathbf{A}(t) \tag{6.31}$$

where $\mathbf{v}$ is the velocity of the Brownian particle, $-\zeta m\mathbf{v}$ is the friction force on it, m is its mass, and $\mathbf{A}(t)$ is its random acceleration, which has the ensemble averages

$$\langle \mathbf{A}(t)\rangle = 0 \tag{6.32a}$$

$$\langle A_i(t)A_j(t')\rangle = \delta_{ij}\frac{2\zeta}{\beta m}\,\delta(t-t'), \qquad i, j = x, y, z \tag{6.32b}$$

The analogy between (6.32) and (6.27) should be clear.

In the case of polymers with excluded volume, however, it is convenient to introduce explicitly the averaging (6.27) in terms of an integration over all functions $\phi(\mathbf{R})$. Such a well-known representation is conveniently introduced by passing from discrete variables to the limit of continuous ones.[10,14,29] Consider the simple integral

$$v = \frac{\int_{-\infty}^{\infty} x^2 \exp(-v^{-1}x^2/2)\, dx}{\int_{-\infty}^{\infty} \exp(-v^{-1}x^2/2)\, dx}$$

which in the case of many variables can be generalized to

$$V_{ij} = \frac{\int_{-\infty}^{\infty} \cdots \int_{-\infty}^{\infty} x_i x_j \exp\left[-\tfrac{1}{2}\sum_{kl} x_k V_{kl}^{-1} x_l\right] \prod_m dx_m}{\int_{-\infty}^{\infty} \cdots \int_{-\infty}^{\infty} \exp\left[-\tfrac{1}{2}\sum_{k,l} x_k V_{kl}^{-1} x_l\right] \prod_m dx_m} \tag{6.33a}$$

where $\mathbf{V}^{-1}$ is the matrix inverse of $\mathbf{V}$; $\mathbf{V}\mathbf{V}^{-1} = \mathbf{1}$ or

$$\sum_j V_{ij}^{-1} V_{jk} = \delta_{ij} \tag{6.33b}$$

If we now associate i, j, etc., in (6.33) with points $\mathbf{R}_i$, $\mathbf{R}_j$, etc., on an arbitrary lattice and take $\phi(\mathbf{R}_i) = x_i$, etc., upon passing to continuous variables (6.33) becomes

$$V(\mathbf{R}, \mathbf{R}') = \mathcal{N} \int \phi(\mathbf{R})\phi(\mathbf{R}') \exp\left[-\tfrac{1}{2}\int d\mathbf{R} \int d\mathbf{R}' \phi(\mathbf{R}) V^{-1}(\mathbf{R}, \mathbf{R}') \phi(\mathbf{R}')\right] \delta\phi \tag{6.34a}$$

where

$$\int V(\mathbf{R}, \mathbf{R}') V^{-1}(\mathbf{R}', \mathbf{R}'') = \delta(\mathbf{R} - \mathbf{R}'') \tag{6.34b}$$

$$\mathcal{N}^{-1} = \int \exp\left[-\tfrac{1}{2}\int d\mathbf{R} \int d\mathbf{R}' \phi(\mathbf{R}) V^{-1}(\mathbf{R}, \mathbf{R}') \phi(\mathbf{R}')\right] \delta\phi \tag{6.34c}$$

and

$$\delta\phi \equiv \prod_{\mathbf{R}} d\phi(\mathbf{R}) \tag{6.34d}$$

Equation (6.34a) is of the required form (6.27b), and it can easily be verified that

$$\mathcal{N} \int \phi(\mathbf{R}) \exp\left[-\tfrac{1}{2}\int d\mathbf{R} \int d\mathbf{R}' \phi(\mathbf{R}) V^{-1}(\mathbf{R}, \mathbf{R}') \phi(\mathbf{R}')\right] \delta\phi \equiv 0 \tag{6.34e}$$

Hence (6.27a) is satisfied. Note that (6.34a and 6.34e) give to

$$P[\phi]\,\delta\phi = \mathscr{N} \exp\left[-\tfrac{1}{2}\int d\mathbf{R} \int d\mathbf{R}' \phi(\mathbf{R}) V^{-1}(\mathbf{R}, \mathbf{R}')\phi(\mathbf{R}')\right] \delta\phi \tag{6.35}$$

the interpretation as "the probability that the random function ϕ lies in the interval $\phi(\mathbf{R})$ to $\phi(\mathbf{R}) + \delta\phi$."

Having introduced the functional integral representation of the averages (6.27), we still must show that (6.25) and (6.26) are valid representations of (6.12). This result is well known and follows from the identity

$$\exp\left\{-\tfrac{1}{2}\int_0^L ds \int_0^L ds' V[\mathbf{r}(s) - \mathbf{r}(s')]\right\}$$

$$\equiv \mathscr{N} \int \delta\phi \exp\left\{-i\int_0^L ds \phi[\mathbf{r}(s)] - \tfrac{1}{2}\iint \phi(\mathbf{r}) V^{-1}(\mathbf{r} - \mathbf{r}')\phi(\mathbf{r}')\, d\mathbf{r}\, d\mathbf{r}'\right\} \tag{6.36}$$

with $\mathscr{N}$ given by (6.34c) and V^{-1} by (6.34d). Equation (6.36) is easily verified by making the transformation

$$\phi'(\mathbf{r}) = \phi(\mathbf{r}) + i\int_0^L ds V[\mathbf{r} - \mathbf{r}(s)], \qquad \delta\phi' \equiv \delta\phi \tag{6.37}$$

in the left-hand side of (6.36) and then changing the dummy integration variable ϕ' to ϕ. Explicitly, substituting (6.36) into (6.12) and switching the order of the $\delta\phi$ and $\mathscr{D}[\mathbf{r}(s)]$ integrations gives

$$G(\mathbf{R}0; L0) = \mathscr{N} \int \delta\phi \exp\left[-\tfrac{1}{2}\int d\mathbf{r} \int d\mathbf{r}' \phi(\mathbf{r}) V^{-1}(\mathbf{r} - \mathbf{r}')\phi(\mathbf{r}')\right]$$

$$\times \int_{\mathbf{r}(0)=\mathbf{0}}^{\mathbf{r}(L)=\mathbf{R}} \mathscr{D}[\mathbf{r}(s)] \exp\left\{-\int_0^L ds\left(\frac{3}{2l}\dot{\mathbf{r}}^2(s) + i\phi[\mathbf{r}(s)]\right)\right\}$$

$$\equiv \int \delta\phi P[\phi] G(\mathbf{R}0; L0[\phi])$$

$$\equiv \langle G([\phi])\rangle_\phi \tag{6.38}$$

with $P[\phi]$ given by (6.35). $G([\phi])$ in (6.38) is then a Wiener integral, which implies the diffusion equation (6.25).

Having verified (6.25)–(6.27), we still must prove that the approximations (6.28) and (6.29) lead to the results stated in Section VIB. If we rewrite (6.38) as

$$G(\mathbf{R}0; L0) = \mathscr{N} \int \delta\phi \exp\left\{\ln G(\mathbf{R}0; L0[\phi]) - \tfrac{1}{2}\iint \phi V^{-1} \phi\right\}$$

$$\equiv \mathscr{N} \int \delta\phi \exp\{-B[\phi]\} \tag{6.39}$$

the dominant field $i\phi_0$ is obtained by taking the solution—the saddle point in function space—to be

$$\left.\frac{\delta B[\phi]}{\delta\phi(\mathbf{R}')}\right|_{\phi=\phi_0} = 0 \tag{6.40}$$

Although $G(\mathbf{R0}; L0[\phi])$ is not known explicitly, we can still obtain its functional derivative with respect to ϕ from the equation of motion (6.25) for $G([\phi])$.[50] The algebra is standard,[50] but it is summarized in Appendix C for convenience. The basic reason for introducing the algebraic details is to note the important result that quantities like $G(\mathbf{R}'\mathbf{R}''; ss'[\phi_0])$, which refer to only a portion of the chain and occur in (6.40), are defined by the equation of motion

$$\left[\frac{\partial}{\partial s} - \frac{l}{6}\nabla_{\mathbf{R}'}{}^2 + i\phi(\mathbf{R}')\right]G(\mathbf{R}'\mathbf{R}''; ss'[\phi]) = \delta(\mathbf{R}' - \mathbf{R}'')\,\delta(s - s') \tag{6.41}$$

and the condition $\phi = \phi_0$. The implication is that only one SCF field $i\phi_0$ is to be used for all $G([\phi])$, even if their indices do not refer to both chain ends. The results are indeed (6.24) as stated. If instead of (6.39) we had

$$\int d\mathbf{R}G(\mathbf{R0}; L0) = \mathcal{N}\int \delta\phi P[\phi]\exp\left\{\ln\int d\mathbf{R}G(\mathbf{R0}; L0[\phi])\right\} \tag{6.42}$$

the final result would be (6.27). Finally, $G([\phi])$ of (6.25), or (6.38), is obviously a Markov process. Therefore, using the definition (6.13), we see that (6.30) is satisfied. The dominant field $i\phi_1$ for G_3 is then the Reiss-like field (6.29). A hierarchy of SCF fields can be obtained, e.g., an $i\phi_2$ which provides the dominant contribution to G_4.

As noted previously, we can in principle treat the fluctuations about ϕ_0. From (6.39) and (6.40),

$$\exp\{-B[\phi]\}$$
$$= \exp\left\{-B[\phi_0] - \tfrac{1}{2}\int d\mathbf{r}\int d\mathbf{r}'K(\mathbf{r}, \mathbf{r}' \mid \mathbf{R}L)\bar{\phi}(\mathbf{r})\bar{\phi}(\mathbf{r}') + \mathcal{O}([\bar{\phi}]^3)\right\} \tag{6.43}$$

where

$$K(\mathbf{r}, \mathbf{r}' \mid \mathbf{R}L) = \left[\frac{\delta^2 B[\phi]}{\delta\phi(\mathbf{r})\,\delta\phi(\mathbf{r}')}\right]_{\phi=\phi_0} \tag{6.44}$$

and

$$\bar{\phi} = \phi - \phi_0 \tag{6.45}$$

If we neglect terms of $\mathcal{O}([\bar{\phi}]^3)$, when (6.43) is substituted into (6.39) the next approximation which includes the fluctuations is

$$G(\mathbf{R0}; L0) \simeq G(\mathbf{R0}; L0[\phi_0])\mathcal{N} \int \delta\phi \exp\left\{-\tfrac{1}{2}\iint \phi K \phi\right\}$$

$$= G(\mathbf{R0}; L0[\phi_0])\{\det VK(\mathbf{rr}' \mid \mathbf{R}L)\}^{-1/2} \tag{6.46}$$

The det in (6.46) implies the Fredholm determinant, whose evaluation would require explicit knowledge of $G([\phi_0])$ for all portions of the chain.

E. Comparison with Diagrammatic Approaches to the Excluded Volume Problem

In many-body theory, it is fashionable to generate "new" approximations by summing infinite classes of the diagrams of perturbation theory[50] (whether or not the approximations give anything useful). The recent diagrammatic approaches to the excluded volume problem[51,52] can be related to the following crude approximation to the simple Edwards field (6.24). Let

$$\left[\frac{\partial}{\partial s} - \frac{l}{6}\nabla_{\mathbf{r}} + V_{\mathrm{SCF}}(\mathbf{rr}'; ss')\right] G_{\mathrm{SCF}}(\mathbf{rr}'; ss') = \delta(\mathbf{r} - \mathbf{r}')\,\delta(s - s') \tag{6.47}$$

with

$$V_{\mathrm{SCF}} = \frac{\displaystyle\int d\mathbf{r}'' \int_0^L ds'' V(\mathbf{r} - \mathbf{r}'') G_{\mathrm{SCF}}(\mathbf{rr}''; ss'') G_{\mathrm{SCF}}(\mathbf{r}''\mathbf{r}'; s''s')}{G_{\mathrm{SCF}}(\mathbf{rr}'; ss')} \tag{6.48}$$

be the approximate SCF theory. This approach completely ignores the requisite lack of translational invariance arising from the specification of the positions of the endpoints. Using as the definition of the distribution function G_0 for the gaussian chain (3.20) and (3.22) (with $V \equiv 0$), we can convert (6.47)–(6.48) to an integral equation

$$G_{\mathrm{SCF}}(\mathbf{rr}'; ss') = G_0(\mathbf{rr}'; ss') + \int d\mathbf{r}'' \int ds'' G_0(\mathbf{rr}''; ss'') \int d\mathbf{r}''' \int ds'''$$

$$\times V(\mathbf{r}'' - \mathbf{r}''') G_{\mathrm{SCF}}(\mathbf{r}''\mathbf{r}'''; s''s''') G_{\mathrm{SCF}}(\mathbf{r}'''\mathbf{r}'; s'''s') \tag{6.49}$$

in the usual manner.[50] Equation (6.49) is the starting point for the diagrammatic analysis. Figure 3 gives a diagrammatic representation of the major quantities where the self-energy is

$$\Sigma_{\mathrm{SCF}}(\mathbf{rr}'; ss') = V(\mathbf{r} - \mathbf{r}') G_{\mathrm{SCF}}(\mathbf{rr}'; ss') \tag{6.50}$$

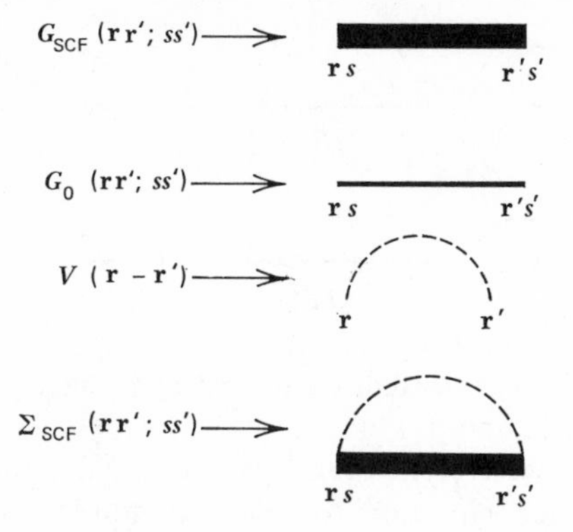

Fig. 3. Diagrammatic representations of G_{SCF}, G_0, V, and Σ_{SCF}. The points $\mathbf{r}(s)$, etc., refer to the ends of the lines and to the vertices.

Using (6.50), we write (6.49) in "matrix notation" as[50]

$$G_{SCF} = G_0 + G_0 \, \Sigma_{SCF} \, G_{SCF} \tag{6.51}$$

and represent it diagrammatically in Fig. 4, where integration over all intermediate vertices is implied as usual.[50] Iterating Fig. 4 into Fig. 3 gives Fig. 5, etc., so that (6.50) contains all possible infinitely nested type structures like those in Fig. 6a where none of the dashed V lines cross, i.e., the terms like those in Fig. 6b are omitted. It can be shown by performing directly the perturbative expansion of the exact (6.12) that the exact Σ should contain both types of Figs. 6a and 6b.

Apart from differences in notation, Figs. 5 and 6a are seen to be identical to the approximation developed by Curro, Blatz, and Pings,[51] when the

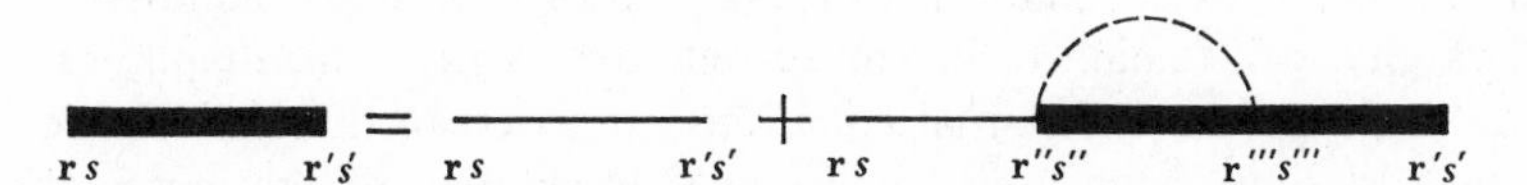

Fig. 4. Diagrammatic representation of the crude SCF approximation (4.3). Integration over all intermediate points (in space and along the chain) is implied.

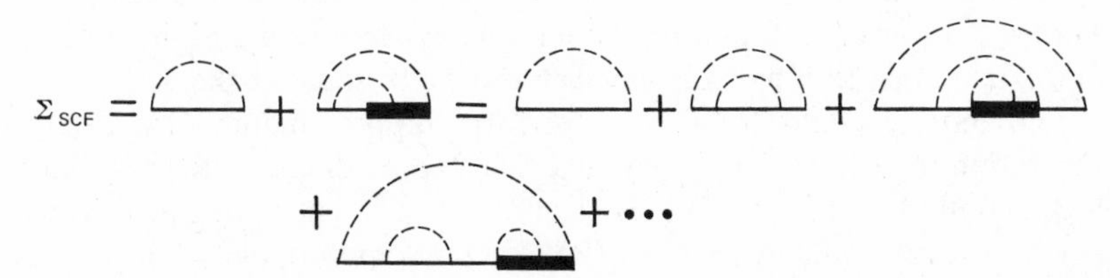

Fig. 5. The iteration of Fig. 4 into the definition of Σ_{SCF} in Fig. 3. Vertices unlabeled for convenience.

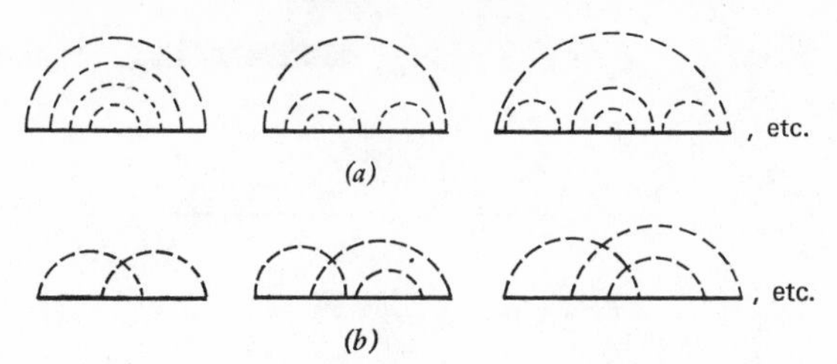

Fig. 6. (*a*) Typical diagrams included in Σ_{SCF} of the crude SCF approximation (6.47)–(6.48). (*b*) Typical diagrams omitted by this approximation.

latter is applied successively, by analogy with the diagrammatic derivation of the Percus-Yevick equation for classical fluids.[6,7] The first term of Fig. 6a is the generic term of the ladder approximation of Chakahisa.[52]

The physical defect of the approximation (6.47) and (6.48), apart from its neglect of the specification of the chain end(s), lies in the fact that the approximate excluded volume field is made up of a sum of contributions from only a portion of the total chain (of contour length L). This arises because the G's in (6.48), which are defined through the inhomogeneous diffusion equations, vanish when $s - s''$ or $s'' - s' < 0$. Thus the actual limits of integration must be $s \geqslant s'' \geqslant s'$.

F. Approximate SCF Solution[19]

In Sections VIC and VID the SCF theory is introduced to exhibit its most fundamental aspects. The mathematical approximations necessary to obtain an explicit solution are not considered. This provides the simplest formulation of the SCF theory of the polymer excluded volume problem, thereby enabling comparisons with other approaches to the problem. The SCF equations (6.21) and (6.24), or higher members of the hierarchy, still remain to be solved self-consistently. These equations are nonlinear integro-differential equations, and we can therefore appreciate Edwards' simplifying approximations. Rather than reiterating Edwards' approximate methods, we indicate an alternate method of solution which relies more heavily on quantum mechanical analogies than the approach given by Edwards. The two methods give identical results (clearly they should!); however, the one presented here gives greater insight into the approximations, leading to better justification of some of them.

In (6.38) we could try first to perform (approximately) the $\mathscr{D}[\mathbf{r}(s)]$ integration to yield $G(\mathbf{R}0; L0[\phi])$ and then (approximately) the remaining $\delta\phi$ integration. From (6.25) it is clear that the Wiener integral is not in general exactly evaluable since $\phi(\mathbf{R})$ is a random variable. One possible simplification arises from the fact that a solution is desired in the asymptotic limit $L \to \infty$. Proceeding by analogy with the semiclassical approach to Feynman path integrals (discussed in Appendix A), in this limit the "action"

$$A(\mathbf{R}; L) = \int_0^L ds \left(\frac{3}{2l} \dot{\mathbf{r}}^2(s) + i\phi[\mathbf{r}(s)] \right) \tag{6.52}$$

is, in general, very large for arbitrary continuous paths $\mathbf{r}(s)$, $0 \leqslant s \leqslant L$, $\mathbf{r}(0) = 0$, $\mathbf{r}(L) = \mathbf{R}$. The fields $i\phi$ must be real; since the whole excluded volume problem is couched in probabilistic language, there is no room for imaginary quantities in the final result. By analogy with action integrals in classical mechanics, $l/3$ is the "particle's mass" and $-i\phi$ is the "potential energy." As $L \to \infty$, the "action" (6.52) is expected to be very large. Thus by analogy with the semiclassical limit in quantum mechanics we look for the classical path(s) that make (6.52) stationary.

Before considering the Euler-Lagrange equations corresponding to (6.52), it is of importance to note that, since $\phi(\mathbf{r})$ is a random variable, there is arbitrariness with respect to its origin. The only reference to any points in space is provided by the two endpoint constraints on $\mathbf{r}(s)$. These conditions destroy the isotropy of space, so the origin for $\phi(\mathbf{r})$ can be taken to be $\mathbf{r} = \mathbf{0}$ or $\mathbf{R}$; we take the first for convenience only. However, this choice has important consequences, since (6.52) describes the "action integral for a particle in three dimensions which passes through the origin of the field." Thus only paths with "zero angular momentum" contribute. The "particle's motion" is then purely radial. More precisely, the system, and hence $\phi(\mathbf{r})$ is of $D_{\infty h}$ symmetry. The "classical motion" should therefore be considered in prolate spheroidal coordinates determined by the vector $\mathbf{R}$. The choice of a spherically symmetric $\phi(\mathbf{r})$ is an approximation that was employed by Edwards.[14] The use of a spherically symmetric $\phi(\mathbf{r})$ is, of course, correct in the interesting case where the chain ends are tied together so $\mathbf{R} = \mathbf{0}$. Thus, if $\phi_\infty(\mathbf{r})$ is the spherically symmetric SCF that is obtained for an infinitely long chain with $\mathbf{R} = 0$, Edwards' approximation follows from the approximation that $G(\mathbf{R}0; L0[\phi_0])$ is equal to the Green's function $G(\mathbf{R0}; L0[\phi_\infty])$ for a large, but finite portion of this infinite chain (see Appendix C). Here we simply consider $\phi(\mathbf{r})$ to be spherically symmetric.

We choose the r-axis to be along the direction of "motion," i.e., parallel to $\mathbf{R}$. The "equation of motion" obtained from (6.52) is[53]

$$\frac{3}{l} \ddot{\mathbf{r}}(s) = i \frac{\partial \phi(r)}{\partial r} \tag{6.53}$$

Equation (6.53) describes the requisite "one-dimensional motion." The two boundary conditions on $\mathbf{r}(s)$ fix the integration constants. There is also the "constant of the motion,[53] the energy" which is

$$E = \frac{3}{2l}[\dot{r}(s)]^2 - i\phi(r) \tag{6.54}$$

It is clear that E depends on the endpoints and the "time" L,

$$E = E(R, L) \tag{6.55}$$

From simple classical mechanics,[53]

$$L = \int_0^R \frac{dr}{\{(2l/3)[E + i\phi(r)]\}^{1/2}} \tag{6.56}$$

implicitly gives this function. This is identical to Edwards' equation (3.6) when it is noted the E of our (6.56) corresponds to his iE.

Evaluating $A(R; L)$ along the "classical path" gives

$$A_{\text{cl}}(R; L) = \int_0^L ds \left\{ -\left(\frac{3}{2l} \dot{\mathbf{r}}_{\text{cl}}^2(s) - i\phi[\mathbf{r}_{\text{cl}}(s)] \right) + \frac{3}{l} \dot{\mathbf{r}}_{\text{cl}}^2(s) \right\}$$

$$= -EL + \int_0^R dr \left\{ \frac{6}{l}[E + i\phi(r)] \right\}^{1/2} \tag{6.57}$$

The contributions from paths $\mathbf{r}(s)$ in the neighborhood of $\mathbf{r}_{\text{cl}}(s)$ are neglected since we are interested only in the dominant terms.

Upon substitution of (6.57) into (6.38) we can proceed with the ϕ integration. By analogy with Section VID, Laplace's method is used to obtain the dominant field ϕ_d:

$$\left\{ \frac{\delta}{\delta\phi} \left[-A_{\text{cl}}(R; L[\phi]) - \frac{l^2}{2v} \int d\mathbf{r}\phi^2(\mathbf{r}) \right] \right\}_{\phi=\phi_d} = 0 \tag{6.58}$$

In (6.58) we have used the simplification (6.2) appropriate to the continuum limit. Moreover, the functional dependence of A_{cl} (and also E) upon ϕ has explicitly been introduced. Note that

$$\frac{\delta A_{\text{cl}}}{\delta\phi} = \frac{\partial A_{\text{cl}}}{\partial E} \frac{\delta E}{\delta\phi} + \left. \frac{\delta A_{\text{cl}}}{\delta\phi} \right)_{\text{explicit}} \tag{6.59}$$

However,

$$\frac{\partial A_{\text{cl}}}{\partial E} \equiv 0 \tag{6.60}$$

[This is essentially the saddle point approximation Edwards uses to obtain (6.56).] If we write $\chi_d = i\phi_d$, (6.59) results in

$$\frac{\theta(R - r)}{\{(2l/3)[E + \chi_d(r)]\}^{1/2}} = \frac{4\pi l^2 r^2}{v} \chi_d(r) \tag{6.61}$$

Equation (6.61) shows that χ_d is zero for r not along the "classical path." [$\theta(x)$ is the usual Heaviside step function which is unity for $x > 0$ and zero otherwise.] Equation (6.61) coincides (apart from differing definitions) with Edwards' equation (3.7). Rather than proceeding as Edwards does to convert (6.61) to an integral equation,[14] we consider the function χ, which is identical to χ_d on the "classical path" but is nonzero elsewhere,

$$\frac{4\pi l^2 r^2 \chi(r)}{v} \equiv \left\{\frac{2l}{3}[E + \chi(r)]\right\}^{-1/2} \tag{6.62}$$

Equation (6.62) is a simple algebraic equation which gives, upon squaring, the equation for $\chi(r)$,

$$E(R; L) = \chi(r) - \frac{3}{2l}\left(\frac{v}{4\pi l^2}\right) r^{-4} \chi^{-2}(r) \tag{6.63}$$

For r small χ is largest, and in this region

$$\chi(r) \simeq \left(\frac{3}{2l}\right)^{1/3} \left(\frac{v}{4\pi l^2}\right)^{2/3} r^{-4/3} \tag{6.64}$$

which is Edwards' solution. As Edwards notes, for large L (and hence R), the dominant contribution must come from $E \approx 0$, in which case (6.64) becomes exact. The reader is referred to Edwards' work for the remaining details which show that γ of (6.4) is 6/5 etc.

It is now of interest to consider the problem with the full $D_{\infty h}$ symmetry. Similarly, the case of $\mathbf{R} = \mathbf{0}$, where the spherical symmetry is indeed correct, should be further investigated to provide an answer to the interesting questions concerning self-avoiding random walks which return to the origin. These problems still remain to be explored.

G. Nonideal Polymer Solutions

Thus far we have discussed the statistical mechanics of individual polymer chains: ideal polymer solutions in which there are only short-range interactions, stiff chains, and chains with excluded volume. The discussion showed how the study of the configurational statistics of polymer chains leads naturally to the use of functional integrals, in particular, those related to Wiener integrals. The Wiener integrals could be evaluated, equivalently, by the calculation of the Green's function for a diffusion equation. However, when long-range interactions are present, the functional integral expression for the thermal Green's function is not equivalent to the solution of a diffusion equation, but instead it is the first member of a hierarchy of equations in the usual many-body sense.

We have not discussed the subject of nonideal polymers in any detail apart from the excluded volume problem. Thus no mention is made of the evaluation of the potential of mean force from the monomer-solvent interaction, and subsequently the evaluation of the osmotic pressure. We refer to the treatment of Yamakawa (Ref. 5, Chapter IV) for this subject and mention only that the osmotic pressure of a polymer solution at finite concentrations is represented as a virial expansion in the polymer concentration.[1-3,5] The second, third, etc., virial coefficients represent the mutual interaction between two, three, etc., polymer chains in solution. Thus the functional integral techniques presented in this review should also be of use in understanding the osmotic pressure of nonideal polymer solutions. We hope that this review will stimulate such studies of this important subject. It should also be mentioned in passing that at the Θ-point the second virial coefficient vanishes.[1] In general, the osmotic pressure π is given by the series

$$\pi = RT\left[\frac{1}{M}C + A_2C + A_3C^3 + \cdots\right] \tag{6.65}$$

where M is the molecular weight of the polymer, C its concentration, and A_2, A_3 are the virial coefficients. At the Θ-point $A_2 \to 0$, and

$$\pi = RT\left[\frac{1}{M}C + A_3C^3 + \cdots\right] \tag{6.66}$$

at Θ-point. Extrapolating $\pi(C)/C$ to zero concentrations enables the determination of M, while the use of Θ-solvent allows the use of higher polymer concentrations without introducing too great a correction for nonideality.

Appendix B. The Hierarchy

The following development is a modification of that given by Feynman for cases analogous to the simple diffusion equation.[12] Consider the functional integral

$$G(\mathbf{R0}; L+\epsilon, 0) = \int_{\mathbf{r}(0)=0}^{\mathbf{r}(L+\epsilon)=\mathbf{R}} \mathscr{D}[\mathbf{r}(s)] \exp\left\{-\int_0^{L+\epsilon} ds \frac{3}{2l}\dot{\mathbf{r}}^2(s) - \tfrac{1}{2}\int_0^{L+\epsilon} ds \int_0^{L+\epsilon} ds' V[\mathbf{r}(s)-\mathbf{r}(s')]\right\} \tag{B.1}$$

where later ϵ will be taken to be an infinitesimal. Divide the interval $0 < s < L + \epsilon$ into two intervals $0 < s < L$ (interval 1) and $L < s < L + \epsilon$ (interval 2), so

$$\mathscr{D}[\mathbf{r}(s)] = \mathscr{D}_1[\mathbf{r}(s)]\, d\mathbf{r}(L)\mathscr{D}_2[\mathbf{r}(s)] \tag{B.2}$$

in obvious notation. Now,

$$\int_0^{L+\epsilon} ds\dot{\mathbf{r}}^2 = \int_0^{L} ds\dot{\mathbf{r}}^2 + \int_L^{L+\epsilon} ds\dot{\mathbf{r}}^2 \tag{B.3}$$

and

$$\int_0^{L+\epsilon} ds \int_0^{L+\epsilon} ds' V[r(s) - r(s')]$$

$$= \int_0^{L} ds \int_0^{L} ds' V + 2\int_0^{L} ds \int_L^{L+\epsilon} ds' V + \int_L^{L+\epsilon} ds \int_L^{L+\epsilon} V$$

$$\equiv V_{11} + 2V_{12} + V_{22} \tag{B.4}$$

The V_{11} and V_{22} represent the chain self-interactions within intervals 1 and 2, respectively, while V_{12} represents the interactions between segments in 1 with those in 2. The presence of V_{12} in (B.4) implies that when (B.2)–(B.4) are substituted into (B.1), the latter does not factor into products of two terms of the form

$$G_M(\mathbf{R0}; L + \epsilon, 0) = \int d\mathbf{R}'' G_M(\mathbf{RR}'', L + \epsilon, L) G_M(\mathbf{R}''\mathbf{0}; L0) \tag{B.5}$$

Equation (B.5) is a Smoluchowski-Chapman-Kolmogorov equation which describes a Markov process (hence the subscript M). Thus (B.1) clearly does not describe a Markov process.

Continuing the standard development, however, the integral over all curves $\mathbf{r}(s)$, $0 \leqslant s \leqslant L + \epsilon$ with $\mathbf{r}(0) = \mathbf{0}$ and $\mathbf{r}(L + \epsilon) = \mathbf{R}$ is identical to the integral over all curves with the same end points and having $\mathbf{r}(L) = \mathbf{R}''$, provided we sum (i.e., integrate) over all possible intermediate points $\mathbf{R}''$:

$$G(\mathbf{R0}; L + \epsilon, 0) = \int d\mathbf{R}'' \int_{\mathbf{r}(0)=\mathbf{0}}^{\mathbf{r}(L+\epsilon)=\mathbf{R}} \mathscr{D}_1[\mathbf{r}(s)]\mathscr{D}_2[\mathbf{r}(s)]\, d\mathbf{r}(L)$$

$$\times\, \delta[\mathbf{r}(L) - \mathbf{R}''] \exp\left\{-\frac{3}{2l}\int_0^L \dot{\mathbf{r}}^2 - \tfrac{1}{2}V_{11}\right\}$$

$$\times \exp\left\{-\frac{3}{2l}\int_L^{L+\epsilon} \dot{\mathbf{r}}^2 - \tfrac{1}{2}V_{22}\right\} \exp\{-V_{12}\} \tag{B.6}$$

Letting ϵ be an infinitesimal,

$$\mathscr{D}_2[\mathbf{r}(s)] \exp\left\{-\frac{3}{2l}\int_L^{L+\epsilon} \dot{\mathbf{r}}^2\, ds\right\} \to \left(\frac{3}{2\pi l\epsilon}\right)^{3/2} \exp\left\{-\frac{3[\mathbf{R} - \mathbf{R}'']^2}{2l\epsilon}\right\} \tag{B.7}$$

In this limit, the V_{22} is of $\mathcal{O}(\epsilon^2)$ and represents the self-interaction of the last segment which is to be omitted by (6.9), so (B.4) becomes

$$\int_0^{L+\epsilon}\int_0^{L+\epsilon} V \rightarrow V_{11} + 2\epsilon \int_0^L ds V[\mathbf{r}(s) - \tfrac{1}{2}(\mathbf{R} + \mathbf{R}'')] \tag{B.8}$$

Substituting (B.7) and (B.8) into (B.6) and expanding the nonsingular terms to $\mathcal{O}(\epsilon)$ gives

$$\begin{aligned} G(\mathbf{R0}; L+\epsilon, 0) = &\int d\mathbf{R}'' \left(\frac{3}{2\pi l\epsilon}\right)^{3/2} \exp\left\{-\frac{3[\mathbf{R} - \mathbf{R}'']^2}{2l\epsilon}\right\} \\ &\times \int_{\mathbf{r}(0)=\mathbf{0}}^{\mathbf{r}(L)=\mathbf{R}''} \exp\left\{-\frac{3}{2l}\int_0^L ds\dot{\mathbf{r}}^2 - \tfrac{1}{2}V_{11}\right\} \\ &\times \left\{1 - \epsilon\int_0^L ds' V[\mathbf{r}(s') - \tfrac{1}{2}(\mathbf{R} + \mathbf{R}'')] + \mathcal{O}(\epsilon^2)\right\} \end{aligned} \tag{B.9}$$

The first term of $\mathcal{O}(\epsilon^0)$ in the functional integral is just the definition of $G(\mathbf{R}''\mathbf{0}; L0)$, whereas if in the second term of $\mathcal{O}(\epsilon)$ we introduce unity in the form

$$1 = \int d\mathbf{R}'\delta[\mathbf{r}(s') - \mathbf{R}']$$

and then invert the orders of the $d\mathbf{R}'$ ds' and $\mathscr{D}[\mathbf{r}(s)]$ integrals, we obtain the definition of $G_3(\mathbf{R}''\mathbf{R}'\mathbf{0}; Ls'0)$ as the functional integral [see (6.13)]. Thus (B.9) becomes

$$\begin{aligned} G(\mathbf{R0}; L+\epsilon, 0) = &\int d\mathbf{R}'' \left(\frac{3}{2\pi l\epsilon}\right)^{3/2} \exp\left\{-\frac{3[\mathbf{R} - \mathbf{R}'']^2}{2l\epsilon}\right\} \\ &\times \left\{G(\mathbf{R}''\mathbf{0}; L0) - \epsilon\int_0^L ds' \int d\mathbf{R}' V[\mathbf{R}' - \tfrac{1}{2}(\mathbf{R} + \mathbf{R}'')] \right. \\ &\left. \times\, G_3(\mathbf{R}''\mathbf{R}'\mathbf{0}; Ls'0) + \mathcal{O}(\epsilon^2)\right\} \end{aligned} \tag{B.10}$$

Letting $\mathbf{R}'' = \mathbf{R} + \boldsymbol{\eta}$, expanding all quantities except the singular $\exp(-3\boldsymbol{\eta}^2/2l\epsilon)$ in a Taylor series in $\boldsymbol{\eta}$ on the right in (B.10), expanding $G(\mathbf{R0}; L + \epsilon, 0)$ in a Taylor series in ϵ, performing the $\boldsymbol{\eta}$ integral, dividing by ϵ, and taking the limit $\epsilon \rightarrow 0$ in the usual manner finally gives the first member of the hierarchy (6.16).

Appendix C. The Saddle Points $\boldsymbol{\phi}_0$ and $\boldsymbol{\phi}_1$

The fundamental results of the following algebra are (1) the exhibition of the explicit dependence of the SCF on the endpoints that is introduced by (6.40) and (2) the demonstration that the same SCF occurs in the equations of motion of both $G_{\mathrm{SCF}}(\mathbf{R0}; L0[\cdots])$ and $G_{\mathrm{SCF}}(\mathbf{RR}'; Ls[\cdots])$, etc.

From (6.39) and (6.40), we have

$$\left[\frac{-1}{G(\mathbf{R0}; L0[\phi])}\frac{\delta G(\mathbf{R0}; L0[\phi])}{\delta\phi(\mathbf{r}')} + \int d\mathbf{R}'' V^{-1}(\mathbf{r}' - \mathbf{R}'')\phi(\mathbf{R}'')\right]_{\phi=\phi_0} = 0 \quad \text{(C.1)}$$

Define $G^{-1}[\phi]$ symbolically by

$$G[\phi]G^{-1}[\phi] = 1 \quad \text{(C.2a)}$$

or explicitly by

$$\int d\mathbf{R}'' \int ds'' G^{-1}(\mathbf{RR}''; ss''[\phi])G(\mathbf{R}''\mathbf{R}'; s''s'[\phi]) = \delta(\mathbf{R} - \mathbf{R}')\, \delta(s - s') \quad \text{(C.2b)}$$

Comparing (C.2b) with the equation of motion (6.25), we infer that

$$G^{-1}(\mathbf{RR}''; ss''[\phi]) = \delta(\mathbf{R} - \mathbf{R}'')\, \delta(s - s'')\left[\frac{\partial}{\partial s} - \frac{l}{6}\nabla_{\mathbf{R}}^2 + i\phi(\mathbf{R})\right] \quad \text{(C.3)}$$

Taking functional derivatives of (C.2a), we find

$$\frac{\delta G^{-1}[\phi]}{\delta\phi(\mathbf{r}')}G[\phi] + G^{-1}[\phi]\frac{\delta G[\phi]}{\delta\phi(\mathbf{r}')} = 0 \quad \text{(C.4)}$$

so that applying $G[\phi]$ on the left results in[50]

$$\frac{\delta G[\phi]}{\delta\phi(\mathbf{r}')} = -G[\phi]\frac{\delta G^{-1}[\phi]}{\delta\phi(\mathbf{r}')}G[\phi] \quad \text{(C.5)}$$

Thus from (C.3)

$$\frac{\delta G^{-1}(\mathbf{RR}''; ss''[\phi])}{\delta\phi(\mathbf{r}')} = i\,\delta(\mathbf{R} - \mathbf{R}'')\, \delta(s - s'')\, \delta(\mathbf{R} - \mathbf{r}') \quad \text{(C.6)}$$

and finally

$$\frac{\delta G(\mathbf{R0}; L0[\phi])}{\delta\phi(\mathbf{r}')} = -i\int ds' G(\mathbf{Rr}'; Ls'[\phi])G(\mathbf{r}'\mathbf{0}; s'0[\phi]) \quad \text{(C.7)}$$

We now note that the Green's function for parts of the chain, on the right in (C.7), *contains the same value of the random field* ϕ *as in* $G(\mathbf{R0}; L0[\phi])$ *and also obeys the diffusion equation*, e.g.,

$$\left[\frac{\partial}{\partial s'} - \frac{l}{6}\nabla_{\mathbf{r}'}^2 + i\phi(\mathbf{r}')\right]G(\mathbf{r}'\mathbf{0}; s'0[\phi]) = \delta(\mathbf{r}')\, \delta(s') \quad \text{(C.8)}$$

Thus substituting (C.8) into (C.1) and multiplying by $\int V(\mathbf{R}' - \mathbf{r}')\, d\mathbf{r}'$ gives (changing dummy indices)

$$\phi_0(\mathbf{R}') = -\frac{i\int d\mathbf{R}''\int ds' V(\mathbf{R}' - \mathbf{R}'')G(\mathbf{R}\mathbf{R}''; Ls'[\phi_0])G(\mathbf{R}''\mathbf{0}; s'0[\phi_0])}{G(\mathbf{R}\mathbf{0}; L0[\phi_0])} \tag{C.9}$$

Because (6.39) *and* (6.40) *imply that* ϕ_0 *is the dominant field* $G(\mathbf{R0}; L0)$, *the end points* $\mathbf{0}0$ *and* $\mathbf{R}L$ *have special significance and appear explicitly in* (C.9). Thus we write

$$G_{\mathrm{SCF}}(\mathbf{R0}; L0[\mathbf{R}L]) \equiv G(\mathbf{R0}; L0[\phi_0]) \tag{C.10a}$$

$$G_{\mathrm{SCF}}(\mathbf{R}''\mathbf{0}; s'0[\mathbf{R}L]) \equiv G(\mathbf{R}''\mathbf{0}; s'0[\phi_0]), \quad \text{etc.} \tag{C.10b}$$

Using (6.29) and substituting (C.10a) into (6.25) and (C.10b) into (C.8) *for the particular value of the random field* $\phi = \phi_0$, gives

$$\left[\frac{\partial}{\partial L} - \frac{l}{6}\nabla_{\mathbf{R}}^2 + V_{\mathrm{SCF}}(\mathbf{R}[\mathbf{R}L])\right]G_{\mathrm{SCF}}(\mathbf{R0}; L0[\mathbf{R}L]) = \delta(\mathbf{R})\,\delta(L) \tag{C.11a}$$

$$\left[\frac{\partial}{\partial s'} - \frac{l}{6}\nabla_{\mathbf{R}''}{}^2 + V_{\mathrm{SCF}}(\mathbf{R}''[\mathbf{R}L])\right]G_{\mathrm{SCF}}(\mathbf{R}''\mathbf{0}; s'0[\mathbf{R}L]) = \delta(\mathbf{R}'')\,\delta(s'), \quad \text{etc.} \tag{C.11b}$$

Hence the same field occurs for both the full Green's function (C.11a) and the partial one (C.11b).

The preceding derivation is easily generalized to deal with the dominant field for $\int d\mathbf{R}G(\mathbf{R0}; L0)$, $G_3(\mathbf{RR}'\mathbf{0}; Ls0)$, G_4, etc.

VII. POLYMERS IN BULK: INTRODUCTION

In addition to being found in solution, polymers also exist in the bulk phase with little or no solvent present. For polymers in bulk, a complete specification of the system is virtually physically meaningless. A statistical description is again required.

It might be argued, for instance, that in measuring the elasticity of a piece of rubber, we obtain the elasticity of a single polymer network and that consequently an ensemble average should not be performed. But in fact most experimental measurements are performed upon a number of identically prepared samples. In the absence of a detailed knowledge of the network structure, the ensemble average becomes necessary. However, measurements performed upon macroscopic samples can be thought to be averages of the properties over small regions of the material. Thus the bulk polymer system can be considered to be composed of macroscopically small, but microscopically large domains, all with differing structures. The measured properties are then "ensemble" averages over these domains. In conclusion, it is necessary to use statistical mechanics to describe polymers in bulk.

Our present theories of rubber elasticity are mostly of the kinetic theory type.[1,2,16] In some of these theories the elasticity of a single chain is multiplied by the number of "effective" chains in the network to provide the total elasticity. In these theories no information concerning the statistical properties of the network structure can be inferred since either no structural aspects are evaluated or a simple structural nature is assumed.[54] These structural properties should be of great interest in furthering our understanding of polymers in bulk. Thus the strain dependence of X-ray scattering measurements from networks, which have heavy atoms at the cross-links and/or the chain ends, should provide some correlation between the structural and elastic properties of polymer networks.

In other theories of rubber elasticity, the network structure is explicitly considered. However, the polymer on the surface is taken to be fixed (according to an affine deformation) upon deformation.[1,2,16] A truly statistical mechanical theory would also treat the surface statistically. More fundamentally, however, in these theories the fixed point character of the surface then completely determines the behavior of the bulk material. This would appear to be nonsense in the thermodynamic limit of infinite volume, unless the "fixed" surface were of finite extent. In this case, the theory is no longer statistical in nature.

The need for a fully statistical mechanical theory of polymers in bulk is more than just a question of rigor or elegance. We mentioned earlier the questions concerning the relation between average structural and elastic properties of polymer networks. More important, the present statistical theories of bulk polymer elasticity fail to account for interactions between different chains in the network. Therefore they cannot be expected to provide a proper molecular basis for an understanding of elasticity at large deformations, of crystallization upon stretching, etc. A description of these phenomena requires a fully statistical mechanical description of the polymer network. In such a description, the observable properties of the system are formally presented in terms of the properties of the constituent polymers and their mutual interactions. Thus all macroscopic properties are expressed in terms of the microscopic properties of the individual polymer chains.

The mathematical problems associated with a fully statistical mechanical description of polymer networks (with interchain interactions possibly present) are staggering in comparison to those posed even by the polymer excluded volume problem. Hence it is expected that a formal rigorous framework for the statistical mechanics of polymers in bulk will provide a fruitful source of new models to treat the phenomena of interest. The rigorous framework also enables the assessment of the range of validity of these models and of the lowest order corrections to them. In summary, the program is a little rigor, some simple models, then hopefully some understanding, etc.[27,29]

The formulation of the foundations of the statistical mechanics of polymers in bulk is just in its infancy. A number of questions of principle still require careful attention. This review can then only be of a state-of-the-art nature, and only a few very simple applications can be considered. In the next section, a simple model is considered which serves as a zeroth-order model for a system of polymers in bulk.[26] This model, which employs the Wiener integral formulation for flexible chains, has great pedagogical value since it introduces some of the formidable problems to be encountered in any statistical mechanical description of polymers in bulk. This model then naturally leads to a discussion of the nature of statistical mechanics for systems with internal constraints (Section IX).

The use of functional integrals in describing the configurational statistics of individual (or a few) polymer chains naturally arises from the chainlike character of the polymers. This mathematical approach bears a strong analogy to the Lagrangian formulation of the quantum mechanical dynamics of single (or a few) particles.[12] It should therefore be expected that the treatment of polymers in bulk should mathematically resemble that of the quantum mechanics of many-body systems. In the quantum many-particle systems, field theoretic approaches have been useful in focusing attention upon the relevant observable quantities. We should expect an analogous situation to hold for the polymer problems. Again for the polymer case the analog of the Lagrangian formulation must be used. This formulation is naturally expressed in terms of functional integrals.[32,55] It is not necessary to assume any knowledge of the usual quantum field theory of many particle systems, much less its Lagrangian form. All of the necessary material can be developed for the polymer system. The analogies are noted for convenience and reference.

VIII. IS THERE A POLYMER ANALOG OF THE PERFECT GAS?

In considering polymers in bulk, it is instructive to reflect upon the successes in the statistical mechanical description of real gases. The ideal gas represents one of the simplest examples from the kinetic theory and the statistical mechanics of fluids. By using kinetic theory arguments, the equation of state of the ideal gas is easily obtained by considering the momentum transferred to the wall by a particle colliding with the wall, and they by multiplying the result by the number of particles in the system.

The model of the ideal gas ignores intermolecular interactions. Given some intermolecular interactions, statistical mechanics must be employed to obtain the equation of state of the real fluid. For fluids of high density, approximations are necessary, but these naturally evolve from the formal, general statistical mechanical framework. For polymers in bulk, the present theories are of the kinetic theory variety since they do not, in principle, relate all macroscopic properties to the constituent molecular properties. In approaching a statistical mechanical theory of polymers in bulk, we can first ask if we can learn from these successes in the statistical mechanics of gases. One way to accomplish this is to consider a bulk polymer situation which simulates as closely as possible ideal gas situations. This model illustrates the following:

1. It requires the solution of a simple Wiener type functional integral.
2. It demonstrates the physical importance of the long-range polymer-polymer interactions in bulk systems.
3. Finally, it serves as an illustration of the difficulties to be encountered in statistical mechanical theories of polymers in bulk, the problems of entanglements and nonergodicity.

Consider a system composed of a large number N of equivalent gaussian polymer chains enclosed in a box of volume $V = B^3$. The individual polymer chains are taken to be "ideal" so all intramolecular long-range interactions can be neglected. The monomer density is

$$\rho = \frac{NL}{lV} \tag{8.1}$$

where L is the contour length of the chains and $n = L/l$ is the effective number of monomers/chain. If ρ is comparable to the density of a dilute gas (say 10^{18} molecules/cm^3), then just as in the case of ideal gases, we can neglect all intermolecular interactions since monomers on different chains are on the average outside of each other's range of force. (We assume either that there is no solvent present or that, if present, it leads to effective monomer-monomer interactions which are not too long range *in real space*.) This physical situation of N basically structureless, non-interacting, gaussian chains in a box is then a direct polymer analog of the ideal gas in ordinary statistical mechanics.

The equation of state $PV = NkT$ of an ideal gas is qualitatively useful for real gases at normal temperatures and pressures because it gives PV/kT correct to an order of magnitude. Even for liquids at normal temperatures and pressures, the ratio is correct to about one to three orders of magnitude.[56] Thus we may ask what is the pressure P exerted by the N polymers in a box, thereby providing an equation of state for "ideal" polymers in a box. Now in posing this question, we really want to know the pressure that arises from the internal degrees of freedom, so it is not necessary to bother with N chains. Consider one very long polymer chain (contour length L) in a box such that the monomer density corresponds to that of an ideal gas. What then is the equation of state?

Having posed this question as an intellectual exercise, we pause to reflect on its physical relevance. The polymer in a box represents a crude zeroth-order approximation to a rubberlike substance. In a polymer system that contains N chains and M cross-links, $M/N \sim 1$–10 so the ratio of number of cross-links per chain to the number of monomers per chain is $\approx 10^{-5}$–10^{-6} for very long chains. Now to lowest order, if chain A cross-links with chain B, chain A does not really "care" whether B is a distant part of the same chain or whether it is part of a different chain. Therefore, in order to obtain a qualitative picture, the ends of the chains can all be attached together to yield a single long, highly self-cross-linked chain. (Then we do not have to worry about the inactive free ends in discussing elasticity.) Experimentally, it turns out that for long enough chains, when there are N chains, a good fraction of the rubber elasticity is present even before the cross-links are introduced. Thus the model of a single long polymer chain in a volume V should be capable, except for the effects of the free ends, of describing this contribution to rubber elasticity. Polymers in bulk are discussed in a subsequent section. We now have only a partial motivation for considering the problem. It may be introduced merely as an illustrative example of the application of functional integral techniques to the statistical mechanics of polymer systems and as an indication of the physical influence of the long range interactions (along the chain).

Since the ideal gas laws work so well, qualitatively, for gases and liquids, it is tempting to try it for the polymer in a box; however, $PV = kT$ (*one molecule*) is not anywhere near the right order of magnitude for a single macroscopic polymer.

In normal classical statistical mechanics, it is assumed that all states which are fixed by the same external constraints, e.g., total volume V, average energy $\langle E \rangle$, average particle number $\langle N \rangle$, are equally probable. All possible states of the system are generated and are assigned weight unity if they are consistent with these constraints and, zero otherwise. Thus in the case of an N-particle system with classical Hamiltonian H_N, the microcanonical ensemble entropy $S(E)$ is obtained from the total number of states $\Omega(E)$ via the definition

$$\Omega(E) = \int_V d\Gamma_N \, \delta(E - H_N) = \exp\left[\frac{S(E)}{k}\right] \tag{8.2}$$

where the integration $d\Gamma_N$ is taken over all of N-particle phase space for which the particles are inside the volume V.

For the polymer in a box, with the neglect of all intermolecular and intramolecular interactions, the configurational partition function just assigns to each polymer configuration an extra weight of unity if it is completely in the box, and zero otherwise. Thus, apart from terms independent of V,

$$S = k \ln \Omega(V) \tag{8.3}$$

where

$$\Omega(V) = \int_V d\mathbf{R} \int_V d\mathbf{R}' G_V(\mathbf{R}, \mathbf{R}'; L) \tag{8.4}$$

gives the total number of chain configurations which are solely in the box. We have included in (8.4) the translational degrees of freedom of the polymer which give rise to the kT/V contribution to the pressure.

For the case in which we consider intramolecular interactions so $l = l(T)$, (8.4) must be interpreted as the partition function

$$Z(V) = \int_V d\mathbf{R} \int_V d\mathbf{R}' G_V(\mathbf{R}, \mathbf{R}'; L) \tag{8.4a}$$

and hence the free energy is

$$A = -kT \ln Z(V) \tag{8.3a}$$

From a comparison of (8.4) with (3.20) (for the potential $V \equiv 0$), the number of chain configurations running from $\mathbf{R}'$, 0 to $\mathbf{R}$, L for which the polymer is solely in the box must be (apart from a normalization factor which is independent of the volume)

$$G(\mathbf{R}, \mathbf{R}'; L) = \mathcal{N} \int_{\mathbf{r}(0)=\mathbf{R}'}^{\mathbf{r}(L)=\mathbf{R}} \exp\left[-\frac{3}{2l}\int_0^L \dot{\mathbf{r}}^2(s)\, ds\right] \delta_{\text{BOX}}[\mathbf{r}(s)] \tag{8.5}$$

The integration $\delta_{\text{BOX}}[\mathbf{r}(s)]$ is over all continuous curves $\mathbf{r}(s)$, which lie entirely within the box, subject to the endpoint conditions at $\mathbf{r}(0)$ and $\mathbf{r}(L)$. $S(V)$ in (8.3) may contain constants which are independent of V. We can then interpret $S(V)$ as the entropy of confinement of a polymer chain to a box.

We could attempt to evaluate (8.5) directly by performing the kind of limiting process which defines the functional integral, e.g., (3.20), and *integrating only over the finite volume of the box*. If this approach is taken, the first gaussian integration gives error functions. The next integration involves the product of the error function and a gaussian whose argument differs from that of the error function. Since this integration is over a finite volume and the argument of the error function takes on very small and very large values in this range, the use of the iterated multiple integral would be extremely cumbersome. Instead, we pursue an approach that preserves the spirit of originally taking the limit in which a discrete chain becomes a continuous chain. Thus (8.5) can be written as an integration over all space by the introduction of the Dirichlet delta function:[33]

$$\Theta(\mathbf{r}) = \begin{cases} 1 & \text{if } \mathbf{r} \in V \\ 0 & \text{otherwise} \end{cases} \tag{8.6}$$

Then (8.5) becomes

$$G(\mathbf{R}, \mathbf{R}'; L) = \int_{\mathbf{r}(0)=\mathbf{R}'}^{\mathbf{r}(L)=\mathbf{R}} \mathscr{D}[\mathbf{r}(s)] \exp\left[-\frac{3}{2l}\int_0^V \dot{\mathbf{r}}^2(s)\, ds\right] \prod_{s=0}^{L} \Theta[\mathbf{r}(s)] \tag{8.7}$$

Since one (zero) to any positive power is also unity (zero), for the case of a discrete chain the product of Dirichlet functions can be taken to be

$$\prod_{s_j} \{\Theta[\mathbf{r}(s_j)]\}^{\Delta s_j/l} = \exp\left\{\frac{1}{l}\sum_{j=1}^{n} \Delta s_j \ln \Theta[\mathbf{r}(s_j)]\right\} \tag{8.8a}$$

which in the limit of a continuous chain ($\Delta s_j \to 0^+$) becomes

$$\exp\left\{\frac{1}{l}\int_0^L ds \ln \Theta[\mathbf{r}(s)]\right\} \tag{8.8b}$$

But (8.8b) has the form of an external potential (the impenetrability of the walls!). When it is substituted into (8.7) we get

$$G(\mathbf{R}, \mathbf{R}'; L) = \int_{\mathbf{r}(0)=\mathbf{R}'}^{\mathbf{r}(L)=\mathbf{R}} \mathscr{D}[\mathbf{r}(s)] \exp\left\{-\frac{3}{2l}\int_0^L \dot{\mathbf{r}}'(s)\, ds + \frac{1}{l}\int_0^L ds \ln \Theta[\mathbf{r}(s)]\right\} \tag{8.9}$$

Comparing (8.9) with (3.20) and (3.21), we immediately see that upon introducing the additional condition (3.42) (dropping the tilde for convenience)

$$\left[\frac{\partial}{\partial L} - \frac{l}{6}\nabla_{\mathbf{R}}^2 - \frac{1}{l}\ln \Theta(\mathbf{R})\right] G(\mathbf{R}, \mathbf{R}'; L) = \delta(L)\,\delta(\mathbf{R} - \mathbf{R}') \tag{8.10}$$

so that G represents a Markov process. Since

$$-\ln \Theta(\mathbf{R}) = \begin{cases} -\ln(1) = 0, & \mathbf{R} \in V \\ -\ln(0) = \infty, & \mathbf{R} \notin V, \end{cases} \tag{8.11a}$$

the "external potential"

$$V(\mathbf{R}) = -\frac{1}{l}\ln \Theta(\mathbf{R}) \tag{8.11b}$$

represents the usual kind of three-dimensional infinite potential well, so (8.10) is the diffusion equation for "a particle in a box." By using (8.11), (8.10) can alternatively be written as

$$\left[\frac{\partial}{\partial L} - \frac{l}{6}\nabla_{\mathbf{R}}^2\right] G(\mathbf{R}, \mathbf{R}'; L) = \delta(L)\,\delta(\mathbf{R} - \mathbf{R}'), \qquad \mathbf{R} \in V \quad (8.13a)$$

$$G(\mathbf{R}, \mathbf{R}'; L) = 0, \qquad \mathbf{R} \notin V \quad (8.13b)$$

Since G must be continuous at the walls, it is necessary for G to approach zero as $\mathbf{R}$ or $\mathbf{R}'$ approaches the wall from within. (Compare the particle in a box in ordinary quantum mechanics.) Thus, expanding G in terms of the complete set of sines inside the box (rather than cosines which individually do not vanish at the walls) we find

$$G(\mathbf{R}, \mathbf{R}'; L) = \prod_{\alpha=x,y,z} \left\{\left(\frac{2}{B}\right) \sum_{n_\alpha=1}^{\infty} \sin\left(\frac{n_\alpha \pi R_\alpha}{B}\right) \sin\left(\frac{n_\alpha \pi R_\alpha'}{B}\right) \exp\left(-\frac{n_\alpha^{\,2}\pi^2 Ll}{6B^2}\right)\right\} \quad (8.14)$$

Substituting (8.14) into (8.4) and performing the trivial integration gives the number of chain configurations

$$\Omega(V) = \left[\frac{2}{B}\sum_{n\,\text{odd}} \left(\frac{2B}{n\pi}\right)^2 \exp\left(-\frac{n\pi^2 Ll}{6B^2}\right)\right]^3 \quad (8.15)$$

There are two simple asymptotic limits; when the polymer's natural size is much smaller than the box, i.e., $(Ll)^{1/2} \ll B$, or when the polymer is larger than the box $(Ll)^{1/2} \gg B$. In the first case, the summand in (8.15) is slowly varying, and the sum is replaced by an integral by taking $k = n\pi/B$, $\Delta k = 2\pi/B$, or

$$\Omega^{1/3} = \frac{16}{\pi}\int_{\pi/B}^{\infty} \frac{dk}{k^2} \exp\left[-\frac{k^2 lL}{6}\right] \quad (8.16a)$$

where the lower limit on the integral must be kept. In the limit of a "small" polymer, (8.16a) can be evaluated by expanding the second integral in

$$\Omega^{1/3} = \frac{16}{\pi}\left\{\int_0^{\infty} \frac{dk}{k^2}\left[\exp\left(-\frac{k^2 Ll}{6}\right) - 1\right] - \int_0^{\pi/B} \frac{dk}{k^2}\left[\exp\left(-\frac{k^2 Ll}{6}\right) - 1\right] + \int_{\pi/B}^{\infty} \frac{dk}{k^2}\right\}$$

to give

$$\Omega = V\left(\frac{16}{\pi}\right)^3 \left\{1 - \frac{\pi}{B}\left(\frac{\pi lL}{6}\right)^{1/2} + \cdots\right\}^3$$

or

$$S(V) = k \ln V - \frac{\pi k}{V^{1/3}}\left(\frac{3\pi lL}{2}\right)^{1/2} + \text{const} + \cdots \quad (8.16b)$$

Since there is no internal energy,

$$A = -TS \quad \text{and} \quad P = -\left(\frac{\partial A}{\partial V}\right)_T = T\left(\frac{\partial S}{\partial V}\right)_T \tag{8.17}$$

so that

$$PV = kT + kT\left(\frac{\pi^3}{6}\right)^{1/2}\left[\frac{lL}{V^{2/3}}\right]^{1/2} + \cdots \tag{8.18}$$

The factor of kT represents the ideal gas law translational degreee of freedom, while the small correction factor represents the fact that the polymer has some internal structure which, in the relatively rare situations where the polymer encounters the wall (its size $\ll$ size of the box), causes some additional pressure.

When $l = l(T)$, (5.3a) and (5.4a) are appropriate to the microcanonical ensemble, and the pressure is

$$P = kT\left(\frac{\partial \ln Z(V, T)}{\partial V}\right)_T = -\left(\frac{\partial A(V, T)}{\partial V}\right)_T \tag{8.17a}$$

Thus the derivation leading to (8.18) is valid in the case in which there are intramolecular interactions and $l(T)$ is temperature dependent.

The more interesting case is $Ll \gg V^{2/3}$, in which the free polymer chain would be larger than the size of the box. In this limit the box encroaches on the freedom of the polymer. From (8.14), the sum is dominated by the $n_\alpha = 1$ term, as

$$\Omega = \frac{2^9}{\pi^6} V \exp\left[-\frac{lL\pi^2}{2V^{2/3}}\right]\left\{1 + \tfrac{3}{4}\exp\left[-\frac{lL\pi^2}{2V^{2/3}}\right] + \cdots\right\} \tag{8.19a}$$

$$S = \text{const.} + k \ln V - \frac{kLl\pi^2}{2V^{2/3}} + \tfrac{3}{4}k \exp\left[-\frac{lL^2\pi^2}{2V^{2/3}}\right] + \cdots \tag{8.19b}$$

$$PV = kT\left[1 + \frac{\pi^2}{3}\frac{Ll}{V^{2/3}}\left(1 + \tfrac{3}{4}\exp\left[-\frac{lL\pi^2}{2V^{2/3}}\right] + \cdots\right)\right] \tag{8.19c}$$

The pressure from translation and the exponential correction are negligible, so we get

$$PV \to kT\frac{\pi^2}{3}\frac{Ll}{V^{2/3}}, \qquad Ll \gg V^{2/3} \tag{8.19d}$$

This defines an effective number of internal degrees of freedom

$$n_{\text{eff}} \equiv \frac{\pi^2}{3}\frac{Ll}{V^{2/3}} \tag{8.20a}$$

or

$$(n_{\text{eff}})^{3/2} \propto \frac{\text{natural volume of free polymer}}{\text{volume of gas}} \tag{8.20b}$$

[The two asymptotic results (8.19) and (8.18) are not more than 10% in error from the complete sum.]

These basic results at first seem quite reasonable. The polymer molecule moves around in the box as a unit until the size of the box becomes comparable to the free size of an unconfined polymer. Then (as a function of L) the pressure increases as the box gets smaller relative to the free size of the polymer. However, as $L \to \infty$ (i.e., $Ll/V^{2/3} \to \infty$), the polymer density is of the form

$$\rho(\mathbf{R}) = C \prod_{\alpha=x,y,z} \sin \frac{\pi R_\alpha}{V^{1/3}} \tag{8.21}$$

which is not the uniform distribution that would be expected from a *real* macroscopic polymer. Furthermore, (8.21) implies that the polymer basically piles up in the middle of the box as $C^{1/3} = \pi N^{1/3}/2V^{1/3}$ so (8.21) is close to N times a delta function. This result can be traced back to the assumption that all configurations of the polymer chain are a priori equally likely provided they lie entirely within the box. Because the chain configuration is described by a random walk, if the chain has a finite probability that any segment be at the wall, then as $L \to \infty$, this configuration will with certainty have a portion outside the box. Thus the only way to keep the polymer inside this (noninteracting) box is to require that the density be zero at the walls. This situation may be contrasted with the case of a Brownian particle in a box. The particle is reflected at the walls, hence there is an impulsive delta function force at the walls, and we do not require continuity across the walls. Therefore, given a long enough time, the particle's trajectory will fill the box.[33]

One consequence of the nonuniformity (8.22) is that the free energy is not an extensive variable, i.e.,

$$A \neq Vf(\rho) \tag{8.22}$$

where ρ is given by (8.1) for $N = 1$; instead, the result here is

$$A = Vf\left(\frac{Ll}{V^{2/3}}\right) \tag{8.23}$$

Equation (8.23) would, of course, be perfectly acceptable for a polymer that was trapped—boxed-in—in some interstitial position, but it is clear that normal polymerized material (apart from minor surface effects) does have uniform density and normal thermodynamic variables like (8.22) with the equality.

But then why is the result (8.21)–(8.22) obtained? Given an enormously long polymer chain, if we wait long enough the chain may take up all configurations, (i.e., the system is ergodic). Then the nonuniform distribution will result (for "inert" walls). But for any chain configuration, there are a large number of knots and entanglements;[24,25] the chain segments cannot pass through each other. It will take too long to reach the state of nonuniform density. Thus, given an initial configuration of the chain, all other configurations that are solely within the box are not accessible; the phase space of chain configurations is partitioned into a large number of disjoint regions. The motion of the polymer is not ergodic. The knots and entanglements represent the fact that the polymer segments have finite extent, and thus are the simplest manifestation of the long-range interactions (along the chain) which have a short range (in space) repulsive character. Thus, in order to discuss qualitatively real polymer systems in bulk, it is necessary to include at least this aspect of the long-range interactions (to get uniform density above, or in the theories of rubber elasticity where the assumption of incompressibility is a reflection of the entanglements, etc.). But the long-range interactions already make the problem equivalent to the full many-body problem. Thus the polymer analog of the perfect gas in classical statistical mechanics is already the full many-body problem!

In Section IX we return to the problem of the polymer in a box to show that if an extremely simple model of the long range interactions is introduced, the usual kind of thermodynamics, (8.22) with equality, is obtained. First we must briefly consider the question of statistical mechanics with internal constraints.[22,24–29]

IX. STATISTICAL MECHANICS WITH CONSTRAINTS: INTRODUCTION

The model of the polymer in a box presented in the preceding section should have provided a zeroth-order picture of a rubber-like substance. However, it turned out that for long enough chains, the model led to nonuniform densities. After finding the source of this difficulty, it was claimed that if we include the fact that the polymer segments are boxed in by their neighbors, so they are not free to move anywhere in the box, then the expected uniform density and extensive free energy $A = Vf(\rho)$ would result. We could imagine forming the very long macroscopic polymer by attaching (chemically) end-to-end a large number N of smaller polymers (contour lengths L_s) which are all much smaller than the size of the box, i.e., $L_s l \ll V^{2/3}$. Thus initially we could assume that the N polymers had a uniform density apart from negligible surface effects. When the large single polymer, $L = NL_s$, is formed, it can be assumed to have uniform density. Although this may or may not reflect the "equilibrium" distribution, it is a consequence of the mode of preparation of the system. Upon formation of the single chain, on the average, this large macroscopic chain will be highly entangled and knotted with itself. Hence the fact that polymer segments cannot pass through each other (the presence of entanglements) implies that the initial uniform distribution cannot become the previous nonuniform density which was derived on the basis of the assumption that all chain configurations that are solely within the box are equally probable. (Except possibly on an astronomical time scale, indeed.) We can, however, prevent this infinite time nonuniform density limit by having the ends of the single macroscopic chain joined together. The chain could then have knots that could never untie. (We are assuming that the temperature is low enough or that there is not enough energy available to break bonds. After all, a piece of rubber breaks only when stretched to very high extensions.)

This situation presents us with a general problem which must be faced in order to have statistical mechanical theories of polymer in bulk: How do we deal with the statistical mechanics of systems which have a large number of internal constraints? This question of principle exists independent of our ability to mathematically describe the knots or entanglements. (See Edwards for the beginnings of a discussion of the matter.[22,24,25,27]) Now we are faced with a system which may have some set of internal constraints which are denoted by the label C. For the polymer in a box C describes the *permanent entanglements*. Given a system which initially has the constraint C, not all other states are accessible, or equally likely. Explicitly the system can *never* be in a state $C' \neq C$ even if this other state C' satisfies the usual external constraints of total volume, average energy, number of particles, etc. A system with internal constraints is definitely not ergodic: a time average over all configurations that the system takes up is definitely not the same as an average over all configurations which assigns to each the equal a priori weight of unity.

A. The Constraints

Consider a statistical mechanical description of the system where Γ represents a point in phase space (or just in configuration space if only the configurational statistics are being discussed). Let $C(\Gamma)$ be a mathematical description of the constraint. For a system with a particular value C of $C(\Gamma)$, the microcanonical ensemble partition function is

$$\Omega_C = \int_V d\Gamma\, \delta[E - H(\Gamma)]\, \delta[C - C(\Gamma)] \tag{9.1}$$

In (9.1) the integration is over the phase space corresponding to a box of volume V, and $H(\Gamma)$ is the Hamiltonian for the system. Thus, as usual,

$$S_C(E, V) = k \ln \Omega_C(E, V) \tag{9.2}$$

is the entropy of the system with constraint C, energy E, and volume V. Alternatively in the canonical ensemble the partition function is

$$Z_C(T, V) = \int_V d\Gamma\, \delta[C - C(\Gamma)] \exp[-\beta H(\Gamma)], \tag{9.3}$$

so that the free-energy of the system with constraint C is

$$A_C(T, V) = -kT \ln Z_C(T, V) \tag{9.4}$$

and its pressure is

$$P_C = -\left(\frac{\partial A_C}{\partial V}\right)_T = kT\left(\frac{\partial \ln Z_C}{\partial V}\right)_T \tag{9.5}$$

However, given a system with possible internal constraints $\{C_j\}$, we do not know which internal constraints are initially present. The system may be described by any of the constraints $\{C_j\}$ with probabilities that reflect the mode of formation of the system. It is therefore necessary to consider a master ensemble of subensembles, each member of which describes a system with internal constraints C. Thus the free-energy, etc., for the system is then taken as some ensemble average over the quantities in (9.4),

$$A(T, V) = \langle A_C(T, V)\rangle_C \tag{9.6}$$

The averaging in (9.6) is yet to be determined.

We can imagine constructing this master ensemble by constructing physical systems a large number of times, N, *in the same manner each time*.[57] The number of times a particular constraint C_i is found is m_{C_i}. For any choice of the $\{m_{C_i}\}$ the number of ways of forming this master ensemble is

$$W_N = \frac{N!}{\prod_{i=1}^{n} (m_{C_i})!} \equiv \frac{N!}{\prod (Np_{C_i})!} \tag{9.7}$$

where

$$p_{C_i} = \frac{m_{C_i}}{N} \tag{9.7a}$$

are the relative frequencies of finding a particular C_i In the limit $N \to \infty$, the quantities in (9.7a) represent probabilities, and the entropy of formation (of "mixing") of the master ensemble is

$$S_{\text{mix}} = k \lim_{N\to\infty} \frac{1}{N} \ln W_N = -k \sum_C p_C \ln p_C \tag{9.8}$$

Stirling's formula has been used in (9.8) and the index i on C has been dropped. However, a given system with constraint C has a large number of other degrees of freedom, which is why it is described by an ensemble as in (9.1)–(9.5). Thus the total entropy of the system, the master ensemble, must be the entropy of "mixing" of (9.8) plus the weighted average of the entropies of the subensembles:

$$S = -k \sum_C p_C \ln p_C + \sum_C p_C S_C \tag{9.9}$$

where

$$\sum_C p_C = 1 \tag{9.10}$$

as appropriate to probabilities.

If the system is formed in "equilibrium," by which we mean that all states are equally probable, then we have

$$p_C = \frac{\Omega_C}{\Omega}, \qquad \text{where} \quad \Omega \equiv \Sigma_C \Omega_C \tag{9.11}$$

By using (9.10), (9.9) becomes

$$S = k \ln \Omega \tag{9.12}$$

which is the usual microcanonical result. For a system that is not formed in equilibrium according to (9.11), (9.12) is not appropriate, and (9.9) must be used.

If the thermodynamic variables are denoted collectively as ξ, then the probabilities of the internal constraints are determined solely by the initial mode of preparation, and if ξ_0 denotes the initial thermodynamic variables,

$$p \equiv p_C(\xi_0) \tag{9.13}$$

Consider what happens when the thermodynamic variables are changed from $\xi_0 \to \xi$: a system with the constraint C upon formation at ξ_0 must also have the constraint C at ξ since the $\{C\}$ represent *permanent* internal constraints. Thus

$$p_C(\xi) = p_C(\xi_0), \qquad \text{all } \xi \tag{9.13a}$$

because the probability of having a given constraint C is established upon the initial formation of the system, i.e., the master ensemble. ($\{C\}$ are imposed constants of the motion.) Each of the subensembles, characterized by a particular C, contains a large number of degrees of freedom. The entropies of the subensembles change upon the transformation of state $\xi_0 \to \xi$. If $S_C(\xi)$ is the entropy of the subensemble C when the thermodynamic variables are ξ, i.e., obtained from (9.1) and (9.2) when a microcanonical ensemble is appropriate, etc., the entropy of the master ensemble is then

$$S(\xi) = -k \sum_C p_C(\xi_0) \ln p_C(\xi_0) + \sum_C p_C(\xi_0) S_C(\xi) \tag{9.14}$$

where the ξ and ξ_0 dependence is used to emphasize those aspects of the system which are fixed upon preparation. [A rubber band has memory of its initial state, the $\{p_C(\xi_0)\}$.] Because we are usually only concerned with entropy changes, the mixing term (9.8) of (9.14) is written as s_0, and by (9.2) for the subensembles at ξ,

$$S(\xi) = k \sum_C p_C(\xi_0) \ln \Omega_C(\xi) + s_0 \tag{9.15}$$

Equivalently from (9.3) and (9.4) for the subensembles

$$A(\xi) = -kT \sum_C p_C(\xi_0) \ln Z_C(\xi) + a_0 \tag{9.16}$$

where a_0 represents the uninteresting free-energy of "mixing" of the initial master ensemble. In general, if we define

$$q_C(\xi) = \frac{\Omega_C(\xi)}{\Omega(\xi)}, \qquad \Omega(\xi) = \sum_C \Omega_C(\xi) \tag{9.17}$$

then

$$q_c(\xi) \neq p_C(\xi) = p_C(\xi_0) \tag{9.18}$$

Thus

$$\begin{aligned} S(\xi) &\neq k \ln \Omega(\xi) \\ A(\xi) &\neq -kT \ln Z(\xi) \end{aligned} \tag{9.19}$$

where

$$Z(\xi) \equiv \sum_C Z_C(\xi) \tag{9.19a}$$

so that we cannot calculate entropy or free energy differences in the usual simple ways from $k \ln [\Omega(\xi)/\Omega(\xi_0)]$ and $-kT \ln [Z(\xi)/Z(\xi_0)]$. Instead it is necessary to deal with the ensemble averages over all constraints (9.14)–(9.16).

But in the usual cases, $\{C\}$ represents an infinite set (whether it is denumerable or nondenumerable is immaterial, since at the present in the case of polymer chains it is not enumerable), and (9.14)–(9.16) imply that we would have to perform the usual kind of statistical mechanics (9.1)–(9.5) for an infinite number of subensembles. Then properties are calculated by the ensemble average over the constraints.

B. Small Changes of State

We can first ask if there are simple approximations that can be used in the case that $\xi - \xi_0$ is small, i.e., are there any representations of (9.14)–(9.16) which in this case do not require us to enumerate each C and then perform the master ensemble average. The answer is, in general, *no*. Polymer systems, glasses, etc., which are typical of systems with internal constraints, need not be formed under the "equilibrium" conditions of

(9.11). However, when (9.11) is applicable, some simplification is introduced.

Suppose (9.11) holds at ξ_0. Then, for instance, using a microcanonical ensemble, where $A = -TS$, (i.e., only configurational entropies are used)

$$P = kT\left[\frac{\partial S(\xi)}{\partial V}\right]_T = kT \sum_C p_C(\xi_0)\left[\frac{\partial \ln \Omega_C(\xi)}{\partial V}\right]_T \tag{9.20a}$$

When energies are not fixed, Ω_C is replaced by Z_C, i.e.,

$$P = -\left[\frac{\partial A(\xi)}{\partial V}\right]_T = kT \sum_C p_C(\xi_0)\left[\frac{\partial \ln Z_C(\xi)}{\partial V}\right]_T \tag{9.20b}$$

in the canonical ensemble where the analog of (9.11) is

$$p_C(\xi_0) = Z_C(\xi_0)/Z(\xi_0) \tag{9.11a}$$

defining the initial equilibrium master ensemble. Since (9.20a) and (9.20b) are formally isomorphic, we deal with only the first case; the second follows immediately.

Substituting (9.11) into (9.20a) and omitting all ξ_0, we find

$$P = \frac{kT}{\Omega} \sum_C \frac{\Omega_C}{\Omega_C(\xi)}\left[\frac{\partial \Omega_C(\xi)}{\partial V}\right]_T \tag{9.21}$$

The initial pressure of the master ensemble upon formation is obtained from (9.21) as the limit $\xi \to \xi_0$, $(V \to V_0)$ so

$$P_0 = \lim_{\xi \to \xi_0} \frac{kT}{\Omega} \sum_C \frac{\Omega_C}{\Omega_C(\xi)}\left[\frac{\partial \Omega(\xi)}{\partial V}\right]_T \tag{9.22a}$$

Since by their definitions (9.1) [or (9.3)],

$$\lim_{\xi \to \xi_0} \frac{\Omega_C}{\Omega_C(\xi_0)} = 1$$

and

$$\begin{aligned} P_0 &= \frac{kT}{\Omega} \lim_{\xi \to \xi_0} \sum_C \left[\frac{\partial \Omega_C(\xi)}{\partial V}\right]_T = \frac{kT}{\Omega} \lim_{\xi \to \xi_0} \left[\frac{\partial}{\partial V} \sum_C \Omega_C(\xi)\right]_T \\ &= \frac{kT}{\Omega} \lim_{\xi \to \xi_0} \left[\frac{\partial \Omega(\xi)}{\partial V}\right]_T = \frac{kT}{\Omega}\left(\frac{\partial \Omega}{\partial V}\right)\bigg|_{V=V_0} = kT\left[\frac{\partial \ln \Omega(\xi_0)}{\partial V_0}\right]_T \end{aligned} \tag{9.22b}$$

where the $\lim_{\xi \to \xi_0}$ and $\partial/\partial V$ have been interchanged. Equation (9.22b) implies that *for the equilibrium initial master ensemble of* (9.11) (*but not in general*) the initial pressure can be evaluated *without enumerating the constraints* C, p_C, or Ω_C; all that is required is Ω, the total number of states which can be evaluated in the usual ways that do not require the specification of

$\{C\}$! For properties which are represented as second derivatives of $S(\xi)$ or $A(\xi)$ with respect to ξ, even the essential simplification as in (9.22b) does not appear when $\xi \to \xi_0$.[58] It is then necessary to include explicitly a description of the internal constraints to explain these "second-order" properties for the initially prepared equilibrium state (e.g., the elastic modulus relationship for infinitesimal deformations). Under any circumstances, apart from quantities like P_0 of (9.22b), it is necessary to consider explicitly the internal constraints in order to have a proper statistical mechanical theory, and hence proper statistical thermodynamics. As an example, in the James and Guth theory of rubber elasticity,[1,2,16] the internal constraints due to the chain entanglements and the network structure are implicitly included by taking the system to be incompressible.

Given that in the general case it is necessary to include the constraints, but with full knowledge that often recourse is made to simple models to describe the net effects of the constraints (many-body effects), some approximations can be introduced which are expected to be valid for small $\xi - \xi_0$. A subensemble labeled C represents a system with a very large number of free degrees of freedom. $\Omega_C(\xi)$ [or similarly $Z_C(\xi)$] of (9.1) does not vary greatly as we range over different constraints C. Thus we expect that $\Omega_C(\xi)$ can be replaced by some average value $\Omega_1(\xi)$ which is determined to minimize the contributions from the fluctuations $\delta_C = \Omega_C(\xi) - \Omega_1(\xi)$. Similarly, using the average $\Omega_1(\xi_0)$ to $\Omega_C(\xi_0)(\Omega_1$ to $\Omega_C)$, we have

$$\Delta S(\xi) = \frac{k}{\Omega} \sum_C \frac{\Omega_C}{\Omega_C(\xi)} \Delta\Omega_C(\xi)$$

$$= \frac{k}{\Omega} \sum_C \frac{[\Omega_1 + (\Omega_C - \Omega_1)]\,\Delta\Omega_C(\xi)}{[\Omega_1(\xi) + (\Omega_C(\xi) - \Omega_1(\xi))]} \tag{9.23a}$$

If we keep terms only through first order in the fluctuations δ and $\Omega_C - \Omega_1$, (9.23a) becomes

$$\Delta S = \frac{k}{\Omega} \frac{\Omega_1}{\Omega_1(\xi)} \sum_C \Delta\Omega_C(\xi) - \frac{k}{\Omega\Omega_1^{\,2}(\xi)} \sum_C \Delta\Omega_C(\xi) \times [\Omega_1\Omega_C(\xi) - \Omega_C\Omega_1(\xi)] + \mathcal{O}(\delta^2) \tag{9.23b}$$

Writing

$$\Delta\Omega = \Delta \sum_C \Omega_C(\xi) = \sum_C \Delta\Omega_C(\xi) \tag{9.23c}$$

as the change in the number of states, and then choosing the ratio $\Omega_1/\Omega_1(\xi)$ to make the first-order term in (9.25b) vanish, we have

$$\Delta S = k \frac{\sum_C \Omega_C \,\Delta\Omega_C(\xi)}{\sum_C \Omega(\xi)\,\Delta\Omega_C(\xi)} \frac{\Delta\Omega(\xi)}{\Omega} + \mathcal{O}(\delta^2) \tag{9.24}$$

for some small change $\Delta\xi$ in the thermodynamic state for a system in thermodynamic state ξ which is not necessarily ξ_0. Note that as $\xi \to \xi_0$ it is found that

$$\Delta S(\xi_0) \to k \frac{\Delta\Omega(\xi_0)}{\Omega(\xi_0)} \tag{9.24a}$$

which is the correct limit in that case. It gives the initial pressure correctly according to (9.22b). Presumably, since only the ratio $\Omega_1/\Omega_1(\xi)$ has been chosen to eliminate the $\mathcal{O}(\delta)$ terms, one of these quantities could be taken to remove the $\mathcal{O}(\delta^2)$ term. Thus, in the case of pressure for instance, from (9.24) (dividing by ΔV and taking limits) we have

$$P(\xi) = \frac{kT \sum_C \Omega_C[\partial\Omega_C(\xi)/\partial V]_T}{\Omega \sum_C \Omega_C(\xi)[\partial\Omega_C(\xi)/\partial V]_T} \left[\frac{\partial\Omega(\xi)}{\partial V}\right]_T \tag{9.25}$$

which reduces to (9.22b) as $\xi \to \xi_0$. [For the canonical ensemble we just use A and Z instead of S and Ω.]

We conclude this introduction to the formal aspects of statistical mechanics with constraints by using an interesting analogy. Given that we can describe the constraints, i.e., $C(\Gamma)$ in (9.1) and (9.2), rather than requiring that the constraint C be met with certainty, we could impose the constraint that $C(\Gamma)$ equal C with the Lagrange multiplier λ. Thus an approximation to (9.3) is taken as

$$\tilde{Z}_C(T, V, \lambda) = \int d\Gamma \exp\left[-\beta H(\Gamma) - \lambda C(\Gamma)\right] \tag{9.26a}$$

where λ is chosen so that

$$-\left[\frac{\partial \ln \tilde{Z}_C}{\partial\lambda}\right]_{T,V} = \langle C(\Gamma)\rangle = C \tag{9.26b}$$

Such an approximation is entirely analogous to the use of gaussian links as an approximation to freely hinged links. The freely hinged links are *constrained* to have length l, whereas the gaussian links only have length l on the rms average!

C. The Polymer in a Box with Constraints

Returning to the problem of the polymer in a box, consider some segment of chain of size Δs_j, which is centered at $\mathscr{R}(s_j)$. The motion of this piece of chain is influenced most by its immediate surroundings, which form a sort of a cage due to the fact that the segment at s_j cannot pass through its surrounding segments. Thus, if the chain initially has a configuration $\mathscr{R}(s)$, at any later time or under some change in external constraints, the net effect of the constraints due to the surroundings (of any particular segment) and the many-body forces is to confine the subsequent configuration $\mathbf{R}(s)$ to be "close" to $\mathscr{R}(s)$. The physical picture presented is therefore in the spirit of a cage theory of liquids, except that in the polymer case the chain connectivity implies that the cage must last much longer and in some respects be permanent.

Thus, given the initial configuration $\mathscr{R}(s)$ and the many-body constraints, we could watch the motion of the chain for a long time as it passes through a set of configurations $\mathbf{R}(s)$ which preserve all the internal constraints. We would then evaluate the probability distribution of $\mathbf{R}(s)$. If this is sensibly independent of the initial distribution $\mathbf{R}(s)$, this distribution for $\mathbf{R}(s)$ can be taken as the probability distribution for the polymer in a box with the many-body constraints. Alternatively, a time-independent procedure can be adopted in which, given an initial configuration $\mathscr{R}(s)$, we determine the probability of some other configuration $\mathbf{R}(s)$ directly. Using this approach, we must now specify or incorporate the constraints.

If the confinement of the segment at s_j is due to segments at s_α, in the discrete picture of the chain the number of configurations in the final state [i.e., for $\mathbf{R}(s)$] is given by

$$G(\{\mathbf{R}(s_j)\}) = \mathscr{N} \exp\left\{-3 \sum_j \frac{[\mathbf{R}(s_j) - \mathbf{R}(s_{j-1})]^2}{2l\,\Delta s_j}\right\} \times \prod_{s_\alpha} \prod_{s_j} C_\alpha[\mathbf{R}(s_j) - \mathscr{R}(s_j)] \quad (9.27)$$

In (9.27), $C_\alpha[\mathbf{R}(s_j) - \mathscr{R}(s_j)]$ represents the constraint on the segment at s_j due to the segment at s_α which tends to keep the new position of the segment at s_j, $\mathbf{R}(s_j)$, "near" its original position $\mathscr{R}(s_j)$. We can always write

$$\prod_{s_\alpha} \prod_{s_j} C_\alpha \equiv \exp\left\{\sum_{s_\alpha} \sum_{s_j} C'_\alpha[\mathbf{R}(s_j) - \mathscr{R}(s_j)]\right\} \quad (9.28)$$

where $C'_\alpha = \ln C_\alpha$. If the constraint is taken to act at the center of the segment s_j but is spread equally over its length Δs_j, (9.28) is equivalently

$$\exp\left\{\sum_{s_\alpha} \sum_{s_j} \frac{\Delta s_j}{l} C'_\alpha[\mathbf{R}(s_j) - \mathscr{R}(s_j)]\right\} \quad (9.29)$$

Now (9.29) is just a formal representation of the full many-body constraints on the system.

In the absence of explicit solutions or representations, we must resort to simple models which are designed to be qualitatively correct. The simplest nontrivial approximation to a many-body problem is a mean field or self-consistent field approximation, and this is the approach to be followed. The constraint $C_\alpha[\cdots s_j \cdots]$ for all α represents the cage which confines $\mathbf{R}(s_j)$ to be "near" $\mathscr{R}(s_j)$. Rather than using the exact confinement field for the segment at s_j, we can use an average confining field. Thus $C'_\alpha[\mathbf{R}(s_j) - \mathscr{R}(s_j)]$ is replaced by its average over all segments s_j and confining segments s_α, i.e., symbolically

$$C'[\mathbf{R}(s) - \mathscr{R}(s)] = \langle C'_\alpha[\mathbf{R}(s_j) - \mathscr{R}(s_j)]\rangle_{s_j, s_\alpha} \tag{9.30}$$

where C' is now independent of s_j and s_α. Replacing the exact constraint in (9.29) by the average one in (9.30), the $\sum_{s_\alpha}$ can be performed immediately. It is proportional to the average number of constraining segments and hence to the monomer density $\rho = L/Vl$. Introducing this factor and passing to the limit of a continuous chain, it is found that the mean field approximation to (9.29) is

$$\exp\left\{-\rho \int_0^L ds C'[\mathbf{R}(s) - \mathscr{R}(s)]\right\} \tag{9.31}$$

We should pause to comment on the expected range of validity of (9.31). When the system is dilute so that the monomer density is similar to that of an ideal gas, the constraints are not very severe. A given segment still has a great deal of freedom (its constraining cage is fairly large relative to its size). Thus in this case of weak constraints we expect (9.31) to be indeed reasonable.

Since C' in (9.31) represents the average constraint and $\mathscr{R}(s)$ is chosen to give a uniform initial distribution, space is isotropic and C' must only depend on $|\mathbf{R}(s) - \mathscr{R}(s)|$. [We neglect any possible differences in the nature of the constraint in directions along the chain from those perpendicular to it.] Hence given the weakness of the constraint we assume that C' in (9.31) is reasonably well behaved so that a Taylor series expansion in $\mathbf{R}(s) - \mathscr{R}(s)$ can be made. The constant term can be absorbed into the normalization, and we keep the lowest order quadratic part since this, of course, can be handled exactly, and at least qualitatively manifests the existence and nature of the constraints, which keeps $\mathbf{R}(s)$ "near" $\mathscr{R}(s)$. Thus with this approximation (9.31) becomes

$$\exp\left\{-\rho C \int_0^L ds[\mathbf{R}(s) - \mathscr{R}(s)]^2\right\} \tag{9.32}$$

where C is a dimensionless constant (undetermined) which may be weakly density dependent and should be temperature dependent if $l = l(T)$.

Now we use this simple mean field theory of Edwards.[27] Given $\mathscr{R}(s)$ initially, the thermal Green's function for $\mathbf{R}(s)$ is given by

$$G(\mathbf{R}, \mathbf{R}'; L[\mathscr{R}]) = \int_{\mathbf{R}(0)=\mathbf{R}'}^{\mathbf{R}(L)=\mathbf{R}} \mathscr{D}[\mathbf{R}(s)] \exp\left\{-\frac{3}{2l}\int_0^L \dot{\mathbf{R}}^2(s)\,ds - \rho C\int_0^L ds[\mathbf{R}(s) - \mathscr{R}(s)]^2\right\} \quad (9.33)$$

where only the distribution of end vectors is considered explicitly. Comparing (9.33) to (9.18), we see that (9.33) corresponds to having a "time"-dependent potential

$$V[\mathbf{R}(s), s] = \rho C[\mathbf{R}(s) - \mathscr{R}(s)]^2 \quad (9.34)$$

because $\mathscr{R}(s)$ is some given fixed function of s. We leave it as an exercise to verify that in this case G of (9.33) satisfies the diffusion equation

$$\left\{\frac{\partial}{\partial L} - \frac{l}{6}\nabla_{\mathbf{R}}{}^2 + C\rho[\mathbf{R} - \mathscr{R}(L)]^2\right\}G(\mathbf{R}, \mathbf{R}'; L[\mathscr{R}]) = \delta(L)\,\delta(\mathbf{R} - \mathbf{R}') \quad (9.35)$$

G in (9.35) depends only on $\mathbf{R} - \mathscr{R}(L)$, etc., so changing variables to $\mathbf{r} = \mathbf{R} - \mathscr{R}(L)$ shows that the resultant Green's function is the solution to

$$\left\{\frac{\partial}{\partial L} - \frac{l}{6}\nabla_{\mathbf{r}}{}^2 + C\rho\mathbf{r}^2\right\}G(\mathbf{r}, \mathbf{r}'; L) = \delta(L)\,\delta(\mathbf{r} - \mathbf{r}') \quad (9.36)$$

which is now independent of $\mathscr{R}$ and is just analogous to the three-dimensional isotropic harmonic oscillator. Instead of using the result quoted in Section VI and derived in Appendix A, the solution to (9.36) is represented in terms of its eigenfunction expansion as

$$G(\mathbf{r}, \mathbf{r}'; L) = \theta(L) \sum_{n_x, n_y, n_z=0}^{\infty} N_{\mathbf{n}} \exp\left(-\frac{\boldsymbol{\xi}^2}{2} - \frac{\boldsymbol{\xi}'^2}{2}\right) H_{\mathbf{n}}(\boldsymbol{\xi}) H_{\mathbf{n}}(\boldsymbol{\xi}') \times \exp\left[-\omega L(n + \tfrac{3}{2})\right] \quad (9.37)$$

where

$$\boldsymbol{\xi} = \left[\frac{6C\rho}{l}\right]^{1/4}\mathbf{r}, \qquad n = n_x + n_y + n_z, \qquad \omega = \left(\frac{2Cl\rho}{3}\right)^{1/2}$$

$N_{\mathbf{n}}$ is the normalization,

$$H_{\mathbf{n}}(\boldsymbol{\xi}) = H_{n_x}(\xi_x) H_{n_y}(\xi_y) H_{n_z}(\xi_z)$$

are Hermite polynomials, and $\theta(L)$ is the step function.

The chain is taken to be closed so that $\mathbf{R} = \mathbf{R}'$. Since $d\mathbf{R} = d\mathbf{r}$, the partition function is (ignoring the walls since the constraint confines each segment to a much smaller volume)

$$\exp\left(\frac{S}{k}\right) = \int G(\mathbf{r}, \mathbf{r}; L)\, d\mathbf{r} = \mathcal{N} \sum_{n=0}^{\infty} \exp\left[-\omega L(n + \tfrac{3}{2})\right] \tag{9.38}$$

or alternatively (9.38) is $\exp(-\beta A)$ if $l = l(T)$, where $\mathcal{N}$ is an overall normalization factor. As $L \to \infty$, the $\mathbf{n} = 0$ term dominates, so in this limit

$$S \to -\frac{3kL}{2}\left[\frac{2C\rho l}{3}\right]^{1/2}, \quad \left(\text{or} -\frac{A}{T}\right) \tag{9.39}$$

and

$$PV = \frac{kTL}{2}\left(\frac{3Cl\rho}{2}\right)^{1/2}\left[1 + \left(\frac{\partial \ln C}{\partial \ln \rho}\right)_T\right] \tag{9.40}$$

where the extra contribution of kT due to the overall translational motion (a factor of V in $\mathcal{N}$) has been omitted. Using $L = Nl$, we have alternatively

$$PV = NkT\left(\frac{3Cl^3}{8}\right)^{1/2} \rho^{1/2}\left[1 + \left(\frac{\partial \ln C}{\partial \ln \rho}\right)_T\right] \tag{9.41}$$

We note that (9.39) represents the free-energy as an extensive quantity $A = Vf(\rho)$. The constraint (9.27)–(9.33) gives the usual type of thermodynamics [independent of the assumed initial $\mathcal{R}(s)$]. However, this result is to be expected since the foregoing model is predicated upon the assumption that normal thermodynamic behavior prevails.

In summary, the model of a polymer in a box has demonstrated that because of the entanglements, etc., we cannot assume that all configurations of the chain(s) which satisfy the external constraints are equally probable (unless of course we have fixed the internal constraints as well). Thus polymer systems are not ergodic because on the time scales of real experiments, phase space is separated into a large number of disjoint regions which are labeled by the constraints C_j.

D. General Changes of State

The formalism of statistical mechanics is generally expressed in terms of sums over states instead of sums over a subset of the degrees of freedom for the system, in this case the constraints C. It is convenient to reconsider the reformulation of Section IXA for the case in which the entropy and free energy are to be expressed in terms of sums over microscopic states. Let an initial microscopic state of the system at ξ_0 be designated as i, while a final microscopic state at ξ is f. The existence of internal constraints implies that i and f must be correlated. There then must be some set of joint probabilities $p(i, f)$ which describes the constraints in terms of the states of the system. We assume that the $\{p(i,f)\}$ are given and we wish to determine the formulation of statistical mechanics in terms of these quantities.

The state i is specified by many degrees of freedom, some of which represent the constraints C. Let $i \equiv C\alpha$, where α denotes all unconstrained degrees of freedom. Similarly, let $f = C\beta$. We assume that the usual rules of statistical mechanics apply to the unconstrained variables α and β. This implies that α and β are totally uncorrelated. Thus $p(i,f)$ must be expressible as

$$p(i,f) = p_C p_C(\alpha) p_C(\beta) \tag{9.42}$$

where p_C is the probability of finding the constraint C, $p_C(\alpha)$ is the probability of having α given C, etc. These probabilities must satisfy the normalizations

$$\sum_C p_C = 1 \tag{9.43a}$$

$$\sum_\alpha p_C(\alpha) = \sum_\beta p_C(\beta) = 1 \tag{9.43b}$$

for probabilities and conditional probabilities, respectively.

Consider the "joint" entropy

$$S_j = -k \sum_{i,f} p(i,f) \ln p(i,f) \tag{9.44}$$

Substituting (9.42) and (9.43) into (9.44), we find

$$S_j = -k \sum_C p_C \ln p_C - k \underset{C}{\quad} p_C \sum_\alpha p_C(\alpha) \ln p_C(\alpha) - k \sum_C p_C \sum_\beta p_C(\beta) \ln p_C(\beta) \tag{9.45}$$

But

$$S_C(\xi) = -k \sum_\beta p_C(\beta) \ln p_C(\beta) \tag{9.46a}$$

is just the definition of the entropy of the subensemble with constraint C in the final state ξ. Similarly,

$$S_C(\xi_0) = -k \sum_\alpha p_C(\alpha) \ln p_C(\alpha) \tag{9.46b}$$

corresponds to the initial constrained entropy. Thus, we have finally that

$$S_j = S(\xi) + S(\xi_0) - s_0 \tag{9.47}$$

where $S(\xi)$ and $S(\xi_0)$ are given by (9.14). In general, we are interested only in entropy differences, and the extra uninteresting constant s_0 in (9.47) is inconsequential. Similarly, $S(\xi_0)$ can be obtained from S_j, apart from this constant, when $\xi = \xi_0$. Thus (9.47) can be used to obtain the thermodynamic properties of the system. If interaction energies are also present, the $p(i, f)$ are to be taken as appropriate to a canonical ensemble. In this case, the entropies (9.44) through (9.47) are replaced by free energies A.

In the canonical ensemble state probabilities are given as the Boltzmann factors $\exp[-\beta H(\Gamma)]$. If similar exponential forms are assumed for $p(i, f)$, then

$$p(i,f) = \exp[-\beta H(\Gamma_0, \Gamma)] \tag{9.48}$$

where Γ_0 and Γ represent the initial and final states of the system in their respective phase spaces. By letting

$$p(i, f; \zeta) \equiv \exp[-\beta(1 + \zeta)H(\Gamma_0, \Gamma)] \tag{9.49}$$

S_j can be obtained from

$$\Lambda(\zeta) = \sum_{i,f} p(i, f; \zeta) \equiv \int d\Gamma_0 \int d\Gamma \exp[-\beta(1 + \zeta)H(\Gamma_0, \Gamma)] \tag{9.50}$$

via

$$S_j = -kT\left(\frac{\partial \Lambda(\zeta)}{\partial \zeta}\right)_{\zeta=1} \tag{9.51}$$

Equation (9.50) is now in the usual form encountered in statistical mechanics. Here, however, each system has twice as many degrees of freedom since it is described both in the initial and final states. Now that statistical mechanics with internal constraints has been considered both in general and with reference to the polymer-in-a-box problem, we turn to the general statistical mechanical treatment of polymers in bulk systems.

X. POLYMERS IN BULK: A FIELD THEORY

The discussions of the excluded volume problem, the polymer in a box, and statistical mechanics with constraints lead quite naturally to a discussion of the statistical mechanics of polymer systems in bulk. Such a system already exhibits the many-body problem in its entirety: excluded volume and the intermolecular and intramolecular interactions which also lead to the constraints. Upon recognition of the complexity of the system, the study of polymers in bulk might be relegated to the disciplines of continuum mechanics or kinetic theory. In the kinetic theory approach the polymer system is viewed in terms of a simple, but physically reasonable, model whose results are usually expressed in terms of some undetermined parameters. These kinetic theories usually have some limited range of validity, but their generalization is not easily accomplished. One approach is to construct slightly differing models and then compare the results of the various models, both among themselves and with experimental results.

The discipline of statistical mechanics is then usually left with the difficult task of providing a molecular (or microscopic) basis for the parameters that occur in the kinetic theories. When success comes in this venture, a search is made for generalizations of these models in order to extend their ranges of validity.

As the systems of the polymers in bulk represent full many-body problems, we do not expect that statistical mechanical treatments will provide exact solutions. Instead, they provide alternate viewpoints, thereby leading to new approximations and models whose ranges of validity may be checked. Furthermore, by using a statistical mechanical approach, the problems associated with describing polymers in bulk can be directly related, by mathematical analogy, to other many-body systems of chemistry and physics. We already noted the strong analogy between the Wiener integrals describing chain configurations and the Feynman path integral approach to quantum mechanics. Similarly, the self-consistent field theory for the polymer excluded volume problem follows from the general type of hierarchy of equations that are well known in other contexts. Since the fundamental basis for the statistical mechanics of polymers in bulk is still being developed, in this section we consider only an introduction to the general techniques. Such an introduction should aid in the understanding of the present literature on the subject as well as the subsequent developments.

As a particular example of the motivation for statistical mechanical approaches to the study of polymers in bulk, we have cited the problem of the description of rubber elasticity. A classic work on the subject is provided by the model of James and Guth (with subsequent refinements) in which attention is focused upon a single segment of chain which is between two cross-links.[1,2,16] Upon deformation of the rubber it is assumed that the polymer segments at the surfaces rigidly follow the deformation (i.e., follow the same transformation as the surface as a whole). Thus it is found that the most probable deformation of any segment under consideration is an affine deformation, i.e., scaling with the surfaces. The elasticity of this chain segment is then easily evaluated. The total elasticity is obtained by multiplying the results by the total number of "effective chains," a quantity that varies from model to model and is usually not independently measurable or determinable. The individual chain segments are taken to be noninteracting. However, the many-body constraints are not entirely forgotten as the rubber is taken to be incompressible (a bulk realization of these constraints). Such a model is, of course, not capable of dealing with situations in which the many-body (intermolecular) interactions become important (as in the case of crystallization upon stretching) or in which the constraints enter in a manner which is not simply expressed in terms of bulk incompressibility (as in the case of stretching to high elongations, etc.). Thus we require a statistical mechanical theory which is also capable, in principle, of treating intermolecular interactions between chains.

In the case of the statistical mechanics of real or ideal dilute polymer solutions, it is only necessary to consider one, two, or maybe three polymer chains (when calculating the second and third virial coefficients for the osmotic pressures).[5] In these cases it is reasonable to consider the configurations of each of the individual chains, expressing physical quantities in terms of averages over the probability distributions for these chain configurations. However, when discussing the properties of polymers in bulk in the thermodynamic limit, where the number of chains $N \to \infty$, the volume $V \to \infty$, but $N/V \to$ constant, it is no longer instructive to view each individual chain separately. Rather, the fundamental quantities are the polymer density throughout space, density-density correlations, etc. This is also the case for classical fluids.[6,7] These quantities would also have a bearing on the light-scattering properties of polymer systems.

Having already demonstrated the analogy between chain configuration probabilities and the probability amplitudes for paths in quantum mechanics of particles, we may again pursue this analogy in order to gain insight into ways of examining polymers in bulk. Actually, the description of polymers in bulk is first developed in an independent, but natural fashion. Then the analogy with the quantum mechanical problem becomes directly apparent. We can then borrow some techniques and representations from nonrelativistic many-body quantum mechanics.[9,10,29,32,50,55,59,60] Some similarities can be noted between the viewpoints taken in the quantum many-body problem and that to be developed for polymers in bulk. In the case of the quantum mechanics of a few particles, the wave function methods, or equivalently Feynman path integrals, focus on the dynamics of individual particles. However, in the many-body limit another viewpoint is taken in which we no longer look at the individual coordinates of each particle. Even an arbitrary number of particles is allowed. Here the basic quantities are propagators for arbitrary generic particles in the system, density-density correlations, etc. Such a physical formulation naturally arises out of the use of the so-called second quantized representation of quantum mechanics.[9,50,55,59,60] Now this viewpoint is just what was said to be necessary in dealing with polymer systems in bulk. The quantum mechanical propagators become, in the polymer case, end-vector probability distribution functions, in particular, Green's functions for the polymer chains. From the strong analogy between the diffusion and Schrödinger equations which has already been exploited, it should be expected that there would be a mode of description of polymers in bulk which would somehow resemble the second quantized form of quantum mechanics.[29] That this representation should exist can be demonstrated in another way.

The use of second quantization in quantum mechanics leads naturally to the introduction of Feynman diagrams.[50,55,59,60] But a box full of cross-linked polymers (a rubber) *is* a box full of Feynman diagrams! Here the abstraction becomes reality. (We would also want to include polymer intermolecular interactions in addition.) Whereas the Feynman diagrams can be written on two-dimensional paper, the polymers exist in three-dimensional space and therefore, as discussed in the preceding section, have topological relations to each other—the knots and entanglements. Thus such a general approach to the statistical mechanics of polymers in bulk automatically leads to the problems of statistical mechanics with constraints. The Feynman diagrams are mentioned to motivate the introduction of the analogy. It should be stressed at the outset that the physical properties, not the diagrams themselves, are the fundamental quantities. The diagrams are mnemonic devices, computational aids, etc., in and for the solution of problems; *they are not ends in themselves*.

Nonrelativistic quantum mechanics is usually expressed in terms of a Hamiltonian representation. In the second quantized version, an operator calculus of creation and annihilation operators is employed. However, as shown in Section VI, polymer-polymer interactions (for a single chain and *ipso facto* for many chains) exist for all "times." The analog of a conserved or approximately conserved Hamiltonian is not apparent for the polymer case. There is a less well known version of the quantum mechanical (second quantized) problem which used the Lagrangian formalism.[10,55] In this case, the creation and annihilation operators become the variables of integration in a functional integration scheme. (After all, this review is focusing upon functional integral techniques in the statistical mechanics of polymer systems!) Thus we are led to a "field theoretic" functional integral representation for polymers in bulk which bears strong (amusingly so) resemblance to the field theoretic formulations of quantum mechanical many-body systems. As these techniques are not generally well known to polymer chemists, or, for that matter, to chemical physicists, we assume no prior knowledge (but reference is made to the strong analogies). The representations can be generated by considering a set of simple mathematical identities. These identities are considered in Section VI in the introduction of gaussian random functions [cf. (6.33) and (6.34)]. It is instructive to consider these functions again.

Let us first consider the case in which the polymer chains are placed only on a lattice, i.e., the polymer bonds can lie only along the bonds of the lattice. It was noted in Section VI that the use of an underlying lattice has been useful in investigations of the polymer excluded volume problem. On a lattice, a polymer with a finite number of bonds can have only a finite number of configurations. For short enough chains, the polymer configurations on a lattice can be explicitly enumerated.[46] It should also be noted that similar "lattice counting" techniques have been employed to discuss critical phenomena by the use of the Ising model.[45] In our case, there is no requirement that the lattice be regular, since ultimately the limit is taken in which all lattice bonds approach zero length and the polymer is then in continuous space.

Consider now the mathematical development which is related to the use of the so-called method of random fields.[61] In the beginning there is the simple integral

$$a = \frac{\int_{-\infty}^{\infty} x^2 \exp(-x^2/2a)\, dx}{\int_{-\infty}^{\infty} dx \exp(-x^2/2a)} \tag{10.1}$$

This identity is then generalized to many variables $\{x_i\}$ by

$$a_i \delta_{ij} = \frac{\int_{-\infty}^{\infty} \cdots \int_{-\infty}^{\infty} x_i x_j \exp\left[-\sum_k x_k^2/2a_k\right] \prod dx_k}{\int_{-\infty}^{\infty} \cdots \int_{-\infty}^{\infty} \exp\left[-\sum_k x_k^2/2a_k\right] \prod dx_k} \tag{10.2}$$

The elements a_k are collected into a matrix $\mathbf{a}$, and the matrix is now allowed to be nondiagonal. In general, if the matrix is A_{ij}, we have

$$A_{ij} B_{ij} = \frac{\int_{-\infty}^{\infty} \cdots \int_{-\infty}^{\infty} x_i B_{ij} x_j \exp\left[-\sum_{kl} \tfrac{1}{2} x_k A_{kl}^{-1} x_l\right] \prod dx_m}{\int_{-\infty}^{\infty} \cdots \int_{-\infty}^{\infty} \exp\left[-\sum_{kl} \tfrac{1}{2} x_k A_{kl}^{-1} x_l\right] \prod_m dx_m} \tag{10.3}$$

where $\mathbf{A}^{-1}$ is the matrix inverse of $\mathbf{A}$, $\mathbf{A}\mathbf{A}^{-1} = \mathbf{1}$, or

$$\sum_j A_{ij}^{-1} A_{jk} = \delta_{ij} \tag{10.3a}$$

Let the quantities A_{ij} be interpreted as the probabilities that a polymer on a lattice have a configuration which runs from the lattice site i to the site j. The variables $\{x_i\}$ are some set of variables which are associated with the lattice sites. More generally, for many polymer chains we can write

$$\frac{\int_{-\infty}^{\infty}\cdots\int_{-\infty}^{\infty} x_i x_j x_k \cdots \exp\left[-\tfrac{1}{2}\mathbf{x}^T\mathbf{A}^{-1}\mathbf{x}\right]\prod dx}{\int_{-\infty}^{\infty}\cdots\int_{-\infty}^{\infty} \exp\left[-\tfrac{1}{2}\mathbf{x}^T\mathbf{A}^{-1}\mathbf{x}\right]\prod dx}$$

$$= \sum_{\substack{\text{all permutations}\\ \text{of indices } i,j,k,\ldots}} A_{ij}A_{kl}A\cdots \tag{10.4}$$

where $\mathbf{x}^T$ denotes the transpose of the column matrix $\mathbf{x}$ and matrix multiplication is implied. Equation (10.4) describes the probability of having chains go through the lattice sites $ijkl, \ldots$, where no specification has been made of which two lattice sites are connected by a given chain (or path) on the lattice. Hence the sum over all permutations of the indices is necessary.

In discussing the configurational statistics of individual chains, we have already considered taking the limit as a discrete chain (or path) approaches a continuous one. So this limit is performed on (10.4). Provided it is independent of the manner in which the limit is taken (thus the use of an arbitrary lattice), the definition of a functional integral is obtained. In this limit the lattice point i is the representation of the point $\mathbf{r}_i$ in real space and becomes continuous. Writing

$$x_i \to \phi(\mathbf{r})\big|_{\mathbf{r}=\mathbf{r}_i}, \qquad \prod_i dx_i \to \prod_{\mathbf{r}} d\phi(\mathbf{r}), \qquad A_{ij} \to A(\mathbf{r}_i, \mathbf{r}_j), \qquad \text{etc.} \tag{10.5}$$

we have an analog of (10.3),

$$A(\mathbf{r}, \mathbf{r}') = \frac{\int \phi(\mathbf{r})\phi(\mathbf{r}') \exp\left[-\frac{1}{2}\int d\mathbf{r}\int d\mathbf{r}'\phi(\mathbf{r})A^{-1}(\mathbf{r},\mathbf{r}')\phi(\mathbf{r}')\right]\delta\phi}{\int \exp\left[-\frac{1}{2}\int d\mathbf{r}\int d\mathbf{r}'\phi(\mathbf{r})A^{-1}(\mathbf{r},\mathbf{r}')\phi(\mathbf{r}')\right]\delta\phi} \tag{10.6}$$

where the continuous matrix $A^{-1}(\mathbf{r}, \mathbf{r}')$ is the inverse of $A(\mathbf{r}, \mathbf{r}')$ by the "matrix multiplication"

$$\int A^{-1}(\mathbf{r}, \mathbf{r}')A(\mathbf{r}', \mathbf{r}'')\, d\mathbf{r}' = \delta(\mathbf{r} - \mathbf{r}'') \tag{10.6a}$$

Note that (10.6) is the same kind of functional integral that is introduced in (6.34) in an entirely different context. It is also shown to represent the variance of a Gaussian random function as encountered in the random forces in Brownian motion.

The simplest case of $A(\mathbf{r}, \mathbf{r}')$ is a "unit matrix" which gives

$$\delta(\mathbf{r} - \mathbf{r}') = \frac{\int \phi(\mathbf{r})\phi(\mathbf{r}') \exp\left[-\frac{1}{2}\int \phi^2(\mathbf{r})\, d\mathbf{r}\right] \delta\phi}{\int \exp\left[-\frac{1}{2}\int \phi^2(\mathbf{r})\, d\mathbf{r}\right] \delta\phi} \tag{10.7}$$

Equation (10.7) can be evaluated directly by first enclosing space in a box of volume B^3, [denoting this integral as $I(B)$] and then taking $B \to \infty$. Analogous to (A.8), consider the linear transformation

$$\phi(\mathbf{r}) = B^{-3/2} \sum_{\mathbf{n}} \phi_{\mathbf{n}} \exp\left(\frac{2\pi i \mathbf{n} \cdot \mathbf{r}}{B}\right) \tag{10.8}$$

which is just a Fourier series. As this is a linear transformation the constant Jacobian factors in the numerator and denominator cancel. Apart from this constant we get

$$\delta\phi \equiv \prod_{\mathbf{r}} d\phi(\mathbf{r}) = \prod_{\mathbf{n}} d\phi_{\mathbf{n}} \tag{10.9}$$

From the development (10.1)–(10.7), $\phi(\mathbf{r})$ for each $\mathbf{r}$ is a real variable, so

$$\phi(\mathbf{r}) = \phi^*(\mathbf{r}) \Rightarrow \phi^*_{\mathbf{n}} = \phi_{-\mathbf{n}} \tag{10.10}$$

Thus $\phi_{\mathbf{n}}$ and $\phi_{-\mathbf{n}}$ can be taken to be independent variables, whereas $\phi^*_{\mathbf{n}}$ and $\phi^*_{-\mathbf{n}}$ are not. Parseval's relation is

$$\begin{aligned} \int d\mathbf{r}\phi^2(\mathbf{r}) &= B^{-3} \sum_{\mathbf{m},\mathbf{m}'} \phi_{\mathbf{m}}\phi_{\mathbf{m}'} \exp\left[\frac{2\pi i(\mathbf{m} + \mathbf{m}') \cdot \mathbf{r}}{B}\right] \\ &= B^{-3} \sum_{\mathbf{m},\mathbf{m}'} \phi_{\mathbf{m}}\phi_{\mathbf{m}'} B^3 \delta_{\mathbf{m},-\mathbf{m}'} = \sum_{\mathbf{m}} \phi_{\mathbf{m}}\phi_{-\mathbf{m}} \end{aligned} \tag{10.11}$$

Therefore

$$\begin{aligned} I(B) = B^{-3} \sum_{\mathbf{n},\mathbf{n}'} \exp\left[\frac{2\pi i(\mathbf{n} \cdot \mathbf{r} + \mathbf{n}' \cdot \mathbf{r}')}{B}\right] \int_{-\infty}^{\infty} \cdots \int_{-\infty}^{\infty} \phi_{\mathbf{n}}\phi_{\mathbf{n}'} \\ \times \exp\left[-\frac{1}{2}\sum_{\mathbf{m}} \phi_{\mathbf{m}}\phi_{-\mathbf{m}}\right] \prod_{\mathbf{k}} d\phi_{\mathbf{k}} \\ \times \left\{\int_{-\infty}^{\infty} \cdots \int_{-\infty}^{\infty} \exp\left[-\frac{1}{2}\sum_{\mathbf{m}} \phi_{\mathbf{m}}\phi_{-\mathbf{m}}\right] \prod_{\mathbf{k}} d\phi_{\mathbf{k}}\right\}^{-1} \end{aligned} \tag{10.12}$$

Let

$$\begin{aligned} \alpha_{\mathbf{m}} &= \tfrac{1}{2}(\phi_{\mathbf{m}} + \phi_{-\mathbf{m}}) \\ \beta_{\mathbf{m}} &= \frac{1}{2i}(\phi_{\mathbf{m}} - \phi_{-\mathbf{m}}) \end{aligned} \tag{10.13}$$

so that

$$d\phi_{\mathbf{m}}\, d\phi_{-\mathbf{m}} = 2\, d\alpha_{\mathbf{m}}\, d\beta_{\mathbf{m}}$$

The factors of two cancel in the numerator and denominator. [Note that (10.13) combines $\phi_{\mathbf{m}}$ and $\phi_{-\mathbf{m}}$ into the symmetric and antisymmetric combinations where some index is taken as **m** and its negative is then $-\mathbf{m}$.] Since $\phi_{\mathbf{m}}\phi_{-\mathbf{m}} = \alpha_{\mathbf{m}}^2 + \beta_{\mathbf{m}}^2$ and this term occurs in the summation in the exponentials of (10.12) once for **m** and once for $-\mathbf{m}$, the exponential is

$$\sum_{\mathbf{m}}{}' [\alpha_{\mathbf{m}}^2 + \beta_{\mathbf{m}}^2]$$

Here the prime indicates that if **m** is included in the summation, $-\mathbf{m}$ is to be omitted. Similarly, the differential is $\prod_{\mathbf{m}}' d\alpha_{\mathbf{m}}\, d\beta_{\mathbf{m}}$. Therefore the gaussian integrals

$$\begin{aligned}\int_{-\infty}^{\infty}\cdots\int_{-\infty}^{\infty} &\{\alpha_{\mathbf{n}}\alpha_{\mathbf{n}'} + i[\alpha_{\mathbf{n}}\beta_{\mathbf{n}'} + \alpha_{\mathbf{n}'}\beta_{\mathbf{n}}] - \beta_{\mathbf{n}}\beta_{\mathbf{n}'}\} \\ &\times \exp\left[-\sum_{\mathbf{m}}{}'(\alpha_{\mathbf{m}}^2 + \beta_{\mathbf{m}}^2)\right] \prod_{\mathbf{k}}{}' d\alpha_{\mathbf{m}}\, d\beta_{\mathbf{k}} \\ &\times \left\{\int_{-\infty}^{\infty}\cdots\int_{-\infty}^{\infty} \exp\left[-\sum_{\mathbf{m}}{}'(\alpha_{\mathbf{m}}^2 + \beta_{\mathbf{m}}^2)\right] \prod_{\mathbf{k}}{}' d\alpha_{\mathbf{m}}\, d\beta_{\mathbf{k}}\right\}^{-1}\end{aligned} \tag{10.14}$$

must be considered. From (10.12) **n** and **n**′ are still free to take on all values. If **n**′ is not one of the chosen variables in the $\sum_{\mathbf{m}}'$ or $\prod_{\mathbf{k}}'$, from (10.13) $\alpha_{-\mathbf{n}'} = \alpha_{\mathbf{n}'}$ and $\beta_{-\mathbf{n}'} = -\beta_{\mathbf{n}'}$ are used. Thus the cross-terms in (10.14) vanish. The only nonzero contributions come from $\mathbf{n}' = \mathbf{n}$ or $-\mathbf{n}$ giving

$$(\tfrac{1}{2} - \tfrac{1}{2}) \qquad \text{or} \qquad (\tfrac{1}{2} - (-1)\tfrac{1}{2}),$$

respectively. Thus only $\mathbf{n}' = -\mathbf{n}$ survives, leaving

$$I(B) = B^{-3}\sum_{\mathbf{n}} \exp\left[\frac{2\pi i\mathbf{n}\cdot(\mathbf{r} - \mathbf{r}')}{B}\right] \to \delta(\mathbf{r} - \mathbf{r}') \tag{10.15}$$

as stated in (10.7). The derivation is much simpler when the expansion of $\phi(\mathbf{r})$ is in terms of real functions, e.g., $\prod_{\alpha=x,y,z} \sqrt{2/B}\cos(n_\alpha \pi r_\alpha/B)$ so there is no bookkeeping to be maintained on the Fourier coefficients, which are now real. It is therefore left as an exercise to prove (10.6) by using the spectral expansion of $A^{-1}(\mathbf{r}, \mathbf{r}')$ in terms of its eigenfunctions $u_n(\mathbf{r})$ and eigenvalues λ_n,

$$A^{-1}(\mathbf{r}, \mathbf{r}') = \sum_n u_n(\mathbf{r})u_n(\mathbf{r}')\lambda_n \tag{10.16}$$

where for convenience A^{-1} is taken to be symmetric and the u_n are real and orthonormal. Hence it is necessary only to show that the expansion

$$\phi(\mathbf{r}) = \sum_n \phi_n u_n(\mathbf{r}) \tag{10.16a}$$

on the right-hand side of (10.6) leads to

$$\sum_n \frac{u_n(\mathbf{r})u_n(\mathbf{r}')}{\lambda_n} \equiv A(\mathbf{r}, \mathbf{r}') \tag{10.16b}$$

Again integration over all functions corresponds to an integration over all generalized Fourier coefficients.

Sometimes, even though only real functions are dealt with, it is convenient to avoid the bookkeeping procedure of (10.13)–(10.14) by the introduction of complex variables $\psi(\mathbf{r})$, $\psi^+(\mathbf{r})$. For instance, when these complex variables are used along with some G^{-1}, the result is

$$G(\mathbf{r}, \mathbf{r}') = \mathcal{N} \int \psi(\mathbf{r})\psi^+(\mathbf{r}') \exp\left[-\int d\mathbf{r} \int d\mathbf{r}' \psi^+(\mathbf{r})G^{-1}(\mathbf{r}, \mathbf{r}')\psi(\mathbf{r}')\right] \delta\psi\, \delta\psi^+ \tag{10.17}$$

The $\delta\psi\, \delta\psi^+$ in (10.17) implies the integration is taken over the real and imaginary parts of ψ and ψ^+. Note the absence of a factor of $\frac{1}{2}$ in the exponent of (10.17). The normalization is again denoted by

$$\mathcal{N}^{-1} = \int \exp\left[-\iint \psi^+ G^{-1} \psi\right] \delta\psi\, \delta\psi^+ \tag{10.17a}$$

For the case $G^{-1}(\mathbf{r}, \mathbf{r}') = \delta(\mathbf{r} - \mathbf{r}')$, the result is easily obtained using

$$\psi(\mathbf{r}) = B^{-3/2} \sum_{\mathbf{n}} \psi_{\mathbf{n}} \exp\left[\frac{2\pi i \mathbf{n} \cdot \mathbf{r}}{B}\right], \qquad \delta\psi = \prod_{\mathbf{n}} \delta\psi_{\mathbf{n}}$$

$$\psi^+(\mathbf{r}') = B^{-3/2} \sum_{\mathbf{n}} \psi_{\mathbf{n}}^+ \exp\left[\frac{-2\pi i \mathbf{n} \cdot \mathbf{r}'}{B}\right], \qquad \delta\psi^+ = \prod_{\mathbf{n}} \delta\psi_{\mathbf{n}}^+$$

Upon formation of

$$\alpha_{\mathbf{n}} = \frac{(\psi_{\mathbf{n}} + \psi_{\mathbf{n}}^+)}{2}, \qquad \beta_{\mathbf{n}} = \frac{(\psi_{\mathbf{n}} - \psi_{\mathbf{n}}^+)}{2i}$$

the term in the exponential is a sum over all $\mathbf{n}$,

$$\sum_{\mathbf{n}} (\alpha_{\mathbf{n}}^2 + \beta_{\mathbf{n}}^2)$$

and the differential is a product over all $\mathbf{n}$, $\prod_{\mathbf{n}} d\alpha_{\mathbf{n}}\, d\beta_{\mathbf{n}}$ with no restrictions as in the case of the real variable $\phi(\mathbf{r})$ above. (The details are again left as an exercise.)

Now that we have established the machinery, we return to study polymer chains. The number of ways of obtaining a random walk, i.e., the number of chain configurations (step length is l) of L/l steps, starting at

$\mathbf{r}'$ and ending at $\mathbf{r}$, for large L/l (or for a gaussian step-length distribution for any L/l) is given by

$$G_0(\mathbf{r}, \mathbf{r}'; L) = \left(\frac{3}{2\pi L l}\right)^{3/2} \exp\left[wL - \frac{3(\mathbf{r} - \mathbf{r}')^2}{2lL}\right] \tag{10.18}$$

Here exp (wl) is the number of ways each link can be added, and kwl is the entropy per link. The quantity w is to be obtained from the study of the configurational statistics of individual polymer chains. If there are intramolecular interactions, then $-kTwl$ is the free energy per link. The end-vector probability distribution $P(\mathbf{r}, \mathbf{r}'; L)$ is then just $\exp(-wL)$ $G_0(\mathbf{r}, \mathbf{r}'; L)$ and is normalized to unity with respect to integration over either $\mathbf{r}$ or $\mathbf{r}'$. G_0 satisfies the differential equation

$$\left[\frac{\partial}{\partial s} - \frac{l}{6}\nabla_{\mathbf{r}}^2 - w\right] G_0(\mathbf{r}, \mathbf{r}'; ss') = \delta(\mathbf{r} - \mathbf{r}')\,\delta(s - s') \tag{10.19}$$

so that its inverse is then

$$G_0^{-1}(\mathbf{r}\mathbf{r}'; ss') = \delta(\mathbf{r} - \mathbf{r}')\,\delta(s - s')\left(\frac{\partial}{\partial s} - \frac{l}{6}\nabla_{\mathbf{r}}^2 - w\right) \tag{10.20}$$

Denote the "Lagrangian density" by

$$\mathscr{L}_0(\mathbf{r}s) = \psi^+(\mathbf{r}s)\frac{\partial \psi(\mathbf{r}s)}{\partial s} - \frac{l}{6}\psi^+(\mathbf{r}s)\nabla_{\mathbf{r}}^2\psi(\mathbf{r}s) - w\psi^+(\mathbf{r}s)\psi(\mathbf{r}s) \tag{10.21a}$$

By analogy with the quantum mechanical problem, the equivalent symmetrized form[55] can be used in which the boundary terms in partial integrations can be taken to vanish:

$$\mathscr{L}_0(\mathbf{r}s) = \frac{1}{2}\left[\psi^+(\mathbf{r}s)\frac{\partial \psi(\mathbf{r}s)}{\partial s} - \frac{\partial \psi^+(\mathbf{r}s)}{\partial s}\psi(\mathbf{r}s)\right] + \frac{l}{6}[\nabla_{\mathbf{r}}\psi^+(\mathbf{r}s)]\cdot[\nabla_{\mathbf{r}}\psi(\mathbf{r}s)] - w\psi^+(\mathbf{r}s)\psi(\mathbf{r}s) \tag{10.21b}$$

Substituting (10.20)–(10.21) into (10.17), we get

$$G_0(\mathbf{r}\mathbf{r}'; ss') = \mathscr{N}\int \psi(\mathbf{r}s)\psi^+(\mathbf{r}'s') \exp\left[-\int d\mathbf{r}\int_0^L ds\,\mathscr{L}_0(\mathbf{r}s)\right]\delta\psi\,\delta\psi^+ \tag{10.22}$$

where

$$\mathscr{N}^{-1} = \int \exp\left[-\int d\mathbf{r}\int_0^L ds\,\mathscr{L}_0(\mathbf{r}s)\right]\delta\psi\,\delta\psi^+ \tag{10.22a}$$

is again the normalization.

From (10.22) it is clear that the variables $\mathbf{r}'s'$ associated with the beginning of the chain segment are the arguments of $\psi^+[\psi^+(\mathbf{r}'s')]$, whereas the end variables $\mathbf{r}s$ are associated with $\psi[\psi(\mathbf{r}s)]$. Complex variables, or *fields*, ψ^+ and ψ have been used so that ψ^+ can denote the start of a chain and ψ the chain end. (It is immaterial which end is chosen to be the start and which is the end. The choice here is consistent with the earlier associations of the arguments of G.) Equations (10.21) are formally analogous to the Lagrangian formulation of the quantum mechanical many-body problem where $\mathscr{L}_0$ is the analog of the Lagrangian density for the particles, s is like the (imaginary) time, and w is like the chemical potential which fixes the number of particles, while $G_0^{-1}G_0 = 1$ is the analog of the Schrödinger equation. This relationship between the polymer and quantum many-body formulations is exploited for convenience. The reader need not be familiar with the latter formulation, since all the necessary material is independently developed here. Numerous texts also are available concerning the quantum mechanical problem. By analogy with the quantum mechanical form for two-body interactions in the action (integral of Lagrangian density),

$$\int dt \int d\mathbf{r} \int d\mathbf{r}' \psi^+(\mathbf{r}t)\psi^+(\mathbf{r}'t)V(\mathbf{r}-\mathbf{r}')\psi(\mathbf{r}'t)\psi(\mathbf{r}t)$$

we would expect polymer-polymer pairwise interactions to introduce a term of the form

$$\int_0^L ds \int_0^L ds' \int d\mathbf{r} \int d\mathbf{r}' \psi^+(\mathbf{r}s)\psi(\mathbf{r}s)V(\mathbf{r}-\mathbf{r}')\psi^+(\mathbf{r}'s')\psi(\mathbf{r}'s') \qquad (10.23)$$

This form can be proven to be correct. Note that the interactions act over all "times." Thus the analog of the Hamiltonian formulation of quantum mechanics is not as convenient as the Lagrangian form since there is no simple analog of the conserved energy.

$\mathscr{L}_0$ could be generalized to describe chains with a slight degree of stiffness by using the appropriate $\mathscr{G}_0^{-1}$, i.e., given any distribution $\mathscr{G}_0$ for end vectors, if $\mathscr{G}_0^{-1}$ exists, (10.22) can be written for that case.

Equation (10.22) describes only a single chain. By analogy with (10.4) an arbitrary number of chains can be considered which start at $\{\mathbf{r}_a s_a, \mathbf{r}_b s_b, \mathbf{r}_c s_c, \ldots\}$ and end at $\{\mathbf{r}_1 s_1, \mathbf{r}_2 s_2, \ldots\}$, making no specification as to which chain beginning goes with which chain end. We have

$$\mathscr{N} \int \psi(\mathbf{r}_1 s_1)\psi(\mathbf{r}_2 s_2)\cdots\psi^+(\mathbf{r}_a s_a)\psi^+(\mathbf{r}_b s_b)\psi^+(\mathbf{r}_c s_c)\cdots \exp(-A_0)\,\delta\psi\,\delta\psi^+$$
$$= \sum_{\substack{\alpha,\beta,\gamma=\text{all} \\ \text{permutations of } a,b,c}} G_0(\mathbf{r}_1\mathbf{r}_\alpha; s_1 s_\alpha)G_0(\mathbf{r}_2\mathbf{r}_\beta; s_2 s_\beta)\cdots \qquad (10.24)$$

where

$$A_0 = \iint \mathscr{L}_0(\mathbf{r}s)\, d\mathbf{r}\, ds \qquad (10.25)$$

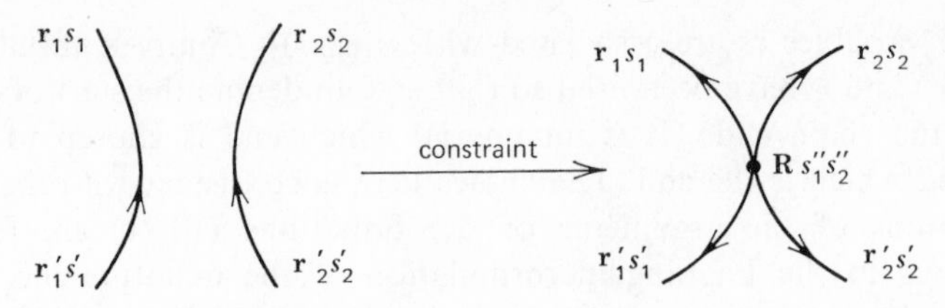

Fig. 7. Schematic representation of cross-link constraint. Chains 1 and 2 with beginnings and ends, respectively at $\mathbf{r}_1's_1'$, $\mathbf{r}_2's_2'$ and $\mathbf{r}_1s_1$, $\mathbf{r}_2s_2$ and required to have a cross-link at $\mathbf{R}$ for s_1'' and s_2'' along the chains.

is the "action" integral. We obtain all permutations because of the lack of connection between specific ends and specific starts. .

Equation (10.24) describes only noninteracting chains. The cross-links are now to be introduced. [The verification of (10.23) can be made by techniques which are similar to those used to introduce the cross-links.] Therefore there must be some mathematical way to require that there be a cross-link. The presence of a cross-link at the point $\mathbf{R}$ and at s_1'' and s_2'' (along chains 1 and 2, respectively) implies that a constraint must be inserted into the integrand of (10.24) to require two chains to pass through the point $\mathbf{R}$ in space for some s_1'' and s_2'' along their chains. The free chains of (10.24) must be constrained to link as in Fig. 7.

There are two simple ways in which this can be accomplished. The first, which is perhaps more familiar, is briefly described, then the second, which is more useful, is treated in more detail. For simplicity, we first consider two chains as in Fig. 7. For two chains (10.24) would have two starts and two ends, and hence two G_0. By analogy with the quantum problem, we want to introduce a start and end at $\mathbf{R}$ for both of the chains:

$$\psi^+(\mathbf{R}s_1'')\psi(\mathbf{R}s_1'')\psi^+(\mathbf{R}s_2'')\psi(\mathbf{R}s_2'') \tag{10.26}$$

Introducing (10.26) into (10.24) with only two starts and two ends leads to

$$G_0(\mathbf{r}_1'\mathbf{R};\, s_1's_1'')G_0(\mathbf{r}_2'\mathbf{R};\, s_2's_2'')G_0(\mathbf{R}\mathbf{r}_1;\, s_1''s_1)G_0(\mathbf{R}\mathbf{r}_2;\, s_2''s_2)$$
$$+ \text{ permutations of ends and beginnings} + \text{unwanted terms} \tag{10.27}$$

Apart from the unwanted terms which contain factors like $G_0(\mathbf{RR};\, s_1''s_1'')$ or $G_0(\mathbf{RR};\, s_2''s_1'')$ and are ultimately removable, (10.27) does indeed describe a cross-link at $\mathbf{R}s_1''s_2''$ along a pair of chains with beginnings and ends at $\mathbf{r}_1's_1'$, $\mathbf{r}_2's_2'$ and $\mathbf{r}_1s_1$, $\mathbf{r}_2s_2$, respectively. There is no specification as to which beginnings and ends go together. If the cross-link is allowed to be distributed randomly in space and along the chains, then this is represented by $\int d\mathbf{R} \int ds_1'' \int ds_2''$ of (10.26). When this integration is applied to (10.27), apart from the unwanted terms, it gives the correct weighting for each individual chain segment. Allowing there to be m cross-links, m factors of the constraint (10.27) can be introduced. It is more convenient, partially in order to remove the unwanted terms, to pass to a grand canonical ensemble where the number of cross-links is arbitrary. Later the given number of cross-links can be reintroduced in the standard manner. Following a procedure which is discussed in more detail later, this would add the term

$$A_1' = -\mu \int_0^L ds \int_0^L ds' \int d\mathbf{R} \psi^+(\mathbf{R}s)\psi(\mathbf{R}s)\psi^+(\mathbf{R}s')\psi(\mathbf{R}s') \qquad (10.28)$$

to the action of (10.25), so that (10.24) would contain $A_1 = A_0 + A_1'$ in order to describe chains with cross-links. ($\mathcal{N}$ must be suitably modified to remove the unwanted terms.) In (10.28) μ is a parameter that is related to the chemical potential which fixed the number of cross-links. Comparing (10.28) with the contribution to the action (10.23) arising from polymer-polymer interactions, we may infer that the cross-linking is equivalent to the *attractive* potential $V(\mathbf{r} - \mathbf{r}') = -\mu\delta(\mathbf{r} - \mathbf{r}')$. Thus the cross-linking constraint is the attractive analog of the excluded volume problem. It remains to be shown that $\mu > 0$, but for large enough μ (large enough number of cross-links, in effect) such an "attractive force" should be sufficient to make the polymer collapse—the phenomenon of syneresis! So the polymer with cross-links but no interactions gives a quartic action ("interactions" acting for all "times") just like the usual many-body systems. However, in the polymer case the strength of the quartic part μ is related to a chemical potential, so solutions are required which are analytic in μ in order to fix the requisite number of cross-links. (We have not yet introduced the entanglements and resultant internal constraints.)

Having briefly discussed a method of introducing the cross-links which utilizes the interpretation of ψ^+ as starting any chain and of ψ as ending any chain, we now turn to a more convenient approach. The cross-links are taken to be introduced by an external field $\phi(\mathbf{r})$, a cross-linking agent. Let there be n noninteracting polymer chains in the presence of an external field $\phi(\mathbf{r})$ such that the number of chain configurations for each chain is given by $G([\phi])$ satisfying

$$\left[\frac{\partial}{\partial s} - \frac{l}{6}\nabla_{\mathbf{r}}^2 + \mu^{1/2}\phi(\mathbf{r}) - w\right] G(\mathbf{r}\mathbf{r}'; ss'[\phi]) = \delta(\mathbf{r} - \mathbf{r}')\,\delta(s - s') \quad (10.29)$$

$\mu^{1/2}$ represents an unspecified "coupling strength." Choosing the action corresponding to (10.29) as

$$A_1 = A_0 + \mu^{1/2} \int d\mathbf{r} \int_0^L ds \phi(\mathbf{r})\psi^+(\mathbf{r}s)\psi(\mathbf{r}s) \qquad (10.30)$$

we see that the n chains have n starts and ends. Thus

$$\mathcal{N}\int \psi\psi\cdots\psi^{+}\psi^{+}\cdots \exp(-A_1)\,\delta\psi\,\delta\psi^{+} = \sum_{\substack{\text{all permutations}\\ \text{of ends}}} \prod_{n\text{ chains}} G([\phi]) \tag{10.31}$$

where the explicit indices are omitted for convenience. Now let the field $\phi(\mathbf{r})$ be a random variable (in particular, gaussian random) so that

$$P[\phi]\,\delta\phi = \mathcal{N}\exp\left[-\frac{1}{2}\int d\mathbf{r}\phi^2(\mathbf{r})\right]\delta\phi \tag{10.32}$$

where of course $\mathcal{N}$ is chosen so $\int P[\phi]\,\delta\phi = 1$. We therefore must verify that

$$\int P[\phi]\,\delta\phi \prod_{n\text{ chains}} G([\phi]) \tag{10.33}$$

introduces an arbitrary number of cross-linkages. These cross-links are distributed at random through space and along the chains with weights that are appropriate to the individual chain segments as in the first term of (10.27). The final sum over permutations in (10.31) then implies the lack of specificity of starts with ends. Having established that (10.33) describes an arbitrary number of linkages, we can then show that the parameter μ can be used to fix the number to be m.

In order to show that (10.33) properly introduces cross-links it is sufficient to consider again only two chains. For each chain the usual perturbation expansion can symbolically be written as a series in powers of ϕ via

$$G([\phi]) = G_0 - \mu^{1/2}\int G_0\phi G_0 + \mu\iint G_0\phi G_0\phi G_0 + \cdots \tag{10.34}$$

where explicitly the first-order term is

$$\int G_0(\mathbf{r}\mathbf{r}'';ss'')\phi(\mathbf{r}'')G_0(\mathbf{r}''\mathbf{r}';s''s')\,d\mathbf{r}''\,ds'' \tag{10.34a}$$

Using the diagrammatic notation of Fig. 8, analogous to that in Section VI, (10.34) becomes Fig. 9. Again the beginnings and ends in Fig. 9 are

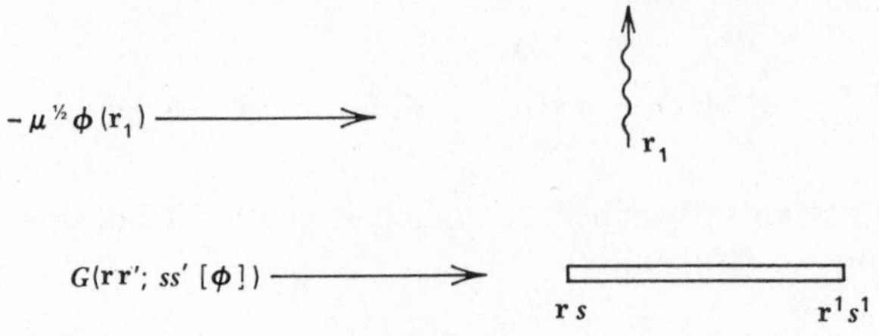

$-\mu^{1/2}\phi(\mathbf{r}_1)$ → $\mathbf{r}_1$

$G(\mathbf{r}\mathbf{r}';ss'[\phi])$ → $\mathbf{r}\,s$ $\mathbf{r}^1 s^1$

Fig. 8. Diagrammatic representation of $\phi(\mathbf{r})$ and $G([\phi])$.

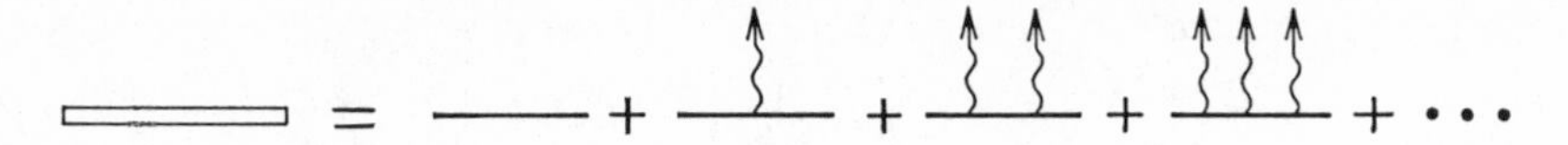

Fig. 9. Expansion of $G([\phi])$ in powers of $\phi \cdot G_0$ as defined in Fig. 3.

to be specified, but sums (i.e., integrals over space and "time") are to be taken over all intermediate points. Taking the first-order term from two different chains, causes the averaging over $\phi(\mathbf{r})$ to give

$$\mathcal{N}\int \delta\phi \exp\left[-\frac{1}{2}\int \phi^2\, d\mathbf{r}\right]\left\{-\mu^{1/2}\int G_0(\mathbf{r}_1\mathbf{r}_1''; s_1 s_1'')\phi(\mathbf{r}_1'')G_0(\mathbf{r}_1''\mathbf{r}_1'; s_1''s_1')\, d\mathbf{r}_1''\, ds_1''\right\}$$

$$\times \left\{-\mu^{1/2}\int G_0(\mathbf{r}_2\mathbf{r}_2''; s_2 s_2'')\phi(\mathbf{r}_2'')G_0(\mathbf{r}_2''\mathbf{r}_2'; s_2''s_2')\, d\mathbf{r}_2''\, ds_2''\right\}$$

$$= \mu\int d\mathbf{R}\int ds_1''\, ds_2'' G_0(\mathbf{r}_1\mathbf{R}; s_1 s_1'')G_0(\mathbf{r}_2\mathbf{R}; s_2 s_2'')$$

$$\times\, G_0(\mathbf{R}\mathbf{r}_1'; s_1''s_1')G_0(\mathbf{R}\mathbf{r}_2'; s_2''s_2') \tag{10.35}$$

The $\delta\phi$ integration trivially gives $\delta(\mathbf{r}_1'' - \mathbf{r}_2'')$. The dummy integration variable has been changed to $\mathbf{R}$. Equation (10.35) gives the correct statistical weight, apart from the factor of μ, for a cross-linkage between two polymer chains when this linkage is randomly distributed. Equation (10.35) is just the "sum" of the statistical weights for having the cross-link at any particular point in space and along the two chains; the chain ends are still fixed in space. It is easily verified that, e.g., the second-order term for a single chain leads to a single self-cross-linkage on that chain, etc.

Note that when (10.31) is expanded in terms of ϕ according to (10.34), the result is a power series in $\mu^{1/2}\phi$; but when (10.33) is averaged over ϕ, all terms with odd powers of ϕ vanish. This results because it requires two ϕ for each linkage. Thus (10.33), if expanded, would be a formal power series in powers of μ. The coefficient of μ^m contains all the terms for which there are m cross-links randomly distributed with the proper weights for each of the chain segments. Therefore for m cross-links we must just take the coefficient of μ^m. This can easily be selected by applying the contour integration

$$m!\oint \frac{d\mu}{\mu^{m+1}}$$

to (10.33), where

$$\oint = \frac{1}{2\pi i} \int_{\mathscr{C}}$$

and $\mathscr{C}$ is a small circle enclosing the origin in the positive direction. Now in an ensemble of n chains, the chains are taken to be distributed at random throughout the volume V; the starts and ends are allowed to be anywhere in the volume V. Thus for n chains and m cross-links randomly distributed (*noninteracting chains only*) (10.32) becomes

$$\exp(S/k) = \oint \frac{m!\, d\mu}{\mu^{m+1}} \mathscr{N} \int \delta\phi \exp\left[-\frac{1}{2}\int \phi^2(\mathbf{r})\, d\mathbf{r}\right] \times \left\{\int d\mathbf{r} \int d\mathbf{r}' G_0(\mathbf{r}\mathbf{r}'; L0[\phi])\right\}^n \quad (10.36)$$

If $l = l(T)$ the left-hand side is taken as $\exp(-\beta A)$ as usual. We could have verified that if the field ϕ, instead of having the probability distribution (10.32), had a distribution like (6.35), then for a single chain, and *ipso facto* for many chains, (10.36) would describe polymer-polymer interactions. Then upon using (10.30) and (10.31) to represent $G([\phi])$ and formally performing the ϕ integral (neglecting for the moment any "unlinked clusters"), the result is that given by (10.23) is analogy with the quantum problem. (We leave the details as an exercise.) However, in order to be sufficiently general to include both the cross-links and polymer-polymer interactions it is desirable to express the ansatz (10.36) in terms of both the ψ and ϕ fields.

If the starts and ends of the chains are distributed randomly throughout V, the ψ and ψ^+ of say (10.31) are taken as ends (at L) and starts (at 0), respectively. The integrand of (10.31) contains

$$\left[\int d\mathbf{r}\psi^+(\mathbf{r}0)\right]^n \left[\int d\mathbf{r}'\psi(\mathbf{r}'L)\right]^n$$

introducing n starts and ends. (We could introduce a distribution in chain lengths, but that is an unnecessary mathematical complication.) It is again convenient—to avoid the need for combinatorial arguments—to pass to a grand canonical ensemble where the number of starts and ends are arbitrary. (They must of course ultimately be the same.) The correct number of chains is fixed by the contour integration as in (10.36). The final result is

$$\exp(-\beta A) = \oint \frac{n!\, d\alpha\, d\alpha^+}{(\alpha\alpha^+)^{n+1}} \oint \frac{d\mu m!}{\mu^{m+1}} \exp\{\Xi(\alpha\alpha^+, \mu, T)\} \quad (10.37)$$

where

$$\exp\{\Xi\} = \int \delta\phi \exp\left[-\frac{1}{2}\int \phi^2(\mathbf{r})\, d\mathbf{r}\right] \mathcal{N}(\phi) \int \delta\psi\, \delta\psi^+ \times \exp\,[-A(\alpha\alpha^+, \mu^{1/2}, T)] \quad (10.38)$$

$$\begin{aligned} A(\alpha\alpha^+, \mu^{1/2}, T) &= -\alpha^+ \int \psi^+(\mathbf{r}0)\, d\mathbf{r} - \alpha \int \psi(\mathbf{r}L)\, d\mathbf{r} \\ &\quad + \int d\mathbf{r} \int_0^L ds \left[\psi^+(\mathbf{r}s)\frac{\partial\psi(\mathbf{r}s)}{\partial s} - \frac{l}{6}\psi^+(\mathbf{r}s)\nabla_{\mathbf{r}}^{\,2}\psi(\mathbf{r}s) - w\psi^+(\mathbf{r}s)\psi(\mathbf{r}s) + \mu^{1/2}\phi(\mathbf{r})\psi^+(\mathbf{r}s)\psi(\mathbf{r}s)\right] \\ &\quad + \beta\int_0^L ds \int_0^L ds' \int d\mathbf{r} \int d\mathbf{r}'\, \psi^+(\mathbf{r}s)\psi(\mathbf{r}s)V(\mathbf{r}-\mathbf{r}')\psi^+(\mathbf{r}'s')\psi(\mathbf{r}'s') \end{aligned} \quad (10.39)$$

or the symmetrized version. $\mathcal{N}(\phi)$ must be chosen to remove all the "unlinked clusters," which are the analogs of the unwanted terms of (10.27). The result can be shown to be

$$\mathcal{N}(\phi)^{-1} = \int \delta\psi\, \delta\psi^+ \exp\,[-A(0, \mu^{1/2}, T)] \quad (10.40)$$

As is usual in statistical thermodynamics, we consider the limit that $n \to \infty$, $m \to \infty$, $V \to \infty$, but n/V and m/V remain constant. In this case the partition function (10.37) is evaluated explicitly by the contour integration when possible, or by employing saddle point integration. If, for instance, the $\alpha\alpha^+$ integration can be evaluated explicitly, the result is

$$\exp(-\beta A) = \oint d\mu \exp\{\Xi(n, \mu, T) + m\ln m - m - m\ln\mu\} \quad (10.41)$$

where Stirling's approximation is used for $\ln m!$. The saddle point μ_0 is obtained from

$$\left[\frac{\partial}{\partial\mu}\{\Xi + m\ln m - m - m\ln\mu\}\right]_{\mu_0} = 0 \quad (10.42a)$$

or

$$\left[\frac{\partial\Xi(n, \mu, T)}{\partial\ln\mu}\right]_{\mu_0} = m \quad (10.42b)$$

which establishes the fact that μ is related to the chemical potential which fixes the number of cross-links.

The result (10.37)–(10.39) provides the partition function for a set of interacting polymer chains with m cross-links, where the chain ends and the cross-links are distributed randomly throughout space. The cross-links are also distributed randomly along the cross-linked chains. This random distribution is the analog of the initial equilibrium distribution considered in Section IX. It is only an idealization of the actual process of cross-linkage. In real experiments, the cross-links are formed over a period of time. Although they may initially be formed with a random distribution, there may be some local changes as a result of these initially formed linkages which cause all subsequently formed cross-links to have a nonrandom distribution. This idealized cross-link preparation, however, leads to the expressions (10.37)–(10.39), which are already as complicated as the general many-body problem. Equation (10.39) is a quartic action (the interaction term is quartic in the ψ). The $\phi\psi^+\psi$ represents the interaction of the polymers with a gaussian random external field ϕ. From the discussions of the excluded volume problem it is clear that having this term is like having another quartic term in the action; the alternate form (10.28) of the cross-link constraint manifests this equivalence. Equation (10.39) differs from the usual quartic field theories by having the linear terms ψ^+ and ψ. (In the usual field theories, these terms would respectively represent sources and sinks, the chain beginnings and ends.) Equations (10.37)–(10.39) are, however, worse than the usual quartic field theories: In order to evaluate the contour integrals of (10.37) directly, or as in (10.42), it is necessary to obtain $\Xi(\alpha\alpha^+, \mu, T)$ as a suitable analytic function. Even in the case of no interchain interactions $[V(\mathbf{r} - \mathbf{r}') = 0]$, this is equivalent to obtaining the solution to a quartic field theory *as an analytic function of the coupling strength.* It is clear then that simple models are to be sought.

It is therefore obvious that (10.37)–(10.39) is, in general, intractable; moreover, *the internal constraints have been omitted.* If free energy differences are to be evaluated for changes in the thermodynamic state from ξ_0 to ξ, account must be taken of the fact that all final states are not accessible from a given initial state. If two chains initially have a chemical cross-link at s and s' when the volume is V_0, then when it is changed to V, these two chains must still have a cross-link at s and s'. Chemical cross-links are not broken and reformed upon deformation. Furthermore, upon deformation chains cannot pass through each other. This is one net effect of the intermolecular excluded volume: it leads to chain entanglements. Thus there are two types of constraint present in the system. One requires the cross-links to be fixed (*along the chains*). The other requires that for a given initial configuration of the network, only those final configurations are accessible which are *topologically equivalent* to the original configuration. The topological equivalence is to be considered in the strictest sense. Two network configurations are topologically equivalent if they can be deformed into each other without requiring that chains pass through each other, and without allowing the cross-linkages to move (along the chains).

The topological questions may in fact be of greater complexity than (10.37)–(10.39) itself. The need for simple confinement models as in Section IX is therefore manifest.

Ignoring the constraints, for the moment, (10.37)–(10.39) can give some information about the bulk polymer system in its initial "equilibrium" state (i.e., a random distribution of cross-links. The initial equation of state can be considered. By considering the Green's function for an added chain via

$$\exp(-\beta A)G(\mathbf{rr}'; ss') = \oint \frac{d\alpha\, d\alpha^{+} n!}{(\alpha\alpha^{+})^{n+1}} \oint \frac{d\mu}{\mu^{m+1}} \int \delta\phi \times \exp\left[-\frac{1}{2}\int \phi^{2}(\mathbf{r})\, d\mathbf{r}\right] \mathcal{N}(\phi) \int \psi(\mathbf{r}s)\psi^{+}(\mathbf{r}'s') \times \exp\left[-A(\alpha\alpha^{+}, \mu^{1/2}, T)\right] \delta\psi\, \delta\psi^{+} \tag{10.43}$$

the structural properties of the network may be examined. (These properties could, in principle, be investigated by X-ray scattering experiments.) In particular, kinetic theories cannot deal with the effects of correlations in the polymer network. These correlations result from the fact that the pattern of cross-linkages in general leads to closed circuits in the polymer network. Thus, because of circuits, by following the chains in a network, it is possible to arrive at an original starting point without traversing the same "chain segment" (the portion of a chain between the two cross-links) twice.

The effects of the simplest circuits upon the initial pressure and the polymer distribution $G(\mathbf{rr}'; ss')$ have been studied.[29] A complete discussion of these topics is beyond the scope of this review; however, a brief summary is of interest. (The following remarks are based on Ref. 29.) If the two points $\mathbf{r}s$ and $\mathbf{r}'s'$ are on the same "chain segment," then $G(\mathbf{rr}'; ss')$ should be similar to the distribution for the free chains $G_0(\mathbf{rr}'; ss')$. If these two points are linked by the network, their correlations are described by some $G_{\text{corr}}(\mathbf{rr}'; ss')$. If they are not connected via the network, $G \to \rho p_{\text{nc}}$, where ρ is the polymer density and p_{nc} is the probability that chains are not connected through the network. $G(\mathbf{rr}; ss')$ exhibits a singularity when m/n reaches a critical value, indicating that the system gels. For these simplest circuits, the pressure is a monotonically decreasing function of m/n, thereby reinforcing the analogy from (10.28) relating the cross-linkage constraint to an effective attractive force.

In order to consider the system in states other than the original one, it is necessary to introduce the constraints. The topological constraints may be introduced by the use of the model discussed in Section IX. This approach has been used to study the phenomenon of syneresis:[29] Beyond the gel point in a highly cross-linked system, the polymerized system is found to separate from the solvent. The treatment of the cross-links themselves as constraints enables a discussion of the portion of the elasticity which is due to the network alone and does not include the effects of entanglements. Such an investigation is presently underway.[62] It involves the generalization of (10.37)–(10.39) along the lines indicated in (9.48)–(9.51). In this case the integration variables ψ and ψ^+ must depend upon the initial $\mathbf{r}$ and final $\mathbf{R}$ positions of the polymer segment at a position s along a given chain.[29,62] The details follow in a manner similar to the development leading to (10.37)–(10.39), so they are not reproduced here. This section is only to be a brief introduction to the subject of the mathematical description of the statistical mechanics of systems of polymers in bulk as the subject is still in its infancy.

Acknowledgments

This review is based in part on lectures presented at the University of Chicago during the winter quarter of 1970. I would like to thank the students in the course for their numerous comments, verbal and otherwise, which have helped to improve the presentation.

Much of the material reviewed was either performed in collaboration with Professor S. F. Edwards or benefited from numerous discussions with him. I am grateful to Pat Gillis for many helpful comments on the original manuscript and for assistance with proofreading, etc.

Acknowledgment is made to the donors of the Petroleum Research Fund, administered by the American Chemical Society, and to a grant from the National Science Foundation for partial support of this research. I have also benefited from the use of facilities provided by Advance Research Projects Agency for materials research at the University of Chicago.

References

1. P. J. Flory, *Principles of Polymer Chemistry*, Cornell University Press, Ithaca, N.Y., 1953.
2. M. V. Volkenstein, *Configurational Statistics of Polymeric Chains*, Interscience, New York, 1963.
3. T. M. Birshtein and O. B. Ptitsyn, *Conformations of Macromolecules*, Interscience, New York, 1966.
4. P. J. Flory, *Statistical Mechanics of Chain Molecules*, Interscience, New York, 1969.,
5. H. Yamakawa, *Modern Theory of Polymer Solutions*, Harper and Row, New York, 1971.
6. S. A. Rice and P. Gray, *The Statistical Mechanics of Simple Liquids*, Interscience, New York, 1965.
7. R. M. Mazo, *Statistical Mechanical Theories of Transport Processes*, Pergamon, New York, 1967.
8. E. W. Montroll, *Comm. Pure Appl. Math.*, **5**, 415 (1952).
9. N. Saitô and M. Namiki, *Prog. Theoret. Phys.* (*Japan*), **16**, 71 (1956).

10. I. M. Gel'fand and A. M. Yaglom, *J. Math. Phys.*, **1,** 48 (1960).
11. S. G. Brush, *Rev. Mod. Phys.*, **33,** 79 (1961).
12. R. P. Feynman and A. R. Hibbs, *Quantum Mechanics and Path Integrals*, McGraw-Hill, New York, 1965.
13. S. F. Edwards, *Proc. Phys. Soc. (London)*, **85,** 613 (1965).
14. S. F. Edwards, *Natl. Bur. Std. (U.S.) Misc. Publ.*, **273,** 225 (1966).
15. K. F. Freed, *J. Chem. Phys.* (in press).
16. L. R. G. Treloar, *The Physics of Rubber Elasticity*, Oxford, 1958.
17. S. F. Edwards, *J. Phys. C.: Solid St. Phys.*, **3,** L30 (1970).
18. S. F. Edwards, *J. Non-Crystalline Solids*, **4,** 417 (1970).
19. E. N. Economou, M. H. Cohen, K. F. Freed, and E. S. Kirkpatrick, in *Amorphous and Liquid Semiconductors*, J. Tauc, Ed., Plenum Press (in press).
20. N. Saitô, K. Takahashi, Y. Yunoki, *J. Phys. Soc. (Japan)*, **22,** 219 (1967).
21. K. F. Freed, *J. Chem. Phys.*, **54,** 1453 (1971).
22. S. F. Edwards, *Disc. Faraday Soc.*, **49,** 43 (1970).
23. S. F. Edwards, *Proc. Phys. Soc. (London)*, **88,** 265 (1966).
24. S. F. Edwards, *Proc. Phys. Soc. (London)*, **91,** 513 (1967).
25. S. F. Edwards, *J. Phys.*, **A1,** 15 (1968).
26. S. F. Edwards and K. F. Freed, *J. Phys.*, **A2,** 145 (1969).
27. S. F. Edwards, *Proc. Phys. Soc. (London)*, **92,** 9 (1967).
28. S. F. Edwards, *J. Phys. C.: Solid St. Phys.*, **2,** 1 (1969).
29. S. F. Edwards and K. F. Freed, *J. Phys. C.: Solid St. Phys.*, **3,** 739, 750, 760 (1970).
30. P.-G. de Gennes, *J. Chem. Phys.*, **48,** 2257 (1968).
31. P.-G. de Gennes, *Rept. Progr. Phys.*, **32,** 187 (1969).
32. S. F. Edwards in *Analysis in Function Space*, W. T. Martin and I. Segal, eds., M.I.T. Press, Cambridge, Mass., 1964, pp. 31, 167.
33. S. Chandrasekhar, *Rev. Mod. Phys.*, **15,** 1 (1943).
34. G. Porod, *J. Polymer Sci.*, **10,** 157 (1953); S. Heine, O. Kratky, and P. J. Schmitz, *Makromol. Chem.*, **44-46,** 682 (1961); J. J. Herman and R. Ullman, *Physica*, **18,** 951 (1952).
35. K. Ito and H. P. McKean, *Diffusion Processes and Their Sample Paths*, Springer, Berlin, 1965; R. E. Mortensen, *J. Stat. Phys.*, **1,** 271 (1969); M. Schilder, *J. Stat. Phys.*, **1,** 475 (1970).
36. J. Popielawski, S. A. Rice and N. Hurt, *J. Chem. Phys.*, **46,** 3707 (1967); H. R. Wilson and S. A. Rice, *J. Chem. Phys.*, **49,** 1697 (1968).
37. R. A. Harris and J. E. Hearst, *J. Chem. Phys.*, **44,** 2594 (1966).
38. P. Bugl and S. Fujita, *J. Chem. Phys.*, **50,** 3137 (1969).
39. F. W. Weigel, *Physica*, **33,** 734 (1967).
40. R. Collins and A. Wragg, *J. Phys. A* (to be published).
41. P. Pechukas, *Phys. Rev.*, **166,** 174 (1969).
42. I. S. Gradshteyn and I. M. Ryzhik, *Tables of Integrals Series and Products*, Academic Press, New York, 1965.
43. C. Garrod, *Rev. Mod. Phys.*, **38,** 483 (1966).
44. S. G. Whittington, *J. Phys.*, **A3,** 28 (1970).
45. M. E. Fisher, *Rept. Progr. Phys.*, **30,** 615 (1967).
46. C. Domb, *Adv. Chem. Phys.*, **15,** 229 (1969).
47. H. Reiss, *J. Chem. Phys.*, **47,** 186 (1967).
48. R. Yeh and I. Isihara, *J. Chem. Phys.*, **51,** 1215 (1969).
49. K. F. Freed and H. P. Gillis, *Chem. Phys. Letters*, **8,** 384 (1971).

50. L. P. Kadanoff and G. Baym, *Quantum Statistical Mechanics*, Benjamin, New York, 1962.
51. J. G. Curro, P. J. Blatz, and C. J. Pings, *J. Chem. Phys.*, **50,** 2199 (1969).
52. Y. Chakahisa, *J. Chem. Phys.*, **52,** 206 (1970).
53. H. Goldstein, *Classical Mechanics*, Addison-Wesley, Reading, Mass., 1950.
54. Some recent work is G. R. Gobson and M. Gordon, *J. Chem. Phys.*, **43,** 705 (1965); M. Gordon, J. A. Love, and D. Pugh, *J. Chem. Phys.*, **49,** 4680 (1968); S. Imai and M. Gordon, *J. Chem. Phys.*, **50,** 3889 (1969).
55. N. N. Bogulubov and D. V. Shirkov, *Introduction to the Theory of Quantized Fields*, Interscience, New York, 1959.
56. H. C. Weber and H. P. Meissner, *Thermodynamics for Chemical Engineers*, Wiley, New York, 1957, Chap. 8.
57. E. T. Jaynes in *Statistical Physics*, Vol. III of *Brandeis Summer Institute Lectures in Theoretical Physics*, Benjamin, New York, 1963, p. 181.
58. S. F. Edwards and K. F. Freed, *J. Phys.*, **C3,** L31 (1970).
59. E. Merzbacher, *Quantum Mechanics*, 2 ed., Wiley, New York, 1970, Chapter 20.
60. A. S. Davydov, *Quantum Mechanics*, Addison-Wesley, Reading, Mass., 1965, Sec. 138.
61. A. J. F. Siegert in *Statistical Physics*, Vol. III of *Brandeis Summer Institute Lectures in Theoretical Physics*, Benjamin, New York, 1963, p. 159; A. J. F. Siegert and D. J. Vezzetti, *J. Math. Phys.*, **9,** 2173 (1968).
62. K. F. Freed, "Statistical Mechanics of Systems with Internal Constraints: Rubber Elasticity," *J. Chem. Phys.* (to be published).

EQUILIBRIUM DENATURATION OF NATURAL AND OF PERIODIC SYNTHETIC DNA MOLECULES

ROGER M. WARTELL* AND ELLIOTT W. MONTROLL

Department of Physics and Chemistry, University of Rochester, Rochester, New York

TABLE OF CONTENTS

* Present address: Department of Biochemistry, University of Wisconsin, Madison, Wisconsin.

I. INTRODUCTION

A. General Background

Since the recognition by Thomas[1] and Rice and Doty[6] that heating a solution of DNA may result in the dissociation of the two-stranded helix, many investigators have studied the helix-coil transformation of naturally occurring and synthetic DNAs. These studies have been undertaken from both the physiochemical and biological viewpoints. The thermally induced transition has been employed to gain information on the molecular bonding and structure of DNA in solution and as a simple means of characterizing the base pair content of natural DNA. The phenomenon has been known by a number of names, among them are denaturation, melting, unwinding, and unzipping.

In this chapter we examine the equilibrium form of the melting transition of DNA. Previous experimental investigations are briefly reviewed, and a comparison is made between a number of theoretical models and calculations. In addition to presenting some new theoretical calculations, we compare several theories with some experimental data.

Over the past decade, substantial literature has accumulated on the helix-coil transition of DNA. Experimental investigations on the transition have been conducted under a variety of solvent conditions. Extremes in *p*H, ion concentration changes, nonpolar solvents all have been examined, and all have a marked effect on the melting transition of a given DNA. These experiments have been useful in delineating the relative importance of different types of bonding (hydrogen bonding, stacking interactions, etc.) on the stability of DNA in solution. Numerous reviews have summarized experimental work in the field.[2–5]

Theoretical investigations of DNA denaturation have attempted to explain the helix-coil transition through molecular models so that both a qualitative and quantitative understanding of the process can be derived. Current studies have been concerned with the unwinding mechanism of DNA,[6,7,13] the thermodynamic energies involved in stabilizing the helix,[8,9] and base distribution information of DNA.[10–14] In recent years several review articles on the theory of denaturation have been published,[10,14,15] and a book[16] has now appeared.

The internal stability of the DNA double helix appears to be due partly to hydrogen bonding between members of the base pairs in the two strands. Each base pair bond, or bonding complex, includes solvent effects, the hydrogen bonds between the complimentary bases (adenine-thymine and guanine-cytosine) and the stabilizing forces resulting from the interaction of adjacent base pairs. The latter "stacking forces" not

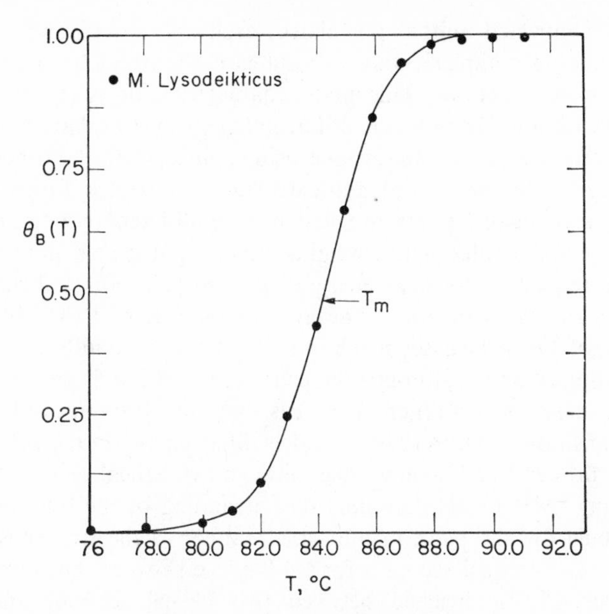

Fig. 1. Helix coil transition of *M.* lysodeikticus DNA in 0.1 SSC (0.015 NaCl + 0.0015*M* sodium citrate *p*H = 7.0).

only confer stability to the base pair bonds but also correlate adjacent base pairs. This correlation, along with other long-range forces, account for the cooperative nature of the helix-to-coil transition.

Experimentally, the DNA denaturation usually is observed by measuring the ultraviolet absorbance of a dilute solution of DNA as a function of temperature. A curve of the fraction of broken bonds versus temperature is commonly obtained by monitoring optical absorbance at 260 millimicrons. An increase in the absorbance of light reflects a change in the electron configuration of the bases due to the breaking of base pair bonds. Assuming the change in absorbance to be proportional to the fraction of broken bonds, we can calculate a melting curve. This curve is often characterized by two parameters: the temperature at which 50% of the bonds are broken, the melting temperature T_m, and the slope at the melting temperature (see Fig. 1). Viscosity measurements of DNA denaturation have been used to provide an estimate of the length of helical segments during unwinding.[10]

B. Theoretical Models

All the theoretical investigations on the equilibrium melting of the DNA-type polymer employ the same basic ideas. A one-dimensional system of N connected units is postulated, each unit corresponding to a base pair bonding complex. Only two states exist for each complex, a broken one or an intact one. This model characterization is equivalent to that in the well-known Ising model of ferromagnetism.[17] In that model the two states are electron spin states, one being spin up and the other spin down at each lattice site. In order to calculate the partition function for the system, statistical weights are assigned to different configurations according to a set of rules. These weights are of a parametric nature, but they can be related to the free energies of base pair bonds. From the partition function we can obtain the average number of bonds broken, average size of the helical segments, etc. A "broken bond" refers to a broken bonding complex, although there are several H bonds per complex.

Basically, there are two different models that have been applied to the helix-coil transition of DNA. These models differ mainly in the treatment of a loop of broken bonds sandwiched between two helical segments. The modified Ising (MI) model considers the unwinding of the interior of a DNA molecule to take place in the same manner as the free ends. The loop entropy (LE) model accounts for a difference between the entropy of unbonded strands sandwiched between two helical sections and the entropy of unbonded strands at the end of the molecule.

In the MI model we assign a statistical weight, or internal partition function to the ith base pair, $\exp(-L_i)$ when it is intact and $\exp(+L_i)$ when it is broken. L_i may depend on the chemical nature of the ith base pair and it may also depend on the types of base pairs adjacent to the ith base pair. The correlation between base pair bonds is expressed as the statistical weight $\exp(U\sigma_i\sigma_{i+1})$ where $\sigma_i = +1$ corresponds to an intact and $\sigma_i = -1$ to a broken bond. U may be interpreted as an average over all base pair interactions rather than as a strictly nearest neighbor interaction. In addition to the statistical weights, the LE model assigns the factor $f(m)$ for a segment of m broken bonds sandwiched between helical segments. The form of $f(m)$ is discussed in a later section.

An alternative notation often used for the statistical weights is that of Zimm and Bragg.[18,19] A statistical weight s_i is assigned to the ith bond if it is intact next to a helical segment, and 1 is assigned if it is broken. The degree of correlation between successive base pairs is specified by a factor σ assigned to each helical segment.

C. Naturally Occurring DNA Molecules

The work of Marmur and Doty[20] has indicated the guanine-cytosine (G–C) pair to be thermally more stable than the adenine-thymine (A–T) base pair. They also demonstrated, to first order, the usefulness of melting curves in obtaining base distribution information. Their observation

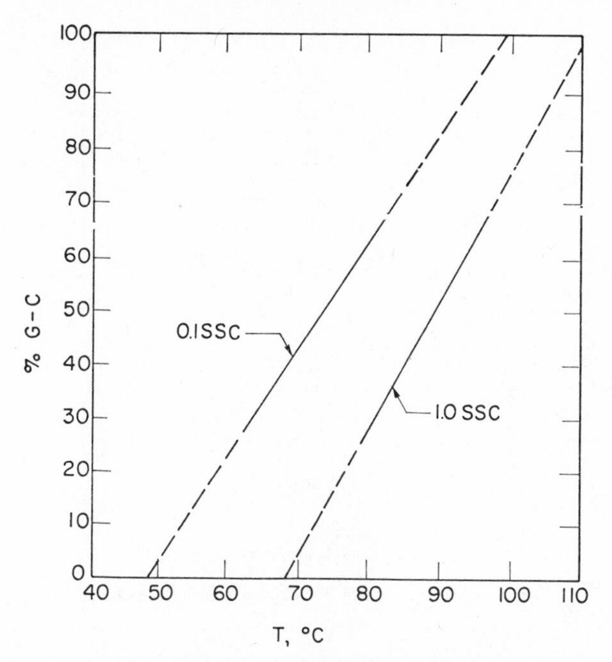

Fig. 2. Dependence of melting temperature, T_m, on %GC, from Marmur and Doty[20] (1.0 SSC) and Mandel[26] et al. (0.1 SSC).

that the melting (50% of bond broken) temperature increases linearly with the fraction of G–C pairs, as shown in Fig. 2, is extremely useful for assaying the composition of naturally occurring DNA from new sources.

Although there are theoretical models which can be applied to the melting of natural DNAs, some difficulty remains in comparing them to experimental melting curves. Since the models are phenomological, certain of the parameters must be evaluated from the experimental curves. With the base sequence of naturally occurring DNAs largely unknown, it is difficult to decipher the effects of base sequence on the melting curves from those of the model or parameters.

The influence of the details of the base sequences on the transition curves has been discussed by several authors.[10–12,14] It has been pointed out that a DNA with long runs of A–T or G–C base pairs will have a broad melting curve when compared to another DNA with the same composition but without long runs of A–T or G–C pairs. The effect can be understood by considering two extreme sequences of chains composed of half A–T and half G–C pairs. All the A–T pairs of sequence **a** are at

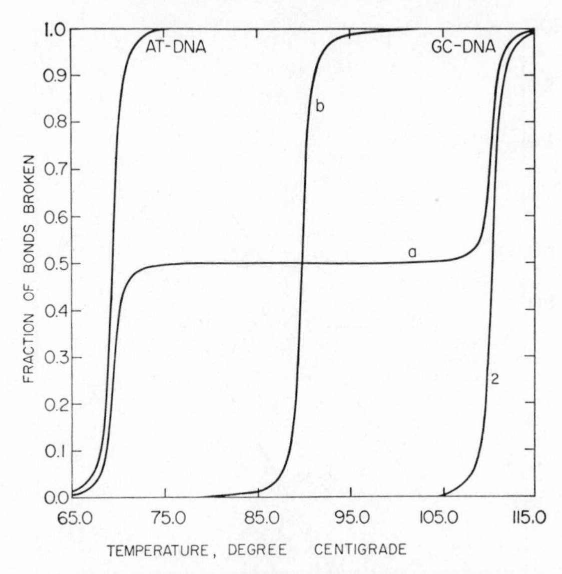

Fig. 3. Melting curves of various DNA molecules.

one end of the chain while all the G–C's are at the other end. Sequence **b**

A A A A G G G G	A G A G
| | | ··· | | | ··· | |	| | | | ···
T T T T C C C C	T C T C
(a)	**(b)**

is an alternating one. In certain media pure AT DNA denatures near 65°C, whereas pure GC DNA does so above 100°C, as indicated schematically in curves 1 and 2 of Fig. 3. The left half block of sequence **a** would denature along curve 1 until practically all AT bonds were broken but little dissociation would penetrate into the GC block until the temperature approached the 100°C level. The melting curve would contain the long, flat region sketched in curve *a* of Fig. 3. As the AT pairs in sequence **a** attempt to break in a medium of increasing temperature, they remain clamped by their intact GC neighbors until the neighbors are overwhelmed by the ATs at a temperature much higher than 65°C. In view of the symmetry of molecule **b**, all GC bonds could be expected to break together in a narrow temperature range yielding the sharp melting curve **b** of Figure 3.

The behavior described for two DNAs with the same A–T/G–C composition but unlike sequences can be observed in a less dramatic fashion in naturally occurring DNAs. Falkow and Cowie[21] have shown bacteriophage λ DNA and E. coli DNA to have measurably different melting curves even though their G–C/A–T ratios are similar. Another pair of DNAs with similar average compositions and rather different transition curves are Bacillus subtilis and Calf thymus DNAs.

D. Synthetic DNA Molecules

In his elegant investigations of the nature of the genetic code, Khorana prepared synthetic DNA molecules [such as poly *d*(A)·poly *d*(T), poly *d*(AT)·poly *d*(AT), etc.] whose base pairs were arranged periodically. This was done to see which DNA triples of bases coded for various amino acids in proteins. When DNA denaturation is studied in such periodic molecules the sequence is known so that in principle the melting curves can be used to determine the parameters used in the various theories. Such molecules are also more appropriate than naturally occurring DNA to check various models.

Studies by Wells and his collaborators[22] and by Langridge[24] indicate that not all of the synthetic DNAs have the same structure. X-ray diffraction has indicated that poly *d*(A–T)·poly *d*(A–T), and poly *d*(A–C)·poly *d*(G–T) form fibrous structures of the Watson Crick type. Poly *d*(A)·poly *d*(T) forms a structure different from but closely resembling the Watson Crick structure. However, poly *d*(G)·poly *d*(C), poly *d*(A–G)·poly *d*(C–T), and poly *d*(I–C)·poly *d*(I–C) appear to be quite unlike the Watson Crick structure. Other physiochemical studies substantiate the X-ray diffraction results.

In addition to the possibility of structural differences between periodic DNAs and natural DNAs, there are other features to be considered. In the polymers poly *d*(A)·poly *d*(T) and poly *d*(A–T)·poly *d*(A–T), the possibility of any A on one strand to bond with any T on the other would permit the two strands to "slide" when loops of broken bonds form as depicted in Fig. 4*b*. The same process can occur with all periodic DNAs, the likelihood of occurrence decreasing with period length. For poly *d*(A–T)·poly *d*(A–T) single-stranded hairpin helices are another type of configuration unlikely in natural DNAs (see Fig. 4*a*). To assess the importance of sliding and hairpin helices on the helix-coil transition of periodic DNAs, we must examine pertinent experiments. We might expect single-stranded branched helices to be an important factor, since entropy considerations would favor such a formation over double-stranded helices. The misalignment or sliding of bases is more difficult to consider. It will depend on the amount of flexibility permitted loops of broken bonds in solution. Qualitatively, this sliding factor would be unimportant for small loops. Furthermore, short-range sliding of bases within a large loop would be expected to dominate over any long-distance sliding of strands.

Baldwin and his collaborators[25] have made a detailed study of poly *d*(A–T)·poly *d*(A–T). Viscosity and optical absorbance were used to

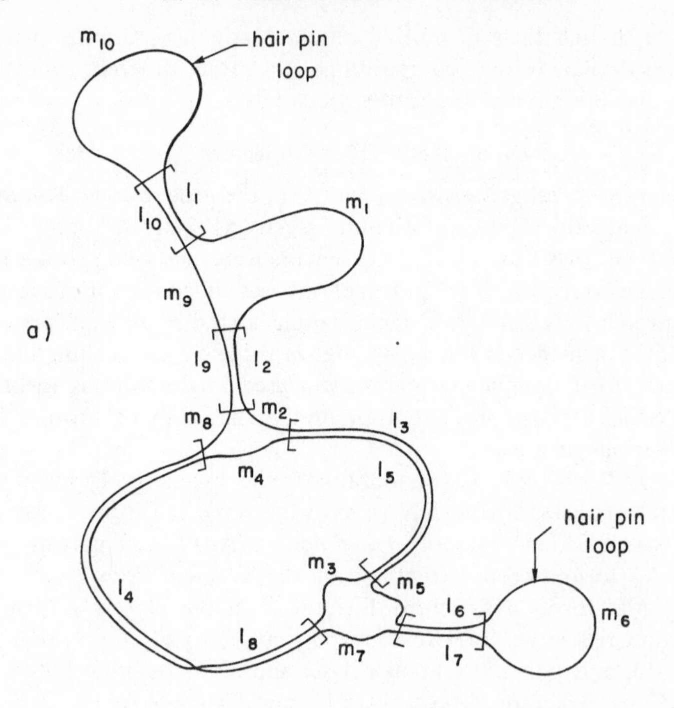

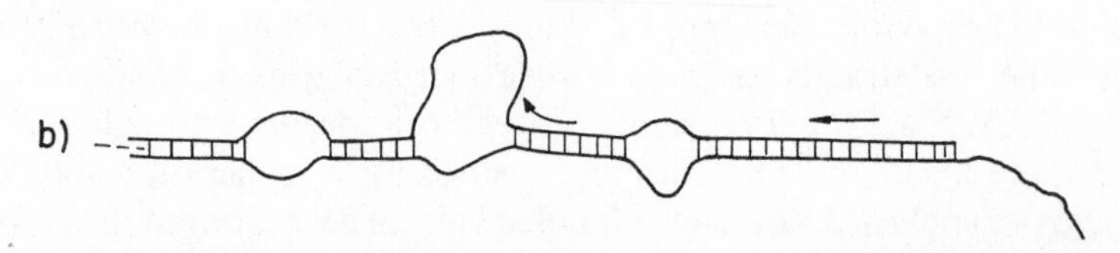

Fig. 4. (*a*) A possible configuration for a circular DNA chain with hairpin helix branching. (*b*) Sketch of strand sliding effect on loops of broken bonds.

monitor the equilibrium melting of the molecule, and optical absorbance was employed in a kinetic study of the melting transition. The results of their investigations indicate that hairpin helices form just before and during the helix-coil transition. The extent of hairpin branching was largest for high salt concentrations and decreased as the salt concentration decreased. These results suggest that hairpin helix formation should be considered in theoretical calculations of polymers like poly *d*(A–T)·poly *d*(A–T). The importance of strand sliding on the melting behavior of periodic DNAs is difficult to assess from experiments on these polymers.

Strand sliding will help to broaden the melting transition of a periodic DNA, but so will several other factors. By including the strand sliding effect in theoretical calculations and comparing them with experiments, its importance might be determined.

There have been a number of other periodic DNAs synthesized by Khorana, Wells, and co-workers in addition to those mentioned. Tri and tetra periodic polymers have been made, and recently poly *d*(G–C)·poly *d*(G–C) has been synthesized. An important question regarding all of the periodic DNAs is whether they can be regarded as reliable models of natural DNA. This is especially true for those periodic DNAs whose X-ray diffraction patterns differ significantly from that of natural DNA. An examination of denaturation studies on some of the polymers, in conjunction with X-ray diffraction results, reveals certain correlations.

The T_m versus %G–C relationship for natural DNAs is a linear one, and it can be extrapolated to 0% and 100% G–C. In the solvent 0.1 SSC (0.015*M* NaCl + 0.0015*M* Na citrate) the extrapolation to a DNA with only A–T base pairs gives a T_m of 49.7°C[26] as shown in Fig. 2. This T_m falls almost precisely between the T_m of poly *d*(A)·poly *d*(T) and that of poly *d*(A–T)·poly *d*(A–T) in the same solvent. Since natural DNAs contain approximately the same amounts of $\overset{A}{\underset{T}{/}}\;\overset{T}{\underset{A}{/}}$ base pair dimers as $\overset{A}{\underset{T}{/}}\;\overset{A}{\underset{T}{/}}$ dimers,[27] this result indicates that the three-dimensional structures of poly *d*(A–T)·poly *d*(A–T) and poly *d*(A)·poly *d*(T) are found in natural DNAs. X-ray diffraction tends to support this conclusion.

Of the two polymers poly *d*(A–C)·poly *d*(G–T) and poly *d*(A–G). poly *d*(C–T), only the first has a T_m value in close agreement to a natural DNA with 50% G–C base pairs. Poly *d*(A–G)·poly *d*(C–T) has a T_m value approximately 6°C less than a natural DNA with 50% G–C base pairs. The stacked base pairs involved in these two synthetic DNAs occur in roughly equal amounts in natural DNAs. Hence, if both polymers were like natural DNA in structure, we would expect their T_ms to be either equidistant from the 50% G–C T_m value or equal to it. An interpretation of the experimental results found is that poly *d*(A–C)·poly *d*(G–T) has a structure similar to natural DNAs, whereas poly *d*(A–G)·poly *d*(G–T) does not. X-ray diffraction results agree with this interpretation. Extrapolating the %G–C versus T_m relation to 100% G–C gives a T_m in 0.1 SSC close to 99.5°C. From the investigation by Wells et al.[22] we can estimate the T_ms of poly *d*(G)·poly *d*(C) and poly *d*(G–C)·poly *d*(G–C) in 0.1 SSC by examining the T_m versus log [Na^+] relations for these polymers. Estimations from the plotted data give a $T_m = 87$°C for poly *d*(G)·poly *d*(C) and $T_m = 98$°C for poly *d*(G–C)·poly *d*(G–C). The diffraction pattern of poly d(G)·poly d(C) is unlike the Watson Crick structure of natural DNA. The diffraction pattern of poly d(G–C)·poly d(G–C) has yet to be obtained. From the T_m value of poly d(G–C)·poly d(G–C), we would expect this polymer to have a Watson Crick structure. The correlation between the denaturation and X-ray diffraction results imply that poly d(G)·poly d(C) and poly d(A–G)·poly d(C–T) have three-dimensional structures not normally found in natural DNAs.

II. MODIFIED ISING MODEL

The modified Ising (MI) model has been applied to the DNA transition by several authors.[11,12,28–30] Although the underlying model employed by these authors has been the same, the methods and approximations made have been different. The notation used in these calculations has also varied, and we will endeavor to show their connection where appropriate. Although the MI model omits the loop entropy factor, its simplicity allows an easy comparison with experiment and gives a basis for further improvements. We first obtain the melting curves of periodic DNAs and then present results for DNAs with a random distribution of A–T and G–C base pairs.

In the formulation of Montroll and Goel I and II,[11,12] a molecule composed of N base pairs in a defined bonding sequence is characterized by the set of parameters $(\sigma_1, \sigma_2, \sigma_3, \ldots, \sigma_N)$ where

$$\sigma_j = \begin{cases} +1 \text{ if the } j\text{th base pair bond is intact} \\ -1 \text{ if the } j\text{th base pair bond is broken} \end{cases}$$

We are seeking the average fraction of base pairs which are broken as a function of temperature, $\theta_B(T)$. For N_{broken}, the number of broken base pair bonds

$$\theta_B(T) = \frac{\langle N_{\text{broken}} \rangle}{N} = N^{-1} \sum_{j=1}^{N} \tfrac{1}{2} \langle (1 - \sigma_j) \rangle \tag{2.1a}$$

where $\langle\ \rangle$ represents the average over all sequence of bonds $(\sigma_1, \sigma_2, \ldots, \sigma_N)$. The numbers $\frac{1}{2}(1 - \sigma)$ are counters of broken bonds since they have the value 0 for intact bonds and 1 for broken ones.

The choice of a model is of course the definition of the weight function

$$F(\sigma_1, \sigma_2, \ldots, \sigma_N)$$

to be used in the averaging process. The simplest form for F is one in which all σ are independent so that

$$F(\sigma_1, \ldots, \sigma_N) = P(\sigma_1)P(\sigma_2) \cdots P(\sigma_N)$$

Each $P(\sigma)$ would have to be normalized so that

$$P(1) + P(-1) = 1$$

A form for $P(\sigma)$ with these properties is the one parameter function of σ,

$$P(\sigma) = \tfrac{1}{2}(\operatorname{sech} J)e^{-\sigma J}$$

The dependence of J on the temperature would have to be such that $\theta_B(T) = \frac{1}{2}$ at the melting temperature T_m. In view of the fact that $\sigma^2 = 1$, this form is equivalent to

$$P(\sigma) = A + B\sigma$$

and is the most general choice that could be made. The average value of σ, $\langle\sigma\rangle$ with this choice is

$$\langle\sigma\rangle = -\tanh J$$

We could assign two values to J, J_1 for an A–T pair and J_2 for a G–C pair. Then for a DNA with n_1N A–T pairs and n_2N G–C pairs,

$$\theta_B(J_1, J_2) = \tfrac{1}{2}\{1 + n_1 \tanh J_1 + n_2 \tanh J_2\} \tag{2.1b}$$

In pure A–T or pure G–C DNA,

$$\theta_B(J_i) = \tfrac{1}{2}(1 + \tanh J_i)$$

If we set $J = a(T - T_M)$, this formula yields melting curves qualitatively similar to the experimental ones. Its great fault is that it omits all correlations between neighbors, so that cooperative effects which depend on base sequences as indicated in Fig. 1 could not be explained.

The next most complicated form that might be chosen for the weight function F is one that would contain correlations between nearest neighbor bond pairs:

$$F(\sigma_1, \ldots, \sigma_N) = Z^{-1}f_1(\sigma_1, \sigma_2)f_2(\sigma_2, \sigma_3)\cdots f_{N-1}(\sigma_{N-1}, \sigma_N)f_N(\sigma_N, \sigma_1)$$

Z^{-1} is a normalizing factor with Z itself called the partition function. This corresponds to a molecule with ends closed to form a ring of base pairs. In an open-chain structure $f_N(\sigma_N, \sigma_1) = 1$.

The most complicated form we could choose for $f(\sigma, \sigma')$ since $\sigma^2 = \sigma'^2 = 1$ (with A, U, B_1, and B_2 parameters that might depend on temperature) is

$$f(\sigma, \sigma') = A \exp(U\sigma\sigma' + B_1\sigma + B_2\sigma')$$

which is equivalent to

$$a + b\sigma + c\sigma' + d\sigma\sigma'$$

This form of $f(\sigma, \sigma')$ is that used in the Ising model of cooperative phenomenon. Other more complicated forms of F could be postulated, say one in value $f(\sigma_i, \sigma_{i+1}; \sigma_{i+2})$. These would require more parameters and would not be warranted by the present state of experimental data. In later sections we discuss models in which longer range effects are introduced in regions in which sequence of broken bonds exists, i.e., entropy factors for the various configurations such regions might have.

When we recognize the four possible base pairs in synthetic DNA analogs, A–T, T–A, G–C, C–G, the most general form for $f(\sigma_j\sigma_{j+1})$ can be written in a Maxwell-Boltzmann manner:

$$f(\sigma_j, \sigma_{j+1}) = \exp\{U_{j,j+1}\sigma_j\sigma_{j+1} - \tfrac{1}{2}K_{j,j+1}(\sigma_j + \sigma_{j+1}) - \tfrac{1}{2}(J_j\sigma_j + J_{j+1}\sigma_{j+1})\} \tag{2.2}$$

Although we could write U, J, and K in terms of free energies, we will, for now, give them a parametric interpretation. $U_{j,j+1}$ represents the correlation between the jth and $(j+1)$st base pairs, $K_{j,j+1}$ the contribution to both jth and $(j+1)$st base pair bonds due to the stacking interactions, and J_j the contributions to the jth bond independent of stacking interactions. It is assumed $J_{(\mathrm{A-T})} = J_{(\mathrm{T-A})}$ and $J_{(\mathrm{G-C})} = J_{(\mathrm{C-G})}$. If we label the four base pairs A–T, T–A, G–C, C–G by the numbers 1, 2, 3, 4, respectively, then $J_j = J_1$ if the jth bond is an A–T, $U_{j,j+1} = U_{12}$ if the jth bond is A–T and the $(j+1)$st T–A, $K_{j,j+1} = K_{14}$ if the jth bond is A–T and the $(j+1)$st C–G, etc. When the subscripts used on U, K are separated by a comma, they (the subscripts) represent the counting number of the sequence. When there is no comma the parameter is associated with the pair of bonds of the type designated by the numbers.

As an introduction to the method of calculating the melting curves for various periodic DNA molecules let us first consider the case of a molecule composed of a sequence of identical base pairs, say poly d(A)·poly d(T) with no complication such as slippage or hairpin loops. The base pair strengths in this case are all the same and are defined as

$$L = J_j + \tfrac{1}{2}(K_{j-1,j} + K_{j,j+1}) = J_1 + K_{11}$$

The quantity to be obtained (postulating closed molecules without ends) is

$$\theta_B(T) = N^{-1}Z^{-1} \sum_{\{\sigma\}=\pm 1} \sum_{k=1}^{N} \tfrac{1}{2}(1 - \sigma_k) \prod_{j=1}^{N} \exp\{U\sigma_j\sigma_{j+1} - \tfrac{1}{2}L(\sigma_j + \sigma_{j+1})\} \tag{2.3a}$$

where the normalizing quantity Z^{-1} is the inverse of the partition function

$$Z = \sum_{\{\sigma\}=\pm 1} \prod_{j=1}^{N} \exp\{U\sigma_j\sigma_{j+1} - \tfrac{1}{2}L(\sigma_j + \sigma_{j+1})\} \tag{2.3b}$$

The summation over $\{\sigma\} = \pm 1$ means over all possible values of the sequence $\sigma_1, \sigma_2 \ldots, \sigma_N$. Since

$$\frac{\partial \log Z}{\partial L} = -\langle \sum \sigma_j \rangle, \qquad \theta_B(T) = \frac{1}{2}\left\{1 + N^{-1}\frac{\partial \log Z}{\partial L}\right\} \tag{2.3c}$$

It is easy to see that

$$Z = \text{trace } F^N = \lambda_1{}^N + \lambda_2{}^N \tag{2.3d}$$

where F is the matrix whose elements are

$$f(\sigma, \sigma') = \exp\{U\sigma\sigma' - \tfrac{1}{2}L(\sigma + \sigma')\}$$

$$F = \begin{pmatrix} e^{U-L} & e^{-U} \\ e^{-U} & e^{U+L} \end{pmatrix} \tag{2.3e}$$

and λ_1 and λ_2 are the two characteristic values of F. Let λ_1 be the larger of the two characteristic values. Then as $N \to \infty$,

$$Z \sim \lambda_1{}^N \qquad \text{or} \quad \log Z = N \log \lambda_1$$

so that as $N \to \infty$

$$\theta_B(T) = \frac{1}{2}\left\{1 + \frac{\partial \log \lambda_1}{\partial L}\right\} \tag{2.4a}$$

Since

$$\lambda_1 = e^U\{\cosh L + (\sinh^2 L + e^{-4U})^{1/2}\} \tag{2.4b}$$

$$\theta_B(T) = \frac{1}{2}\left(1 + \frac{s}{[s^2 + e^{-4U}]^{1/2}}\right)$$

$$\to \begin{cases} \tfrac{1}{2}(1 + s/|s|) & \text{as} \quad U \to \infty \\ \tfrac{1}{2}(1 + \tanh L) & \text{as} \quad U \to 0 \end{cases} \tag{2.4c}$$

where

$$s = \sinh L \qquad \text{and} \quad c = \cosh L$$

The proper choice of the temperature dependence of U and L will be considered later, but let us now postulate that U is temperature independent and that with a constant a

$$L = a(T - T_m) \tag{2.4d}$$

At the melting temperature T_m, the slope of the melting curve is

$$\frac{\partial \theta_B}{\partial T} = \tfrac{1}{4}a \exp 2U \tag{2.4e}$$

which means that the slope of the melting curve increases with the strength of the neighbor correlation parameter U.

Detailed calculations using these ideas have been made for a number of periodic DNA molecules in MGII. Some of these will be sketched in a later section of this chapter.

There is an alternative way of deriving (2.3d) which we present here because it is a more natural way to analyze loop entropy models and the "hairpinning" effect. Let us consider a DNA molecule composed of a respected sequence of identical base pairs and in a bonding state such that there is a set of runs of l_1 successive bound base pairs, followed by m_1 successive broken bond base pairs, l_2 bound, m_2 broken, . . . , l_n bound and m_n broken as exhibited in Fig. 5. We can assign a weight to this sequence,

$$g(l_1)f(m_1)g(l_2)f(m_2)\cdots g(l_n)f(m_n) \tag{2.5a}$$

where $g(l)$ is the weight of a sequence of l successive bonded base pairs and $f(m)$ that of a sequence of m unbonded pairs. This sequence corresponds to the boundary condition that the left end bond (on base pair 1) always remains intact, whereas that at the right end (base pair N) always is broken. Other boundary conditions are discussed in Appendix A. If we are interested in very long molecules, the boundary conditions at the ends of the chain are unimportant. However, an important property of the integers l_i and m_i is that the sum over all base pairs, those intact and broken, must be the total number of pairs N:

$$\sum_{j=1}^{n}(m_j + l_j) = N, \qquad 0 < n < \left[\frac{N}{2}\right] \tag{2.5b}$$

$[N/2]$ being the largest integer less than or equal to $\frac{1}{2}N$. The partition function is then the sum over all products of the form (2.5a) which satisfy the restrictions (2.5b),

$$Z_N = \sum_n \sum_{\{l_j\}} \sum_{\{m_j\}} \prod_{j=1}^{n} g(l_j)f(m_j) \tag{2.5c}$$

The supplementary condition (2.5b) can be neglected at the expense of

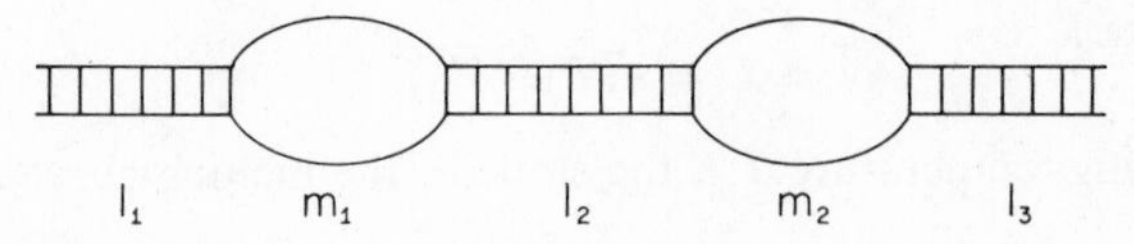

Fig. 5. Representation of l_1 intact, m_1 broken, l_2 intact, m_2 broken, . . . , for a DNA configuration.

introducing a delta function into the summation:

$$Z_N = \sum_{n}^{[N/2]} \sum_{l=1}^{\infty} \sum_{m=1}^{\infty} \delta\left(N - \sum_{j=1}^{n} [l_j + m_j]\right)$$

$$\times \exp\left\{\beta\left(N - \sum_{j=1}^{n} [l_j + m_j]\right)\right\} \prod_{k=1}^{n} g(l_k) f(m_k) \quad (2.6)$$

where we interpret

$$\sum_{l=1}^{\infty} \equiv \sum_{l=1_1}^{\infty} \sum_{l_2=1}^{\infty} \cdots \sum_{l_n=1}^{\infty}$$

and similary for $\sum_{m=1}^{\infty}$. The parameter β is to be chosen so that certain integrals which are to be calculated later converge. Of course the exponential term has the value 1 because only those partitions of N for which the multiplier of β vanish are allowed by the delta function. The restrictions have been taken off the summations over l_j and m_j.

The summations themselves are most easily carried out by using the Fourier integral representation of the delta function so that

$$Z_N = \frac{1}{2\pi} \int_{-\infty}^{\infty} d\alpha e^{N(\beta + i\alpha)} \sum_{n=1}^{[N/2]} \prod_{j=1}^{n} \{\Phi_j \psi_j\} \quad (2.7a)$$

where

$$\Phi_j = \sum_{l_j=1}^{\infty} g(l_j) \exp\{-(\beta + i\alpha) l_j\} \equiv \Phi(\beta + i\alpha) \quad (2.7b)$$

$$\psi_j = \sum_{m_j=1}^{\infty} f(m_j) \exp\{-(\beta + i\alpha) m_j\} \equiv \psi(\beta + i\alpha) \quad (2.7c)$$

It should be noted that since β is arbitrary, a value can be chosen for it so that Φ and ψ converge and $|\Phi\psi| < 1$. Since all Φ_j and ψ_j are identical and since the summation over n can be extended to ∞ when N is very large, we finally obtain

$$Z_N = \frac{1}{2\pi} \int_{-\infty}^{\infty} d\alpha e^{N(\beta + i\alpha)} \sum_{n=1}^{\infty} (\Phi\psi)^n \quad (2.8a)$$

$$= \frac{1}{2\pi} \int_{-\infty}^{\infty} d\alpha e^{N(\beta + i\alpha)} \{(\Phi\psi)^{-1} - 1\}^{-1} \quad (2.8b)$$

It was shown in MGII that when the free end boundary condition is chosen (each end of the chain can be either bonded or unbonded), then

$$Z_N^{(\mathrm{F.E})} = \frac{1}{2\pi} \int_{-\infty}^{\infty} d\alpha e^{N(\beta + i\alpha)} \{(\Phi\psi)^{-1} - 1\}^{-1} \{2 + \Phi^{-1} + \psi^{-1}\} \quad (2.9)$$

As we shall see, when N is large the difference in end or boundary conditions makes no difference in $\log Z_N$. If the integrand has poles the integral can be calculated as the sum of all residues of the poles.

To proceed further we must postulate some form for the weights $g(l)$ and $f(m)$. At this point we accept the MI form for these; later, when considering the loop entropy models, other forms will be analyzed. In the nearest neighbor MI case we set

$$g(l) = \exp\{(U - L)l - 2U\} \tag{2.10a}$$

and

$$f(m) = \exp\{(U + L)m - 2U\} \tag{2.10b}$$

As can be seen from (2.3b), if a patch of l bonds are intact there are $(l - 1)U$. Since the connection with the first neighbor to the left and to the right of the bond patch contribute $-U$, we find the term $(l - 2)U$ in the exponent of $g(l)$, and similarly for $f(m)$. Incidentally, we can write $f(m)$ as the square of

$$k(m) = \exp\{\tfrac{1}{2}(U + L)m - U\} \tag{2.11}$$

where $k(m)$ is to be interpreted as the weight factor for a *single* strand of length m of the double-stranded unbound segment.

We note from (2.7a) and (2.7b) that

$$\Phi(\beta + i\alpha) = \frac{\exp[-(U + L + \beta + i\alpha)]}{1 - \exp(U - L - \beta - i\alpha)} = \frac{\exp - (U + L)}{\lambda - \exp(U - L)} \tag{2.12a}$$

$$\psi(\beta + i\alpha) = \frac{\exp - (U - L)}{\lambda - \exp(U + L)} \tag{2.12b}$$

where we have chosen β such that $(U + |L| - \beta) < 0$ and defined

$$\lambda \equiv \exp(\beta + i\alpha) \tag{2.13}$$

The poles of the integrand of (2.8b) and (2.9) correspond to the value of λ for which

$$(\Phi\psi)^{-1} - 1 = 0 \tag{2.14a}$$

or

$$\lambda^2 - 2\lambda e^U \cosh L + 2 \sinh 2U = 0 \tag{2.14b}$$

which is just the characteristic equation obtained from F in (2.3e).

We find [in accordance with (2.4b)] that

$$\lambda_\pm = e^U\{\cosh L \pm (\sinh^2 L + e^{-4U})^{1/2}\}$$

so that the poles are at

$$\alpha_\pm = i(\beta - \log \lambda_\pm)$$

on the upper positive imaginary axis. It is more convenient to calculate the partition function (2.8b) as a contour integral in the variable λ. Then (2.8b) becomes

$$Z_N = \frac{e^{-2U}}{(\lambda_+ - \lambda_-)} \frac{1}{2\pi i} \oint_C \lambda^{N-1} \left\{ \frac{1}{\lambda - \lambda_+} - \frac{1}{\lambda - \lambda_-} \right\} d\lambda \qquad (2.15a)$$

where the contour C encircles over the two poles λ_+ and λ_- in the counter-clockwise direction. Hence, since λ_+ [$\equiv \lambda_1$ of (2.4b)] $> \lambda_-$, as $N \to \infty$,

$$Z_N = e^{-2U}(\lambda_+ - \lambda_-)^{-1}[\lambda_+^{N-1} - \lambda_-^{N-1}] \sim \left[\frac{e^{-2U}}{(\lambda_+ - \lambda_-)} \right] \lambda_+^{N-1} \qquad (2.15b)$$

so that as $N \to \infty$,

$$N^{-1} \log Z_N \sim \log \lambda_1$$

When we treat the loop entropy model and the "hairpin" branches we use the preceding ideas but give an alternative form to the weight function $f(m)$ for unbonded regions.

A. Periodic DNAs without Branching

Since not all of the synthetic DNA analogs necessary for a complete comparison with this model are available, certain assumptions will be made. In this section we write $U_{j,j+1}$ independent of the types of base pair, i.e., $U_{11} = U_{12} = U_{24} = \cdots U$. Finally, after reviewing the formulas from MGII to compare with experimental results, an equivalent thermodynamic formulation used by Strassler[31] and Goel et al.[9] will be presented. This formulation enables a connection to be made between the parameters and thermodynamic quantities. The synthetic polynucleotides for which theoretical calculations will be presented are poly d(A)·poly d(T), poly d(A–T)·poly d(A–T) and poly d(A–C)·poly d(G–T). The structure of these polymers appears to be found in natural DNAs and they were obtainable for experimental comparison.

If we define

$$L_{j,j+1} = K_{j,j+1} + \tfrac{1}{2}(J_j + J_{j+1}) \qquad (2.16b)$$

and

$$M_{j,j+1} = \tfrac{1}{2}(J_j - J_{j+1}) \qquad (2.16c)$$

then from (2.2)

$$Z_N = \sum_\sigma \prod_{j=1}^{N} \exp \{U\sigma_j\sigma_{j+1} - \tfrac{1}{2}L_{j,j+1}(\sigma_j + \sigma_{j+1}) - \tfrac{1}{2}M_{j,j+1}(\sigma_j - \sigma_{j+1})\} \qquad (2.16d)$$

The base pair strength L_j for the jth base pair will be given by

$$L_j = J_j + \tfrac{1}{2}(K_{j-1,j} + K_{j,j+1}) = \tfrac{1}{2}(L_{j-1,j} + L_{j,j+1}) + \tfrac{1}{2}(M_{j,j+1} - M_{j-1,j}) \tag{2.17}$$

From (2.16)

$$Z_N = \text{Tr}\,(F_{1,2}F_{2,3}F_{3,4}\cdots F_{N,1}) \tag{2.18a}$$

where

$$F_{i,j} = \begin{pmatrix} \exp\,(U - L_{i,j}) & \exp\,(-U - M_{i,j}) \\ \exp\,(-U + M_{i,j}) & \exp\,(U + L_{i,j}) \end{pmatrix} \tag{2.18b}$$

The fraction of broken bonds is obtained for a one-component DNA from (2.16) and (2.1):

$$\theta_B(T) = \frac{1}{2}\left(1 + \frac{1}{N}\frac{\partial \log Z_N}{\partial L_{jj}}\right)$$

For poly d(A)·poly d(T), all the $L_{j,j+1} = L_{11}$ and $M_{j,j+1} = M_{11} = 0$. From (2.17) L_{11} is the base pair bond strength for an A–T in poly d(A)·poly d(T). The partition function for this DNA will be

$$Z_N = \text{Tr}\,F_{11}{}^N \simeq \lambda_1{}^N \tag{2.19a}$$

where λ_1 is the largest eigenvalue of

$$F_{11} = \begin{pmatrix} \exp\,(U - L_{11}) & \exp\,(-U) \\ \exp\,(-U) & \exp\,(U + L_{11}) \end{pmatrix} \tag{2.19b}$$

The fraction of broken base pairs is given by

$$[\theta_B(T)]_{d\text{A}:d\text{T}} = \frac{1}{2}\left\{1 + \frac{\sinh L_{11}}{[\sinh^2 L_{11} + \exp\,(-4U)]^{1/2}}\right\} \tag{2.20}$$

The form of L_{11}, which will be apparent from physical arguments, is assumed to be

$$L_{11} = a_{11}(T - T_{11}) \tag{2.21}$$

where T_{11} is the melting temperature of poly(A)·poly d(T), T the temperature, and a_{11} a constant. We assume for this and all other polymers that U is a constant.

For poly d(A–T)·poly d(A–T), neglecting hairpin helix branching, we find two $L_{j,j+1}$s, L_{12}, and L_{21}. If we average over both of these values,

$$\bar{L}_{12} = \tfrac{1}{2}(L_{12} + L_{21}) \tag{2.22a}$$

we obtain from (2.17) the base pair strength for an A–T bond in poly d(A–T)·poly d(A–T). Although we have two matrices, F_{12} and F_{21}, it has been shown that an equation identical to (2.20) results for this polymer with L_{11} replaced by $\bar{L}_{12}$. Assuming $L_{12} = L_{21}$,

$$\bar{L}_{12} = a_{12}(T - T_{12}) \tag{2.22b}$$

with T_{12} the melting temperature of poly d(A–T)·poly d(A–T) and a_{12} a constant.

For the period-two, two-component poly d(A–C)·poly d(G–T) we have two alternating matrices,

$$F_{14} = \begin{Bmatrix} \exp(U - L_{14}) & \exp(-U - M_{14}) \\ \exp(-U + M_{14}) & \exp(U + L_{14}) \end{Bmatrix} \tag{2.23}$$

and

$$F_{41} = \begin{Bmatrix} \exp(U - L_{41}) & \exp(-U - M_{41}) \\ \exp(-U + M_{41}) & \exp(U + L_{41}) \end{Bmatrix} \tag{2.24}$$

Averaging over the two L_{ij}, $\bar{L}_{14} = \frac{1}{2}(L_{14} + L_{41})$, we get the average base pair strength of the A–T and G–C in the polymer. From Equations (41) and (62) of MGII, we find

$$2[\theta_B(T)]_{d\text{AC}:d\text{TG}} = 1 + \tfrac{1}{2}\sinh(2\bar{L}_{14})\{\sinh^2 \bar{L}_{14} + e^{-4U}\cosh^2 M_{14}\} \times \{\cosh^2 \bar{L}_{14} + e^{-4U}\sinh^2 M_{14}\}^{-1/2} \tag{2.25}$$

As before,

$$\bar{L}_{14} = a_{14}(T - T_{14}) \tag{2.26}$$

with T_{14} the melting temperature of poly d(A–C)·poly d(T–G). We assume M_{14} is a constant. Note that $M_{14} = M_{23} = M_{13} = M_{24}$ and $M_{ij} = -M_{ji}$.

In order to relate the parameters with thermodynamic quantities we return to (2.16). Taking the polynucleotide poly d(A)·poly d(T) as an example, we have the parameters L_{11}, M_{11}, and U. Multiplying the partition function by a constant (exp NP), we find

$$Z = \sum_{\sigma} \prod_{j=1}^{N} \exp\{U\sigma_j\sigma_{j+1} - \tfrac{1}{2}L_{11}(\sigma_j + \sigma_{j+1}) + P\} \tag{2.27}$$

The constant P has been employed to normalize the partition function to 1 when all bonds are broken. An equivalent form of (2.27) is

$$Z = \sum_{\sigma} \prod_{j=1}^{N} \exp\left\{\frac{\frac{1}{4}(1 + \sigma_j)(1 + \sigma_{j+1})\,\Delta G_S}{RT} + \frac{\frac{1}{4}[(1 + \sigma_j) + (1 + \sigma_{j+1})]\,\Delta G_B}{RT}\right\} \tag{2.28}$$

where

$$U = \frac{\Delta G_S}{4RT} \tag{2.29a}$$

$$L_{11} = \frac{-(\Delta G_B + \Delta G_S)}{2RT} \tag{2.29b}$$

$$P = \frac{\Delta G_S + 2\,\Delta G_B}{4RT} \tag{2.29c}$$

Within the framework of this model we can interpret ΔG_B as the free energy required to break a base pair bond when the neighboring bonds are both broken. ΔG_S is the additional stacking free energy necessary to break a bond when one adjacent base pair bond is intact and the other broken. It follows that $L_{11} = -\frac{1}{2}RT$ times the free energy change in the reaction of breaking a base pair next to an intact one. Writing this free energy in terms of enthalpy and entropy,

$$L_{11} = -\left(\frac{\Delta G_S + \Delta G_B}{2RT}\right) = \frac{-\Delta G}{2\,RT} = \frac{-(\Delta H - T\,\Delta S)}{2RT} \tag{2.30a}$$

at the melting temperature T_m, for this model

$$\Delta H = T_m \Delta S \tag{2.30b}$$

so that if we assume that ΔH and ΔS are constant in the melting region,

$$L_{11} = \frac{\Delta H}{2RTT_m}(T - T_m) \tag{2.31a}$$

and thus in this region

$$a_{11} \simeq \frac{\Delta H}{2RT_m^{\ 2}} \tag{2.31b}$$

where ΔH is the enthalpy change at T_m from the breaking of a base pair next to an intact one. This relation holds in a similar way for the other a_{ij}.

For periodic DNAs that have short periods, say of length six or less, we could go through the algebraic manipulations to obtain an analytic expression for $\theta_B(T)$ such as (2.20) and (2.25). However, as period lengths increase to 16 or 32, the length of the expression for $\theta_B(T)$ becomes rather long and tedious to calculate. For a periodic DNA that has a long period, the calculation can be simplified by numerically multiplying l two-by-two matrices where l is the period length. As an example, we treat the following DNA with period six:

$$\begin{array}{c} \text{A T G A G G A T} \\ |\ |\ |\ |\ |\ |\ |\ |\ \cdots \\ \text{T A C T C C T A} \end{array}$$

The partition function Z_N is

$$\begin{aligned} Z_N &= \mathrm{Tr}\,(F_{12}F_{23}F_{31}F_{13}F_{33}F_{31}F_{12}F_{23}\cdots) \\ &= \mathrm{Tr}\,(F_{12}F_{23}F_{31}F_{13}F_{33}F_{31})^{N/6} \end{aligned} \tag{2.32}$$

Defining the matrix

$$G_{12\cdots 31} \equiv F_{12}F_{23}F_{31}F_{13}F_{33}F_{31} \tag{2.33}$$

we have as $N \to \infty$

$$Z_N = \mathrm{Tr}\, G_{12\cdots 31}^{N/6} \simeq \lambda_{\max}^{N/6} \tag{2.34}$$

where $\lambda_{\max}$ is the largest eigenvalue of

$$\lambda^2 - \lambda\, \mathrm{Tr}\, G_{12\cdots 31} + \det G_{12\cdots 31} = 0 \tag{2.35}$$

Choosing a U independent of base pair types, we find

$$\det G_{12\cdots 31} = (2 \sinh 2U)^6$$

and the fraction of bonds broken is given by

$$\theta_B(T) = \frac{1}{2}\left\{1 + \frac{\frac{1}{2}\partial\, \mathrm{Tr}\, G_{12\cdots 31}}{6[(\frac{1}{2}\, \mathrm{Tr}\, G_{12\cdots 31})^2 - (2 \sinh 2U)^6]^{1/2}}\right\} \tag{2.36}$$

with

$$\partial \equiv \sum_{\{ij\}} \frac{\partial}{\partial L_{ij}}$$

The sum is over all of the L_{ij} values found within the period length. Operationally we can obtain $\mathrm{Tr}\, G_{12\cdots 31}$ and $\partial\, \mathrm{Tr}\, G_{12\cdots 31}$ from (2.33) through numerical matrix multiplication. The extension to a DNA of period l is made by replacing 6 with l in (2.36) and carrying out the appropriate multiplications of matrices.

B. Periodic DNAs with Branching

The formulation of the theory of denaturation already given omits the possibility of "hairpin" helix branching, which is likely in synthetic DNAs such as poly d(A–T)·poly d(A–T). To include this effect, the partition function calculation can be accomplished by a slight modification of its integral representation in Eq. (2.8). Let us regard poly d(A–T)·poly d(A–T) as a long single strand of bases. In order to avoid free ends (which can be taken into account with a little more effort), we consider the ends of the double helix "ladder" to be tied together as a Moebius strip as in Fig. 4*a*. If there are N base pairs, the single strand of double length will have $2N$ single bases. A typical sequence of bonded and unbonded bases, as given in Fig. 4*a*, includes several hairpin regions.

A weight $h(l)$ is given to a bonded segment of l base units and $k(m)$ to an unbonded segment of m base units. On this basis, the weight of the configuration of Fig. 4 is

$$h(l_1)k(m_1)h(l_2)k(m_2)h(l_3)k(m_3)\cdots h(l_{10})k(m_{10}) \tag{2.37}$$

Notice that every bonded segment is connected to another bonded segment so that the factor of the preceding expression, which corresponds to the bonded segments, is

$$[h(l_1)h(l_{10})][h(l_2)h(l_9)][h(l_3)h(l_5)][h(l_4)h(l_8)][h(l_7)h(l_6)] \tag{2.38}$$

Since $l_1 = l_{10}$, $l_2 = l_9, \ldots l_5 = l_3$ in our example, we can write the bonded factors of (2.37) as

$$g(l_1)g(l_2)\cdots g(l_5), \qquad \text{with} \quad g(l) \equiv [h(l)]^2 \tag{2.39}$$

Since the h and k factors alternate in (2.37), the number of k factors must be the same as the number of h factors.

Several special features of $k(m)$ and $g(l)$ are evident in poly d(A–T)· poly d(A–T). The alternating A and T sequence restricts the l and m according to the types of base pair or base at the ends of the bound and unbound segments. A sequence of l intact base pairs which begins with an A–T (type 1) pair and ends with a T–A (type 2) pair must have l an even number. A sequence of m unbound bases which begins with an A type base and ends with an A type base is restricted to odd values for m, etc.

We insert two subscripts into the weight for l intact base pairs, $g_{pq}(l)$. Here pq denote the first and last base pair types of the sequence. For certain (pq), there are sets of l values not allowed. Specifically,

$$\begin{aligned} g_{11}(l) = g_{22}(l) = 0, &\qquad l = \text{even} \\ g_{12}(l) = g_{21}(l) = 0, &\qquad l = \text{odd} \end{aligned} \tag{2.40}$$

Since the two different bases of this polymer belong to both base pair types, we do not account for the restriction on $k(m)$ in the same manner as $g_{pq}(l)$. Instead, we note that the number of unbound bases in a closed loop must be even. This restriction results from the specific A with T binding to form base pairs. The hairpin loops will necessarily have an even number of bases. Other loops, with more than one helical segment branching from them, are also restricted to an even number of unbound bases. We can describe this restriction by combining the weights $k(m)$ in pairs, i.e., $[k(m_1)k(m_9)]$. If we have a hairpin region such as $k(m_{10})$ of

Fig. 4*a*, we can combine this weight with one of the three unbonded segments from the loop (m_2, m_4, m_8). One of these three *m*s must be even, and the other two will be either both odd or both even. In this way we could have

$$[k(m_1)k(m_9)][k(m_2)k(m_4)][k(m_3)k(m_5)][k(m_6)k(m_7)][k(m_8)k(m_{10})] \tag{2.41a}$$

For poly d(A–T)·poly d(A–T) we define a weight for two unbonded strands as

$$k(m_1)k(m_2) \equiv j(m_1 + m_2) = (m - 1) \exp \{-2U + \tfrac{1}{2}(U + \bar{L}_{12})m\}$$
$$m \equiv m_1 + m_2 \text{ even} \tag{2.41b}$$

The factor $(m - 1)$ counts the number of ways of arranging the m bases among the two strands with at least one unbonded base on each strand. We can rewrite the factors of (2.41a) as

$$j(m_1)j(m_2)j(m_3)j(m_4)j(m_5) \tag{2.41c}$$

A poly d(A–T)·poly d(A–T) molecule with $2n$ alternating bonded and unbonded single stranded segments will have the weight

$$g_{pq}(l_1)g_{pq}(l_2) \cdots g_{pq}(l_{n/2})j(m_1)j(m_2) \cdots j(m_{n/2})$$

The partition function Z_N for a chain of N base pairs is

$$Z_N = \sum_{n=1}^{N} \sum_{\{l\}} \sum_{\{m\}} (10) \left[\prod_{v=1}^{n/2} \begin{pmatrix} g_{11}(l_v) & g_{12}(l_v) \\ g_{21}(l_v) & g_{22}(l_v) \end{pmatrix} \begin{pmatrix} j(m_v) & j(m_v) \\ j(m_v) & j(m_v) \end{pmatrix} \right] \begin{pmatrix} 1 \\ 0 \end{pmatrix}$$
$$+ (01) \left[\prod_{v=1}^{n/2} \begin{pmatrix} g_{11}(l_v) & g_{12}(l_v) \\ g_{21}(l_v) & g_{22}(l_v) \end{pmatrix} \begin{pmatrix} j(m_v) & j(m_v) \\ j(m_v) & j(m_v) \end{pmatrix} \right] \begin{pmatrix} 0 \\ 1 \end{pmatrix} \tag{2.42a}$$

The sets of $\{l\}$ and $\{m\}$ must obey

$$2N = \sum_{v=1}^{n/2} m_v + 2l_v, \qquad 0 < n < N$$

Inserting

$$\delta\left[2N - \left(\sum_{v=1} m_v + 2l_v\right)\right] \quad \text{and} \quad \exp\left\{\beta\left[2N - \left(\sum_{v=1}^{n/2} m_v + 2l_v\right)\right]\right\} = 1$$

into (2.42a) and using the Fourier representation of the delta function, we find in a manner similar to Eq. (2.7)

$$Z_N = \frac{1}{2\pi} \int_{-\infty}^{\infty} d\alpha e^{2N(\beta + i\alpha)} \sum_{n}^{N} \sum_{l}^{\infty} \sum_{m}^{\infty} \left[(1 \quad 0)P_{n/2} \begin{pmatrix} 1 \\ 0 \end{pmatrix} + (0 \quad 1)P_{n/2} \begin{pmatrix} 0 \\ 1 \end{pmatrix} \right] \tag{2.42b}$$

where

$$P_{n/2} = \prod_{\nu=1}^{n/2} j(m_\nu)G(l_\nu) \exp\{-(\beta + i\alpha)m_\nu\}$$

$$G(l_\nu) = \exp\{-(\beta + i\alpha)2l_\nu\}\begin{pmatrix} g_{11}(l_\nu) & g_{12}(l_\nu) \\ g_{21}(l_\nu) & g_{22}(l_\nu)\end{pmatrix}\begin{pmatrix} 1 & 1 \\ 1 & 1\end{pmatrix}$$

Including the l and m sums in $P_{n/2}$,

$$Z_N = \frac{1}{2\pi}\int_{-\infty}^{\infty} d\alpha \exp[2N(\beta + i\alpha)] \sum_n^N \left[(1 \quad 0)\chi^{n/2}\begin{pmatrix}1\\0\end{pmatrix} + (0 \quad 1)\chi^{n/2}\begin{pmatrix}0\\1\end{pmatrix}\right] \tag{2.42}$$

$$\chi \equiv \psi\begin{pmatrix} \phi_{11} & \phi_{12} \\ \phi_{21} & \phi_{22}\end{pmatrix}\begin{pmatrix} 1 & 1 \\ 1 & 1\end{pmatrix}$$

with

$$\psi = \sum_{m=2,4}^{\infty} j(m)\lambda^{-m} \tag{2.42d}$$

$$\phi_{ij} = \sum_{l=1}^{\infty} g_{ij}(l)\lambda^{-2l} \tag{2.42e}$$

$$\lambda \equiv \exp(\beta + i\alpha)$$

We can select β such that ψ and ϕ_{ij} converge and

$$|\chi_{ij}| < 1$$

The weights $j(m)$ and $g_{ij}(l)$ are given by (2.40), (2.41b), and

$$g_{11}(l) = g_{22}(l) = \exp[-2U + (U - \bar{L}_{12})l], \qquad l \text{ odd} \tag{2.43a}$$

$$g_{12}(l) = g_{21}(l) = \exp[-2U + (U - \bar{L}_{12})l] \qquad l \text{ even} \tag{2.43b}$$

After inserting these weights into the expressions (2.42d), (2.42e) yields

$$\psi = \frac{(\exp[2\bar{L}_{12}] + \lambda^2 \exp[-U + \bar{L}_{12}])}{(\lambda^2 - \exp[U + \bar{L}_{12}])^2}$$

$$\phi_{11} = \phi_{22} = \frac{\lambda^2 \exp[-(U + \bar{L}_{12})]}{\lambda^4 - \exp[2(U - \bar{L}_{12})]}$$

$$\phi_{12} = \phi_{21} = \frac{\exp[-2\bar{L}_{12}]}{\lambda^4 - \exp[2(U - \bar{L}_{12})]}$$

and

$$\chi = \psi(\phi_{11} + \phi_{12})\begin{pmatrix} 1 & 1 \\ 1 & 1\end{pmatrix}$$

Since

$$\begin{pmatrix} 1 & 1 \\ 1 & 1 \end{pmatrix}^{n/2} = \begin{pmatrix} 2^{n/2-1} & 2^{n/2-1} \\ 2^{n/2-1} & 2^{n/2-1} \end{pmatrix}$$

we obtain as $N \rightarrow \infty$ in n sum of (2.42c),

$$Z_N = \frac{1}{2\pi} \int_{-\infty}^{\infty} d\alpha e^{2N(\beta+i\alpha)} \{[2\psi(\phi_{11} + \phi_{12})]^{-1/2} - 1\}^{-1} \qquad (2.43c)$$

The integral is evaluated by the method of residues with the poles of the integrand given by

$$[2\psi(\phi_{11} + \phi_{12})]^{-1/2} - 1 = 0$$

or

$$(\lambda^2 - e^{U+\bar{L}_{12}})(\lambda^2 - e^{U-\bar{L}_{12}})^{1/2} - \sqrt{2}\, e^{-U}(\lambda^2 + e^{U+\bar{L}_{12}})^{1/2} \qquad (2.43d)$$

We can evaluate Z_N by changing the variable of integration to λ. Then in the limit of $N \rightarrow \infty$, $Z_N \sim \lambda_{\max}^{2N}$ where $\lambda_{\max}$ is the largest root of (2.43d).

The fraction of broken base pairs is given by

$$\theta_B = \frac{1}{2}\left[1 + \frac{N^{-1}\partial \log \lambda_{\max}^{2N}}{\partial \bar{L}_{12}}\right] \qquad (2.43e)$$

C. Unbranched Periodic DNAs with Strand Sliding

In the earlier description of unbranched periodic DNA molecules (Section IIB), no attempt was made to account for flexible features of unbonded regions. Two effects may be considered for unbonded regions in periodic DNA molecules: (1) the possibility of strand sliding as the molecule unwinds, and (2) the different configurational entropy assigned to broken base pairs at an open end and broken base pairs between bonded base pairs. In this section we consider only the first of these two factors, strand sliding.

We employ the integral equation approach of Eq. (2.8) and treat poly d(A)·poly d(T) as the example. Again, for mathematical simplicity, a circular two-stranded molecule is assumed with all A on one strand and all T on the other. For $N \rightarrow \infty$ the use of this closed boundary condition makes no difference in the final result.

As the molecule unwinds, various configurations occur which we can describe as l_1 bound base pairs, followed by m_2 unbound base pairs, then l_2 bound, m_2 unbound, etc. If we include sliding of strands, the number of bases in the unbounded regions can be odd as well as even. Hence we must count the number of unbonded bases in a loop rather than base pairs. By letting $g(l_\mu)$ be the weight for a region of l_μ bound base pairs and $K(m_\mu)$ be the weight for loop with m_μ unbonded **bases**, the weight function for a particular set of $\{l_1, m_1, l_2, m_2, \ldots, l_n, m_n\}$ is

$$g(l_1)K(m_1)g(l_2)K(m_2) \cdots g(l_n)K(m_n)$$

The condition that must hold for all sets of l and m is

$$N = \sum_{\mu=1}^{n}(l_\mu + \tfrac{1}{2}m_\mu), \qquad 0 < n < \left[\frac{N}{2}\right]$$

The partition function for this polymer is

$$Z_N = \sum_{n=1}^{N/2} \sum_{\{l_\mu\}} \sum_{\{m_\mu\}} \prod_{\mu}^{n} g(l_\mu)K(m_\mu) \tag{2.44a}$$

Inserting the factors

$$\delta\left[N - \sum_{\mu=1}^{n}\left(l_\mu + \tfrac{1}{2}m_\mu\right)\right] \quad \text{and} \quad \exp\{\beta[N - \textstyle\sum(l_\mu + \tfrac{1}{2}m_\mu)]\}$$

into (2.44a), we have, as in (2.8),

$$Z_N = \tfrac{1}{2}\pi \int_{-\infty}^{\infty} d\alpha e^{N(\beta+i\alpha)} \sum_{n=1}^{N/2} \prod_{\mu}^{n} \phi(l_\mu)\psi(m_\mu) \tag{2.44b}$$

with

$$\phi(l_\mu) = \sum_{l_\mu=1}^{\infty} g(l_\mu)e^{-(\beta+i\alpha)l_\mu} \equiv \phi(\beta + i\alpha) \tag{2.44c}$$

$$\psi(m_\mu) = \sum_{m_\mu=1}^{\infty} K(m_\mu)e^{-1/2(\beta+i\alpha)m_\mu} \equiv \psi(\beta + i\alpha) \tag{2.44d}$$

For poly d(A)·poly d(T) the weights $g(l)$ and $K(m)$ are

$$g(l) = \exp\{-2U + (U - L_{11})l\} \tag{2.45a}$$

and

$$K(m) = \exp\{-2U + \tfrac{1}{2}(U + L_{11})m\} \tag{2.45b}$$

and thus

$$\phi = \frac{\exp(-U - L_{11})}{\lambda - \exp[U - L_{11}]} \tag{2.46a}$$

$$\psi = \frac{\exp[-\frac{1}{2}(3U - L_{11})]}{(\lambda^{1/2} - \exp[\frac{1}{2}(U + L_{11})]} \tag{2.46b}$$

$$\lambda \equiv \exp(\beta + i\alpha)$$

As $N \to \infty$ in the n sum of (2.44b),

$$Z_N = \frac{1}{2\pi}\int_{-\infty}^{\infty} d\alpha \exp[N(\beta + i\alpha)]\{(\phi\psi)^{-1} - 1\}^{-1} \tag{2.47a}$$

We can evaluate Z_N by the method of residues as demonstrated in Eq. (2.8). Changing the integration variable to λ, we obtain a clockwise circular contour in λ space. From (2.47a) and (2.46),

$$Z_N = \frac{1}{2\pi i}\oint d\lambda\lambda^{N-1}\{(\lambda - \exp[U - L_{11}])(\lambda^{1/2} - \exp[\tfrac{1}{2}(U + L_{11})])$$
$$\times \exp[\tfrac{1}{2}(5U + L_{11})] - 1\}^{-1} \quad (2.47b)$$

In the limit of $N \to \infty$, $Z_N \sim \lambda_{max}^N$ where λ_{max} is the largest root of

$$(\lambda - \exp[U - L_{11}])(\lambda^{1/2} - \exp[(U + L_{11})]) = \exp[-\tfrac{1}{2}(5U + L_{11})]$$

$\theta_B(T)$ is derived from (2.48a)

$$\theta_B(T) = \frac{1}{2}\left[1 + N^{-1}\frac{\partial \log Z_N}{\partial L_{11}}\right] \quad (2.48b)$$

D. DNAs with a Random Distribution of Two Bond Types

In order to compare a theoretical model with the melting curves of natural DNAs, it is clear we must treat nonperiodic distributions of A–T and G–C base pairs. There have been a number of calculations which have considered random or Markov distributions of two base pair components. Poland and Scheraga[14] have compared several approximate calculations with the results of Lehman and McTague.[28] Fink and Crothers[32] made a comparison between several theoretical approaches. In this section we review and compare the theories of Montroll and Yu,[33] Vedenov and Dykhne,[30] and Lehman and McTague.

1. *Theory of Montroll and Yu*

The paper by Montroll and Yu makes use of a general formalism for the investigation of the properties of lattices with two types of component, type I and type II. The formalism is in the spirit of the Mayer cluster integral approach. Let us consider a pure type I component lattice with the effect of the type II components imposed upon it. Type II components are considered first as single units, then as pairs, triples, etc. For the DNA problem, the pure lattice is a one-dimensional string of A–T (or G–C) base pairs, with the G–C (or A–T) base pairs acting as the type II components.

The formalism proceeds as follows. Let us consider a pure A–T DNA, N base pairs long, with m G–C base pairs placed at the positions $(r_1, r_2, \ldots, r_m)$. We are interested in obtaining a thermodynamic property $F(m) = F(r_1, r_2, \ldots, r_m)$. This function will depend on the number and arrangement of G–C pairs.

We define a set of differences in terms of the functions $F(0)$, $F(r_1)$, $F(r_1r_2), \ldots$, etc.:

$$\Delta(r_1) \equiv F(r_1) - F(0) \tag{2.49a}$$

$$\Delta(r_1r_2) \equiv F(r_1r_2) - F(r_1) - F(r_2) + F(0) \tag{2.49a}$$

$$\begin{aligned}\Delta(r_1r_2r_3) \equiv{} & F(r_1r_2r_3) - F(r_1r_2) - F(r_1r_3) - F(r_2r_3) \\ & + F(r_1) + F(r_2) + F(r_3) - F(0)\end{aligned} \tag{2.49c}$$

$$\begin{aligned}\Delta(r_1r_2r_3r_4) \equiv{} & F(r_1r_2r_3r_4) - F(r_1r_2r_3) - F(r_1r_2r_4) - F(r_2r_3r_4) \\ & - F(r_1r_3r_4) + F(r_1r_2) + F(r_1r_3) + F(r_1r_4) + F(r_2r_3) \\ & + F(r_2r_4) + F(r_3r_4) - F(r_1) - F(r_2) - F(r_3) - F(r_4) \\ & + F(0), \text{ etc.}\end{aligned} \tag{2.49d}$$

With a small number of G–Cs in an A–T lattice, the Δs will be small. We can write the $F(r_1, \ldots, r_m)$s in terms of the Δs as

$$F(r_1) = F(0) + \Delta(r_1) \tag{2.50a}$$

$$F(r_1r_2) = F(0) + \Delta(r_1) + \Delta(r_2) + \Delta(r_1r_2) \tag{2.50b}$$

$$\cdot \quad \cdot \quad \cdot \quad \cdot \quad \cdot \quad \cdot$$

$$\begin{aligned}F(r_1 \cdots r_m) \equiv F(m) = {} & F(0) + \sum_{j=1}^{m} \Delta(r_j) + \sum_{j=1}^{m-1} \sum_{k=j+1}^{m} \Delta(r_jr_k) \\ & + \sum_{j=1}^{m-2} \sum_{k=j+1}^{m-1} \sum_{l=j+2}^{m} \Delta(r_jr_kr_l) + \cdots\end{aligned} \tag{2.51}$$

Since $F(m)$ is some averaged property of the system, let us assume we can express it through a linear operation on the logarithm of the partition function of the system,

$$F(r_1r_2 \cdots r_m) = \mathcal{O} \log Z(r_1, r_2, \ldots, r_m) \tag{2.52}$$

As in the case of F, we define $Z(0)$ as the partition function of a pure A–T lattice, $Z(r_j)$ the partition function of an A–T lattice with a single G–C pair at r_j, and so on for $Z(r_jr_k)$, $Z(r_jr_kr_l)$, etc. It follows from (2.52) that

$$F(0) = \mathcal{O} \log Z(0) \tag{2.53a}$$

$$F(r_j) = \mathcal{O} \log Z(r_j) \tag{2.53b}$$

$$F(r_jr_k) = \mathcal{O} \log Z(r_jr_k), \text{ etc.} \tag{2.53c}$$

We can now express the expansion (2.51) for $F(m)$ in terms of the partition functions for lattices with one G–C, two G–Cs, three G–Cs, etc.

From (2.49), (2.51), and (2.53),

$$\begin{aligned}F(m) &= \mathcal{O}\log Z_N(m)\\ &= \mathcal{O}\log Z(0) + \sum_{j=1}^{m}\mathcal{O}\left[\log\left\{\frac{Z(r_j)}{Z(0)}\right\}\right]\\ &\quad+\sum_{j=1}^{m-1}\sum_{k=j+1}^{m}\mathcal{O}\left[\log\left\{\frac{Z(r_jr_k)Z(0)}{Z(r_j)Z(r_k)}\right\}\right]\\ &\quad+\sum_{j=1}^{m-2}\sum_{k=j+1}^{m-1}\sum_{l=j+2}^{m}\mathcal{O}\left[\log\left\{\frac{Z(r_jr_kr_l)Z(r_j)Z(r_k)Z(r_l)}{Z(r_jr_k)Z(r_jr_l)Z(r_kr_l)Z(0)}\right\}\right]+\cdots\end{aligned}\tag{2.54}$$

We can make certain simplifying assumptions in our treatment of a random distribution of A–T and G–C bonds in this model. It will be assumed that the contribution of the stacking interactions to the jth bond strength is independent of the adjacent bond types. Thus we have, from (2.17), two bond strengths,

$$L_1 = J_1 + K_1 \quad\text{and}\quad L_3 = J_3 + K_3 \tag{2.55}$$

We adjoin one subscript to K since its value is independent of its neighbors. As before we define

$$L_1 = a_1(T - T_1), \qquad L_3 = a_3(T - T_3) \tag{2.56}$$

where T_1 is the T_m of a "natural" DNA containing only A–T bonds, and T_3 is the T_m of a DNA with just G–C type bonds. It is assumed further that the correlation parameter U is independent of bond types. We can also develop the present formalism when the correlation parameter U is dependent on the bond types.

The partition function for the particular sequence of base pairs shown in Fig. 6 is

$$\begin{aligned}Z_N(r_1, r_2, \ldots, r_m)\\ &= \sum_\sigma \exp\{-[L_3\sigma_1 + L_1\sigma_2 + L_1\sigma_3 + L_3\sigma_4 + L_1\sigma_5 + \cdots\}\\ &\qquad\times\exp\{U(\sigma_1\sigma_2 + \sigma_2\sigma_3 + \cdots \sigma_N\sigma_1)\}\end{aligned}\tag{2.57}$$

We can consider this DNA as having m G–C bonds placed in a perfect lattice of A–T bonds. The G–C bonds are placed at the first bond or rung, another n_1 rungs above the first, the third n_2 rungs above the second, etc.

Before we proceed with the systematic theory, we note two asymptotic results, one for $U \to 0$ and the other for $U \to \infty$. The $U = 0$ case has already been given in Eq. (2.1b). The fraction of bonds broken at temperature T is

$$\theta_B(T) = \tfrac{1}{2}\{1 + c_1 \tanh L_1 + c_2 \tanh L_3\} \tag{2.57a}$$

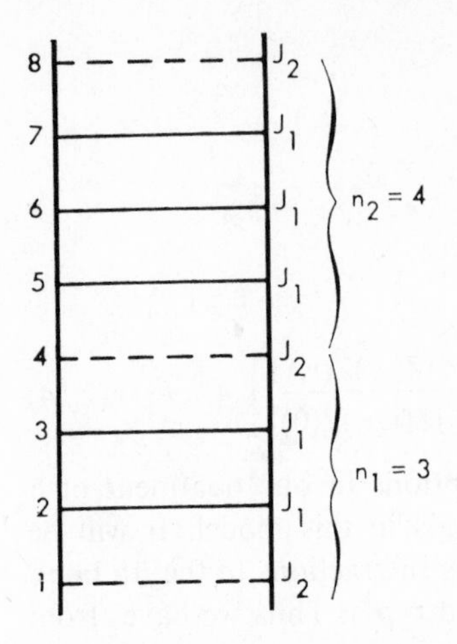

Fig. 6. Schematic representation of a DNA with two bond types. J_2 represents a G-C bond, J_1 corresponds to an A-T bond.

c_1 and c_2 being, respectively, the fraction of AT and GC pairs. The quantities L_1 and L_3 are related to the temperature by (2.56). It is known that as $U \to \infty$,

$$\theta_B(T) = \begin{cases} 0 & \text{if} \quad c_1L_1 + c_2L_3 < 0 \\ 1 & \text{if} \quad c_1L_1 + c_2L_3 > 0 \end{cases} \tag{2.57b}$$

If L_1 and L_3 are related to the temperature by (2.56), the critical temperature that separates the two cases of (2.57b) is

$$T_c = \frac{c_1a_1T_1 + c_2a_3T_3}{a_1c_1 + a_3c_2} \tag{2.57c}$$

Generally the runs of A–T bonds in a molecule can be grouped together by employing the relation of a matrix to its eigenvalues and eigenvectors. From the matrix

$$\begin{pmatrix} e^{-L_1+U} & e^{-U} \\ e^{-U} & e^{L_1+U} \end{pmatrix} \tag{2.58}$$

with eigenvalues λ_1 and λ_2 and eigenvectors written as $[\psi_j(1), \psi_j(-1)]$, we have

$$\exp\,[U\sigma_1\sigma_2 - \tfrac{1}{2}L_1(\sigma_1 + \sigma_2)] = \sum_{j=1}^{2} \lambda_j \psi_j(\sigma_1)\psi_j(\sigma_2)$$

$$\sum_{\sigma_2=1} \exp\,\{U\sigma_1\sigma_2 - \tfrac{1}{2}L_1(\sigma_1 + \sigma_2) + U\sigma_2\sigma_3 - \tfrac{1}{2}L_1(\sigma_2 + \sigma_3)\}$$
$$= \sum_{j=1}^{2} \lambda_j^{\,2} \psi_j(\sigma_1)\psi_j(\sigma_3), \quad \text{etc.}$$

Detailed properties of the eigenvalues and eigenvectors of (2.58) are given in Montroll and Goel, I.

Defining

$$F(\sigma, \sigma'; n) = \sum_{j=1}^{2} \lambda_j^n \psi_j(\sigma)\psi_j(\sigma') \tag{2.59}$$

we can rewrite (2.57) as

$$Z_N(r_1 \cdots r_m) = \sum_{\sigma} \{\exp - \varepsilon(\sigma_1 + \sigma_2 + \cdots + \sigma_m)\} \times F(\sigma_1, \sigma_2; n_1)F(\sigma_2, \sigma_3; n_2) \cdots F(\sigma_m, \sigma_1; n_m) \tag{2.60}$$

where $\varepsilon \equiv L_3 - L_1$. To obtain (2.60) we have incorporated all of the sums over the σ_i which are A–T bonds into the $F(\sigma, \sigma'; n)$ term and renumbered the G–C bonds by letting the first G–C be σ_1, the second σ_2, the third σ_3, etc. Using (2.59), we find

$$Z_N(r_1 \cdots r_m) = \sum_{j_1=1}^{2} \sum_{j_2=1}^{2} \cdots \sum_{j_m=1}^{2} \lambda_{j_1}^{n_1} \lambda_{j_2}^{n_2} \cdots \lambda_{j_m}^{n_m} (j_1, j_2)(j_2, j_3) \cdots (j_m, j_1) \tag{2.61a}$$

where

$$\begin{aligned} (j, k) &= \sum_{\sigma=\pm 1} \psi_j(\sigma) e^{-\varepsilon\sigma} \psi_k(\sigma) \\ (1, 1) &= \psi_1^2(1)e^{-\varepsilon} + \psi_1^2(-1)e^{\varepsilon} \\ (2, 2) &= \psi_2^2(1)e^{-\varepsilon} + \psi_2^2(-1)e^{\varepsilon} \end{aligned} \tag{2.61b}$$

The employment of the formula (2.54) requires the calculation of the partition function of an AT chain in the absence of GC pairs, with one GC pair, with two, etc. In the absence of GC bonds, we use (2.3). As $N \to \infty$

$$Z(0) \simeq \lambda_1^N$$

We introduce the abbreviations

$$\omega \equiv \frac{\lambda_2}{\lambda_1} \quad \text{and} \quad a_{ij} \equiv \frac{(i, j)}{(1, 1)} \tag{2.62}$$

into the formulas for $Z(r)$, $Z(r_1 r_2)$, etc. When a single GC pair is placed into a chain of AT pairs, we have from (2.61a) $n_1 = N$ (since the molecule is assumed circular) and

$$Z(r) = \sum_{j=1}^{2} \lambda_j^N (j, j) \sim (1, 1)\lambda_1^N h_1 \quad \text{as} \quad N \to \infty \tag{2.63}$$

We have introduced the h_1 to conform to a pattern that will develop. For two G–Cs at the j_1th and j_2th sites, chosen from the set $1, 2, 3, \ldots, m$

and separated by n rungs or bonds,

$$Z(r_{j_1}r_{j_2}) = \sum_{j_1=1}^{2} \sum_{j_2=1}^{2} \lambda_{j_1}{}^{n} \lambda_{j_2}^{N-n} (j_1, j_2)(j_2, j_1)$$

Letting $t_1 = n$, $t_2 = N - n$, we have

$$\begin{aligned} Z(r_{j_1}, r_{j_2}) &= (1, 1)^2 \lambda_1^N + (1, 2)(2, 1)[\lambda_1^{t_1} \lambda_2^{t_2} + \lambda_1^{t_2} \lambda_2^{t_1}] + (2, 2)^2 \lambda_2^N \\ &= (1, 1)^2 \lambda_1{}^N [1 + a_{12} a_{21} (\omega^{t_1} + \omega^{t_2}) + \omega^N a_{22}{}^2] \end{aligned}$$

As $N \to \infty$, since $\omega < 1$, we can neglect the ω^N-term inside the brackets and the smaller member of the pair ω^{t_1} and ω^{t_2},

$$Z(r_{j_1}, r_{j_2}) = (1, 1)^2 \lambda_1{}^N [1 + a_{12} a_{21} \omega^t] = (1, 1)^2 \lambda_1{}^N h_2(t) \tag{2.64}$$

For three G–Cs at site j_1, j_2, and j_3, chosen from our m possible sites, and separated by n_1 and n_2 rungs,

$$Z(r_{j_1}, r_{j_2}, r_{j_3}) = \sum_{j_1=1}^{2} \sum_{j_2=1}^{2} \sum_{j_3=1}^{2} \lambda_{j_1}{}^{n_1} \lambda_{j_2}{}^{n_2} \lambda_{j_3}^{N-n_1-n_2} (j_1, j_2)(j_2, j_3)(j_3, j_1)$$

Letting $t_1 = n_1$, $t_2 = n_2$, $t_3 = N - n_1 - n_2$, we have

$$\begin{aligned} Z(r_{j_1}, r_{j_2}, r_{j_3}) = (1, 1)^3 \lambda_1{}^N [1 &+ a_{12} a_{21} (\omega^{t_1} + \omega^{t_2} + \omega^{t_3}) \\ &+ a_{12} a_{22} a_{21} (\omega^{t_1+t_2} + \omega^{t_1+t_3} + \omega^{t_2+t_3}) + a_{22}{}^3 \omega^N] \end{aligned}$$

As $N \to \infty$ for $0 < t_1 < N\alpha_1$, $0 < t_2 < N\alpha_2$, $0 < \alpha_j < 1, j = 1, 2$ we have $t_3 > N(1 - \alpha_1 - \alpha_2) \to \infty$ and

$$\begin{aligned} Z(r_{j_1}, r_{j_2}, r_{j_3}) &= (1, 1)^3 \lambda_1{}^N [1 + a_{12} a_{21} (\omega^{t_1} + \omega^{t_2}) + a_{12} a_{22} a_{21} \omega^{t_1+t_2}] \\ &= (1, 1)^3 \lambda_1{}^N h_3(t_1, t_2) \end{aligned} \tag{2.65}$$

with

$$h_3(t_1 t_2) = \frac{h_2(t_1) h_2(t_2)}{h_1} \left\{ 1 + \frac{a_{12} a_{21} (a_{22} - a_{21} a_{12}) \omega^{t_1+t_2}}{h_2(t_1) h_2(t_2)} \right\}$$

If we continue this process, by inserting more G–Cs into the AT lattice, we find for n G–Cs,

$$Z(r_{j_1}, r_{j_2}, \ldots, r_{j_n}) = (1, 1)^n \lambda_1{}^N h_n(t_1, t_2, t_3, \ldots, t_{n-1}) \tag{2.66a}$$

where each t_i is the distance between successive G–C bonds and

$$\begin{aligned} h_n(t_1, t_2 \cdots t_{n-1}) &= \frac{h_{n-1}(t_1 \cdots t_{n-2}) h_{n-1}(t_2 \cdots t_{n-1})}{h_{n-2}(t_2 \cdots t_{n-2})} \\ &\quad \times \left\{ 1 + \frac{a_{12} a_{21} (a_{22} - a_{12}{}^2)^{n-2} \omega^{t_1+t_2+\cdots t_{n-1}}}{h_{n-1}(t_1 \cdots t_{n-2}) h_{n-1}(t_2 \cdots t_{n-1})} \right\} \end{aligned} \tag{2.66b}$$

The fraction of bonds broken is

$$\theta_B(T) = \tfrac{1}{2}(1 + N^{-1}\partial \log Z_N)$$

$$\partial \equiv \frac{\partial}{\partial L_1} + \frac{\partial}{\partial L_3} \tag{2.67a}$$

From (2.54), with the operator $\mathcal{O} \equiv \partial$, we have

$$\theta_B(T) = \tfrac{1}{2}(S_0 + S_1 + S_2 + S_3 + \cdots) \tag{2.67b}$$

with

$$S_0 = 1 + N^{-1}\,\partial \log Z(0) = 1 + \partial \log \lambda_1 \tag{2.68a}$$

$$S_1 = \left(\frac{m}{N}\right) \partial \log \left\{\frac{Z(r)}{Z(0)}\right\} = \left(\frac{m}{N}\right) \partial \log (1, 1) \tag{2.68b}$$

$$\begin{aligned} S_2 &= \frac{1}{N} \sum_{r_1} \sum_{r_2} \partial \log \left\{\frac{Z(r_1, r_2)Z(0)}{Z(r_1)Z(r_2)}\right\} \\ &= N^{-1} \sum_{t=1}^{N-1} g(t)\, \partial \log \left\{\frac{h_2(t)}{h_1}\right\} \end{aligned} \tag{2.68c}$$

$$\begin{aligned} S_3 &= N^{-1} \sum\sum\sum \partial \log \left\{\frac{Z(r_j r_k r_l)Z(r_j)Z(r_k)Z(r_l)}{Z(r_j r_k)Z(r_j r_l)Z(r_k r_l)Z(0)}\right\} \\ &= N^{-1} \sum_{t_1=1}^{N-2} \sum_{t_2=1}^{N-2} g(t_1, t_2)\, \partial \log \left\{\frac{h_3(t_1 t_2)}{h_2(t_1)h_2(t_2)h_2(t_1 + t_2)}\right\}, \quad \text{etc.} \end{aligned} \tag{2.68d}$$

Here $g(t)$ is the number of pairs of G–C bonds which are separated by t bonds of either type, $g(t_1, t_2)$ is the number of triples of G–C with t_1 bonds of either type separating the first and the second and t_2 bonds separating the second and third, etc.

It is convenient to define

$$U_1(t) \equiv h_2(t)$$

$$U_2(t_1 t_2) \equiv \frac{h_3(t_1 t_2)}{h_2(t_1)h_2(t_2)}$$

and, generally,

$$U_n(t_1 t_2 \cdots t_n) \equiv \frac{h_{n+1}(t_1 t_2 \cdots t_n)h_{n-1}(t_2 t_3 \cdots t_{n-1})}{h_n(t_1 \cdots t_{n-1})h_n(t_2 \cdots t_n)} \tag{2.69}$$

Then

$$S_2 = N^{-1} \sum_t g(t)\, \partial \log U_1(t) \tag{2.70a}$$

$$S_3 = N^{-1} \sum_{t_1} \sum_{t_2} g(t_1 t_2)\, \partial \log \left(\frac{U_2(t_1 t_2)}{U_1(t_1 + t_2)}\right), \quad \text{etc.} \tag{2.70b}$$

The derivation of the general form for S_n is given in References 33 and 34.

When the sequence, or the statistics of the sequence is known, we can determine the form of $g(t)$, $g(t_1t_2)$, $g(t_1t_2t_3)$, etc. We restrict ourselves to a random distribution of G–Cs in concentration $c = m/N$ imbedded in an AT matrix. Given a G–C pair, the probability that another one is t bonds away is c. Thus with Nc the total number of G–Cs, the expected number of pairs of G–C separated by t bonds of either type will be, for N large,

$$g(t) = Nc^2 \tag{2.71a}$$

We find in a similar fashion

$$g(t_1t_2) = Nc^3 \tag{2.71b}$$

$$g(t_1t_2t_3) = Nc^4 \tag{2.71c}$$

As $N \to \infty$ we find from (2.67b) that $\theta_B(T)$, the fraction of bonds broken, becomes

$$\begin{aligned}\theta_B(T) = \tfrac{1}{2}\Bigg\{1 &+ \partial \log \lambda_1 + c\,\partial \log (1,1) + c^2 \sum_{t=1}^{\infty} \partial \log U_1(t) \\ &+ c^3 \sum_{t_1=1}^{\infty} \sum_{t_2=1}^{\infty} \partial \log \left[\frac{U_2(t_1t_2)}{U_1(t_1+t_2)}\right] \\ &\times c^4 \sum_{t_1}^{\infty} \sum_{t_2}^{\infty} \sum_{t_3}^{\infty} \log \left[\frac{U_3(t_1t_2t_3)U_1(t_1+t_2+t_3)}{U_2(t_1+t_2, t_3)U_2(t_1, t_2+t_3)}\right] + \cdots \Bigg\}\end{aligned}$$

This sum can be simplified by collecting all the terms involving $\partial \log U_1(t)$ from each sum, i.e., $\partial \log U_1(t)$, $\partial \log U_1(t_1 + t_2)\partial \log U_1(t_1 + t_2 + t_3) \cdots$. Then doing the same thing for terms of the form $\partial \log U_2(t_1t_2)$, i.e., $\partial \log U_2(t_1t_2)$, $\partial \log U_2(t_1 + t_2, t_3)$, $\partial \log U_2(t_1, t_2 + t_3) \ldots$, etc. If this process is continued, we obtain the expression

$$\begin{aligned}\theta_B(T) = \frac{1}{2}\Bigg[1 &+ \partial \log \lambda_1 + c\,\partial \log (1,1) \\ &+ \sum_{j=1}^{\infty} c^{j+1} \sum_{t_1=1}^{\infty} \cdots \sum_{t_j=1}^{\infty} (1-c)^{t_1+t_2+\cdots+t_j-j}\,\partial \log U_j(t_1 \cdots t_j)\Bigg]\end{aligned} \tag{2.72}$$

We have treated a DNA molecule as a matrix of AT pairs with a concentration c of GCs randomly imbedded in it. Clearly we can reverse the role of the ATs and GCs. In the temperature range $T < \frac{1}{2}(T_1 + T_3)$, the convergence of (2.72) is more rapid when the basic matrix is composed of only AT pairs, whereas at the range $T > \frac{1}{2}(T_1 + T_3)$, the basic matrix of GC pairs yields the more rapid convergence.

For a given j, the contribution to (2.72) will be negligible beyond a value $R_{j(\max)}$ where

$$R_j = \sum_{i=1}^{j} t_i$$

To collect the largest terms for a given $R_{j(\max)}$ we can rearrange the sum in (2.72) such that only the terms where $\sum_{i=1}^{j} t_i \leqslant R_{j(\max)}$ are collected. Defining

$$y_j = R_{j(\max)} - (j-1)$$

we will call

$$Y_1 = y_j$$
$$Y_2 = y_j + 1 - t_1$$
$$Y_3 = y_j + 2 - t_1 - t_2$$
$$\vdots$$
$$Y_l = y_j + (l-1) - \sum_{i=1}^{l-1} t_i$$

Then the sums in (2.72) can be written as

$$\sum_{j=1}^{\infty} c^{j+1} \sum_{t_1=1}^{Y_1} \sum_{t_2=1}^{Y_2} \cdots \sum_{t_j=1}^{Y_j} (1-c)^{\Sigma_{}^{j} t_i - j}\, \partial \log U_j(t_1 t_2 \cdots t_j)$$

Although this procedure helps speed the convergence of the series, its efficiency diminishes rapidly with increasing values of the correlation parameter U. When $U > 2.0$ and $c = 0.50$, the convergence of (2.72) is slow. The reason is that as U increases, the terms of $\partial \log U_j(t_1 \cdots t_j)$ which contribute most to the series of (2.72) are found to be those with successively higher values of j.

As pointed out by Montroll and Yu in the regime of strong correlation, in very high values of U the melting curve of a random sequence of two components should be approximated by a long periodic sequence. We can select periodic sequences which have the same number of pairs of Bs separated by one bond, two bonds, three bonds, etc., as in the random case. The closer the periodic DNA approximates the random case, the smaller will be their differences in melting curves.

Consider a DNA molecule with a random sequence 50% A–T (or A) and G–C (or B). In this random chain the number of pairs of B separated by any integral number of bonds is $N/4$ for an N unit chain. A periodic DNA molecule of period 16 which has $N/4$ B as a nearest neighbor to another B, $N/4$ pairs of B as second neighbors, as well as the required $N/4$, third, fourth, fifth, and sixth neighbors is

BBBAAAABAABABBA¦BBAAAAB ----

Seventh neighbor Bs do not appear in the required number $N/4$. A molecule of period 32 which has the required $N/4$ pairs of nearest, next-nearest, and all longer ranged pairs up to order 11 is

AAAAABABABBABAABBBBBABBBAAABAABB¦AAAAABA—

Twelfth-order neighbor pairs do not appear with the correct frequency.

2. *Lehman-McTague Theory*

Lehman and McTague have treated both single-molecule and ensemble averaged properties of DNA using the MI model. Their notation follows that of Zimm and Bragg. Since it differs from that previously employed, we give the relationship between the two notations and briefly outline the approach of Lehman and McTague.

In the Zimm and Bragg notation, the statistical weight s_m is assigned if the mth base pair is bonded next to a helical segment and 1 is assigned if it is broken. The end of each helical section along the molecule is given a statistical factor σ. For a two-component DNA, s_A and s_B are assigned to the A–T and G–C base pairs, respectively. The two notations are connected through the relations

$$\begin{aligned} U &= -\tfrac{1}{4}\log\sigma \\ L_A &= -\tfrac{1}{2}\log s_A \\ L_B &= -\tfrac{1}{2}\log s_B \end{aligned} \tag{2.73}$$

The partition function for a specific sequence of N base pairs can be written as

$$Z_N = (1, 1)\left\{\prod_{m=1}^{N}\begin{pmatrix} s_m & \sigma s_m \\ 1 & 1 \end{pmatrix}\right\}\begin{pmatrix} 1 \\ 0 \end{pmatrix} \tag{2.74}$$

The fraction of broken bonds is obtained from

$$\theta_B(T) = 1 - N^{-1}\left(\frac{\partial}{\partial \log s_A} + \frac{\partial}{\partial \log s_B}\right)\log Z_N \tag{2.75}$$

An algorithm for a finite length DNA follows from (2.74) by letting

$$\begin{pmatrix} U_n \\ V_n \end{pmatrix} = \begin{pmatrix} s_n & \sigma s_n \\ 1 & 1 \end{pmatrix}\begin{pmatrix} U_{n-1} \\ V_{n-1} \end{pmatrix} \tag{2.76}$$

Since

$$W_n = \frac{U_n}{V_n} = s_n \frac{(\sigma + W_{n-1})}{(1 + W_{n-1})} \tag{2.77}$$

and

$$V_n = U_{n-1} + V_{n-1} = V_{n-1}(1 + W_{n-1})$$

it follows that

$$\log Z_N = \log \{(1 + W_N)V_N\} = \log (1 + W_N) + \sum_{n=2}^{N} \log \left(\frac{V_n}{V_{n-1}}\right)$$

$$= \sum_{n=1}^{N} \log (1 + W_n) \tag{2.78}$$

with

$$(U_0, V_0) = (1, 0) \qquad \text{and} \qquad (U_1, V_1) = (s_1, 1)$$

To obtain the melting curve for an N base pair DNA, we apply (2.75) to (2.78), to find that

$$\theta_B(T) = 1 - N^{-1} \sum_{n=1}^{N} (1 + W_n)^{-1} X_n \tag{2.79}$$

where X_n is defined by

$$X_n = \left(\frac{s_A \partial}{\partial s_A} + \frac{s_B \partial}{\partial s_B}\right) W_n = W_n + \frac{s_n(1 - \sigma) X_{n-1}}{(1 + W_{n-1})^2}$$

The preceding algorithm for $\theta_B(T)$ has been derived by Lehman and McTague, and by Vedenov et al.,[29] using a different notation.

To obtain the ensemble average of $\log Z$ for a large number of DNA molecules, each containing N base pairs, Lehman and McTague presented two iterative techniques. We review one of them, the Fredholm integral equation approach for a random sequence of two units, in the limit as $N \to \infty$.

The starting point is the integral

$$F(a, b) \equiv 2\int_0^\infty U^{-1}(e^{-aU^2} - e^{-bU^2})\, dU = \lim_{\varepsilon \to 0} \int_\varepsilon^\infty z^{-1}(e^{-az} - e^{-bz})\, dz$$

$$= \lim_{\varepsilon \to 0} \left\{\int_{a\varepsilon}^\infty e^{-x}\, d\log x - \int_{b\varepsilon}^\infty e^{-x}\, d\log x\right\}$$

$$= \lim_{\varepsilon \to 0} \left\{-e^{-a\varepsilon} \log \varepsilon a + e^{-b\varepsilon} \log \varepsilon b \right.$$

$$\left. + \int_{a\varepsilon}^\infty (\log x) e^{-x}\, dx - \int_{b\varepsilon}^\infty (\log x) e^{-x}\, dx\right\}$$

Since the two integrals to the right converge as $\varepsilon \to 0$ and therefore cancel each other, we have

$$F(a, b) = \log b - \log a$$

Hence, if we set $a = y$ and $b = y + W_n$, we find

$$\log (y + W_n) = \log y + 2\int_0^\infty U^{-1}e^{-yU^2}(1 - e^{-U^2W_n})\, dU \tag{2.80}$$

An integral over the Bessel function of purely imaginary argument $I_1(x)$ is also useful at this point. It is obtained from the integral formula

$$\int_0^\infty I_\nu(at) \exp(-p^2t^2)\, dt = \left(\frac{\pi^{1/2}}{2p}\right) \exp\left(\frac{a^2}{8p^2}\right) I_{\nu/2}\left(\frac{a^2}{8p^2}\right)$$

which is derived from page 394 of Watson's *Treatise on Bessel Function.* Since

$$I_{1/2}(z) = \frac{1}{2}\left(\frac{2}{\pi z}\right)^{1/2}(e^z - e^{-z})$$

it follows that

$$\exp kU^2 = 1 + 2U\int_0^\infty I_1(2UV) \exp\left(\frac{-V^2}{k}\right) dV \tag{2.81}$$

Lehman and McTague have combined (2.80) and (2.81) in an elegant way to give an algorithm for the calculation of the partition function in the case in which the two base pair components are distributed randomly. Let

$$C_N(y) \equiv N^{-1}\sum_{n=1}^{N} \log (y + W_n) \tag{2.82}$$

Then, from (2.80),

$$C_N(y) \equiv \log y + 2\int_0^\infty U^{-1}e^{-yU^2}N^{-1}\left\langle \sum_{n=1}^{N} [1 - \exp(-U^2W_n)] \right\rangle dU$$

$$= \log y + 2\int_0^\infty U^{-1}[1 - R_N(U)] \exp(-yU^2)\, dU \tag{2.83a}$$

with

$$R_N(U) \equiv N^{-1}\left\langle \sum_{n=1}^{N} \exp(-U^2W_n) \right\rangle \tag{2.83b}$$

Since [from (2.77)]

$$W_n = s_n - \frac{(1 - \sigma)s_n}{(1 + W_{n-1})} \tag{2.84}$$

we can combine (2.83b) and (2.84) to yield

$$R_N(U) = N^{-1}\left\langle \sum_{n=1}^{N} [\exp(-U^2 s_n)] \exp\left\{\frac{-U^2\alpha_n{}^2}{(1+W_{n-1})}\right\}\right\rangle$$

$$= N^{-1}\left\langle \sum_{n=1}^{N} [\exp(-U^2 s_n)] \right.$$

$$\left. \times \left\{1 + 2U\alpha_n \int_0^\infty I_1(2UV\alpha_n) \exp[-(1+W_{n-1})V^2]\, dV\right\}\right\rangle \quad (2.85a)$$

where

$$\alpha_n = [s_n(1-\sigma)]^{1/2} \quad (2.85b)$$

Now let us suppose that our base pairs are of two types, A and B, and that those of type j exist in concentration c_j. Then, if they are randomly distributed, the state of the nth is independent of the $(n-1)$st (as far as type is concerned). Hence, if the two values of s_n are s_A and s_B, then as $N \to \infty$, $R(U)$ is the solution of the Fredholm type of integral equation

$$R_\infty(U) = c_A \exp\{-U^2 s_A\} + C_B \exp\{-U^2 s_B\} + \int_0^\infty K(U, V) R_\infty(V)\, dV \quad (2.86a)$$

where

$$K(U, V) = 2 \sum_{j=A,B} c_j \alpha_j U I_1(2UV\alpha_j) \exp\{-(U^2 s_j{}^2 + V^2)\} \quad (2.86b)$$

The integral equation has been solved numerically and the results substituted into (2.83a) to obtain $C_N(y)$. Our required partition function, being $C_N(1)$ as $N \to \infty$, is then

$$C_\infty(1) = 2\int_0^\infty U^{-1}\{1 - R_\infty(U)\} \exp(-U^2)\, dU \quad (2.87)$$

The fraction of bonds broken is finally obtained by numerical differentiation from (2.75), with $N^{-1} \log Z_N \sim C_N(1)$.

3. *Vedenov and Dykhne Theory*

Unlike either of the previous theories, the calculations by Vedenov and Dykhne for a DNA with a random distribution of two components results in a closed analytical expression. This compactness, however, was obtained at the expense of several approximations other than letting $N \to \infty$.

The procedure is to write $\log Z_N$ as a sum of N terms recursively related, and then express the sum as an integral over an appropriate distribution function. The distribution function was obtained, after several approximations, from a Fokker-Planck differential equation. The result for $\log Z_N$ is given by

$$\log Z_N = U + [(1-c)L_1 + cL_3] \coth \left\{ \frac{(1-c)L_1 + cL_3}{2(L_1 - L_3)^2 c(1-c)/U} \right\} \tag{2.88}$$

As before, c is the concentration of G–C base pairs. The notation actually employed by Vedenov and Dykhne differed slightly from that given here. The variables U_1, U_2, V used in their paper correspond to L_1, L_3, U, respectively. $\theta_B(T)$ is obtained from Eq. (2.88):

$$\theta_B(T) = [1 + \coth \alpha - \alpha \operatorname{csch}^2 \alpha] \tag{2.89a}$$

with

$$\alpha = \frac{2U[(1-c)L_1 + cL_3]}{[(L_1 - L_3)^2 c(1-c)]} \tag{2.89b}$$

A comparison of the three theories is given in Table I. The Lehman-McTague result was obtained by computer generating a random sequence

TABLE I

Comparison of Fraction of Bonds Broken for a DNA with a Random Distribution of A-T and G-C Pairs

$c_{GC} = 0.500$ $\quad a = e^{-2U}$

$T_1 = 69.3°C$ $\quad T_2 = 110.3°C$

Temp. (°C)	Montroll and Yu	Lehman and McTague	Vedenov and Dykhne
		$U = 1.50$	
77.8	0.072	0.070	0.084
81.8	0.179	0.176	0.174
85.8	0.329	0.326	0.316
89.8	0.500	0.499	0.500
93.8	0.671	0.674	0.683
97.8	0.821	0.823	0.826
101.8	0.928	0.930	0.916
		$U = 2.20$	
81.8	0.012	0.016	0.000
83.8	0.036	0.047	0.000
85.8	0.105	0.111	0.006
87.8	0.251	0.264	0.086
89.8	0.500	0.496	0.500
91.8	0.749	0.740	0.914
93.8	0.895	0.887	0.994
95.8	0.965	0.949	1.000

of 5000 base pairs, evaluating $\theta_B(T)$ for this sequence, repeating the process ten times and averaging the results. As can be seen, while the Vedenov-Dykhne approximation differs by a small amount from the other results for $U = 1.50$; it becomes highly inaccurate for $U = 2.20$.

III. LOOP ENTROPY MODEL

The model just described accounts for the basic interactions needed in a cooperative transition but fails to include several factors. Correlations between base pairs more remote than nearest ones was not explicitly considered. Since we would expect the long-range correlations to damp out rapidly, this is probably not too important. Another factor that has been omitted in the previous model is the loop entropy. This is the configurational entropy which remains in unbonded strands after connecting bonds are broken. This entropy will depend on the size of the unbonded strands or loop and the flexibility of the DNA strands.

As mentioned earlier, the flexibility of DNA strands in solution is not well known. Most studies assume that the strand configurations are generated by a random walk process. As a first estimate of the loop entropy, Zimm[19] and others have used the Jacobson-Stockmayer[35] calculations for the probability of ring closure as described by an unrestricted random walk process. Several authors[36–38,40] recently have related the loop entropy to the probability of ring closure for a restricted random walk on lattices. In these studies, the random walker is not allowed to return to a point he has visited except the origin. The results of numerical computations indicate the probability of ring closure relative to an open walk varies as j^{-k} with j the number of steps and k a constant. Calculations on different lattices indicate that k is closer to 2 than the $\frac{3}{2}$ value in the unrestricted walk. Although these calculations are a more realistic estimate of the loop entropy than the Jacobson-Stockmayer treatment, there are solvent effects and structural aspects of the polymer strands to consider. These factors would tend to reduce the number of configurations available to both free ends and interior loops, perhaps by different amounts. For these reasons it is best to regard the j^{-k} factor as a physically reasonable phenomonological one.

A. Periodic DNAs without Branching

Some of the first theoretical work on DNA denaturation employed the LE model. Gibbs and DiMarzio[39] and Zimm[20] made use of the Jacobson and Stockmayer factor in melting curve calculations for one-component DNAs. The statistical weights s and σ as defined earlier were applied to these calculations. The loop entropy factor was included by defining the statistical weight for a loop of j broken base pairs as

$$\frac{\sigma}{(j+1)^{3/2}}, \quad j > 0 \tag{3.1}$$

In an effort to apply this model to synthetic periodic DNAs, Crothers and Zimm[8] included the possibility that the strands may slide with respect to each other. Assuming that all possible loop configurations are equally likely, Crothers and Zimm used

$$\frac{\sigma(j+1)}{(j/2+1)^{3/2}}, \quad j > 0 \tag{3.2}$$

as the statistical weight for a ring of j unbonded bases for the poly d(A)· poly d(T) type polymer.

Klotz[40] extended the work of Crothers and Zimm by using an arbitrary value k in place of the exponent $\frac{3}{2}$ in the loop entropy factor. Thus in analogy to (3.2), we now have

$$\frac{\sigma(j+1)}{(j/2+1)^{k}} \tag{3.3}$$

The fraction of broken bonds versus temperature is obtained from the foregoing theories through the use of the quasigrand partition function,

$$\Xi = \sum_{n=1} Z_n a^n$$

a^n being the weight of the reference state (all bonds broken) of the molecule and Z_n the partition function for the molecule. As $n \to \infty$, thermodynamic properties of Z_n derivable from a can be obtained most easily from Ξ. The details of this method are discussed by Lifson and Zimm.[41] Through the preceding techniques, we can calculate thermodynamic properties of a one-component DNA molecule in the LE case.

1. *One-Component DNA*

An alternate method of considering configurational entropy factors is through the Fourier integral representation of the microcanonical distribution described earlier. We will show how that formalism can be applied to a two-component, period-two DNA molecule as well as the one-component DNA molecule.

We start by considering the simpler one-component case, say poly d(A)· poly d(T). As in the derivation of (2.8), we consider the various configurations of the molecule as sequences of intact and broken bonds, the first l_1 bonds being intact, the next m_1 being broken, l_2 intact, m_2 broken, etc. The weight function for a particular arrangement of intact and broken bonds is

$$g_1(l_1)f_1(m_1)g_2(l_2)f_2(m_2)\cdots g_n(l_n)f_n(m_n) \tag{3.4}$$

with $g_i(l_i)$ the weight of l_i intact bonds, and $f_i(m_i)$ the weight of m_i broken bonds. The partition function of a molecule containing N base pairs is the sum of all terms such as (3.4) over all configurations. Neglecting end effects, we have

$$Z_N = \sum_n \sum_{\{l_j\}} \sum_{\{m_j\}} \prod_{j=1}^{n} g(l_j)f(m_j) \tag{3.5}$$

where $\{l_j\}$ and $\{m_j\}$ represent the sets of l_j and m_j consistent with

$$\sum_{j=1}^{n}(l_j + m_j) = N, \qquad 0 < n < [N/2] \tag{3.6}$$

By introducing the condition (3.6) through a delta function in (3.5), and by introducing the factor

$$\exp\{[N - \sum^{n}(m_j + l_j)]\} = 1,$$

we obtain, as before,

$$Z_N = \frac{1}{2\pi}\int_{-\infty}^{\infty} e^{N(\beta+i\alpha)}\{(\Phi\Psi)^{-1} - 1\}^{-1}\,d\alpha \tag{3.7}$$

where β is an arbitrary constant and

$$\Phi = \sum_{l=1}^{\infty} g(l)\lambda^{-l} \tag{3.8a}$$

$$\Psi = \sum_{m=1}^{\infty} f(m)\lambda^{-m} \tag{3.8b}$$

and

$$\lambda \equiv e^{\beta+i\alpha} \tag{3.8c}$$

β is chosen such that Φ and Ψ are convergent and $|\Phi\Psi| < 1$. The integral can be determined by the method of residues providing the integrand has singularities.

To apply this formalism to the LE model, we need the forms of $f(m)$ and $g(l)$. For poly d(A)·poly d(T), we could take

$$f(m) = (m+1)^{-k}e^{-2U_{11}}\exp(U_{11} + L_{11})m \tag{3.9a}$$

as the weight of a bubble of m broken bonds. The factor $(m+1)^{-k}$ is the configurational entropy of a bubble of m broken bonds sandwiched between helical sections relative to m broken bonds at free ends. For $2m$

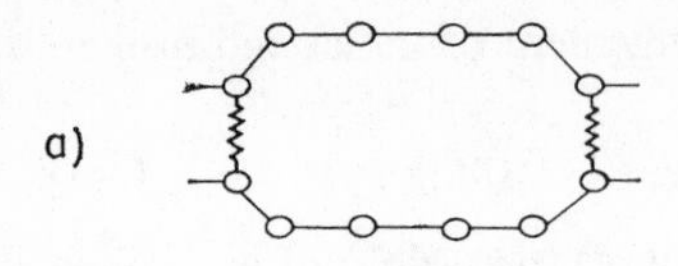

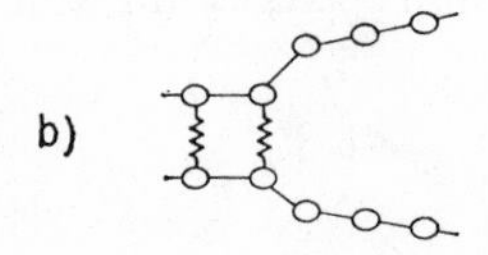

Fig. 7. Sketch of (*a*) a loop of broken bonds, and (*b*) a free end.

unbonded bases in a loop, there are $(2m + 2)$ strand linkages (Fig. 7). Assigning 2^{-k} as the probability of the open-ended configuration b of Fig. 7, we have $(m + 1)^{-k}$. The weight of a sequence of intact bonds, as already given in (2.10a), is

$$g(l) = e^{-2U_{11}} \exp{(U_{11} - L_{11})l} \tag{3.9b}$$

A set of configurations not considered in writing (3.9a) is that due to strand slippage. If we include this effect, the number of bases in a loop of broken bonds can be odd as well as even. Thus we must count the number of bases in the bubble factor, rather than base pairs. Returning to (3.4), we introduce $d(m')$, the weight for a closed loop of m' unbonded bases. In analogy to (3.5), we now have

$$Z_N = \sum_n \sum_{\{l_j\}} \sum_{\{m_j'\}} \prod_{j=1}^{n} g(l_j)\, d(m_j') \tag{3.10a}$$

with m' now being understood as the number of bases in a loop. In place of (3.6) we have

$$N = \sum_{j=1}^{n} l_j + \sum_{j=1}^{n} \tfrac{1}{2} m_j', \qquad 0 < n < \left[\frac{N}{2}\right] \tag{3.10b}$$

If we insert the delta function

$$\delta\left[N - \sum_{j=1}^{n} (l_j + \tfrac{1}{2} m_j')\right]$$

into (3.7a), as well as

$$\exp\left\{\left[N - \sum_{j=1}^{n} (l_j + \tfrac{1}{2} m_j')\right]\right\}$$

we obtain

$$Z_N = \frac{1}{2\pi} \int_{-\infty}^{\infty} e^{N(\beta + i\alpha)} \{(\Phi\Psi)^{-1} - 1\}^{-1}\, d\alpha \tag{3.11a}$$

with

$$\Phi = \sum_{l=1}^{\infty} g(l)\lambda^{-l} \tag{3.11b}$$

$$\Psi = \sum_{m'=1}^{\infty} d(m')\lambda^{-m/2} \tag{3.11c}$$

Since we have changed our counting of states only in unbonded regions, (3.9b) is applicable to both (3.8a) and (3.11b). For $d(m')$, we must consider the effect of strand slippage. A loop of m' unbonded bases will have $m' + 1$ ways of placing the bases on two strands. Although not all of these configurations are equally likely, we will take $(m' + 1)/(\frac{1}{2}m' + 1)^k$ as the configuration entropy for a loop of m' broken bases. Then we set

$$d(m') = (m' + 1)(1 + \tfrac{1}{2}m')^{-k} \exp\{-2U_{11} + \tfrac{1}{2}m'(U_{11} + L_{11})\} \tag{3.12}$$

Placing (3.12) into (3.11c), we obtain

$$\psi = 2^k e^{-2U_{11}}[r^{-2}\{\mathrm{T}(r, k-1) - \mathrm{T}(r, k)\} - 2^{-k}] \tag{3.13}$$

where

$$r \equiv \lambda^{-1/2} \exp \tfrac{1}{2}(U_{11} + L_{11}) \quad \text{and} \quad T(r, k) \equiv \sum_{m'=1}^{\infty} \frac{r^{m'}}{m'^k}$$

From (3.11b) and (3.9b),

$$\Phi = \frac{\{\exp - (U_{11} + L_{11})\}}{(\lambda - \exp[U_{11} - L_{11}])} \tag{3.14}$$

In order to insure the convergence of Φ and Ψ, we choose $\beta > U_{11} + |L_{11}|$. Thus $|r| < 1$ and $\mathrm{T}(r, k)$ is convergent.

Poles of the integrand of (3.11a) are the roots of

$$(\Phi\Psi)^{-1} - 1 = 0 \tag{3.15}$$

The function $\mathrm{T}(r, k)$ is known as the Truesdell function.[42] For noninteger values of $k > 0$, a rapidly convergent series for $\mathrm{T}(r, k)$ can be written when $r \simeq 1$:

$$\mathrm{T}(r, k) = (-\log r)^{k-1}\Gamma(1 - k) + \sum_{n=0}^{\infty} \frac{(\log r)^n \zeta(k - n)}{n!} \tag{3.16}$$

$\zeta(x)$ is Riemann's zeta function of x, and $\Gamma(x)$ is the gamma function of x. We can approximate $\mathrm{T}(r, k)$ quite well for $1 > r > 0.9$ by the first two terms of 3.16. Hence

$$\mathrm{T}(r, k) \simeq (-\log r)^{k-1}\Gamma(1 - k) + \zeta(k), \qquad k \neq \text{integer} \tag{3.17a}$$

For $k = 1$, we have (when $0 < r < 1$)

$$\mathrm{T}(r, 1) = -\log(1 - r) \tag{3.17b}$$

and, for $k = 2$, using the dilogarithm function and Debye function,

$$\mathrm{T}(r, 2) \simeq \frac{\pi^2}{6} + \log r\{1 - \log(1 - r)\} \tag{3.17c}$$

From (3.15), the poles of the integrand are the roots of

$$\lambda - \exp(U_{11} - L_{11}) - e^{-2U_{11}}\{2^k r^{-2}[\mathrm{T}(r, k-1) - \mathrm{T}(r, k)] - 1\} \times \exp\{-(U_{11} + L_{11})\} = 0 \tag{3.18}$$

which satisfy the convergence condition on (3.13), $|\lambda| > \exp(U_{11} + |L_{11}|)$. Evaluating the contour integral in λ space, we find that as $N \to \infty$, $Z_n \sim \lambda_{\max}^N$ where $\lambda_{\max}$ is the largest root of (3.18) and is obtained numerically as a function of L_{11}. The fraction of broken bonds can be derived from

$$\theta_B(T) = \frac{1}{2}\left(1 + \frac{\partial \log \lambda_{\max}}{\partial L_{11}}\right) \tag{3.19}$$

The results for $\theta_B(T)$ from this formulation and that of Crothers and Zimm were identical to four places for $a_{11} = 0.02$, $k = \frac{3}{2}$, and $2 < U_{11} < 3$.

2. *Two-Component DNA*

In order to analyze the melting behavior of poly d(A–C)·poly d(G–T) with the LE model, we must outline a scheme for finding the partition function of DNA whose bases are regularly distributed with period 2. In this case the statistical weights for a sequence of intact or broken bonds depend on the types of base pairs at the terminal ends of the sequence as well as the number of bonds in the sequence. The statistical weight for a sequence of l intact base pairs starting with a p-type bond and ending with a q-type bond is given as $g_{pq}(l)$. As before, the four types of bonds A–T, T–A, G–C, C–G are represented by the numbers 1, 2, 3, 4. For a sequence of m broken bonds, starting with a p and ending with a q bond, we have, similarly, $f_{pq}(m)$. Certain sets of pq are not allowed for even and odd values of l and m. For the concrete example of poly d(A–C)·poly d(T–G), we have

$$\begin{aligned} g_{11}(l) &= f_{11}(m) = 0, && l, m \text{ even} \\ g_{44}(l) &= f_{44}(m) = 0, && l, m \text{ even} \\ g_{41}(l) &= g_{14}(l) = f_{14}(m) = f_{41}(m) = 0, && l, m \text{ odd} \end{aligned} \tag{3.20}$$

Physically, this simply means a sequence starting and ending with an A–T bond cannot have an even number of bonds in the sequence, etc.

The partition function Z_N for this period-two DNA is obtained by the same procedure which we used in the one-component case.

As in the case of a one-component DNA molecule, the partition function of the copolymeric DNA which has an A–T bond on one end and a G–C bond on the other is

$$Z = \frac{1}{2\pi}\int_{-\infty}^{\infty} d\alpha\xi^N \sum_{n=1}^{[N/2]}\sum_{l=1}^{\infty}\sum_{m=1}^{\infty}\left[(1,0)P_n\begin{pmatrix}1\\0\end{pmatrix} + (1,0)P_{n-1}G(l_n)\begin{pmatrix}0\\1\end{pmatrix}\right.$$

$$\left. + (0,1)P_n\begin{pmatrix}0\\1\end{pmatrix} + (0,1)F(m_1)P_{n-1}\begin{pmatrix}1\\0\end{pmatrix}\right] \quad (3.21)$$

where

$$P_n = \prod_{j=1}^{n} G(l_j)F(m_j) \quad (3.22a)$$

$$G(l_j) \equiv \xi^{-l_j}\begin{pmatrix} g_{11}(l_j) & g_{14}(l_j) \\ g_{41}(l_j) & g_{44}(l_j)\end{pmatrix} \quad (3.22b)$$

$$F(m_j) \equiv \xi^{-m_j}\begin{pmatrix} f_{44}(m_j) & f_{41}(m_j) \\ f_{14}(m_j) & f_{11}(m_j)\end{pmatrix} \quad (3.22c)$$

$$\xi \equiv \exp\{\beta + i\alpha\} \quad (3.22d)$$

and as before we have introduced the delta function to insure the supplementary condition (3.6) and have used its Fourier transform representation. It may be noted that the choice of the row and column matrices in (3.21) has been made such that the DNA molecule has an A–T bond on one end and a G–C bond on the other, and we have used the fact that the chain starting with broken bonds and ending with intact bonds is equivalent to the chain starting with intact bonds and ending with broken bonds, but A–T and G–C bonds at the ends interchanged. If the chain ends have some other kind of bonds, the row and column vectors have to be chosen correspondingly. By taking the factors involving the l_j and m_j terms inside of the matrices, then performing the appropriate summation, and, finally, combining products of identical terms, we find that

$$Z_n = \frac{1}{2\pi}\int_{-\infty}^{\infty} \xi^N\, d\alpha \sum_{n=1}^{[N/2]}\left[(1,0)(\Phi\Psi)^n\begin{pmatrix}1\\0\end{pmatrix} + (1,0)(\Phi\Psi)^n\begin{pmatrix}0\\1\end{pmatrix}\right.$$

$$\left. + (0,1)(\Phi\Psi)^n\begin{pmatrix}0\\1\end{pmatrix} + (0,1)(\Phi\Psi)^n\begin{pmatrix}1\\0\end{pmatrix}\right] \quad (3.23)$$

with

$$\Phi \equiv \begin{pmatrix}\Phi_{11} & \Phi_{14}\\ \Phi_{41} & \Phi_{44}\end{pmatrix}, \qquad \Psi = \begin{pmatrix}\Psi_{44} & \Psi_{41}\\ \Psi_{14} & \Psi_{11}\end{pmatrix} \quad (3.24a)$$

$$\Phi_{pq} = \sum_{l=1}^{\infty} g_{pq}(l)\xi^{-l}, \qquad \Psi_{pq} = \sum_{m=1}^{\infty} f_{pq}(m)\xi^{-m} \quad (3.24b)$$

We have made several simplifications in the passage from (3.21) to (3.23). We have replaced $P_{n-1}G(l_n)$ by P_n and $F(m_1)P_{n-1}$ by P_n. This is justified as $N \to \infty$ since they correspond to making slight changes in conditions at the ends of our chain. As has been observed in the preceding section such changes do not change $N^{-1} \log Z_N$ as $N \to \infty$.

As before, we can choose β such that Φ and Ψ exist, and $|(\Phi\Psi)_{ij}| < 1$. We can write the components of the matrix $\Phi\Psi$ in terms of its eigenvalues δ_k $(k = 1, 2)$ and eigenvectors $[\psi_k(1), \psi_k(-1)]$ and the eigenvectors of its transpose matrix $[\phi_k(1), \phi_k(-1)]$. From the relation of a matrix to its eigenvalues and eigenvectors we have

$$[(\Phi\Psi)^n]_{ij} = \sum_{k=1}^{2} \delta_k^{\,n} \psi_k(i)\phi_k(j) \tag{3.25a}$$

The normalization condition is chosen as

$$\sum_{i=\pm 1} \phi_l(i)\psi_m(i) = \begin{cases} 1, & l = m \\ 0, & l \neq m \end{cases} \tag{3.25b}$$

Using (3.25) into (3.23), we obtain as $N \to \infty$,

$$Z_N = (2\pi)^{-1} \int_{-\infty}^{\infty} \xi^N \frac{d\alpha\{(1 - \delta_2)\Omega(\delta_1) + (1 - \delta_1)\Omega(\delta_2)\}}{(1 - \delta_1)(1 - \delta_2)} \tag{3.26}$$

with

$$\Omega(\delta_k) = \delta_k[1 + \psi_k(1)\phi_k(-1) + \psi_k(-1)\phi_k(1)]$$

The poles of the integral are given by

$$(1 - \delta_1)(1 - \delta_2) = 0 \tag{3.27}$$

We can evaluate the contour integral in α space by the method of residues as before. As $N \to \infty$, $\log Z_N \sim N \log \xi_{\max}$ where $\xi_{\max}$ is the largest root of (3.27).

The statistical weights $g_{pq}(l)$ and $f_{pq}(m)$ will determine $\Phi\Psi$ and hence the δ_k. For poly d(A–C)·poly d(T–G) we have three parameters; the correlation parameter between stacked base pairs $u_{14} = \frac{1}{2}(U_{14} + U_{41})$, the averaged base pair strength parameter $\bar{L}_{14} = \frac{1}{2}(L_{14} + L_{41})$, and M_{14}. The statistical weights for sequences of intact bonds are given by (3.20) and

$$g_{14}(l) = \exp\{u_{14} - \bar{L}_{14})l - 2u_{14}\}, \qquad l \text{ even} \tag{3.28a}$$

$$g_{11}(l) = \exp\{(u_{14} - \bar{L}_{14})l - 2(u_{14} + \tfrac{1}{2}M_{14})\}, \qquad l \text{ odd} \tag{3.28b}$$

$$g_{44}(l) = \exp\{(u_{14} - \bar{L}_{14})l - 2(u_{14} - \tfrac{1}{2}M_{14})\}, \qquad l \text{ odd} \tag{3.28c}$$

When we include the loop entropy factor $(m + 1)^{-k}$ and the sliding factor as in the one-component case in the weight for an unbonded segment,

$f_{pq}(m)$, it is found from (3.20) that

$$f_{14}(m) = (2m + 1)(m + 1)^{-k} \exp\{-2u_{14} + m(u_{14} + \bar{L}_{14})\}, \quad m \text{ even} \qquad (3.29a)$$

$$f_{11}(m) = (2m + 1)(m + 1)^{-k} \exp\{-2u_{14} + m(u_{14} + \bar{L}_{14}) + M_{14}\}, \quad m \text{ odd} \qquad (3.29b)$$

$$f_{44}(m) = (2m + 1)(m + 1)^{-k} \exp\{-2u_{14} + m(u_{14} + \bar{L}_{14}) - M_{14}\}, \quad m \text{ odd} \qquad (3.29c)$$

Substituting the preceding relations into (3.24), we obtain

$$\Phi_{11} = \frac{\xi \exp\{(u_{14} - \bar{L}_{14}) - (2u_{14} + M_{14})\}}{\xi^2 - \exp\{2(u_{14} - \bar{L}_{14})\}} \qquad (3.30a)$$

$$\Phi_{14} = \frac{\{\exp - 2\bar{L}_{14}\}}{\{\xi^2 - \exp 2(u_{14} - \bar{L}_{14})\}} = \Phi_{41} \qquad (3.30b)$$

$$\Phi_{44} = \frac{\xi \exp\{(u_{14} - \bar{L}_{14}) - (2u_{14} - M_{14})\}}{\xi^2 - \exp\{2(u_{14} - \bar{L}_{14})\}} \qquad (3.30c)$$

and

$$\Psi_{11} = \frac{2^{-k}\xi \exp(M_{14})}{\exp(3u_{14} + \bar{L}_{14})}[4\mathrm{T}(r^2, k - 1) - \mathrm{T}(r^2, k)] \qquad (3.31a)$$

$$\Psi_{41} = \Psi_{14} = r^{-1} \exp(-2u_{14}) \times [2\mathrm{T}(r, k - 1) - \mathrm{T}(r, k) - r - 2^{-k}\{4\mathrm{T}(r^2, k - 1) - \mathrm{T}(r^2, k)\}] \qquad (3.31b)$$

$$\Psi_{44} = 2^{-k}\xi \exp\{-M_{14} - 3u_{14} - \bar{L}_{14}\}[4\mathrm{T}(r^2, k - 1) - \mathrm{T}(r^2, k)] \qquad (3.31c)$$

with

$$r \equiv \xi^{-1} \exp(u_{14} + \bar{L}_{14})$$

The eigenvalues of $\Phi\Psi$ are

$$\delta_\alpha = A - (-1)^\alpha B, \qquad \alpha = 1, 2 \qquad (3.32)$$

$$A = \tfrac{1}{2}[\Phi_{11}\Psi_{44} + \Phi_{44}\Psi_{11} + 2\Phi_{14}\Psi_{14}]$$

$$B = \tfrac{1}{2}\{[\Phi_{11}\Psi_{44} - \Phi_{44}\Psi_{11}]^2 + 4(\Phi_{14}\Psi_{44} + \Phi_{44}\Psi_{14})(\Phi_{11}\Psi_{14} + \Phi_{14}\Psi_{11})\}^{1/2}$$

Since $Z_N \sim \xi_{\max}^N$ as $N \to \infty$, we must evaluate $\xi_{\max}$ from (3.27), (3.30), (3.31), and (3.32). The fraction of broken bonds is obtained from

$$\theta_B(T) = \frac{1}{2}\left[1 + \frac{N^{-1}\partial \log Z_N}{\partial \bar{L}_{14}}\right] \qquad (3.33)$$

For both the one- and two-component DNAs described by the LE model, the form of L_{ij} is slightly changed from that used in the MI model for similar DNAs [(2.21) and (2.26)]. For the MI model $L_{ij} = 0$ at $T = T_m$ for the unbranched periodic DNAs. This is not exactly true in the LE model due to the inclusion of the loop entropy factor. Instead, we have

$$L_{ij} = a_{ij}(T - T_{cm}) \tag{3.34}$$

where T_{cm} is very close to the T_m of the polymer ($\pm 1C$) but not equal to it.

B. Periodic DNAs with Branching

A proper theoretical examination of poly d(A–T)·poly d(A–T) employing our LE model must also account for hairpin helix branching. There have been three calculations that have included both branching and loop entropy for poly d(A–T)·poly d(A–T), we briefly review those approaches here.

Go[43] extended the method of the quasigrand (QG) partition function of Crothers and Zimm to consider the branched tree structures. The melting behavior of poly d(A–T)·poly d(A–T) as a single and a double-stranded molecule was investigated. The difference between the melting curves for these two cases was insignificant when physically reasonable parameter values were used. All possible branched structures were included in the QG partition function by considering broken bond loops of zeroth order, first order, second order, all the way to infinite order. A zeroth-order loop has one helical segment emanating from it, a first-order loop two helical segments, etc. An infinite-order loop has all possible types of helical segment emanating from it. For a loop of j links, the statistical weight is given by

$$\sigma(2/j)^{-3/2} \tag{3.35}$$

Hijmans[44] derived a formula for the melting curve of poly d(A–T)·poly d(A–T) also based on the model of Crothers and Zimm. All possible configurations for the copolymer were counted by regarding the structure (Fig. 4a) as a rooted tree in the graph theoretical sense. The statistical weights were assigned in the following manner. For each helical section of j consecutive base pairs, the weight was given by

$$\sigma s^j(2)^{3/2} \tag{3.36}$$

A loop of j unboned bases with f helices branching from it was assigned a weight

$$[j!/f!(j-f)!](j-1)^{-3/2}/j \tag{3.37}$$

and an open chain of $j + 1$ bases with f helices branching from it was assigned the weight

$$j!/f!(j-f)! \tag{3.38}$$

DeGennes[45] also studied the poly d(A–T)·poly d(A–T) copolymer, deriving equations for the melting curve and radius of gyration. For a single strand of j unbonded bases starting at $\bar{r}$ and ending at $\bar{r}'$ (in real space), the factor

$$G_j(\bar{r}, \bar{r}') = \left(\frac{3}{2\pi j l_c^2}\right)^{3/2} \exp\left\{\frac{-3\,|\bar{r} - \bar{r}'|}{2jl_c^2}\right\} \tag{3.39}$$

was assigned. l_c is a measure of a coil link. Equation (3.39) is the probability that an unrestricted random walker starting at $\bar{r}$ will, after j steps, be at $\bar{r}'$. For each helical section of $2j$ bases extending from $\bar{r}$ to $\bar{r}'$, the weight is given by

$$\Gamma_j(\bar{r}, \bar{r}') = \left(\frac{c\sigma s^j}{4\pi R^2}\right) \delta(R - jl_h), \qquad R \equiv |\bar{r} - \bar{r}'| \tag{3.40}$$

where c and l_h are constants. The partition function was obtained using Fourier and Laplace transform techniques, integrating the statistical weight for all configurations of a molecule N bases long over all space. An interesting aspect of DeGennes' work was his calculation of the radius of gyration of the branched poly d(A–T)·poly d(A–T). A minimum in the radius of gyration was shown to exist on the low-temperature side of the melting point. This result is suggested by the viscosity curves of poly d(A–T)·poly d(A–T).

Although Go, Hijmans, and DeGennes all use the unrestricted random walk in their loop entropy factor, each uses it in a different way. Let us consider a loop of j unbonded bases—j_1 on one strand and j_2 on the other—between two helical segments. Whereas Go and Hijmans consider the entire j bases in assigning the loop factor, DeGennes has a factor $(j_1)^{-3/2}(j_2)^{-3/2}$.

C. DNAs with a Random Distribution of A–T and G–C Pairs

Several melting curve calculations have been presented using the LE model for a two-component DNA with a random distribution of components. Crothers[13] and Poland and Scheraga[16] have obtained melting curves for a two-component DNA by essentially numerical means. In both cases a coarse graining approximation was employed to minimize the numerical counting of possible DNA configurations. "Coarse graining" groups base pairs together into blocks and assumes each block acts as a unit. Frank-Kamenetskii and Frank-Kamenetskii[46] have also treated a random distribution of two components in the LE model. In this treatment, however, an average loop size was taken for every loop of broken bonds. For a heteropolymer, this approximation may lead to significant errors. In this section an algorithm will be introduced which can produce exactly the partition function for a finite length DNA of specific sequence.

To construct the algorithm for the melting curve, we use a procedure similar to that employed by Lehman and McTague. Given a specific sequence of N base pairs, with N_A A–T and N_B G–C, we can divide Z_N into the following four groupings:

1. Z_N^{hh}. All configurations starting and ending with a bonded or a helical base pair.
2. Z_N^{hc}. All configurations starting with a helical base pair and ending with a broken base pair.
3. Z_N^{cc}. All configurations starting and ending with a coil base pair.
4. Z_N^{ch}. All configurations starting with a coil or broken base pair and ending with a helical base pair.

Clearly

$$Z_N = Z_N^{hh} + Z_N^{hc} + Z_N^{ch} + Z_N^{cc} \tag{3.41}$$

Using the Zimm-Bragg notation we can write Z_N as

$$Z_N = \sum_{\{m_A m_B h\}} s_A^{m_A} s_B^{m_B} \sigma^h \prod_{\{j\}} f(j) \tag{3.42}$$

where m_A and m_B are the number of A–T and G–C base pairs, respectively, which are bonded and h is the number of helical segments. As before, $f(j)$ is the entropy factor for a section of j broken bonds bounded by helical segments. We assume $f(j) = (j + 1)^{-k}$. $\{m_A m_B h\}$ and $\{j\}$ are all the sets of values consistent for the polymer. The fraction of broken bonds is

$$\theta_B(T) = 1 - N^{-1}\left(\frac{\partial}{\partial \log s_A} + \frac{\partial}{\partial \log s_B}\right) \log Z_N \tag{3.43}$$

If we make the transformation $m_A = N_A - n_A$, $m_B = N_B - n_B$ where n_A, n_B are the number of broken A–T and G–C bonds, we have

$$Z_N = s_A^{N_A} s_B^{N_B} \sum s_A^{-n_A} s_B^{-n_B} \prod f(j) = s_A^{N_A} s_B^{N_B} Q_N \tag{3.44}$$

where Q_N is the partition function for the same molecule assigning s_A^{-1} for a broken A–T bond, s_B^{-1} for a broken G–C bond, and 1 for any intact bond. The fraction of broken bonds in this case becomes

$$\theta_B(T) = N^{-1}\left(\frac{\partial}{\partial \log s_A^{-1}} + \frac{\partial}{\partial \log s_B^{-1}}\right) \log Q_N \tag{3.45}$$

The formulation of (3.44) and (3.45) form a useful check of the numerical calculations and helps overcome the computational difficulty of very large numbers in the early regions of the transition.

The algorithm for Z_N is obtained from induction and hence we can show the cases $N = 2$, $N = 3$, $N = 4$, in detail. Consider the case $N = 2$.

The configurations are

We have

$$Z_2^{hh} = s_2 s_1, \qquad Z_2^{ch}(1) = s_1$$
$$Z_2^{hc} = s_2\sigma, \qquad Z_2^{cc}(2) = 1$$

$Z_N^{ch}(i)$ denotes the configurations of Z_N^{ch} which have i broken bonds looking in from the left. A similar definition holds for $Z_N^{cc}(i)$ and Z_N^{cc}. For $N = 3$ we have

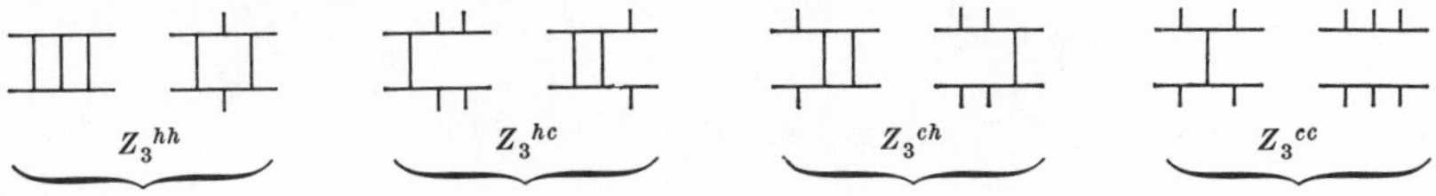

Z_3^{hh} Z_3^{hc} Z_3^{ch} Z_3^{cc}

where

$$Z_3^{hh} = s_3 s_2 s_1 + s_3(2)^{-k}\sigma s_1, \qquad Z_3^{ch}(1) = s_2 s_1, \qquad Z_3^{ch}(2) = s_1$$
$$Z_3^{hc} = s_3 s_2\sigma + s_3\sigma, \qquad Z_3^{cc}(1) = \sigma s_2, \qquad Z_3^{cc}(3) = 1$$

Now Z_3 can be written as

$$Z_3 = (1 \quad 1)\begin{pmatrix} Z_3^{hh} & Z_3^{hc} \\ \sum_{i=1}^{2} Z_3^{ch}(i) & \sum_{i=1}^{3} Z_3^{cc}(i) \end{pmatrix}\begin{pmatrix} 1 \\ 1 \end{pmatrix}$$
$$= (1 \quad 1)\begin{pmatrix} s_3 & \sigma s_3 0 \\ 1 & 1 \end{pmatrix}\begin{pmatrix} Z_2^{hh} & Z_2^{hc} \\ \sum_{i=1}^{1} Z_2^{ch}(i) & \sum_{i=1}^{2} Z_2^{cc}(i) \end{pmatrix}\begin{pmatrix} 1 \\ 1 \end{pmatrix} \tag{3.46}$$

with 0 an operator with the properties

$$0\left[\sum_{i=1}^{1} Z_2^{ch}(i)\right] = \sum_{i=1}^{1} (i+1)^{-k} Z_2^{ch}(i) \tag{3.47a}$$

$$0\left[\sum_{i=1}^{2} Z_2^{cc}(i)\right] = \sum_{i=1}^{1} (i+1)^{-k} Z_2^{cc}(i) + Z_2^{cc}(2) \tag{3.47b}$$

For $N = 4$ we have the configurations

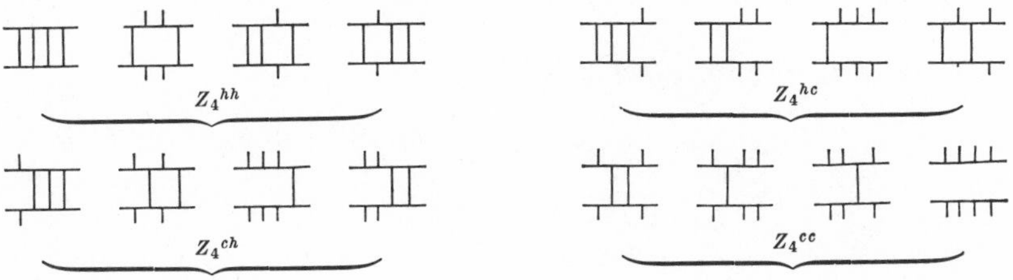

Z_4^{hh} Z_4^{hc}

Z_4^{ch} Z_4^{cc}

where

$$Z_4^{hh} = s_4 s_3 s_2 s_1 + s_4 \sigma (3)^{-k} s_1 + s_4 s_3 \sigma (2)^{-k} s_1 + s_4 (2)^{-k} \sigma s_2 s_1$$
$$Z_4^{hc} = s_4 s_3 s_2 \sigma + s_4 s_3 \sigma + s_4 \sigma + s_4 (2)^{-k} \sigma s_2 \sigma$$
$$Z_4^{ch}(1) = s_3 s_2 s_1 + s_3 (2)^{-k} \sigma s_1, \qquad Z_4^{ch}(2) = s_2 s_1, \qquad Z_4^{ch}(3) = s_1$$
$$Z_4^{cc}(1) = s_3 s_2 \sigma + s_3 \sigma, \qquad Z_4^{cc}(2) = s_2 \sigma, \qquad Z_4^{cc}(4) = 1$$

We find

$$\begin{pmatrix} Z_4^{hh} & Z_4^{hc} \\ \sum_{i=1}^{3} Z_4^{ch}(i) & \sum_{i=1}^{4} Z_4^{cc}(i) \end{pmatrix} = \begin{pmatrix} s_4 & \sigma s_4 0 \\ 1 & 1 \end{pmatrix} \begin{pmatrix} Z_3^{hh} & Z_3^{hc} \\ \sum_{i=1}^{2} Z_3^{ch}(i) & \sum_{i=1}^{3} Z_3^{cc}(i) \end{pmatrix} \tag{3.48}$$

and if we continue the process to higher N we find, in general,

$$\begin{pmatrix} Z_N^{hh} & Z_N^{hc} \\ \sum_{i=1}^{N-1} Z_N^{ch}(i) & \sum_{i=1}^{N} Z_N^{cc}(i) \end{pmatrix} = \begin{pmatrix} s_N & \sigma s_N 0 \\ 1 & 1 \end{pmatrix} \begin{pmatrix} Z_{N-1}^{hh} & Z_{N-1}^{hc} \\ \sum_{i=1}^{N-2} Z_{N-1}^{ch}(i) & \sum_{i=1}^{N-1} Z_{N-1}^{cc}(i) \end{pmatrix} \tag{3.49}$$

with

$$\begin{aligned} 0\left[\sum_{i=1}^{N-2} Z_{N-1}^{ch}(i)\right] &= \sum_{i=1}^{N-2} (i+1)^{-k} Z_{N-1}^{ch}(i) \\ 0\left[\sum_{i=1}^{N-1} Z_{N-1}^{cc}(i)\right] &= Z_{N-1}^{cc}(N-1) + \sum_{i=1}^{N-2} (i+1)^{-k} Z_{N-1}^{cc}(i) \end{aligned} \tag{3.50a}$$

and

$$\left.\begin{aligned} Z_N^{ch}(1) &= Z_{N-1}^{hh}, \qquad Z_N^{ch}(i) = Z_{N-1}^{ch}(i-1) \\ Z_N^{cc}(1) &= Z_{N-1}^{hc}, \qquad Z_N^{cc}(i) = Z_{N-1}^{cc}(i-1) \end{aligned}\right\} \quad i > 1 \tag{3.50b}$$

The partition function for a specified sequence of base pairs is thus

$$Z_N = G_N + H_N = (1 \quad 1) \prod_{m=3}^{N} \begin{pmatrix} s_n & \sigma s_n 0 \\ 1 & 1 \end{pmatrix} \begin{pmatrix} G_{n-1} \\ H_{n-1} \end{pmatrix} \tag{3.51}$$
$$G_n \equiv Z_n^{hh} + Z_n^{hc}$$
$$H_n \equiv \sum_{i=1}^{n-1} Z_n^{ch}(i) + \sum_{i=1}^{n} Z_n^{cc}(i)$$

The fraction of broken bonds is

$$\theta_B(T) = 1 - N^{-1}(\partial G_N + \partial H_N)(G_N + H_N)^{-1}$$
$$\partial \equiv \frac{\partial}{\partial \log s_{\mathrm{A}}} + \frac{\partial}{\partial \log s_{\mathrm{B}}} \tag{3.52}$$

For $\theta_B(T)$ we need, in addition to (3.50) and (3.51), the derivative relations

$$\partial 0\left(\sum_{i=1}^{N-2} Z_{N-1}^{ch}(i)\right) = \sum_{i=1}^{N-2} (i+1)^{-k}\, \partial Z_{N-1}^{ch}(i) \tag{3.53a}$$

$$\partial Z_N{}^{hh} = s_N\left\{Z_{N-1}^{hh} + \partial Z_{N-1}^{hh} + \sigma 0\left[\sum_{i=1}^{N-2} Z_{N-1}^{ch}(i)\right] + \sigma 0\left[\sum_{i=1}^{N-2} \partial Z_{N-1}^{ch}(i)\right]\right\} \tag{3.53b}$$

$$\partial Z_N{}^{ch}(1) = \partial Z_{N-1}^{hh}, \qquad \partial Z_N{}^{ch}(i) = \partial Z_{N-1}^{ch}(i-1), \qquad i > 1 \tag{3.53c}$$

$$\partial Z_N{}^{cc}(1) = \partial Z_{N-1}^{hc}, \qquad \partial Z_N{}^{cc}(i) = \partial Z_{N-1}^{cc}(i-1), \qquad i > 1 \tag{3.53d}$$

$$\partial Z_N{}^{hc} = s_N\left\{Z_{N-1}^{hc} + \partial Z_{N-1}^{hc} + \sigma 0\left[\sum_{i=1}^{N-1} Z_{N-1}^{cc}(i)\right] + \sigma 0\left[\sum_{i=1}^{N-1} \partial Z_{N-1}^{cc}(i)\right]\right. \tag{3.53e}$$

The preceding algorithm can be used to find the exact solution of $\theta_B(T)$ for a specified sequence of base pairs. However, the amount of time necessary to perform the calculation is large. To obtain a melting curve for $N = 1000$ requires about 16 minutes on the IBM 360-65. In keeping with the economic pressures of the day, we can drastically reduce the execution time for larger N by making use of the coarse graining approximation. This also provides a check on the accuracy of the coarse graining approximation.

In the coarse graining approximation, a DNA N base pairs long is divided into N/M blocks, each block containing M base pairs. Each block is treated as a single unit with its properties derived from an average over its components. Some fine graining is restored in blocks that border on bonded and unbonded regions.

Consider a typical configuration of such blocks with the right end a terminal end of the molecule

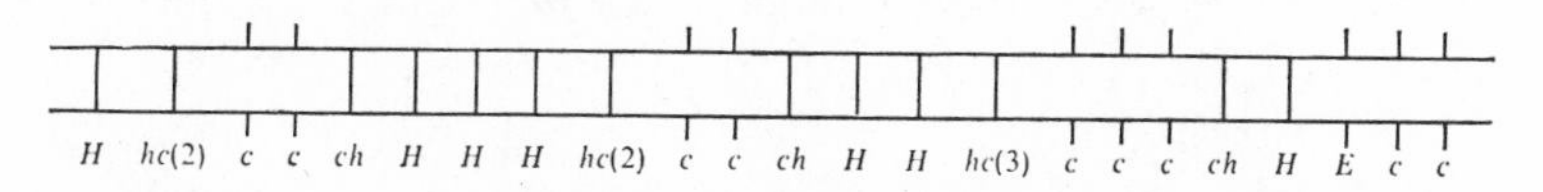

We define the blocks in the following manner. An H block is a run of M intact bonds, a C block is a run of M broken bonds, a ch block is a helical border block with a C to the left and an H or the end to the right, $hc(i)$ represents a helical border block with $i$$C$s and a ch to the right, and an E block is a coil border block with an H or ch block to the left and all Cs or the end to the right.

The statistical weights of the blocks in terms of s_A, s_B, and σ are as follows:

Block Type	Statistical Weight
H	$\langle s_m \rangle^M \equiv \mathcal{s}_m \quad \text{with} \quad \langle s_m \rangle = s_A^{f_A(m)} s_B^{1-f_A(m)}$
ch	$\sum_{j=1}^{M} \langle s_m \rangle^j = \frac{\langle s_m \rangle(\mathcal{s}_m - 1)}{(\langle s_m \rangle - 1)} \equiv \xi_m$
E	$\sigma \sum_{j=0}^{M-1} \langle s_m \rangle^j = \frac{\sigma(\mathcal{s}_m - 1)}{(\langle s_m \rangle - 1)} = \frac{\sigma \xi_m}{\langle s_m \rangle}$
$hc(i)$	$\left(\frac{\sigma}{\xi_{m-i-1}}\right) \sum_{l=1}^{M} \sum_{j=1}^{M} \langle s_m \rangle^l \frac{\langle s_{m-i-1} \rangle^j}{[(i+2)M - j - l + 1]^k} \equiv \sigma \Lambda_m(i)$
C	1

In this scheme $\ell_A(m)$ is the fraction of A–T bonds in the mth block. For $hc(i)$ we make a further approximation employed by Poland and Scheraga. Treating the two sums independently and using averages of l and j, we have

$$\Lambda_m(i) \simeq \sigma \xi_m / [(i+2)M + 1 - \partial \xi_m / \xi_m - \partial \xi_{m-i-1} / \xi_{m-i-1}]^k \tag{3.54}$$

The partition function is obtained in a manner similar to the exact calculation. In this case we have N_b blocks with M base pairs each, and the running subscripts, n, go from 1 to N_b. In order not to count a border block twice, we have to divide Z_n^{hh} and Z_n^{hc} into two groupings,

$$Z_n^{hh\,(\text{or } hc)} = Z_{nH}^{hh\,(\text{or } hc)} + Z_{nL}^{hh\,(\text{or } hc)} \tag{3.55}$$

$Z_{nH}^{hh(\text{or } hc)}$ is the statistical weight, s.w., of the hh (or hc) configurations where the nth block is an H block, and $Z_{nL}^{hh\,(\text{or } hc)}$ is the s.w. of hh (or hc) configurations where the nth block is an $hc(i)$ block. Examining the cases $n = 1, 2, 3, \ldots$, we find

$$\begin{bmatrix} Z_{nH}^{hh} + Z_{nL}^{hh} & Z_{nH}^{hc} + Z_{nL}^{hc} \\ \sum_{i=1}^{n-1} Z_n^{ch}(i) & \sum_{i=1}^{n} Z_n^{cc}(i) \end{bmatrix} = \begin{bmatrix} \mathcal{s}_n & \sigma \mathcal{O}_n \\ F_n & 1 \end{bmatrix} \begin{bmatrix} Z_{(n-1)H}^{hh} + Z_{(n-1)L}^{hh} & Z_{(n-1)H}^{hc} + Z_{(n-1)L}^{hc} \\ \sum_{i=1}^{n-2} Z_{n-1}^{ch}(i) & \sum_{i=1}^{n-1} Z_{n-1}^{cc}(i) \end{bmatrix} \tag{3.56}$$

where the operators $\mathcal{O}_n$ and F_n are defined by

$$\mathcal{O}_n\left[\sum_{i=1}^{n-2} Z_{n-1}^{ch}(i)\right] \equiv \sum_{i=1}^{n-2} \Lambda_n(i) Z_{n-1}^{ch}(i) \tag{3.57a}$$

$$\mathcal{O}_n\left[\sum_{i=1}^{n-1} Z_{n-1}^{cc}(i)\right] \equiv \frac{\mathcal{s}_n \xi_{n-1}}{\langle s_{n-1}\rangle} + \sum_{i=1}^{n-2} \Lambda_n(i) Z_{n-1}^{cc}(i) \tag{3.57b}$$

$$F_n[Z_{(n-1)H}^{hh} + Z_{(n-1)L}^{hh}] \equiv \left(\frac{\xi_{n-1}}{\mathcal{s}_{n-1}}\right) Z_{(n-1)H}^{hh} + Z_{(n-1)L}^{hh} \tag{3.57c}$$

further,

$$Z_{nH}^{hh} = \mathcal{s}_n[Z_{(n-1)H}^{hh} + Z_{(n-1)L}^{hh}], \qquad Z_{nL}^{hh} = \sigma \mathcal{O}_n \sum_{i=1}^{n-2} Z_{n-1}^{ch}(i) \tag{3.58a}$$

$$Z_{nH}^{hc} = \mathcal{s}_n[Z_{(n-1)H}^{hc} + Z_{(n-1)L}^{hc}] + \frac{\sigma \mathcal{s}_n \xi_{n-1}}{\langle s_{n-1}\rangle} \tag{3.58b}$$

$$Z_{nL}^{hc} = \sigma \mathcal{O}_n \sum_{i=1}^{n-2} Z_{n-1}^{cc}(i) \tag{3.58c}$$

The partition function is thus given by

$$Z_N = Z_{(N_b \times M)} = (1 \quad 1) \prod_{n=3}^{N_b} \begin{bmatrix} \mathcal{s}_n & \sigma \mathcal{O}_n \\ F_n & 1 \end{bmatrix} \begin{bmatrix} V_{n-1} \\ W_{n-1} \end{bmatrix} \tag{3.59}$$

with

$$V_n = Z_n^{hh} + Z_n^{hc}, \qquad W_n = Z_n^{ch} + Z_n^{cc}$$

For the fraction of broken bonds we have

$$\theta_B(T) = 1 - \frac{N^{-1}(\partial V_{N_b} + \partial W_{N_b})}{V_{N_b} + W_{N_b}} \tag{3.60}$$

The recursive relations for the terms of ∂V_{N_b} and ∂W_{N_b} must be obtained. These relations follow in a straightforward manner from (3.52) and (3.58).

In Fig. 8 are the results of an investigation on the effect of block size on the melting curve of a specific sequence DNA generated by a random number generator. Other authors who have produced melting curves using algorithmic methods have usually employed a block size of 50 base pairs. As can be seen, there is a significant dependence of the melting curve on the block size taken for a finite length DNA. For a random distribution of two base pair strengths, it is not until $M = 10$ that the convergence of the melting curve to $M = 1$ becomes clear.

Crothers as well as Poland and Scheraga have pointed out that a linear map of the melting DNA can be obtained theoretically. This is of particular interest with relation to electron microscope pictures of melting DNA,

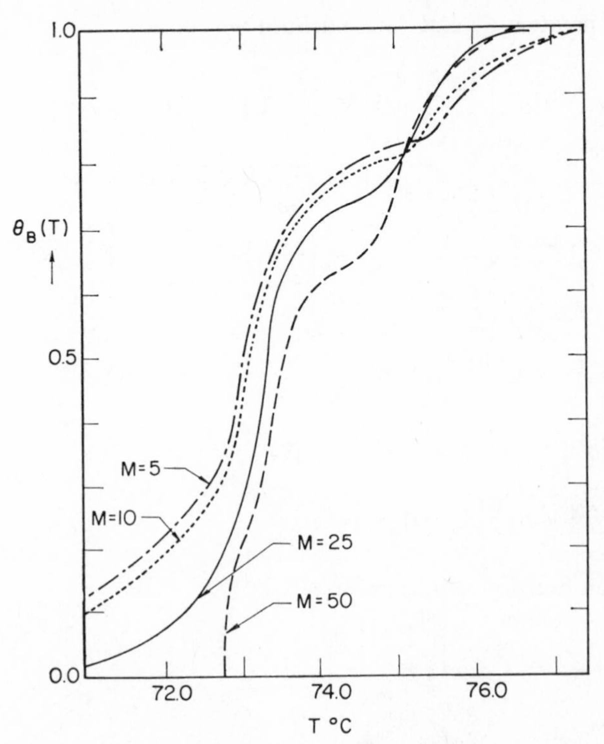

Fig. 8. Comparison of coarse graining approximation for same random sequence using block sizes of $M = 5, 10, 25, 50$. $N = 5000$, $T_1 = 49.5$, $T_2 = 99.5$, $a_1 = 0.0194$, $a_2 = 0.0168$.

and more will be said on this in the discussion section. To get the probability that the jth bond (or jth block) is broken we need to numerically evaluate

$$\theta_j = 1 - \frac{N^{-1}(\partial^j V_{N_b} + \partial^j W_{N_b})}{V_{N_b} + W_{N_b}} \tag{3.61}$$

$$\partial^j \equiv \frac{\partial}{\partial \log s_j} \quad \text{for} \quad j\text{th bond}$$

$$\partial^j \equiv \frac{\partial}{\partial \log \mathscr{s}_j} \quad \text{for} \quad j\text{th block}$$

IV. COMPARISON OF THEORIES WITH EXPERIMENT

A. Periodic DNAs

The melting curves for poly d(A)·poly d(T), poly d(A–T)·poly d(A–T), and poly d(A–C)·poly d(G–T) are shown in Figs. 9, 11, and 12, respectively. The theoretical results for these periodic sequences were compared with these curves. Using the MI models, we have for poly d(A)·poly d(T) the parameters a_{11}, T_{11} and U_{11}; for poly d(A–T)·poly d(A–T) a_{12}, T_{12}, and u_{12}; and for poly d(A–C)·poly d(T–G), a_{14}, T_{14}, u_{14}, and M_{14}. For the LE model, we have additionally the loop entropy exponent k.

The T_{ij} are obtained from the experimental melting temperatures. The constants a_{ij} can be obtained from Eq. (2.31b), if we know the enthalpy change of bond breaking. The calorimetric experiments of Scheffler and Sturtevant[47] for poly d(A–T)·poly d(A–T) indicate an enthalpy change of approximately $\Delta H = 8.0$ kcal/mole of base pairs for the bond breaking of this polymer in 0.1 SSC. This gives $a_{12} = 0.01966$. To evaluate the other a_{ij}, it is assumed that the entropy change in bond breaking is independent of the base pair types. Equations (2.30), (2.31b), and the appropriate T_m determine a_{11} and a_{14}.

Poly d(A)·poly d(T) was examined first. Using $a_{11} = 0.0192$ and $T_{11} = 53.25$°C, the parameter U_{11} was determined by comparing the theories with the experimental curve. For the MI model [Eq. (2.20)] a value of $U_{11} = 2.95$ gave a very good fit (Fig. 9). An equally good fit was made for the MIS model with $U_{11} = 3.10$. The values correspond to stacking free energies of $\Delta G_s = 7.7$ kcal/mole base pairs and 8.1 kcal/mole base pairs, respectively. The LE model has two unknown parameters for poly d(A)·poly d(T), k and U_{11}. If we choose $k = \frac{3}{2}$, a good fit can be

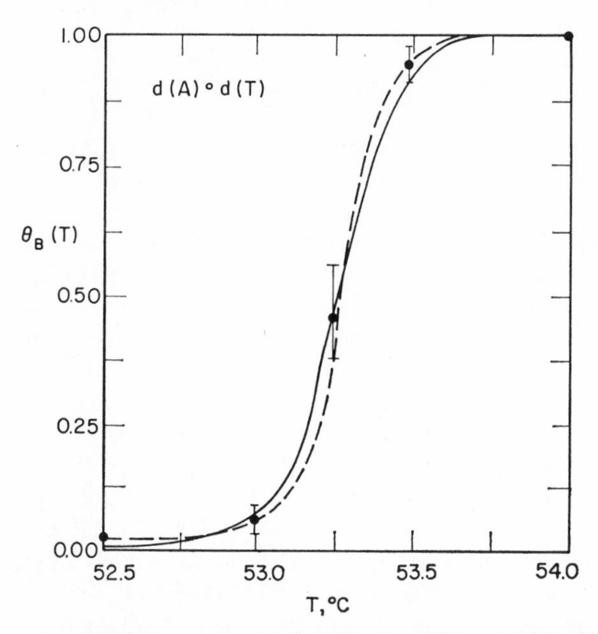

Fig. 9. Experimental melting curve of poly d(A)·poly d(T) in 0.1 SSC (Φ). Theoretical melting curves; MI model – – –, LE model ———.

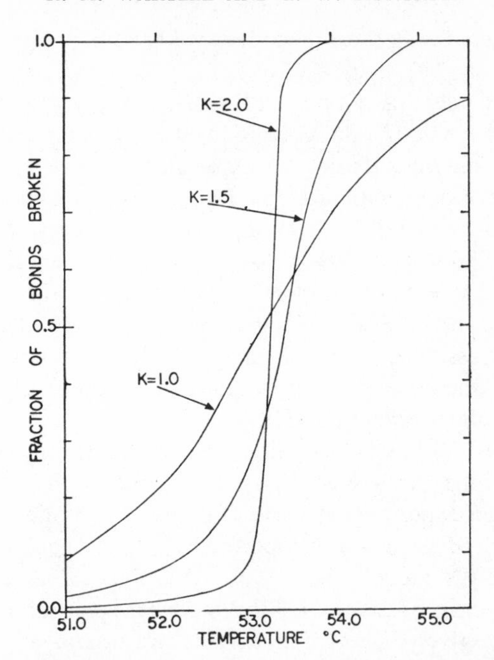

Fig. 10. Three melting curves using the LE model for poly d(A)·poly d(T). For $k = 1.0$, $a_{11} = 0.019$, $U_{11} = 1.70$; $k = 1.5$, $a_{11} = 0.019$, $U_{11} = 1.70$; $k = 2$, $a_{11} = 0.019$, $U_{11} = 1.70$.

made with $U_{11} = 2.68$. This U value corresponds to a $\Delta G_s \simeq 7.0$ kcal/mole base pairs.

The melting curve is quite sensitive to the choice of k, as demonstrated in Fig. 10. Generally, as k increases, the melting curve becomes more asymmetric; the initial phase of the transition flattens out and the later portion steepens. This behavior persists even if adjustments are made in U_{11} or a_{11}. It has also been described by Klotz. Since the asymmetry of the curve depends on k, an estimate of the upper limit of k can be made by comparing theory with experiment. It was found that a good fit could not be made to poly d(A)·poly d(T) data when $k > 2$.

Using the U_{11} values obtained from poly d(A)·poly d(T), the three models were compared with the melting curve of poly d(A–T)·poly d(A–T). The theoretical calculations included the hairpin helix effects. Equation (2.43e) is applicable to both the MI and MIS models and was employed in both cases. Hijman's equation (42) was used for the LE model.

As can be seen from Fig. 11, the MIS and LE models agree well with the experimental points. The fit for the MI model ($u_{12} = 2.95$) is not as good. Theoretical curves produced from DeGennes' model of

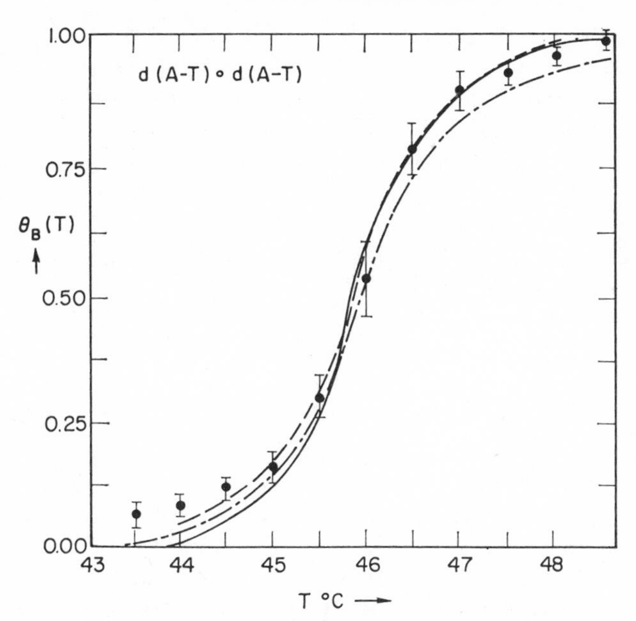

Fig. 11. Melting curve of poly d(A-T)·poly d(A-T) in 0.1 SSC (Φ). Theoretical curves; MI model (— – —), MIS model (——), LE model (- - - -).

poly d(A–T)·poly d(A–T) are sharper than the LE model curves when the same u_{12} are employed (not shown).

A comparison of the various theories with the experimental results for poly d(A–C)·poly d(T–C) was carried out employing one additional parameter, M_{14}. Since all the experimental data are required to establish the value of M_{14}, the theory using it can only be tested when more periodic molecules are studied. Another factor which differentiates poly d(A–C)· poly d(T–G) from the other polymers discussed is its shorter length. Whereas the previous polymers are composed of approximately 2000 base pairs, the poly d(A–C)·poly d(G–T) samples employed are about 400–500 base pairs long. At this length, end effects and dissociation of strands become increasingly important. The end effects can be estimated by evaluating exactly the melting curves for an N base pair long DNA using the algorithmic methods presented in Section IIIC (LE model) and the Lehman-McTague method (MI model). A straightforward modification of Eqs. (2.79) and (3.53) can be made to include the factor

$$\exp\{-\tfrac{1}{2}M_{j,j+1}(\sigma_j - \sigma_{j+1})\}$$

These modifications enable the algorithmic models to include four rather than two different components or base pair types. Appendix B shows the modifications needed to change from the two-component to four-component models. The dissociation of the helix into two strands has been considered by several authors.[16,20,48,49] We do not include it in our present comparison since it would add parameters which are not yet well known. Instead we make the assumption that dissociation will have a small effect on the early part of the transition and compare theoretical curves with the experimental curve in the region $\theta_B < 0.250$.

The parameter M_{14} was defined earlier as the difference between self-energy parameters J_1 and J_3,

$$M_{14} = \frac{(J_1 - J_3)}{2}$$

In the region of the transition we will postulate M_{14} to be a constant. Theoretical curves for poly *d*(A–C)·poly *d*(T–G), 450 base pairs long were obtained with the LE model algorithm for a specific sequence, modified to include M_{14}, [Eq. (3.52)]. The parameters employed were $a_{14} = 0.0181$, $T_{14} = 73.6$, $u_{14} = 2.68$, and various M_{14}. $M_{14} = 3.10$ gave a good fit for $\theta_B < 0.250$ as shown in Fig. 12. Also included in this figure is the theoretical curve for the four-component MI model algorithm with $M_{14} = 3.10$ and $u_{14} = 2.95$. Neither of the theoretical curves fits the data in the late stages of the transition. This is not surprising since it is in this region that dissociation effects will be strongest. Varying u_{14} or M_{14} by $\pm 10\%$ changes

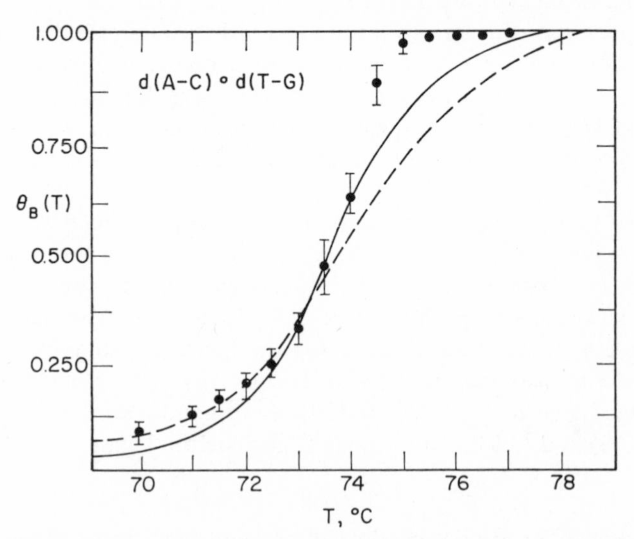

Fig. 12. Melting curve of poly *d*(A-C)·poly *d*(T-G) in 0.1 SSC (Φ). Theoretical curves; MI model (– – –), LE model (——).

the theoretical curves slightly, but it does not alter the comparison relative to the experimental points.

B. Natural DNAs

In Figs. 1, 14, and 15 are the experimental curves of four naturally occurring DNA samples. Figure 15 illustrates the influence the base sequence can have on the shape of the melting curve. The G–C concentration of Bacillus subtilis and Calf thymus DNAs are almost identical, yet their melting curves are significantly different.

Theoretical curves are obtained in the following manner. For the MI and MIS models, the Lehman-McTague algorithm has been employed with $U = 2.95$ and $U = 3.10$. The melting curves are obtained by randomly generating a hypothetical DNA 5×10^3 base pairs long. Increasing the number of units to 10^4 made little difference in the resultant curves. Theoretical curves using the LE model are produced from (3.60) by randomly generating 5×10^3 unit sequences with $U = 2.68$ and $k = \frac{3}{2}$. Coarse graining is employed with a block size of $M = 5$. A random number generator is employed with p_a the probability of having an A–T bond, $1 - p_a$ the probability of having a G–C bond. The values of G–C concentration when given are the actual values of the sequences generated.

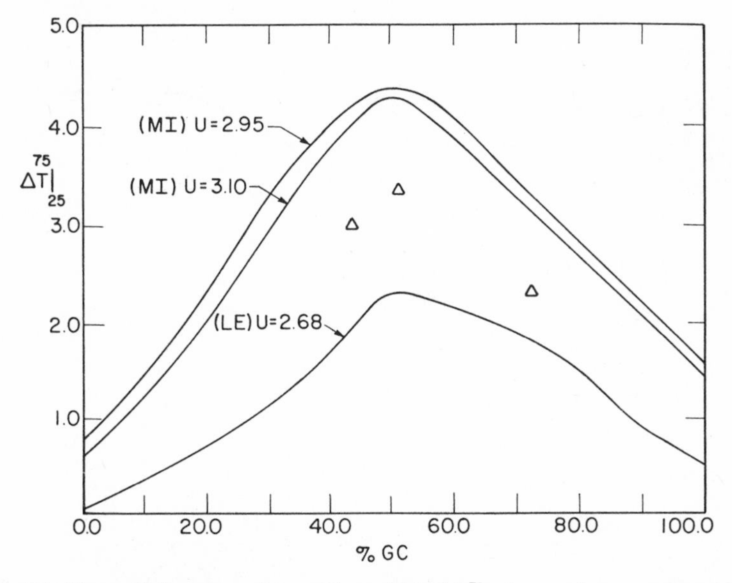

Fig. 13. Theoretical curves of transition width ΔT_{25}^{75} of a random sequence DNA versus %GC for the three models shown. Experimental points (Δ) are from bacterial DNA transition curves in 0.1 SSC.

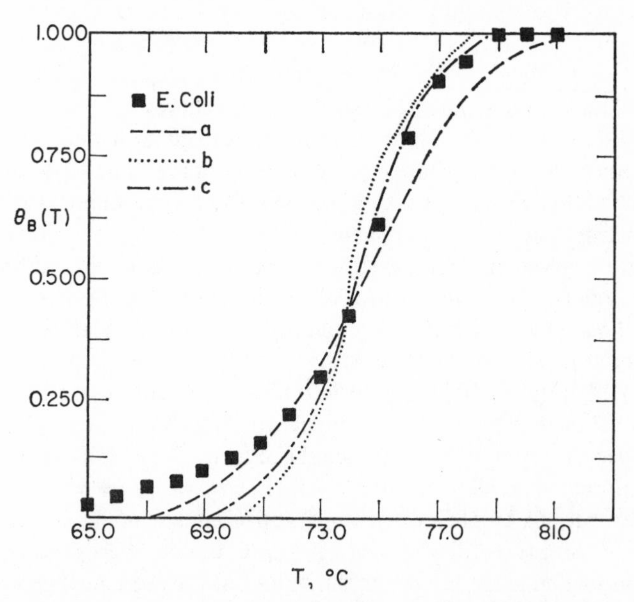

Fig. 14. Transition curve of E. coli DNA (■) and theoretical curves (a), (b), and (c)

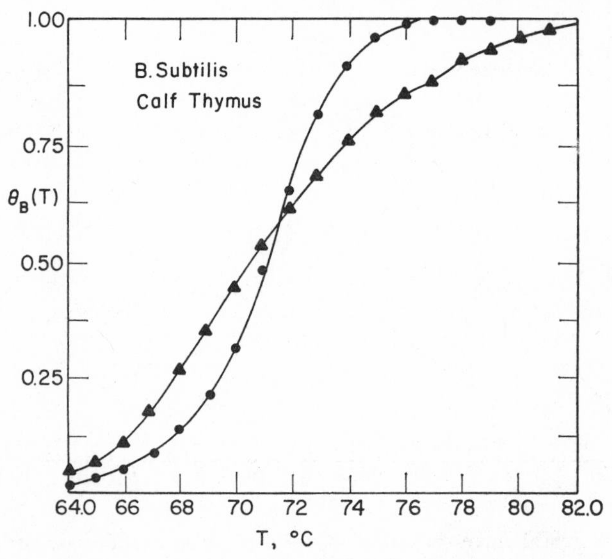

Fig. 15. Transition curves for B. subtilis DNA (●) and Calf thymus DNA (▲) in 0.1 SSC.

As a measure of the width of the transition curve we use $\Delta T_{0.25}^{0.75}$, the temperature interval between $\theta_B = 0.250$ and 0.750. Figure 13 shows a plot of $\Delta T_{0.25}^{0.75}$ versus %G–C for randomly generated DNA sequences using the three models. Although the models gave essentially the same transition widths for the periodic DNAs, they are quite different for the random sequence molecules. The difference between the MI and MIS models is obviously due to different U values. The higher U for the MIS case results from the inclusion of strand sliding in the periodic DNAs. Strand sliding has also been considered in the LE model. For the periodic DNAs, the effective k value is $\frac{1}{2}$ and, for random sequence molecule, $\frac{3}{2}$. It should be mentioned that the curves of Fig. 13 are not to be considered highly accurate since only single melting curves were generated at 10% G–C intervals. The main features of Fig. 13, the hierarchy of transition widths for the models, is valid. By generating five curves with the same a priori probability of $p_a = 0.50$, we find standard deviations in $\Delta T_{0.25}^{0.75}$ of ± 0.19°C for the LE model and ± 0.085°C for the MI model. Experimental points are from the three bacterial DNAs.

A comparison between theoretical transition curves and the experimental melting curve of E. coli DNA molecules is presented in Fig. 14. Theoretical curves were obtained by averaging the melting curves calculated from five sequences generated with the same initial instructions. Curves a, b, and c correspond respectively to a random sequence DNA using the MI model, a random sequence DNA using the LE model, and a block sequence DNA using the LE model.

The initial instruction for a random sequence DNA is that each base pair has the probability p_a of being an A–T, and $(1 - p_a)$ of being a G–C. For a block sequence DNA, we divide an N-unit DNA into K blocks with L_i base pairs in the ith block:

$$\sum_{i=1}^{K} L_i = N$$

The lengths L_i are not necessarily equal. Each block is assigned a probability p_i of having an A–T base pair. Within each block, a sequence of A–Ts and G–Cs is randomly generated according to its p_i. This technique has been introduced to simulate the frequency distribution of E. coli DNA segments found by Yamagisha.[50]

From buoyant density centrifugation with mercury, Yamagisha has obtained a histogram of the relative number of segments, $N_s(y)$ versus y %GC content (segment length 1.4×10^6 daltons). The y values ranged from about 39% to 57%GC. The histogram was converted to a frequency distribution $P(y)$ by normalizing to 1 and used to simulate a block sequence DNA. Blocks in the generated sequence were assigned a $p_i = y$ with lengths L_i proportional to $P(y)$. The block order and p_i and L_i values are listed in Table II. Blocks were arranged such that for any 500 base pairs the average %GC would be close to that of the entire molecule.

TABLE II

Block number	Length	p_i
1	250	0.525
2	170	0.390
3	100	0.568
4	100	0.437
5	100	0.560
6	90	0.440
7	170	0.556
8	110	0.450
9	270	0.546
10	120	0.457
11	350	0.541
12	160	0.464
13	450	0.5333
14	170	0.470
15	210	0.477
16	480	0.517
17	270	0.485
18	470	0.509
19	330	0.495
20	400	0.500
21	250	0.525

From Fig. 14 we observe the transition of E. coli DNA is wider than the random sequence DNA in the LE case. When block heterogeneity is introduced into the theoretical sequence, curve c, it brings theoretical and experimental curves into closer agreement. $\Delta T|_{0.25}^{0.75}$ for curve c is ~0.6°C larger than for curve b. Compared to the random sequence molecule of the MI model, we find the E. coli transition the steeper of the two curves. When block heterogeneity is introduced into the MI calculation in the same manner as for the LE case, we again find a broadening of the transition over the random sequence DNA. The change in the transition width in this case, enlarges the difference between theory and experiment.

V. DISCUSSION

The comparison between theoretical models and poly d(A)·poly d(T), and poly d(A–T)·poly d(A–T) indicates that hairpin helix structures must be considered in the latter polymer. If hairpin helices are omitted in the models [e.g., Eq. (2.20)], predicted transition curves for poly *d*(A–T)·poly *d*(A–T) are much too narrow. When the strand-sliding effect is included in the theoretical models, the same correlation parameter, U provides good agreement with experiment for both polymers. Although this indicates that strand sliding may be important, it is not an overwhelming piece of evidence.

We might also argue that since the A–T bond strength differs between poly *d*(A–T)·poly *d*(A–T) and poly *d*(A)·poly *d*(T), the correlation forces may also differ. In this case the simple MI model with $U_{11} = 2.95$ and $U_{12} = 3.10$ would describe the experimental transitions. A further examination of a variety of other periodic DNA molecules should provide a more conclusive check of the parameter values and models.

The comparison of theory with experiment for naturally occurring DNAs can also provide some information on the validity of the various models. A randomly generated sequence of A–T and G–C base pairs is a reasonable first approximation to the base pair sequence of an actual DNA molecule from an organism. We can expect, however, that natural DNA molecules will have a more heterogeneous base pair distribution due to information coding requirements. Thus we should find naturally occurring DNAs with a broader transition curve than a theoretical random sequence DNA. From Fig. 14 we observe this to be definitely the case for the LE model. In the MI case the random sequence DNA had a broader transition than the bacterial DNAs. The MIS model's random sequence molecules also have broader transitions than the bacterial DNAs. Since some virulent bacteriophage DNAs have steeper transition curves than bacterial DNAs with similar base compositions, the LE model appears to best describe both periodic and naturally occurring DNAs. Another factor strengthening this conclusion is the increase in transition width caused by block heterogeneity. Including the segment distribution of E. coli DNA brings the theoretical LE calculation closer to the experimental transition. For the MI and MIS calculations, the opposite effect occurs.

Although the preceding arguments favor the LE model, they are not conclusive regarding the k value. The value of $k = \frac{3}{2}$ used in the LE calculations is probably a good estimate with the precise value being in the range $1.0 < k < 2.0$. An experiment which may be useful in establishing k would be to measure the amount of unwinding from the ends relative to the interior of a periodic DNA molecule. The MI model ($k = 0$) predicts unwinding occurs equally fast from the ends and from loop formation. The LE model predicts unwinding to occur mainly from the ends, the relative amount of free ends and interior loop contribution depending on k. Electron microscopy may provide a method of measuring the relative amounts of unbonded base pairs at free ends and in the interior as a function of temperature. Experiments of this nature have been used in making denaturation maps of A–T-rich regions in natural DNAs.

The comparison of an accurate theoretical transition theory with the electron microscope pictures of unwinding DNA molecules provides an interesting though difficult possibility. Crothers[13] has briefly discussed this and alluded to the problem of preserving the equilibrium condition of DNA molecules in solution when they are transferred to a solid matrix for microscopy. Equation (3.61) might provide a regional map of the uncoiling DNA in the same manner as the denaturation maps. Which region unwinds will depend on the base pair composition of neighboring regions as well as the composition of the region in question. The main application of denaturation maps obtained by Inman,[57] Doerfler and Kleinschmidt,[58] and others, has been to locate A–T-rich regions in natural DNAs. Electron microscope pictures are taken of partially uncoiled DNA in the presence of formaldehyde. When this procedure is carried out at different temperatures in the melting zone, a map of the unwinding DNA is obtained. The segments of the molecules which melt out at lower temperatures are assumed to be A–T-rich segments.

Although A–T-rich segments are expected to unwind early in the helix-coil transition, the temperature at which they uncoil will clearly be influenced by their length, base pair sequence, and the base pair composition of adjoining segments. Theoretical models of the unwinding process can provide quantitative estimates on the influence of these factors and hence have potential in obtaining base pair distribution information. A precondition for a theory-experiment comparison of this nature is the maintenance of structural equilibrium of DNA during the experiment.

Acknowledgments

The authors wish to express appreciation to Dr J. Maniloff for allowing unrestricted use of his laboratory facilities, for his interest, and for his advice Also, we gratefully acknowledge useful discussions with Dr. N. S. Goel and the generous gift of poly *d*(A–C)·poly *d*(T–G) from Dr. F. N. Hayes.

References

1. R. Thomas, *Biochim. Biophys. Acta*, **14**, 231 (1954).
2. W. Szybalski, *Thermobiology*, A. H. Rose, ed., Academic Press, London, 1967, Chap. 4.
3. G. Felsenfeld and H. Miles, *Ann. Rev. Biochem.*, **36**, 407 (1967).
4. J. Marmur, R. Round, and C. Schildkraut, *Progr. Nucleic Acid Res.*, **1**, 231 (1963).
5. A. Michelson, J. Massoulie, and W. Guschbauer, *Progr. Nucleic Acid Res.*, **6**, 83 (1967).
6. S. A. Rice and P. Doty, *J. Amer. Chem. Soc.*, **79**, 3937 (1957).

7. E. Simon and B. H. Zimm, *J. Stat. Phys.*, **1,** 41 (1969).
8. D. M. Crothers and B. H. Zimm, *J. Mol. Biol.*, **9,** 1 (1964).
9. N. S. Goel, N. Fukuda, and R. Rein, *J. Theor. Biol.*, **18,** 350 (1968).
10. Yu. S. Lazurkin, M. D. Frank-Kamenetskii, and E. N. Trifonov, *Biopolymers*, **9,** 1253 (1970).
11. E. W. Montroll and N. S. Goel, *Biopolymers*, **4,** 855 (1966).
12. N. S. Goel and E. W. Montroll, *Biopolymers*, **6,** 731 (1968).
13. D. M. Crothers, *Biopolymers*, **6,** 1391 (1968).
14. D. Poland and H. A. Scheraga, *Biopolymers*, **7,** 887 (1969).
15. D. M. Crothers, *Accts. Chem. Res.*, **2,** 225 (1969).
16. D. Poland and H. A. Scheraga, *Theory of Helix Coil Transition in Biopolymers*, Academic Press, New York, 1970.
17. G. Newell and E. W. Montroll, *Rev. Mod. Phys.*, **25,** 353 (1953).
18. B. H. Zimm and J. K. Bragg, *J. Chem. Phys.*, **31,** 526 (1959).
19. B. H. Zimm, *J. Chem. Phys.*, **33,** 1349 (1960).
20. J. Marmur and P. Doty, *J. Mol. Biol.*, **5,** 109 (1962).
21. S. Falkow and D. Cowie, *J. Bacteriol.*, **96,** 777 (1968).
22. R. D. Wells, J. E. Larson, R. C. Grant, C. R. Cantor, and B. Shortle, *J. Mol. Biol.* **54,** 465 (1970), and papers cited therein.
23. M. Riley, B. Maling, and M. J. Chamberlin, *J. Mol. Biol.*, **20,** 359 (1966).
24. R. Langridge, *J. Cell Physiol.*, **74,** 1 (1969).
25. H. C. Spatz and R. L. Baldwin, *J. Mol. Biol.*, **11,** 213 (1965), and papers in this series.
26. M. Mandel, L. Igambi, J. Bergendahl, M. L. Dodson, and E. Scheltgen, *J. Bacteriol.* **101,** 333 (1970).
27. J. Josse, A. D. Kaiser and A. Kornberg, *J. Biol. Chem.*, **236,** 864 (1961).
28. (a) G. W. Lehman and J. P. McTague, *J. Chem. Phys.*, **49,** 3170 (1968); (b) G. W. Lehman, *Statistical Mechanics*, T. A. Bak, Ed., Benjamin, New York, 1967, p. 204.
29. A. A. Vedenov, A. M. Dyhkne, A. D. Frank-Kamenetskii, and M. D. Frank-Kamenetskii, *Mol. Biologiya*, **1,** 313 (1967).
30. A. A. Vedenov and A. M. Dykhne, *Zh. Ekop. Teor. Fiz.*, **55,** 357 (1968).
31. S. Strassler, *J. Chem. Phys.*, **46,** 1037 (1967).
32. T. R. Fink and D. M. Crothers, *Biopolymers*, **6,** 863 (1968).
33. E. W. Montroll and L. Yu, *Proc. Irvine Conf. Localized Excitations in Solids*, R. F. Wallis, Ed., Plenum Press, New York, 1968.
34. L. Yu, Ph.D. Dissertation, University of Maryland (1967).
35. H. Jacobson and W. H. Stockmayer, *J. Chem. Phys.*, **18,** 1600 (1950).
36. M. Fisher, *J. Chem. Phys.*, **45,** 1469 (1966).
37. M. F. Sykes, *J. Chem. Phys.*, **39,** 410 (1963).
38. C. Domb, J. Gillis, and G. Wilmers, *Proc. Phys. Soc. London*, **85,** 625 (1965).
39. J. H. Gibbs and E. A. DiMarzio, *J. Chem. Phys.*, **30,** 271 (1959).
40. L. C. Klotz, *Biopolymers*, **7,** 265 (1969).
41. S. Lifson and B. H. Zimm, *Biopolymers*, **1,** 15 (1963).
42. C. Truesdell, *Ann. Math.*, **46,** 144 (1945).
43. M. Go, *J. Phys. Soc. Japan* **23,** 597 (1967).
44. J. Hijmans, *J. Chem. Phys.*, **47,** 5116 (1967).
45. P. G. DeGennes, *Biopolymers*, **6,** 715 (1968).
46. A. D. Frank-Kamenetskii and M. D. Frank-Kamenetskii, *Mol. Biologiya*, **3,** 375, (1969).
47. I. Scheffler and J. Sturtevant, *J. Mol. Biol.*, **42,** 577 (1969).
48. D. M. Crothers, N. R. Kallenbach, and B. H. Zimm, *J. Mol. Biol.*, **11,** 802 (1965).
49. J. Applequist and V. Damle, *J. Chem. Phys.*, **39,** 2719 (1963).

50. H. Yamagisha, *J. Mol. Biol.*, **49,** 603 (1970).
51. J. Marmur, *J. Mol. Biol.*, **3,** 208 (1961).
52. E. J. King, *Biochemical J.*, **26,** 292 (1932).
53. A. H. Brown, *Arch. Biochem.*, **11,** 269 (1946).
54. W. Mejbaum, *Z. Physiol. Chem.*, **258,** 117 (1939).
55. O. H. Lowry, N. J. Rosebrough, A. L. Farr, and R. J. Randall, *J. Biol. Chem.*, **193,** 265 (1951).
56. J. Eigner and P. Doty, *J. Mol. Biol.*, **12,** 549 (1965).
57. R. B. Inman, *J. Mol. Biol.*, **28,** 103 (1967).
58. W. Doerfler and A. K. Kleinschmidt, *J. Mol. Biol.*, **50,** 579 (1970).

APPENDIX A. END EFFECTS FOR LONG DNA CHAINS

In the Fourier integral representation of the partition function, the boundary or end conditions of the DNA molecule have corresponded to the first base pair always remaining intact and the Nth base pair always being broken. We now describe more realistic end conditions and show that, for long molecules, the choice of ends makes no difference on $\log Z_N$.

Let us consider a straight DNA molecule with N identical base pairs. The following four end conditions cover all possible bonding states or configurations of the molecule:

$$\left.\begin{matrix}\text{(i)}\\ \text{(ii)}\end{matrix}\right\}\ \text{first base pair intact, } N\text{th base pair}\left\{\begin{matrix}\text{broken}\\ \text{intact}\end{matrix}\right.$$

$$\left.\begin{matrix}\text{(iii)}\\ \text{(iv)}\end{matrix}\right\}\ \text{first base pair broken, } N\text{th base pair}\left\{\begin{matrix}\text{broken}\\ \text{intact}\end{matrix}\right.$$

As in the derivations of (2.8) and (3.5), we consider the various configurations of the molecule as alternating runs of intact and broken base pair bonds.

A configuration with end conditions (i) is described as having the first l_1 bonds intact, the next m_1 bonds broken, l_2 intact, m_2 broken, . . . , l_n intact, m_n broken. The weight function for this arrangement of intact and broken bonds is

$$g_1(l_1)f_1(m_1)g_2(l_2)f_2(m_2)\cdots g_n(l_n)f_n(m_n) = \prod_{j=1}^{n} g_j(l_j)f_j(m_j) \quad \text{(A.1a)}$$

with $g_i(l_i)$ the weight of l_i intact bonds and $f_i(m_i)$ the weight of m_i broken bonds. All configurations with end conditions (i) are counted when we sum over all permitted l_j and m_j.

For configurations with end conditions (ii), the weight functions have the form

$$\frac{\left[\prod_{j=1}^{n} g_j(l_j)f_j(m_j)\right]}{f_n(m_n)} \quad \text{(A.1b)}$$

Division by $f_n(m_n)$ insures that the last bond is always intact.

In a similar manner, the weight functions for configurations with end conditions (iii) and (iv) are written as

$$\frac{\left[\prod_{j=1}^{n} f_j(m_j)g_j(l_j)\right]}{g_n(l_n)} \tag{A.1c}$$

and

$$\left[\prod_{j=1}^{n} f_j(m_j)g_j(l_j)\right] \tag{A.1d}$$

respectively. The partition function of the molecule is the sum over all terms such as (A.1),

$$Z_N = \sum_{n=1}^{N} \sum_{\{l_j\}} \sum_{\{m_j\}} \prod_{j=1}^{n} g(l_j)f(m_j)[1 + f_n^{-1}(m_n) + g_n^{-1}(l_n) + 1] \tag{A.2}$$

where $\{l_j\}$ and $\{m_j\}$ are the sets of l_j and m_j consistent with

$$N = \sum_{j=1}^{n}(l_j + m_j), \qquad 0 < n < \left[\frac{N}{2}\right] \tag{A.3}$$

Introducing the condition (A.3) through a delta function in (A.2) and inserting the factor

$$1 = \exp\{\beta[N - \sum^{n}(m_j + l_j)]\}$$

in (A.2), we find in a manner similar to (2.8b)

$$Z_N = \left(\frac{1}{2\pi}\right)\int_{-\infty}^{\infty} d\,\alpha e^{N(\beta+i\alpha)}\{(\phi\psi)^{-1} - 1\}^{-1}[2 + \phi^{-1} + \psi^{-1}] \tag{A.4}$$

with

$$\phi = \sum_{l=1}^{\infty} g(l)\lambda^{-l} \tag{A.5a}$$

$$\psi = \sum_{m=1}^{\infty} f(m)\lambda^{-m} \tag{A.5b}$$

$$\lambda \equiv e^{\beta+i\alpha} \tag{A.5c}$$

Equation (A.4) is the result for free end boundary conditions as shown in MGII.

We next consider the nearest neighbor MI case to obtain $\log Z_N$. With

$$g(l) = \exp\{(U - L)l - 2U\}$$
$$f(m) = \exp\{(U + L)m - 2U\}$$

we have from (A.5)

$$\phi = \frac{[\exp - (U + L)]}{[\lambda - \exp (U - L)]} \tag{A.6a}$$

$$\psi = \frac{[\exp - (U - L)]}{[\lambda - \exp (U + L)]} \tag{A.6b}$$

The poles of the integrand of (A.4) correspond to the values of λ which obey

$$(\phi\psi)^{-1} - 1 = 0$$

or

$$\lambda_{1,2} = e^U[\cosh L \pm (\sinh^2 L + e^{-4U})^{1/2}] \tag{A.7}$$

The integral (A.4) can be transformed to a contour integral in the variable λ. The contour C is a circle of radius e^β enclosing the two poles $\lambda_{1,2}$. The integral becomes

$$Z_N = \left[\frac{2e^{-U}(\lambda_1 - \lambda_2)^{-1}}{2\pi i}\right]\int_C \lambda^{N-1}[(\lambda - \lambda_1)^{-1} - (\lambda - \lambda_2)^{-1}] \times [\lambda \cosh L - 2 \sinh U]\, d\lambda \tag{A.8a}$$

or

$$Z_N = 2e^{-U}(\lambda_1 - \lambda_2)^{-1} \times [\lambda_1^{N-1}(\lambda_1 \cosh L - 2 \sinh U) - \lambda_2^{N-1}(\lambda_2 \cosh L - 2 \sinh U)] \tag{A.8b}$$

As $N \to \infty$,

$$Z_N \sim 2e^{-U}(\lambda_1 - \lambda_2)^{-1}\lambda_1{}^N(\cosh L - 2\lambda_1{}^{-1}\sinh U)$$

and

$$\log Z_N \sim \lambda_1{}^N \tag{A.9}$$

This is the same result obtained with the end conditions of (2.15).

APPENDIX B. FOUR-COMPONENT SPECIFIC SEQUENCE ALGORITHMS

In the two-component algorithm of Lehman and McTague, there are two statistical factors s_A and s_B corresponding to the A–T (or T–A) and G–C (or C–G) base pairs, respectively. s_A and s_B are related to the bond strengths L_1 and L_3 by Eq. (2.73). We now consider the nearest-neighbor contribution to the bond strength, K_{ij}, to depend on the types of bonds involved. The most general case allows four different bond types A–T, T–A, G–C, C–G and 10 values of K_{ij}, i.e., K_{11}, K_{12}, K_{14}, K_{41}, ... The bond strength L_i of the ith bond, and hence s_i, now depends on the $i \pm 1$ bond types as well as the ith bond type.

For the case of poly d(A–C)·poly d(T–G), we have two alternating bond strengths, L_1 and L_4. The two s_j from (2.17), (2.73), and the equation preceding (3.28a),

$$s_1 = \exp\{-2L_1\} = \exp\{-2(\bar{L}_{14} + M_{14})\} \tag{B.1a}$$

$$s_4 = \exp\{-2L_4\} = \exp\{-2(\bar{L}_{14} - M_{14})\} \tag{B.1b}$$

If we wished to include a bond-type dependence on U, the correlation parameter, then σ_i, the parameter for the Lehman-McTague calculation, would depend on the ith and $(i - 1)$st bond type:

$$\sigma_i = \exp\{-4U_{i,i-1}\} \tag{B.2}$$

For poly d(A–C)·poly d(T–G), we have averaged over the two U_{ij}s, $u_{14} = \frac{1}{2}(U_{14} + U_{41})$, and thus

$$\sigma = \exp\{-4u_{14}\} \tag{B.3}$$

The partition function takes on the same form:

$$Z_N = \prod_{m=1}^{N} \begin{pmatrix} s_m & s_m\sigma \\ 1 & 1 \end{pmatrix} \begin{pmatrix} 1 \\ 0 \end{pmatrix} \tag{B.4}$$

For the LE algorithm the modification to the four-component case follows in exactly the same manner. For poly d(A–C)·poly d(T–G), there are two alternating matrices in the s_js given by (B.1) and a single σ related to U by (B.3).

We should mention that the Zimm-Bragg notation actually describes a slightly different model than the notation used for the MI model in Section II. In the limit of large N, both models give the same result. However, for finite N, assigning a σ to only one end of each helical segment is not the same as assigning $\sigma^{1/2}$ to both ends. A matrix in Zimm-Bragg notation analogous to (2.18b) is

$$Z_N = \prod_{m=1}^{N} \begin{pmatrix} s_m & s_m\sigma^{1/2} \\ \sigma^{1/2} & 1 \end{pmatrix} \begin{pmatrix} 1 \\ 0 \end{pmatrix} \tag{B.5}$$

with (B.1) and (B.2) remaining the same.

APPENDIX C. EXPERIMENTAL MATERIALS AND PROCEDURES

Experimental melting curves of three synthetic types of DNA molecule and four naturally occurring types were obtained. Poly d(A)·poly d(T) and poly d(A–T)·poly d(A-T) were purchased from Biopolymers Inc.;

poly d(A–C)·poly d(G–T) was the generous gift of Dr. F. N. Hayes. Calf thymus DNA was purchased from Sigma Chemical Co. and Cal. Bio. Chem. Co. The three bacterial DNAs Escherichia coli B, Bacillus subtilis 168, and Micrococcus lysodeikticus were isolated by a modified version of the Marmur[51] procedure. After lysis, RNAse I (Cal. Bio. Chem. Co., final concentration 50 μg/ml) was incubated with the cell lysate for 10 min at 37°C, and then followed by a pronase digestion (Cal. Bio. Chem. Co., 100 μg/ml) for 30 min at 37°C. Distilled phenol buffered with Tris SDS (0.1M Tris buffer + 1% sodium dodecyl sulfate + 0.1M NaCl pH 9.0) was used in the place of chloroform-isoamyl alcohol as a protein denaturing agent. The solvent 0.1SSC (0.015M NaCl + 0.0015M sodium citrate pH = 7.0) was employed for the melting curve studies. 1.0 SSC, a tenfold concentration of 0.1 SSC, was used in the sedimentation velocity measurements, and to obtain melting curves of the DNAs. All DNAs were dialysed against two or three changes of the solvent being employed.

The natural DNA samples were characterized by ϵ_p^{260}, their molar extinction coefficients at 260 mu based on phosphorus,[52] and their RNA[53,54] and protein[55] content. All the samples had values within 6100–6800 cm²/mol P and less than 3% RNA or protein. DNA concentrations were measured via the net absorbance at 260 mμ assuming $\mathrm{Abs}_{260:25°\mathrm{C}}$ = 0.200 per 10 μg/ml of DNA.

The synthetic DNAs were characterized by their absorbance spectrum and the T_m of their melting curves in 0.1 SSC and 1.0 SSC. The values of these parameters agreed with those in the literature.[20,22,23]

Sedimentation coefficients $S_{20°\mathrm{C},w}$ were obtained at a concentration of 20 μg/ml in 1.0 SSC. A Spinco model E ultracentrifuge was used at a speed of 44,000 rpm using ultraviolet optics. The centrifuge cell was fitted with a standard 12 mm Kel-F centerpiece. The three synthetic DNAs gave the following values: poly d(A)·poly d(T) $S_{20°\mathrm{C},w} = 10.4$S; poly d(A–T)·poly d(A–T) $S_{20°\mathrm{C},w} = 10.9$S; and poly d(A–C)·poly d(G–T) $S_{20°\mathrm{C},w} =$ 6.5S. These values correspond to molecular weights of 1.18×10^6, 1.02×10^6, and 2.40×10^5 daltons, respectively, using the relationship[56] $S_{20°\mathrm{C},w} = 0.116\ (\mathrm{M.W.})^{0.325}$. The sedimentation coefficient of Calf thymus DNA measured 15.3S, and that of E. coli B 22.2S.

Melting curves were obtained using a Gilford 240 spectrophotometer. Readings were taken at 5-mμ intervals from 245–265 mμ with a fixed slit width of 0.10 mm. The cuvette chamber was thermally controlled by a Haake circulating bath which pumped a water-ethylene glycol mixture through thermoplates on the sides of the chamber. Temperature of the solvent blank was measured with a platinum resistance probe. The probe was placed into the cuvette containing the solvent blank through a specially made teflon stopper. The leads from the probe were led out of the top of a high-walled chamber cover to a bridge circuit (Gilford model 210) which provided a voltage output proportional to temperature. The temperature was read on a Varian x,x recorder.

Before the run the DNA solutions and solvent blank were bubbled with helium and pipetted into 3-ml cuvettes with 1-cm lightpaths. Concentrations were approximately 20 ug/ml. A few drops of mineral oil were placed on top of all solutions and the cuvettes stoppered.

In the region of the helix coil transition, bath temperature was raised slowly so that measurements could be taken under equilibrium conditions at 0.5-C intervals. Readings at the five wavelengths were taken manually and repeated to check equilibrium at each temperature.

The temperature at which all bonds are intact T_i is defined as 10°C before any significant absorbance rise. The temperature T_f at which all bonds are broken is defined as the temperature such that for $T > T_f$ $(e_T/e_{Ti} - e_{T-1}/e_{Ti}) < 0.005$ where e_T is the extinction coefficient at temperature T. Six runs were taken for each polymer in 0.1 SSC. After correcting the relative absorbance for thermal expansion, the fraction of bonds broken was calculated at each wavelength assuming the absorbance change proportional to fraction bonds broken. These values were then averaged. It was noted that all wavelengths used gave nearly identical melting curves. The standard deviation errors were caused almost entirely by the run-to-run deviations.

THE LATTICE VIBRATIONS OF MOLECULAR SOLIDS*

O. SCHNEPP AND N. JACOBI

Department of Chemistry, University of Southern California, Los Angeles, California

CONTENTS

* This work was supported by a grant from the National Science Foundation and from the the Army Research Office-Durham.

I. INTRODUCTION

Molecular solids contain units which are tightly bound groups of atoms. This molecular unit is recognizable by its resemblance to the gas phase entity. The binding energy between these molecular units is generally much smaller (by a factor of 10) than that of the interatomic binding energies. It is a direct result of this difference that the internal vibrations of a molecule are at much higher frequencies (typically 1000 cm^{-1}) than the relative motions of the molecules in the solid. These latter motions are called lattice vibrations and typically have frequencies of the order of 100 cm^{-1} or less.

Since the internal molecular frequencies are so much higher than the lattice vibrational frequencies, the two classes of motion are mostly separable. It is then possible to discuss the lattice vibrations separately, assuming the molecule to be a rigid body. The motions of these molecular units can be divided into two types: translatory displacements of the centers of mass and orientational or librational displacements during which the centers of mass remain fixed. The importance of such lattice motions in molecular solids was recognized early by Cruikshank (1958) who was concerned with their effects on the determination of crystal structures by X-ray diffraction. This author made, in fact, the first attempts to deduce the frequencies of lattice vibrations from the X-ray experimental results. This application is still of great interest (Pawley, 1967; Schomaker and Trueblood, 1968).

More recently, interest in the lattice vibrations of molecular solids has centered around the elucidation of intermolecular potential functions. If a pair potential is assumed, it can be tested by calculating the observables by application of the appropriate lattice dynamics. Dows (1962) was the first to attempt a calculation of lattice vibrational frequencies from an assumed potential. He treated solid ethylene and used a model which represented the pair interaction by repulsions between hydrogen atoms on neighboring molecules.

The lattice dynamics of solids consisting of atoms or ions is a well developed field in physics. The work of Born and Huang (1954) is the classic which deals with the theory of small vibrations as applied to periodic solids. Another standard reference on the subject is by Ziman (1960). However, a new element is introduced when molecular solids are treated. Librational motions occur only in these solids and in some ionic solids which contain polyatomic ions.

The total number of external degrees of freedom in a molecular solid, including translations and librations, is $6NZ$ ($5NZ$ for linear molecules). Here N is the total number of unit cells in the crystal and Z the number of molecules or sites in the unit cell. Of these, three modes have zero frequency; these are the acoustic modes at $\mathbf{q} = 0$.

Direct observation and measurement of lattice vibrations were made possible by the development of modern instrumentation in both Raman and far infrared spectroscopy. These optical techniques are restricted to the study of lattice modes at $\mathbf{q} = 0$ (where $\mathbf{q}$ is the wave vector). They are a very small fraction of all $6NZ$ modes of the solid. However, because of the symmetry selection rules which apply at $\mathbf{q} = 0$, it is possible to make definitive assignments of the observed lines. Early investigators used Raman spectroscopy before the advent of the laser. Kastler and Rousset (1941) investigated solid naphthalene and Fruehling (1950) solid benzene. The first far infrared studies of lattice vibrations of molecular solids are due to the group of Gebbie who applied interferometry or Fourier spectroscopy to the problem. Anderson, Gebbie, and Walmsley (1964) and Walmsley and Anderson (1964; also Anderson and Walmsley, 1964) studied the far infrared spectra of the hydrogen halide, halogen and carbon dioxide crystals.

Inelastic neutron scattering is a most powerful tool for the study of lattice dynamics. In particular, coherent scattering by single crystals allows, in principle, the measurement of all modes of the crystal as a function of $\mathbf{q}$, the wave vector in the Brillouin zone. The extensive investigation of a molecular solid, hexamethylenetetramine, has recently been reported by Dolling and Powell (1970). It is to be expected that our knowledge of intermolecular potentials will be enriched greatly by such studies. Incoherent scattering experiments also are profitable inasmuch as they measure the total frequency distribution. This distribution has usually prominent peaks, which do not necessarily coincide with the $\mathbf{q} = 0$ modes and therefore provide additional observables for comparison with theory.

The lattice vibrations of molecular solids have received brief consideration in reviews dealing with the infrared spectra of these solids (Dows, 1963, 1965, 1966). The review by Schnepp (1969) provides a good summary of the field. A recent review by Venkataraman and Sahni (1970) of the lattice dynamics of complex crystals contains much subject matter related to the present review. A number of good reviews are available on the lattice motions of ionic, covalent, and metallic solids (Mitra and Gielisse, 1964; Martin, 1965; Cochran and Cowley, 1969).

In Section II of this review we discuss the different forms of classical lattice dynamical treatments which have been applied to molecular solids. The applications to specific systems and comparison of results with experiment will then be taken up. In Section III we give a short treatment of quantum lattice dynamics, which has been developed to deal with "quantum solids" as helium and hydrogen. Classical approaches in the harmonic approximation fail for these systems. In Section IV, intensities of infrared and Raman spectra in the lattice vibration region are discussed. A group theoretical appendix has been added for the reader who is not familiar with this aspect.

The review is not meant to be exhaustive. Rather, theory and experiment as they relate to a number of important systems are treated. The lattice dynamics of the rare gas crystals are discussed only as they relate to quantum lattice dynamics. For the interested reader, good and comprehensive reviews on this subject are available (Horton, 1968).

II. CLASSICAL LATTICE DYNAMICS

A. The Intermolecular Potential Model

In any lattice dynamical treatment a potential of the solid Φ must be formulated. If the solid has a very high degree of symmetry, it is possible to express the vibrational frequencies in terms of a limited number of parameters or "force constants." It may then be possible to determine the force constants from measured frequencies. This approach is, in fact, usual in the analysis of molecular vibrations (Wilson, Decius, and Cross, 1955). It has been used by Cochran and Pawley (1964) in their treatment of hexamethylene-tetramine. However, the number of parameters generally is prohibitively large compared to the number of available observable frequencies. As a result, it is necessary to assume an explicit potential Φ which contains a very limited number of parameters, and to calculate the frequencies of the system from it. Such an approach is also desirable since it leads to an understanding of the physical content of the nature of the potential function.

It has generally been assumed that the potential of the solid can be written as a sum over pair potentials:

$$\Phi = \frac{1}{2} \sum_{ll'} \sum_{kk'} v(lk, l'k') \tag{2.1}$$

Here $v(lk, l'k')$ is the contribution to the potential by the pair of molecules lk, $l'k'$ where l designates the unit cell and k the site. This assumption is probably valid in most cases and there has been some discussion of this question in the literature (Dymond and Alder, 1968; Williams, Schaad, and Murrell, 1967).

Two general forms have been used for the pair potential v. The first was introduced by Walmsley and Pople (1964) in their treatment of the $\mathbf{q} = 0$ lattice frequencies of solid CO_2. It consists of a 6–12 Lennard-Jones term (between molecular centers) and an orientation-dependent term in the form of a quadrupole-quadrupole interaction. The second form, which has found wide application, consists of a sum over atom-atom interactions, summed over the nonbonded atoms of the two molecules. This type of pair potential function was developed for organic molecules and was used to account for the crystal structures of these systems. Kitaigorodskii (1966) determined the parameters for such potentials in this way. Dows (1962) first applied such a potential to the calculation of the librational lattice modes frequencies of solid ethylene using hydrogen-hydrogen repulsion terms as given by de Boer (1942). The usual form of this type of potential contains 6-exponential atom-atom terms:

$$-\frac{A}{R^6} + B \exp(-CR) \tag{2.2}$$

where R is the distance between the interacting atoms. Coulson and Haigh (1963) listed parameters for hydrogen-hydrogen interactions and Williams (1967) derived parameters for H–H, C–H, and C–C potential terms from a large number of crystal properties including structures, elastic constants, and sublimation energies of nine aromatic hydrocarbons. In some instances 6–12 atom-atom terms have been used, as in the application to the lattice dynamics of α–N_2 (Kuan, Warshel, and Schnepp, 1969).

B. Treatment of Translational Motions

1. *General*

The lattice dynamics of translational motions has been formulated in a number of ways. The classical formulation is due to Born and Huang (1954) and will be described briefly. It uses the Lagrangian formalism in terms of local displacement coordinates and then assumes plane wave solutions. A second approach also makes use of the Lagrangian method but the dynamical variables are now "symmetry coordinates" belonging to irreducible representations of the infinite translation group reflecting the periodicity of the solid. Finally, it will be shown that we can derive the Hamiltonian in terms of these symmetry coordinates from the Lagrangian by following the usual procedures of classical mechanics (Goldstein, 1950). A treatment beginning with the Hamiltonian is given by Ziman (1960).

2. *Classical Treatment*

We consider here the three degrees of translational freedom of each molecule. The component of the displacement vector $\mathbf{u}$ of a molecule at site k in unit cell l is $u_j(lk)$ where j spans x, y, z. The kinetic energy T_{trans} is then given by

$$T_{\text{trans}} = \frac{m}{2} \sum_{jlk} \dot{u}_j^2(lk) \tag{2.3}$$

The potential energy Φ is expanded in a power series in the displacements and only the second-order (harmonic) term is retained; the first-order term vanishes as a condition of equilibrium since the u_j are all independent coordinates:

$$\begin{aligned}\Phi_{\text{trans}} &= \frac{1}{2} \sum_{ij} \sum_{ll'} \sum_{kk'} \Phi_{ij}(lk, l'k') u_i(lk) u_j(l'k') \\ &= \frac{1}{2} \sum_{ij} \sum_{ll'} \sum_{kk'} \left[\frac{\partial^2 \Phi}{\partial u_i(lk)\, \partial u_j(l'k')} \right]_0 u_i(lk) u_j(l'k')\end{aligned} \tag{2.4}$$

In the preceding equation $\Phi_{ij}(lk, l'k')$ is defined.

We now formulate the Lagrangian L,

$$L = T - \Phi \tag{2.5}$$

and standard procedure (Goldstein, 1950) using (2.5) after substitution of (2.3) and (2.4) leads to the equations of motion:

$$m\ddot{u}_i(lk) = -\sum_{j}\sum_{l'k'} \Phi_{ij}(lk, l'k')u_j(l'k') \tag{2.6}$$

It should be noted that i, j span only x, y, z in (2.6) since we are concerned with the translational degrees of freedom only in this section. Following Born and Huang (1954) we now assume a plane wave solution:

$$u_i(lk) = U_i(\mathbf{q}k) \exp i(\mathbf{q} \cdot \mathbf{R}(lk) - \omega(\mathbf{q})t) \tag{2.7}$$

Here $\mathbf{U}(\mathbf{q}k)$ is the amplitude which is independent of the unit cell index l, ω is the frequency, and $\mathbf{q}$ is the wave vector which is a vector in reciprocal space and designates the modes of the system. Its components (q_x, q_y, q_z) have values

$$q_i = \frac{2\pi}{L_i} n_i = \frac{2\pi}{N_i a} n_i \tag{2.8}$$

where L_i is the length of the solid in the ith dimension, N_i is the number of unit cells in this dimension, and a is the cell constant. The n_i are integers $0, 1, \ldots, (N_i - 1)$. Thus q_i assumes values from 0 to $2\pi/a$. In practice it is found convenient to use values from $-\pi/a$ to $+\pi/a$ since modes of q_i differing by $2\pi/a$ turn out to be identical. In addition, a mode of q_i has the same frequency as the mode of $-q_i$ and therefore we need only allow q_i to vary between 0 and π/a to obtain all modes, but they are all doubly degenerate.

Substituting (2.7) in the equations of motion (2.6), we obtain

$$-m\omega^2(\mathbf{q})u_i(lk) = -\sum_{j}\sum_{l'k'} \Phi_{ij}(lk, l'k')u_j(l'k')$$

and dividing throughout by $-\exp i[\mathbf{q} \cdot \mathbf{R}(lk) - \omega(\mathbf{q})t]$ we obtain

$$\begin{aligned} m\omega^2(\mathbf{q})U_i(\mathbf{q}k) &= \sum_{j}\sum_{l'k'} \Phi_{ij}(lk, l'k')U_j(\mathbf{q}k') \exp i\mathbf{q} \cdot \{\mathbf{R}(l'k') - \mathbf{R}(lk)\} \\ &= \sum_{j}\sum_{k'} U_j(\mathbf{q}k') \sum_{l'} \Phi_{ij}(lk, l'k') \exp i\mathbf{q} \cdot \mathbf{R}(l'k'; lk) \\ &= \sum_{j}\sum_{k'} U_j(\mathbf{q}k')M_{ij}(\mathbf{q}, kk') \end{aligned} \tag{2.9}$$

In (2.9) we use $R(l'k'; lk)$ to designate the vector connecting molecules $(l'k')$ and (lk) and $M_{ij}(\mathbf{q}, kk')$ is there defined. It should be noted that cell index l is arbitrary and serves only as origin or reference point for the lattice sum over l'.

The secular equation is now

$$|M_{ij}(\mathbf{q}, kk') - m\omega^2(\mathbf{q})\delta_{ij}\delta_{kk'}| = 0 \tag{2.10}$$

The secular equation, as indicated in (2.10), can be written separately for each $\mathbf{q}$ since the translational symmetry of the solid guarantees the factoring, as will be explicitly demonstrated in Section IIB.3. The dimension of (2.10) is given by the number of components of the molecular displacement coordinates (2.3) (if we consider translations only) $D(i)$ times the number of sites in the primitive unit cell $D(k)$. Equation (2.10) has $D(i) \cdot D(k)$ roots of ω^2, which are all real.

3. *Lattice Dynamics in Terms of Translationally Symmetric Coordinates*

In order to make use of the translational symmetry of the solid, we use the transformation (2.11):

$$u_j(\mathbf{q}k) = N^{-1/2} \sum_l \exp i[\mathbf{q} \cdot \mathbf{R}(lk)] u_j(lk) \tag{2.11}$$

We thereby factor the problem, as we see next (Landsberg, 1969).

The kinetic energy T_{trans} is given by [see (2.3)]:

$$\begin{aligned} T_{\text{trans}} &= \frac{m}{2} \sum_{jlk} \dot{u}_j^2(lk) \\ &= \frac{m}{2N} \sum_{jk} \sum_{\mathbf{q}\mathbf{q}'} \sum_{l} \dot{u}_j(\mathbf{q}k) \dot{u}_j(\mathbf{q}'k) \exp i[-(\mathbf{q} + \mathbf{q}') \cdot \mathbf{R}(lk)] \end{aligned} \tag{2.12}$$

where we have used the inverse transformation of (2.11). We now note that we can perform the sum over l and use the well known result given by (2.13),

$$\sum_l \exp i[-(\mathbf{q} + \mathbf{q}') \cdot \mathbf{R}(lk)] = N\delta_{\mathbf{q}+\mathbf{q}',0} \tag{2.13}$$

This sum therefore vanishes except for the case $\mathbf{q} + \mathbf{q}' = 0$, when every unit cell makes the same contribution of unity, giving N. Thus we set $\mathbf{q}' = -\mathbf{q}$ and obtain

$$\begin{aligned} T_{\text{trans}} &= \frac{m}{2} \sum_{\mathbf{q}} \sum_{jk} \dot{u}_j(\mathbf{q}k) \dot{u}_j(-\mathbf{q}k) \\ &= \frac{m}{2} \sum_{\mathbf{q}} \sum_{jk} \dot{u}_j(\mathbf{q}k) \dot{u}_j^*(\mathbf{q}k) \end{aligned} \tag{2.14}$$

Here we use the relation $u_j(-\mathbf{q}k) = u_j^*(\mathbf{q}k)$, which is obvious from the definition (2.11). Clearly, (2.14) shows that the kinetic energy is still diagonal in the new coordinates (2.11).

The potential energy Φ_{trans} as given by (2.4) is now also transformed using (2.11):

$$\Phi = \frac{1}{2}\sum_{ij}\sum_{ll'}\sum_{kk'}\Phi_{ij}(lk, l'k')u_i(lk)u_j(l'k')$$
$$= \frac{1}{2N}\sum_{ij}\sum_{ll'}\sum_{kk'}\Phi_{ij}(lk, l'k')\sum_{\mathbf{q}\mathbf{q}'}u_i(\mathbf{q}k)u_j(\mathbf{q}'k')$$
$$\times \exp i[-\mathbf{q}\cdot\mathbf{R}(lk) - \mathbf{q}'\cdot\mathbf{R}(l'k')] \tag{2.15}$$

Equation (2.15) can be rewritten in the following form:

$$\Phi = \frac{1}{2N}\sum_{ij}\sum_{kk'}\sum_{\mathbf{q}\mathbf{q}'}u_i(\mathbf{q}k)u_j(\mathbf{q}'k')\sum_{l}\exp i[-(\mathbf{q}+\mathbf{q}')\cdot\mathbf{R}(lk)]$$
$$\times \sum_{l'}\Phi_{ij}(lk, l'k')\exp i[-\mathbf{q}'\cdot\{\mathbf{R}(l'k') - \mathbf{R}(lk)\}] \tag{2.16}$$

The last sum in (2.16) is actually independent of l for the following reason. Here $\Phi_{ij}(lk, l'k')$ depends on $(l' - l)$, or on the distance between the unit cells considered, rather than on the actual position of any one of the unit cells in the lattice. The exponential argument depends on this difference only. In other words, the value of the sum over l' will be the same, no matter what the value of l is set to be. We can therefore set $l = 0$ on some reference unit cell. Then we write

$$F_{ij}(\mathbf{q}kk') = \sum_{l'}\Phi_{ij}(lk, l'k')\exp i[-\mathbf{q}\cdot\{\mathbf{R}(l'k') - \mathbf{R}(0k)\}] \tag{2.17}$$

As a result of this conclusion we find that the summation over l in (2.16) can be executed without further difficulty using (2.13), and we obtain the result

$$\Phi = \frac{1}{2}\sum_{\mathbf{q}}\sum_{ij}\sum_{kk'}u_i(\mathbf{q}k)u_j^*(\mathbf{q}k')F_{ij}(\mathbf{q}kk')$$
$$= \frac{1}{2}\sum_{\mathbf{q}}\sum_{ij}\sum_{kk'}u_i(\mathbf{q}k)u_j^*(\mathbf{q}k')$$
$$\times \sum_{l'}\Phi_{ij}(0k, l'k')\exp i[-\mathbf{q}\cdot\{\mathbf{R}(l'k') - \mathbf{R}(0k)\}] \tag{2.18}$$

It is clear from (2.18) that the potential energy has been decoupled into noninteracting parts, each designated by a specific value of $\mathbf{q}$, due to the transformation (2.11).

We could also formulate T and Φ directly in terms of the "symmetry coordinates" (2.11) by using a procedure analogous to that for molecular vibrations. First it should be pointed out that a symmetry coordinate

(2.11) can be obtained in the usual way by applying the group theoretical Wigner projection operator (Wigner, 1931) to one of the $u_i(lk)$. The symmetry operations of the translation group (involving cyclic boundary conditions) are all translations by combinations of multiples of primitive vectors of the lattice and the coefficients $\exp i[\mathbf{q} \cdot \mathbf{R}(lk)]$ make up the character of representation $\mathbf{q}$. Since then the $u_i(\mathbf{q}k)$ are symmetry coordinates, we expect to be able to separate the problem by representations $\mathbf{q}$ (Wilson, Decius, and Cross, 1955) and this is just what has been shown.

The $u_i(\mathbf{q}k)$ form a set of independent coordinates if the $u_i(lk)$ are such a set since they are related by the unitary transformation defined by (2.11). We note, however, as a new feature, that the $u_i(\mathbf{q}k)$ are complex. Both the kinetic energy T [from (2.14)] and the potential energy Φ [from (2.18)] are hermitian and therefore the Lagrangian L derived from these forms is hermitian as required. In tensor notation we obtain

$$L = \sum_{\mathbf{q}} L(\mathbf{q}) = \frac{1}{2} \sum_{\mathbf{q}} [m\dot{\mathbf{u}}^{+}(\mathbf{q})\dot{\mathbf{u}}(\mathbf{q}) - \mathbf{u}^{+}(\mathbf{q})\mathbf{F}(\mathbf{q})\mathbf{u}(\mathbf{q})] \tag{2.19}$$

where $\mathbf{u}(\mathbf{q})$ is a vector whose components are $u_i(\mathbf{q}k)$ and the components of $\mathbf{F}(\mathbf{q})$ are identified in (2.17).

We now derive the Lagrangian equations of motion keeping in mind that $u_j(\mathbf{q}k)$ and $u_j^*(\mathbf{q}k)$ are treated as independent coordinates.

We calculate

$$\begin{aligned} \frac{d}{dt}\frac{\partial L}{\partial \dot{u}_i(\mathbf{q}k)} &= \frac{d}{dt}[\tfrac{1}{2}m\dot{u}_i^*(\mathbf{q}k)] = \tfrac{1}{2}m\ddot{u}_i^*(\mathbf{q}k) \\ \frac{\partial L(\mathbf{q})}{\partial u_i(\mathbf{q}k)} &= -\frac{1}{2}\sum_{jk'} F_{ij}(\mathbf{q}kk')u_j^*(\mathbf{q}k') \end{aligned} \tag{2.20}$$

By substituting from (2.20) into the Lagrangian equation of motion

$$\frac{d}{dt}\frac{\partial L}{\partial \dot{u}} - \frac{\partial L}{\partial u} = 0$$

we obtain the equations of motion of the system

$$m\ddot{u}_i^*(\mathbf{q}k) + \sum_{jk'} F_{ij}(\mathbf{q}kk')u_j^*(\mathbf{q}k') = 0 \tag{2.21}$$

In (2.21), as in (2.6), the indices i, j run only over x, y, z as long as we treat translational degrees of freedom only.

We now seek periodic solutions to (2.21) with the same frequency for all components of $u(\mathbf{q})$:

$$\begin{aligned} u_i(\mathbf{q}k, t) &= U_i(\mathbf{q}k) \exp[i\omega(\mathbf{q})t] \\ u_i^*(\mathbf{q}k, t) &= U_i^*(\mathbf{q}k) \exp[-i\omega(\mathbf{q})t] \end{aligned} \tag{2.22}$$

where $U_i(\mathbf{q}k)$ is the amplitude. Substitution of (22) in the equations of motion (21) gives

$$-m\omega^2(\mathbf{q})U_i^*(\mathbf{q}k) + \sum_{jk'} F_{ij}(\mathbf{q}, kk')U_j^*(\mathbf{q}k') = 0 \tag{2.23}$$

which leads to the expected secular equation

$$|F_{ij}(\mathbf{q}kk') - m\omega^2(\mathbf{q})\delta_{ij}\delta_{kk'}| = 0 \tag{2.24}$$

The secular equation (2.24) is identical with (2.10) since $F_{ij}(\mathbf{q}, kk')$ is identical to $M_{ij}(\mathbf{q}, kk')$, as can be seen by comparing (2.17) to (2.9).

To test the validity of the Lagrangian treatment in terms of the complex coordinates (2.11), we easily determine that we obtain the identical result as above by deriving the equation of motion

$$\frac{d}{dt}\frac{\partial L}{\partial \dot{u}^*} - \frac{\partial L}{\partial u^*} = 0$$

4. *The Hamiltonian of the Lattice Dynamics Problem*

We consider the coordinates $u_i(\mathbf{q}k)$ and $u_i^*(\mathbf{q}k)$ and calculate their conjugate momenta $p_i(\mathbf{q}k)$, $p_i^*(\mathbf{q}k)$:

$$\begin{aligned} p_i(\mathbf{q}k) &= \frac{\partial L(\mathbf{q})}{\partial \dot{u}_i(\mathbf{q}k)} = \frac{m}{2}\,\dot{u}_i^*(\mathbf{q}k) \\ &= \frac{m}{2}\,N^{-1/2}\sum_l \exp\,[-i\mathbf{q}\cdot\mathbf{R}(lk)]\dot{u}_i^*(lk) \end{aligned} \tag{2.25}$$

The final result is obtained by substituting for $u_i^*(\mathbf{q}k)$ from (2.11). Similarly we calculate $p_i^*(\mathbf{q}k)$. It is indeed found that the coordinate $u_i(\mathbf{q}k)$ and its conjugate momentum $p_i(\mathbf{q}k)$ given by (2.11) and (2.25), respectively, satisfy the commutation relation as required by quantum mechanics (Ziman, 1960, p. 27).

The Hamiltonian is given (Goldstein, 1950) by

$$H(p, u, t) = \sum_i \dot{u}_i p_i - L(u, \dot{u}, t) \tag{2.26}$$

In our derivation we must remember to include in the sum over i both $\dot{u}_i p_i$ and $u_i^* p_i^*$ for each value of i since the u_i and u_i^* were treated as independent coordinates in this sense. We then obtain

$$\begin{aligned} \sum_i \dot{u}_i p_i &= \frac{m}{2}\sum_{\mathbf{q}}\sum_{jk} [\dot{u}_j(\mathbf{q}k)\dot{u}_j^*(\mathbf{q}k) + \dot{u}_j^*(\mathbf{q}k)\dot{u}_j(\mathbf{q}k)] \\ &= m\sum_{\mathbf{q}}\sum_{jk} \dot{u}_j^*(\mathbf{q}k)\dot{u}_j(\mathbf{q}k) \end{aligned} \tag{2.27}$$

We then use (2.27) and (2.19) to obtain from (2.26)

$$H = \sum_{\mathbf{q}} \sum_{ik} \left[m\dot{u}_i^*(\mathbf{q}k)\dot{u}_i(\mathbf{q}k) - \frac{m}{2}\dot{u}_i^*(\mathbf{q}k)\dot{u}_i(\mathbf{q}k) \right] + \frac{1}{2}\sum_{\mathbf{q}}\sum_{ij}\sum_{kk'} F_{ij}(\mathbf{q}kk')u_i^*(\mathbf{q}k)u_j(\mathbf{q}k') \quad (2.28)$$

Equation (2.28) can, of course, be decomposed into a sum of terms, each for a specific value of $\mathbf{q}$:

$$H = \sum_{\mathbf{q}} H_{\mathbf{q}}$$

where $H_{\mathbf{q}}$ is given by

$$H_{\mathbf{q}} = \sum_{ik} \frac{m}{2}\dot{u}_i^*(\mathbf{q}k)u_i(\mathbf{q}k) + \frac{1}{2}\sum_{ij}\sum_{kk'} F_{ij}(\mathbf{q}kk')u_i^*(\mathbf{q}k)u_j(\mathbf{q}k')$$

$$= \sum_{ik} \frac{1}{2m} p_i^*(\mathbf{q}k)p_i(\mathbf{q}k) + \frac{1}{2}\sum_{ij}\sum_{kk'} F_{ij}(\mathbf{q}kk')u_i^*(\mathbf{q}k)u_j(\mathbf{q}k') \quad (2.29)$$

This is, in fact, the Hamiltonian assumed by Ziman (1960, p. 29). The derivation of the secular equation or dynamical equation of the problem from H is given there.

5. *Molecular Solids*

We now consider the application of the methods described in Sections IIB.2 and IIB.3 to molecular solids. We limit our discussion to the special case when translational lattice motions are separable from other degrees of freedom, specifically angular motions or "librations." Such a separation is valid only for centro-symmetric solids and then only at a limited number of points in the Brillouin zone, i.e., the zone center (Γ-point) and some points on the zone boundary. In all other cases, interactions must be considered.

To a good approximation, the molecule can be assumed to be rigid since intermolecular binding is much weaker than intramolecular binding. Then the translational motions of the molecule can be treated by the described methods. The molecule is considered as a single mass point with the total mass concentrated at its center of gravity.

If, on the other hand, each atom is considered as a mass point with three translational degrees of freedom, then a treatment along the lines of Sections IIB.2 and IIB.3 includes all vibrational degrees of freedom, both external (translational and librational lattice modes) and internal (intramolecular vibrations). In this case also all interactions, both intermolecular and intramolecular, must be included in the potential function Φ. We discuss such treatments further in Section IIC.

C. Treatments Including Librational Motions

1. *General*

The lattice dynamics of librational motions is much less developed than that for translational motions. The first full treatments are due to Cochran and Pawley (1964) and Pawley (1967), although the treatment of these degrees of freedom at $\mathbf{q} = 0$ was included in the matrix formulation of Shimanouchi, Tsuboi, and Miyazawa (1961) and the work on solid CO_2 by Walmsley and Pople (1964). More recently Schnepp and Ron (1969) and Kuan, Warshel, and Schnepp (1969) carried out calculations on solid α–N_2, and Suzuki and Schnepp (1971) on solid CO_2.

In this section we describe all of these formulations and others reported in the literature and discuss them critically. First, however, the validity of treating librations in terms of the harmonic oscillator problem will be considered.

2. *Libration as Harmonic Oscillator*

An angular small displacement in one dimension (θ) can be reduced immediately to an harmonic oscillator problem. If we call the displacement coordinate u_θ, the linear displacement element ds can be expressed in terms of u_θ and the distance from the center r:

$$ds = r\, du_\theta$$

and the kinetic energy T takes the form

$$T = \frac{m}{2} r^2 \dot{u}_\theta{}^2 = \tfrac{1}{2} I \dot{u}_\theta{}^2$$

where I is the moment of inertia. If now the potential Φ is expanded in a power series in u_θ to second order, we have a problem identical to that in a linear displacement. This is also true for the quantum mechanical problem, which has been widely discussed in connection with internal rotations in molecules (Pitzer, 1953; Lin and Swalen, 1959).

An entirely new degree of complexity is introduced into the problem of the angular oscillator if two degrees of freedom are allowed. This is the case of the angular oscillations of a linear molecule in an angle-dependent potential. It has been implicit in the treatments reported to date that the solutions of this problem can be assumed to be those of the harmonic oscillator. Two questions arise. First, assuming a purely harmonic potential, is the problem that of a harmonic oscillator for which the classical solutions of this problem may be used? Here the question concerns the separability of the kinetic energy. Second, how does anharmonicity of the potential affect the results?

We consider first the classical treatment of the two-dimensional rotation problem in the conventional spherical angular coordinates θ, ϕ, with the corresponding displacement coordinates u_θ, u_ϕ (Schnepp and Ron, 1969). This is equivalent to the motion of a particle, constrained to the surface of a sphere of constant radius r. The square of the element of displacement is given by

$$ds^2 = r^2\, du_\theta^{\,2} + r^2 \sin^2\theta\, du_\phi^{\,2} \tag{2.30}$$

where θ is measured from the lattice fixed Z-axis to the molecular figure axis, subtended at the center of mass of the molecule. We use

$$\theta = \theta_0 + u_\theta$$

and obtain from (2.30) the expression for the kinetic energy of the librational motion:

$$T_{\text{lib}} = \frac{I}{2} \sum_{lk} [\dot{u}_\theta^{\,2}(lk) + \sin^2\theta(lk)\dot{u}_\phi^{\,2}(lk)] \tag{2.31}$$

The potential energy is assumed to be harmonic in the angular displacements as in (2.4) where now i spans θ, ϕ and j spans x, y, z, θ, ϕ. On applying the Lagrangian treatment, we find that the appearance of the variable factor $\sin^2\theta(lk)$ in T_{lib} [Eq. (2.31)] as coefficient of $\dot{u}_\theta^{\,2}(lk)$ causes some problems. It has been common to assume the amplitude of the vibration to be small enough to justify expansion of $\sin^2\theta(lk)$ about the equilibrium position and to retain only the first term, which is constant, e.g., $\sin^2\theta_0(lk)$. This procedure amounts to a harmonic approximation for the kinetic energy. The validity of the procedure depends sensitively, from the classical point of view, on the amplitude of the vibration since generally first-order corrections occur. First-order correction vanishes for $\theta_0 = 0°$ or $90°$, but we may expect that a physical result will not be affected by a rotation of coordinate system. Estimates indicate (Cahill and Leroi, 1969) that the root mean square librational amplitudes are of the order of 4° for solid CO_2 and N_2O and 10° for solid α-N_2 and α-CO. For naphthalene and anthracene the parallel values are of the order of 1° (Suzuki, Yokoyama, and Ito, 1967). These amplitudes lead to estimates of the first-order correction term of the order of 10% for CO_2 and N_2O, 30% for α-N_2 and α-CO, and 3% for naphthalene and anthracene. Clearly, it is impossible to conclude that the infinitesimal displacement limit applies accurately for the linear molecules just considered. On the other hand, if we try to retain the nonconstant factor $\sin^2\theta(lk)$, the Lagrangian problem becomes intractable.

In view of the foregoing discussion, it is relevant to investigate the quantum mechanical formulation of the librational problem. No serious attempt has yet been made to treat the librational *lattice* problem quantum mechanically, although Martin and Walmsley (1969) have discussed one possible approach. However, Devonshire (1935) and Sauer (1966) have given a detailed treatment of a linear molecule (or two-dimensional rigid rotor problem in u_θ, u_ϕ) in an octahedral potential. We are here interested in deriving criteria for the applicability of the harmonic oscillator result and therefore discuss the high barrier limit treated by these authors.

The octahedral Devonshire potential has the form

$$V = -K[P_4^0(\cos\theta) + \tfrac{1}{168}P_4^4(\cos\theta)\cos 4\phi]$$
$$= -\frac{K}{8}[3 - 30\cos^2\theta + 35\cos^4\theta + 5\sin^4\theta\cos 4\phi] \qquad (2.32)$$

This potential has six equivalent minima at the intersection of the coordinate axes with the sphere centered at the origin, or at $\theta = 0$, π and at $\theta = \pi/2$, $\phi = 0$, $\pm\pi/2$ with a value $V = -K$ The maxima are at $\theta = \cos^{-1}(\pm 1\sqrt{3})$, $\phi = \pm\pi/4$ or $\pm 3\pi/4$ with a value $V = +\frac{2}{3}K$. Thus the depth of the six wells is $\frac{5}{3}K$ and the constant K determines the barrier height. The quantum mechanical problem was solved by assuming trial functions which are linear combinations of spherical harmonics of suitable symmetry and a variational calculation was carried out using this basis. Devonshire had to limit his basis set in view of the lack of adequate computing facilities, but Sauer included terms up to $l = 12$. It is clear from both Sauer's and Devonshire's results that in the limit of large K the lowest state of the system has a degeneracy of 6 arising from the 6 equivalent wells of the potential. As K decreases, this degeneracy is removed due to interaction between different wells. Similarly, for very large K the first excited state becomes twelvefold degenerate, reflecting the two degrees of freedom u_θ, u_ϕ but for the excited state this limit is reached at appreciably higher K than for the lowest state. We follow Devonshire's nomenclature and define $k = K/B$ where B is the rotational constant. The energy W is expressed in units of B, $W = E/B$. In the high-barrier limit Devonshire obtains the following for the energy W (in units of B) as a function of the quantum number n:

$$\Delta W = \frac{E}{B} = -k + 2(n+1)(5k)^{1/2} - (\tfrac{3}{2}n^2 + 3n + \tfrac{11}{4}) \qquad (2.33)$$

We obtain from this energy equation the excitation energy W for the transition $n = 0$ to $n = 1$:

$$\Delta W = 2(5k)^{1/2} - \tfrac{9}{2} \qquad (2.34)$$

On inspection of Sauer's results, which we shall accept as "exact," we find that the sixfold degeneracy of the ground state is good to 10% for $k = 30$ and to 1% for $k = 40$. The twelvefold degeneracy of the $n = 1$ state is good to 15% for $k = 40$, to 10% for $k = 50$, to 5% for $k = 60$, to 3% for $k = 70$, and to 1.5% for $k = 80$. Therefore we note that (2.34) will give a realistic representation of the excitation energy only for cases corresponding to $k = 60$ and larger, and ideally for k larger than 80. Comparison with Sauer's results also shows that (2.34) is accurate to 3% for $k = 80$ and to 4% for $k = 70$ We then conclude that (2.34) is useful for at least $k = 60$–80.

The next step is a comparison between (2.34) and a classical harmonic oscillator solution for the potential V given in (2.32). We develop V in a power series up to second order in u_θ, u_ϕ about the well minimum at $\theta = \pi/2$, $\phi = 0$:

$$V = -K + 5Ku_\theta^2 + 5Ku_\phi^2 \tag{2.35}$$

The kinetic energy is given by (2.31), setting $\sin^2\theta = \sin^2\theta_0 = 1$ for $\theta_0 = \pi/2$. Therefore

$$T = \frac{I}{2}(\dot{u}_\theta^2 + \dot{u}_\phi^2) \tag{2.36}$$

The secular equation is then

$$\begin{vmatrix} \left(\dfrac{\partial^2 V}{\partial u_\theta^2}\right)_0 - I\omega^2 & \left(\dfrac{\partial^2 V}{\partial u_\theta\,\partial u_\phi}\right)_0 \\ \left(\dfrac{\partial^2 V}{\partial u_\theta\,\partial u_\phi}\right)_0 & \left(\dfrac{\partial^2 V}{\partial u_\phi^2}\right)_0 - I\omega^2 \end{vmatrix} = 0 \tag{2.37}$$

and using the values of the second derivatives at the minimum position implied in (2.35),

$$\left(\frac{\partial^2 V}{\partial u_\theta^2}\right)_0 = \left(\frac{\partial^2 V}{\partial u_\phi^2}\right)_0 = +10K; \qquad \left(\frac{\partial^2 V}{\partial u_\theta\,\partial u_\phi}\right)_0 = 0$$

we have a simple diagonal problem with two identical roots:

$$\omega_1 = \omega_2 = \left(10\frac{K}{I}\right)^{1/2}$$

which reduces to the one-quantum excitation energy in units of B:

$$\Delta W = \frac{h\nu}{B} = 2(5k)^{1/2} \tag{2.38}$$

Comparison between (2.38) and the high barrier limit excitation energy equation (2.34) leads first to the satisfying conclusion that the angular oscillator problem excitation energy does indeed approach the classical harmonic oscillator value for very high barriers, i e., for

$$2(5k)^{1/2} \gg \frac{9}{2}, \qquad k \gg 2$$

We note that for $k = 80$ the correction of $\frac{9}{2}$ constitutes 11% and for $k = 1000$ this correction still amounts to 3.2%.

TABLE I
Librational parameters for various molecular solids

Substance	B (cm^{-1})	$\tilde{\nu}_{\text{lib}}$ (cm^{-1})	$\tilde{\nu}_{\text{lib}}/B$	k[a] (lowest)
C_2H_2	1.18	22, 33[b]	18.6, 28	26.5
HCl	10.59	218, 290[c]	20.6, 27.4	31
HBr	8.47	200, 269[c]	23.6, 31.8	39
α-N_2	2.01	50, 70[d]	25, 35	45
CO	1.93	50[e]	26	46
CO_2	0.39	72, 92[e]	185, 236	1,900
N_2O	0.42	69, 81[e]	162, 193	1,300
C_2N_2	0.16	43, 57[f]	270, 356	3,800
C_6H_6	0.19	60, 80[g]	316, 420	4,500
Cl_2	0.24	83, 100[h]	346, 416	6,000
OCS	0.20	92[h]	460	10,800

[a] The constant k is defined by (2.34).
[b] Ito, Yokoyama, and Suzuki, 1970.
[c] Ito, Suzuki, and Yokoyama, 1969.
[d] Brit, Ron, and Schnepp, 1969.
[e] Cahill and Leroi, 1969.
[f] Richardson and Nixon, 1968.
[g] Ito and Shigeoka, 1966.
[h] Suzuki, Yokoyama, and Ito, 1969.

We now examine actual instances. In Table I we have listed the parameters pertinent to the preceding discussion for a number of solids; the lowest two librational frequencies are considered. We see that CO_2, N_2O, Cl_2, C_2N_2, OCS, and benzene all have k larger than 1000, and many of them several times that value. We conclude therefore that the harmonic oscillator result is probably quite accurate for the librations in these solids. On the other hand, α-N_2, CO, C_2H_2, and the hydrogen halides (HCl, HBr) have low values of k and therefore the harmonic oscillator results may be inaccurate. It is difficult, on the other hand, to assess the errors in such treatments for the latter group of solids since we must expect the inaccuracy for a lattice to be different from that for a single angular oscillator.

It is of interest to note that the identical result (2.38) is obtained for the librations by a treatment using cartesian displacement coordinates for the atoms of a diatomic molecule (Suzuki, 1970). Although we do not face the problems of separability of coordinates in the kinetic energy in this case, the assumption of "harmonicity" of the potential energy presumably causes the same inaccuracies. In other words, it is immaterial which coordinates are used but in terms of Eulerian angle displacement coordinates the "harmonic approximation" is not confined to the potential energy but also appears in the kinetic energy.

3. *Cartesian Coordinate Treatment*

As pointed out in Section IIB, it is possible to approach the lattice dynamics problem of a molecular crystal by choosing the cartesian displacement coordinates of the atoms as dynamical variables (Pawley, 1967). In this case, all vibrational degrees of freedom of the system are included, i.e., translational and librational lattice modes (external modes) as well as intramolecular vibrations perturbed by the solid (internal modes). It is then obviously necessary to include all intermolecular and intramolecular interactions in the potential function Φ. For the intramolecular part a force field derived from a molecular normal coordinate analysis is used. The force constants in such a case are calculated from the measured vibrational frequencies. The intermolecular part of Φ is usually expressed as a sum of terms, each representing the interaction between a pair of atoms on different molecules, as discussed in Section IIA.

In general, the interactions between atoms in the same molecule are much stronger than those between atoms on different molecules and therefore it is usually relatively simple to recognize the external or lattice vibrations. Further identification and, in particular, characterization of translational or librational motions, where applicable, is contingent on an analysis of the eigenvectors. A calculation, as described here, is usually quite long and requires extensive use of computers. In principle, however, there is no difficulty in carrying it out throughout the Brillouin zone.

Several authors have reported that they introduced a minor correction to the X-ray determined crystal structure to conform to the equilibrium conditions based on the assumed potential function of the type of Eq. (2.2). Such corrections are usually within the accuracy of the structure determination (± 0.05 Å).

It seems of interest and it will be useful for reference in the following sections to discuss one aspect of a lattice dynamical calculation based on a potential as given in (2.2). It is easily ascertained that the interatomic distances R are not linearly dependent on the cartesian coordinates of the atoms x_α; similarly, the corresponding changes in interatomic distances r_l have nonlinear dependence on the atomic cartesian displacements x_l. Further, the r_l are not independent displacement coordinates and they are, in fact, highly redundant. Therefore $\partial\Phi/\partial r_l$ evaluated at the equilibrium configuration does not vanish for all l. On the other hand, the atomic cartesian displacements x_α are independent variables. The potential Φ is then expanded in the x_α to second order [Eq. (2.4)] and the force constants $\Phi_{\alpha\beta}{}^c$ in this expansion are given by (2.39) (the superscript c denotes the cartesian displacement basis):

$$\begin{aligned}\Phi_{\alpha\beta}{}^c &= \left(\frac{\partial^2\Phi}{\partial x_\alpha\, \partial x_\beta}\right)_0 \\ &= \sum_{kl}\left(\frac{\partial^2\Phi}{\partial r_k\, \partial r_l}\right)_0 \frac{\partial r_k}{\partial x_\alpha}\frac{\partial r_l}{\partial x_\beta} + \sum_l \left(\frac{\partial\Phi}{\partial r_l}\right)_0 \frac{\partial^2 r_l}{\partial x_\alpha\, \partial x_\beta}\end{aligned} \tag{2.39}$$

The second term on the right-hand side of (2.39) is often, but not always small. We shall return to discuss its importance in the next section, when we describe the "matrix method," since this term is neglected in that method.

4. *Matrix Treatment*

As already mentioned, Shimanouchi, Tsuboi, and Miyazawa (1961) developed a matrix treatment of lattice dynamics for molecular solids which is analogous to the molecular vibration *F–G* matrix formulation of Wilson, Decius, and Cross (1955). This method was applied to the $\mathbf{q} = 0$ lattice motions of solid benzene by Harada and Shimanouchi (1966, 1967) and was subsequently used by many other authors (e.g., Bernstein, 1970).

In this formulation the secular equation takes the form

$$|\mathbf{G\Phi} - \mathbf{E}\lambda| = 0 \tag{2.40}$$

where $\mathbf{\Phi}$ is the potential energy (force constant) matrix and $\mathbf{G}$ the inverse kinetic energy matrix in the displacement coordinates chosen. $\mathbf{E}$ is the unit matrix of dimension equal to that of $\mathbf{\Phi}$ and $\mathbf{G}$ and $\lambda = 4\pi^2c^2\tilde{\nu}^2$. For a crystal the matrices $\mathbf{\Phi}$ and $\mathbf{G}$ can be transformed into semidiagonal form by making use of the translational symmetry, as discussed previously. It is advantageous to use again the cartesian displacements of the *atoms* as a basis since the $\mathbf{G}$ matrix is then diagonal. Moreover, as discussed in the preceding section, in this representation all motions are included, i.e., both the internal vibrations of the molecules and the translational and librational lattice motions. It is thus possible to couple internal and external motions in this treatment. The potential function used is the same as given in Eq. (2.2). We shall now briefly sketch the procedure and discuss its implications and the approximations intrinsic to the method. We refer the reader for the details of the calculation to the references noted.

The potential energy is written in terms of "internal" coordinates or here explicitly atom-atom distances, and it takes the form of a sum of terms as in (2.2). All these distances (including those within the molecule and those between different molecules) form the components of a column vector **R**. The harmonic force constant matrix in terms of the corresponding displacements **r** is $\mathbf{\Phi}^i$. As already noted, however, it is advantageous to use cartesian displacement coordinates since the kinetic energy matrix $\mathbf{G}^c$ is then diagonal. It is therefore necessary to express also the potential Φ in terms of the cartesian displacements (vector **x**) and to write the force constant matrix $\mathbf{\Phi}^c$ accordingly:

$$\Phi = \tfrac{1}{2}\mathbf{r}^+\mathbf{\Phi}^i\mathbf{r} = \tfrac{1}{2}\mathbf{x}^+\mathbf{\Phi}^c\mathbf{x} \tag{2.41}$$

In order to do this we require a transformation:

$$\mathbf{r} = \mathbf{B}\mathbf{x} \tag{2.42}$$

Comparison between (2.41) and (2.42) yields the relation between the force constant matrices in the two coordinate systems:

$$\mathbf{\Phi}^c = \mathbf{B}^+\mathbf{\Phi}^i\mathbf{B} \tag{2.43}$$

To use this scheme we require then a *linear* transformation **B** between the internal displacement coordinates **r** (usually atom-atom distance changes in terms of which the potential is expanded to second order) and the cartesian displacement coordinates of the atoms **x**. However, the relation between these coordinates is not linear and therefore an approximation must be made at this point. Shimanouchi, Tsuboi, and Miyazawa (1961) expand the change in distance between atoms a and b, r_{ab}, from its equilibrium value $R_{\mathrm{ab}}{}^0$ to first order in the cartesian displacement of the atoms as given in (2.44):

$$\begin{aligned} r_{\mathrm{ab}} &= R_{\mathrm{ab}} - R_{\mathrm{ab}}{}^0 \\ &= \frac{1}{R_{\mathrm{ab}}{}^0}\left[X_{\mathrm{ab}}{}^0(x_{\mathrm{b}} - x_{\mathrm{a}}) + Y_{\mathrm{ab}}{}^0(y_{\mathrm{b}} - y_{\mathrm{a}}) + Z_{\mathrm{ab}}{}^0(z_{\mathrm{b}} - z_{\mathrm{a}})\right] \end{aligned} \tag{2.44}$$

In this equation $X_{\mathrm{ab}}{}^0$, $Y_{\mathrm{ab}}{}^0$, $Z_{\mathrm{ab}}{}^0$ are the components of the interatomic equilibrium distance $R_{\mathrm{ab}}{}^0$ and x_{a}, y_{a}, z_{a} are the cartesian displacements of atom a. Equation (2.44) defines the linear transformation matrix **B** of (2.42) and (2.43). We next examine the consequences of this approximation. We refer back to (2.39) where the exact $\alpha\beta$-element of $\mathbf{\Phi}^c$ is given, and find that we can rewrite the first term on the right-hand side of (2.39) to conform with (2.43), but the second term does not appear in (2.43):

$$\Phi_{\alpha\beta}{}^{c} = \sum_{kl} B_{\alpha k}^{+} \Phi_{kl}{}^{i} B_{l\beta} + \sum_{l} \left(\frac{\partial \Phi}{\partial r_l}\right)_0 \frac{\partial^2 r_l}{\partial x_\alpha \, \partial x_\beta} \tag{2.45}$$

The second term of (2.39) and (2.45) vanishes only in the approximation of (2.44) and (2.42), i.e., of linear dependence of the **r** on the **x**.

The extent of the contribution the second term in (2.45) is then decisive for the accuracy of the matrix formulation. Investigations have shown that this term affects the $\mathbf{q} = 0$ or optic mode frequencies to greatly different degrees in different cases. For solid CO_2, it was found (Suzuki and Schnepp, 1971) that the translational modes are changed by less than 1% except for the lowest frequency mode, which is changed by 5%. On the other hand, the librational modes are affected by between 3 and 8%. [Note that this separation between translations and librations is only valid at the Γ and R points of the Brillouin zone of space group $Pa3$ (T_h^6).] For α-N_2 the error introduced in the translational modes is less than 2% in all cases but for the librations the error ranges from 25 to 44%. It should be noted that the classical harmonic treatment is in any case subject to question for librational modes of α-N_2 but this is not the case for solid CO_2 (see Section IIC.2). No accurate data are available for aromatics, but it may be assumed that the error in these cases is similar to those for solid CO_2.

It is important to note that most treatments using the matrix method for the lattice dynamical treatment of molecular solids adjust the parameters of the potential model to fit observed frequencies. In this case, the formal error in the treatment is minimized in its importance. It may be of importance only inasmuch as the same potential model is assumed to be useful to interpret other physical properties such as gas phase second virial coefficients. It may also be important when one attempts to determine parameters for intermolecular potentials in terms of atom-atom interactions which are general to a class of molecules rather than specific to one substance. It is also evident that the error will be very serious in every case involving low-frequency librations (as for α-N_2). The reason for the small effect of the first derivative term in many cases is as follows. Both repulsive and attractive terms of the potential usually contribute significantly to the first derivative but their contributions have opposite sign and cancel. By comparison, the contribution to the second derivative of the potential is usually much larger for the repulsive potential term than for the attractive term (Shimanouchi, 1970).

Piseri and Zerbi (1968) extended the matrix method to include treatment throughout the Brillouin zone. They define the symmetry coordinates (symmetry with respect to translation) as defined by our (2.11) where the x_α or r_l are substituted for $u_j(lk)$. The transformation matrix $\mathbf{B}$ is then also $\mathbf{q}$ dependent and has the form

$$\mathbf{B}(\mathbf{q}k) = N^{-1/2} \sum_l \exp i[\mathbf{q} \cdot \mathbf{R}(lk)]\mathbf{B}(lk) \tag{2.46}$$

The secular equation is then again of the form of (2.40), except that $\mathbf{\Phi}^i$ and $\mathbf{\Phi}^c$ are $\mathbf{q}$-dependent.

5. *Direction Cosine Displacement Coordinates*

We now turn our attention to methods aimed at the investigation of the librational motions of the molecule (assumed rigid) directly and explicitly. In these methods displacement coordinates are chosen which describe the librations only without involving the translations of the center of gravity of the molecule or its internal degrees of freedom. The normal coordinates may, however, be mixtures of translational and librational degrees of freedom or even may represent interactions between internal molecular and lattice vibrations. Such interactions occur due to mixed translational and librational displacement terms in the potential expansion [Eq. (2.4)]. These interactions generally must be considered, and only in special cases and at very special points in the Brillouin zone can translational and librational modes be separated. If a solid has centers of inversion and molecular centers of symmetry coincide with these, then at $\mathbf{q} = 0$ the g- and u-character of vibrational modes is preserved. Since translations are odd and librations even to inversion, the separation can be effected in such cases. Carbon dioxide, α-nitrogen, benzene, naphthalene, and anthracene are examples of such structures. Our discussion here concerns the treatment of librational modes only and assumes their separability. However, the treatment can always be expanded to take account of nonseparability by expanding it to include translational displacement coordinates in the basis. Also, translation-libration interaction matrix elements must in general be calculated before the secular equation can be solved. These interaction terms will be discussed in Section IID.

Walmsley and Pople (1964) discussed the lattice dynamics of solid CO_2. In their description of the librations, they chose as displacement coordinates the changes in the direction cosines of the molecular axis in the crystal (space-fixed) coordinate system. If $\Lambda_j(lk)$ is the direction cosine for the symmetry axis of a molecule at site k in unit cell l with respect to crystal coordinate axis j (x, y, or z), then we call $\lambda_j(lk)$ the change of the direction cosine such that

$$\Lambda_j(lk) = \Lambda_j^0(lk) + \lambda_j(lk), \qquad j = x, y, z \tag{2.47}$$

where the superscript 0 designates the equilibrium orientation. These displacement coordinates can again be symmetrized with respect to translation in the solid:

$$\lambda_j(\mathbf{q}k) = N^{-1/2} \sum_l \exp i[\mathbf{q} \cdot \mathbf{R}(lk)]\lambda_j(lk) \tag{2.48}$$

in analogy to (2.11). We now note that a linear molecule has only two independent angular degrees of freedom, whereas we have used three displacement coordinates ($j = x, y, z$). We are therefore dealing with a problem involving redundancy. Walmsley and Pople (1964) restricted their treatment to $\mathbf{q} = 0$.

The CO_2 solid contains four distinctly oriented molecules per primitive unit cell and therefore at $\mathbf{q} = 0$ there are eight angular degrees of freedom and twelve displacement coordinates $\lambda_j(lk)$, or there are four redundant coordinates. The redundancy conditions are obtained from the conditions of normalization, i.e., the sum of the squares of the direction cosines must be unity both before and after displacement

$$\sum_j (\Lambda_j^0)^2(lk) = \sum_j [\Lambda_j^0(lk) + \lambda_j(lk)]^2 = 1 \tag{2.49}$$

Equation (2.49) immediately yields the following conditions:

$$\sum_j [2\Lambda_j^0(lk)\lambda_j(lk) + \lambda_j^2(lk)] = 0 \tag{2.50}$$

Using the inverse transformation of (2.48) and setting all $\lambda_j(\mathbf{q}k)$ to zero for $\mathbf{q} \neq 0$, we obtain (2.51). We note that $\Lambda_j^0(k)$ is independent of l.

$$\sum_j [2\Lambda_j^0(k)\lambda_j(\mathbf{0}k) + N^{-1/2}\lambda_j^2(\mathbf{0}k)] = 0 \tag{2.51}$$

For the particular example of solid CO_2, the molecular axes of the four sublattices are oriented parallel to the four body diagonals of the cube, and the direction cosines are therefore as follows:

$$\begin{aligned}
&k = 1 \quad \Lambda_x^0 = \frac{1}{\sqrt{3}}, \quad \Lambda_y^0 = \frac{1}{\sqrt{3}}, \quad \Lambda_z^0 = \frac{1}{\sqrt{3}} \\
&k = 2 \quad \Lambda_x^0 = \frac{1}{\sqrt{3}}, \quad \Lambda_y^0 = -\frac{1}{\sqrt{3}}, \quad \Lambda_z^0 = -\frac{1}{\sqrt{3}} \\
&k = 3 \quad \Lambda_x^0 = -\frac{1}{\sqrt{3}}, \quad \Lambda_y^0 = \frac{1}{\sqrt{3}}, \quad \Lambda_z^0 = -\frac{1}{\sqrt{3}} \\
&k = 4 \quad \Lambda_x^0 = -\frac{1}{\sqrt{3}}, \quad \Lambda_y^0 = -\frac{1}{\sqrt{3}}, \quad \Lambda_z^0 = \frac{1}{\sqrt{3}}
\end{aligned} \tag{2.52}$$

Using (2.52) in (2.51), we obtain the set of redundancy conditions for $\mathbf{q} = 0$:

$$\begin{aligned} \frac{2}{\sqrt{3}}(l_1 + m_1 + n_1) + N^{-1/2}(l_1^{\,2} + m_1^{\,2} + n_1^{\,2}) &= 0 \\ \frac{2}{\sqrt{3}}(l_2 - m_2 - n_2) + N^{-1/2}(l_2^{\,2} + m_2^{\,2} + n_2^{\,2}) &= 0 \\ \frac{2}{\sqrt{3}}(-l_3 + m_3 - n_3) + N^{-1/2}(l_3^{\,2} + m_3^{\,2} + n_3^{\,2}) &= 0 \\ \frac{2}{\sqrt{3}}(-l_4 - m_4 + n_4) + N^{-1/2}(l_4^{\,2} + m_4^{\,2} + n_4^{\,2}) &= 0 \end{aligned} \tag{2.53}$$

In (2.53) $l_1 = \lambda_x(\mathbf{0}1)$, $m_1 = \lambda_y(\mathbf{0}1)$, $n_1 = \lambda_z(\mathbf{0}1)$, etc., where the $\mathbf{0}$ in the parentheses signifies $\mathbf{q} = 0$ [see (2.48)]. Clearly, the redundant coordinates are $(l_1 + m_1 + n_1)$, $(l_2 - m_2 - n_2)$, $(-l_3 + m_3 - n_3)$, $(-l_4 - m_4 + n_4)$, which vanish to first order in small displacements.

The use of redundant coordinates requires extensive modification of the lattice dynamical procedure. It is, however, often worth the additional complication to use redundancies if this facilitates the formulation of symmetry coordinates. When the Wigner projection operator (Wigner, 1931) is used to build such symmetry coordinates, it is necessary to first understand the results of the application of all symmetry operations of the applicable group to the displacement coordinates chosen. This is indeed relatively straightforward for the direction cosine displacement coordinates and therein lies their principal value. These coordinates transform like axial vectors in contrast to cartesian coordinates, which transform like polar vectors.

We now turn our attention to the procedure necessary to take into account redundancies. These have been discussed by Wilson, Decius, and Cross (1955, pp. 171–173) for molecular vibration problems and by Walmsley and Pople (1964) and Oliver and Walmsley (1968) for the lattice dynamics problem. Since the method has been applied to solid CO_2 (Walmsley and Pople, 1964), we shall continue referring to this system as illustrative example.

In the development of the potential Φ [Eq. (2.4)] only second-order terms appear since the first derivatives of the potential are assumed to vanish as condition of equilibrium. However, this is true only for independent displacement coordinates. Generally,

$$\Phi = \sum_i \Phi_i u_i + \frac{1}{2}\sum_{ij} \Phi_{ij} u_i u_j \tag{2.54}$$

and the condition for equilibrium requires that the work for a virtual displacement from the equilibrium position vanishes:

$$\delta\Phi = \sum_i \Phi_i u_i = 0 \tag{2.55}$$

If now the u_i are independent, it follows that $\Phi_i = 0$ for all i. If, on the other hand, we use m variables, of which n are independent, then there exist $m - n$ constraints in the form of redundancy conditions connecting the coordinates u_i:

$$R^a(\ldots, u_i, \ldots) = 0; \qquad a = 1, \ldots, m - n \tag{2.56}$$

In the present case (for $\mathbf{q} = 0$) we use twelve displacement coordinates $\lambda_j(\mathbf{0}k)$ of which eight are independent, as already discussed, and the four relations of (2.53) represent the redundancy conditions R^a of (2.56). In this case the Φ_k for the four redundant coordinates will not be zero in the potential expansion (2.54). In fact, Walmsley and Pople (1964) construct symmetry coordinates

$$S_i = \sum_{jk} T_{i,jk}\lambda_j(\mathbf{0}k) \tag{2.57}$$

and four of these turn out to be linear combinations of the first-order redundant coordinates appearing in (2.53). The remaining eight S_i are linearly independent coordinates. Therefore, that part of the potential energy expansion that depends on the librational coordinates takes the form

$$\Phi = \sum_{h=9}^{12} \Phi_h S_h + \frac{1}{2} \sum_{i,j=1}^{12} \Phi_{ij} S_i S_j \tag{2.58}$$

The next step involves expressing the redundant S_h in terms of the independent set S_i ($i = 1, \ldots, 8$). This results in expressions of the form

$$\begin{aligned} S_h &= \sum_{p,q=1}^{12} a_{hpq} S_p S_q \\ &= \sum_{p,q=1}^{8} a_{hpq} S_p S_q + O(S^4); \qquad h = 9, 10, 11, 12 \end{aligned} \tag{2.59}$$

Equation (2.59) represents the redundancy conditions (2.53) after suitable transformations; (2.53) relates linear displacements to second-order terms in the displacements. Substitution of the four equations (2.59) into (2.58) results in an expansion of the potential in second-order terms of the independent coordinates only:

$$\Phi = \frac{1}{2} \sum_{p,q=1}^{8} G_{pq} S_p S_q \tag{2.60}$$

where we identify G_{pq} by substituting (2.59) in (2.58) and differentiating twice. Explicitly we then obtain

$$\begin{aligned} G_{pq} &= \Phi_{pq} + 2\sum_{h=9}^{12} a_{hpq}\Phi_h \\ &= \Phi_{pq} + \sum_{h=9}^{12} \frac{\partial^2 S_h}{\partial S_p\, \partial S_q} \cdot \frac{\partial \Phi}{\partial S_h} \end{aligned} \tag{2.61}$$

The frequencies are then obtained from the standard secular equation

$$|G_{pq} - I\omega^2\delta_{pq}| = 0$$

where I is the moment of inertia and ω the angular frequency. The derivatives Φ_h and Φ_{pq} can be expressed in terms of the displacement coordinates of individual molecules $\lambda_j(lk)$ by making use of (2.57) and (2.48), noting that the matrix of the transformation $\mathbf{T}$ is an orthogonal matrix:

$$\Phi_h = \frac{\partial \Phi}{\partial S_h} = N^{1/2} \sum_{jk} T_{h,jk} \left[\frac{\partial \Phi}{\partial \lambda_j(lk)}\right]_0$$

$$\begin{aligned} \Phi_{pq} &= \frac{\partial^2 \Phi}{\partial S_p\, \partial S_q} \\ &= \sum_{ij}\sum_{kk'} T_{p,ik} T_{q,jk'} \sum_{l'} \left[\frac{\partial^2 \Phi}{\partial \lambda_i(lk)\, \partial \lambda_j(l'k')}\right]_0 \end{aligned}$$

where all derivatives are evaluated at the equilibrium configuration.

Walmsley and Pople (1964) carried out a calculation using cosine displacement coordinates for the $\mathbf{q} = 0$ modes of a linear molecule, CO_2. Oliver and Walmsley (1968) showed how the method may be applied, in principle, to a nonlinear molecule with three angular degrees of freedom. In this case there are nine direction cosines (three for each molecular axis) but only three independent degrees of freedom, or six redundancies per molecule in the primitive unit cell. The method has not been applied to calculations for $\mathbf{q} \neq 0$ modes. For such a case the redundancy conditions (2.59) would in general be functions of $\mathbf{q}$ and a method would be required to remove the redundancies automatically in a computer calculation. This appears to be possible for molecules of three librational degrees of freedom according to the method described by Oliver and Walmsley (1968), but it is not clear from this work how it could be best done for the case of linear molecules with two angular degrees of freedom. It appears that the most arduous task in a calculation using redundant coordinates is to cast the redundancy conditions in the form of (2.59). The redundant coordinates must be expressed as second-order expressions in the dependent coordinates to make it possible to use the procedure described here.

Coll and Harris (1970) have described another method of dealing with the librations of a crystal of linear molecules by use of direction cosine displacement coordinates. These authors eliminate the z-component (λ_z) at the beginning of the treatment and thereby avoid the redundancy problem.

6. *Infinitesimal Rotation Coordinates*

The rotational kinetic energy of a rigid body can be expressed in terms of rotational displacements u_α, u_β, u_γ where the subscripts refer to rotations about the x,y,z-coordinate axes, respectively:

$$T_{\text{lib}} = \frac{1}{2}\sum_i I_i \dot{u}_i^2, \qquad i = \alpha, \beta, \gamma \tag{2.62}$$

Such displacement coordinates have been used as a basis in the analysis of the lattice dynamics of librational motions (Cochran and Pawley, 1963; Pawley, 1967). Oliver and Walmsley (1968) have pointed out that u_α, u_β, u_γ are independent to first order only and are often called "infinitesimal" rotation coordinates for that reason. We shall here discuss the use of these coordinates in some detail as applied to a linear molecule having two angular degrees of freedom.

Considering first the kinetic energy, we note that (2.62) is an approximation. When we consider a rotation by α about the x-axis followed by a rotation about the space-fixed (equilibrium configuration) y-axis, we find that the corresponding transformation is given by

$$\begin{aligned} x &= r \cos\alpha \sin\beta \\ y &= -r \sin\alpha \\ z &= r \cos\alpha \cos\beta \end{aligned} \tag{2.63}$$

and the square of the infinitesimal displacement is given by

$$ds^2 = r^2\, du_\alpha^2 + r^2 \cos^2\alpha\, du_\beta^2 \tag{2.64}$$

resulting in the kinetic energy expression

$$T_{\text{lib}} = \frac{I}{2}\sum_{lk} [\dot{u}_\alpha^2(lk) + \cos^2\alpha(lk)\dot{u}_\beta^2(lk)] \tag{2.65}$$

As in our discussion of (2.31), it is usual to expand the nonconstant coefficient $\cos^2\alpha(lk)$ about the equilibrium value of α which is $\alpha = 0°$ or $\cos\alpha = 1$ for the case where the molecular figure axis at equilibrium is taken as the z-axis and the angles α and β are measured from this equilibrium axis. Equation (2.65) then takes the form to be expected from (2.62). It is, of course, clear that the displacements u_α, u_β are defined in terms of the local axes appropriate for each molecule (lk).

When using a potential $\Phi(R_i)$ as given by a sum of terms as in (2.2), which is a function of distances R_i between pairs of atoms on different molecules, it is necessary to expand Φ up to second order in the translational displacements $u_x(lk)$, $u_y(lk)$, $u_z(lk)$ of the molecular centers of mass and also to the same order in the librational coordinates $u_\alpha(lk)$, $u_\beta(lk)$. Again, as in (2.39), if we designate by r_i the changes in R_i, we obtain for a typical coefficient of a second-order term in $u_\alpha(lk)$, $u_{\alpha'}(l'k')$

$$\begin{aligned}\Phi_{\alpha\alpha'} &= \frac{\partial^2\Phi}{\partial u_\alpha(lk)\,\partial u_{\alpha'}(l'k')}\\ &= \sum_{ij}\left(\frac{\partial^2\Phi}{\partial R_i\,\partial R_j}\right)_0 \frac{\partial r_i}{\partial u_\alpha(lk)}\cdot\frac{\partial r_j}{\partial u_{\alpha'}(l'k')}\\ &\quad+\sum_j\left(\frac{\partial\Phi}{\partial R_j}\right)_0 \frac{\partial^2 r_j}{\partial u_\alpha(lk)\,\partial u_{\alpha'}(l'k')}\end{aligned} \tag{2.66}$$

The derivatives of the r_i with respect to the $u_\alpha(lk)$ are now calculated by making use of the derivatives of the cartesian components $x_i y_i z_i$ of the r_i which can be derived from the transformations (2.63). Here the cartesian components are assumed to be in the coordinate system appropriate to molecule (lk) and then the only nonvanishing derivatives up to second order are given in (2.67):

$$\begin{gathered}\left(\frac{\partial x}{\partial u_\beta(lk)}\right)_0 = r\\ \left(\frac{\partial y}{\partial u_\alpha(lk)}\right)_0 = -r\\ \left(\frac{\partial^2 z}{\partial u_\alpha^2(lk)}\right)_0 = \left(\frac{\partial^2 z}{\partial u_\beta^2(lk)}\right)_0 = -r\end{gathered} \tag{2.67}$$

Here again the molecular figure axis is taken as parallel to the z-axis and the sign convention of rotational displacements is such that it is positive for a right-handed screw advancing in the direction of the positive coordinate axis (a right-handed coordinate frame is assumed).

It should be pointed out that inclusion of the second-order derivatives in (2.67) is important and their neglect may give rise to serious errors. In actual calculations it is necessary to consider the displacements of two molecules which contribute to a particular r_i such that

$$\mathbf{r}_i = \mathbf{r}_i(lk) + \mathbf{r}_i(l'k')$$

Then the components of each $r_i(lk)$ in the coordinate system of molecule (lk) must be used in (2.66), as already stated.

Cochran and Pawley (1963) and Pawley (1967) chose to calculate the first derivatives of the potential analytically and the second derivatives numerically. In this way they avoided the more complicated analytical expression (2.66) while including the terms arising from the expansion of the displacements to second order.

7. *Eulerian Angle Displacement Coordinates*

Schnepp and Ron (1969) and Kuan, Warshel, and Schnepp (1969) treated the lattice dynamics of solid α-N_2 using Eulerian angles θ, ϕ to describe the librations. Suzuki and Schnepp (1971) applied a similar calculation to solid CO_2. The kinetic energy T_{lib} has already been discussed in Section IIC.1 [see (2.31)]. The authors mentioned replaced $\sin^2 \theta(lk)$ appearing in T_{lib} by the equilibrium values $\sin^2 \theta_0(lk)$, all angles being measured relative to the space-fixed crystal axes. The potential was again a sum of pair potentials, each term being a function of the distance between a pair of atoms on different molecules $[\Phi = \sum_i v_i(R_i)]$. The x-component X_i of R_i, is given by the relation

$$X_i(lk, l'k') = \frac{a}{2} + u_x(lk) - u_x(l'k')$$
$$+ h \sin \theta(lk) \cos \phi(lk) - h \sin \theta(l'k') \cos \phi(l'k') \quad (2.68)$$

Here the subscript i designates a particular pair of atoms, one in molecule (lk), the other in molecule $(l'k')$. In this crystal structure ($Pa3 - T_h^6$), the molecular centers occupy face-centered cubic positions and the cubic cell parameter is a. The $u_x(lk)$ are, as before, the cartesian displacements of the molecular center of mass. The letter h designates the half distance between atoms (or the distance of the atom from the molecular center of mass). This distance h is also used as an adjustable parameter of the potential and then $2h$ is the distance between the two "centers of interaction" of the molecule. The u_θ are related to the instantaneous orientations θ by

$$\theta = \theta_0 + u_\theta$$

It is clear from a consideration of (2.68) that the nonlinear dependence of X_i on θ, ϕ requires again the inclusion of terms similar to the last term in (2.66). Moreover, terms like $\partial^2 X_i(lk, l'k')/\partial u_\theta^2(lk)$ do not vanish, in contrast to $\partial^2 X_i(lk, l'k')/\partial u_x^2(lk)$. Such terms appear in the calculation of $\partial^2\Phi/\partial u_j^2$ where $j = \theta, \phi$.

The equations of motion here are

$$I\ddot{u}_\theta(lk) = -\sum_j \sum_{l'k'} \Phi_{\theta j}(lk, l'k')u_j(l'k')$$

$$I \sin^2 \theta_0(lk)\ddot{u}_\phi(lk) = -\sum_j \sum_{l'k'} \Phi_{\phi j}(lk, l'k')u_j(l'k')$$

The index j runs over x, y, z, θ, ϕ. These equations are analogous to (2.6) except for the factor $\sin^2 \theta_0(lk)$ in the ϕ-equation. This factor appears then in the rows of the secular determinant designated by u_ϕ and as a result the determinant loses its symmetric structure. This difficulty is readily overcome by defining a new displacement coordinate u_ω:

$$u_\omega = \sin \theta_0 u_\phi$$

Alternatively, the secular determinant can be transformed by multiplying appropriate rows and dividing appropriate columns by $\sin \theta_0$.

It was pointed out in Section IIC.5 that the understanding of the behavior of coordinates under symmetry operations is of great importance. Such an understanding is essential for the construction of symmetry coordinates and the choice of suitable coordinates is, indeed, often determined by such considerations. Schnepp and Ron (1969) therefore investigated the transformation of the u_θ, u_ω under the operations of the factor group of α-N_2, T_h^6. These transformations were found to be non-linear and were summarized in Table 2 of their paper up to second order in the displacements. For the construction of symmetry coordinates it is sufficient to use the transformations to first order only since these coordinates must hold for arbitrarily small displacements. The transformations to second order are useful when it is desired to find relations between elements of the dynamical equation based on symmetry. The

TABLE II

Transformation Properties of Librational Displacement Coordinates u_θ, u_ω and Molecular Positions and Orientations for the α-N_2 or CO_2 structure ($T_h^6 - Pa3$)

A. Molecular Positions and Orientations

Type	Orientation	Specific molecule	Location
1	1 1 1	1	(0 0 0)
2	1, −1, −1	2_1	($\frac{1}{2}$ $\frac{1}{2}$ 0)
		2_2	($\frac{1}{2}$ $-\frac{1}{2}$ 0)
		2_3	($-\frac{1}{2}$ $-\frac{1}{2}$ 0)
		2_4	($-\frac{1}{2}$ $\frac{1}{2}$ 0)
3	−1, 1, −1	3_1	(0 $\frac{1}{2}$ $\frac{1}{2}$)
		3_2	(0 $\frac{1}{2}$ $-\frac{1}{2}$)
		3_3	(0 $-\frac{1}{2}$ $-\frac{1}{2}$)
		3_4	(0 $-\frac{1}{2}$ $\frac{1}{2}$)
4	−1, −1, 1	4_1	($\frac{1}{2}$ 0 $\frac{1}{2}$)
		4_2	($\frac{1}{2}$ 0 $-\frac{1}{2}$)
		4_3	($-\frac{1}{2}$ 0 $-\frac{1}{2}$)
		4_4	($-\frac{1}{2}$ 0 $\frac{1}{2}$)

TABLE II (continued)

B. Transformations (to second order only)

1. Inversion i located at molecule j
   $u_\theta(j) = u_\theta(j)$
   $u_\omega(j) = u_\omega(j)$
2. Screw diads (positive displacement)
   (a) C_2^x (at $y = \frac{1}{4}$, $z = 0$)
   $u_\theta(1) = -u_\theta(2_1), \quad u_\omega(1) = -u_\omega(2_1)$
   $u_\theta(3_1) = -u_\theta(4_1), \quad u_\omega(3_1) = -u_\omega(4_1)$
   (b) C_2^y (at $z = \frac{1}{4}$, $x = 0$)
   $u_\theta(1) = -u_\theta(3_1), \quad u_\omega(1) = -u_\omega(3_1)$
   $u_\theta(2_1) = -u_\theta(4_1), \quad u_\omega(2_1) = -u_\omega(4_1)$
   (c) C_2^z (at $x = \frac{1}{4}$, $y = 0$)
   $u_\theta(1) = u_\theta(4_1), \quad u_\omega(1) = u_\omega(4_1)$
   $u_\theta(2_1) = u_\theta(3_1) \quad u_\omega(2_1) = u_\omega(3_1)$
3. Triads located on molecule of type 1
   (a) $C_3^1(1 \to 1, 3_1 \to 2_1)$

$$u_\theta{}^1(1) = -\tfrac{1}{2}u_\theta(1) + \frac{\sqrt{3}}{2} = u_\omega(1) + \frac{3\sqrt{2}}{16}u_\theta{}^2(1) + \frac{3\sqrt{2}}{16}u_\omega{}^2(1) + \frac{3\sqrt{6}}{8}u_\theta(1)u_\omega(1)$$

$$u_\omega{}^1(1) = -\frac{\sqrt{3}}{2}u_\theta(1) - \tfrac{1}{2}u_\omega(1) + \frac{\sqrt{6}}{8}u_\theta{}^2(1) + \frac{\sqrt{6}}{4}u_\omega{}^2(1)$$

$$u_\theta(2_1) = -\tfrac{1}{2}u_\theta(3_1) + \frac{\sqrt{3}}{2}u_\omega(3_1) - \frac{3\sqrt{2}}{16}u_\theta{}^2(3_1) - \frac{3\sqrt{2}}{16}u_\omega{}^2(3_1) - \frac{3\sqrt{6}}{8}u_\theta(3_1)u_\omega(3_1)$$

$$u_\omega(2_1) = -\frac{\sqrt{3}}{2}u_\theta(3_1) - \tfrac{1}{2}u_\omega(3_1) + \frac{\sqrt{6}}{8}u_\theta{}^2(3_1) - \frac{\sqrt{6}}{4}u_\omega{}^2(3_1)$$

   (b) $C_{-3}^1(1 \to 1, 4_1 \to 2_1)$

$$u_\theta{}^1(1) = -\tfrac{1}{2}u_\theta(1) - \frac{\sqrt{3}}{2}u_\omega(1) + \frac{3\sqrt{2}}{16}u_\omega{}^2(1) + \frac{3\sqrt{2}}{16}u_\omega{}^2(1) - \frac{3\sqrt{6}}{8}u_\theta(1)u_\omega(1)$$

$$u_\omega{}^1(1) = \frac{\sqrt{3}}{2}u_\theta(1) - \tfrac{1}{2}u_\omega(1) + \frac{\sqrt{6}}{8}u_\theta{}^2(1) - \frac{\sqrt{6}}{4}u_\omega{}^2(1)$$

$$u_\theta(2_1) = \tfrac{1}{2}u_\theta(4_1) + \frac{\sqrt{3}}{2}u_\omega(4_1) - \frac{3\sqrt{2}}{16}u_\theta{}^2(4_1) + \cdots$$

$$u_\omega(2_1) = -\frac{\sqrt{3}}{2}u_\theta(4_1) + \tfrac{1}{2}u_\omega(4_1) - \frac{\sqrt{6}}{8}u_\theta{}^2(4_1) + \cdots$$

---

transformations of u_θ, u_ω can be derived by considering the behavior of the cartesian coordinates of an end atom of the linear molecule and using the usual transformations between these cartesian coordinates and the angular coordinates θ, ϕ. Table 2 of Schnepp and Ron (1969) is reproduced in Table II.

8. *Extended Point-Mass Model*

Rafizadeh and Yip (1970) have recently proposed a new approach for the treatment of librations in connection with their treatment of solid hexamine (hexamethylenetetramine). The potential is assumed to be a function of the "net" molecular displacements $\mathbf{\Delta}(lk)$ only. The ith cartesian component of net displacement includes translational and librational motions and is defined as the sum of the center-of-mass translational displacement component $u_i(lk)$ and the displacement component due to an angular displacement $\boldsymbol{\theta} \times \mathbf{R}$. The molecule is labeled (lk) as before. The defining relation is given in (2.69):

$$\Delta_i(lk) = u_i(lk) + [\boldsymbol{\theta}(lk) \times \mathbf{R}(lk)]_i \tag{2.69}$$

The vector $\mathbf{R}(lk)$ is a characteristic feature of the model and its components are parameters of the potential model. This vector can be thought of as defining the "effective size" of the molecule; it determines the position of a point mass relative to the molecular center of mass. The translational displacement $\mathbf{u}(lk)$ and the angular displacement $\boldsymbol{\theta}(lk)$ produce then the net displacement $\mathbf{\Delta}(lk)$.

The potential is expanded up to second order in terms of the $\Delta_i(lk)$. It is then possible to express all force constants (translational, librational, and interaction) in terms of the translational force constants only. However, the components of the vectors $\mathbf{R}(kl)$, $\mathbf{R}(k'l')$ also appear in these expressions. If, for example, we consider the problem of one molecule per primitive unit cell, the dynamical equation is of dimension six and there are, in general, 36 force constants. In the extended point-mass approximation considered here, on the other hand, there are only nine force constants, as would appear in the dynamical equation containing translational motions only. In addition, there are the components of $\mathbf{R}(l)$, $\mathbf{R}(l')$ (for the interacting molecules l, l') which represent six further parameters. If we regard the force constants as parameters, the total number of these is 15 in this model, as compared to 36 in the general case. Actually, the symmetry of the dynamical matrix further reduces this number in both cases.

It appears that the extended point-mass treatment is useful inasmuch as it requires relations between the force constants and thereby the total number of parameters is reduced. Thus far it has been assumed that the force constants are all independent parameters. If, on the other hand, a pair potential function is specified which contains a limited number of parameters, and all force constants can be derived from this function, then the number of independent parameters is only as the number of parameters of the potential function. In such a case, which is more usual, it is not yet clear if the treatment has advantages. At this time, the experience with the treatment is too limited to draw any useful conclusion in this regard. It is, however, evident that this model does not emphasize the shape of the molecule. In the case of the hexamine molecule, which is very close to spherical, the shape is probably not a factor, but this may not always be the case. It should also be pointed out that the treatment is not immediately applicable to a linear molecule solid since the resulting redundancy would have to be carefully considered.

D. Translation-Libration Interactions and Solution of the Lattice Dynamics Problem

In all lattice dynamics treatments for librational degrees of freedom discussed in Section IIC, interactions between these coordinates and translations must be considered. As already pointed out, interaction matrix elements vanish only at the zone center and some very special points on the zone boundary and this only for centro-symmetric solids. The first calculation of the dispersion curves for a molecular solid throughout the Brillouin zone was carried out by Cochran and Pawley (1964) for hexamethylenetetramine (hexamine). Once the librational displacement coordinates have been defined and a potential function chosen, the interaction force constants $\Phi_{i\alpha}$ can be calculated. The subscripts refer to the displacement coordinate components, where we use i to designate a translational displacement component u_i and α to designate a librational displacement component u_α. The corresponding dynamical matrix elements are $M_{i\alpha}$, analogous to the M_{ij} defined in (2.9). In general, the matrix elements are complex, and the matrix is hermitian.

If the crystal is centrosymmetric, some simplifications result. Referring to the designations used in (2.9), we choose the reference molecule (lk) which is now located at a center of inversion for the whole solid, and consider the interaction of this molecule with two other molecules $(l'k')$ and $(l''k')$ which are related by the operation of inversion. Then, clearly,

$$\mathbf{R}(l'k'; lk) = -\mathbf{R}(l''k'; lk)$$

Considering a translation-translation matrix element M_{ij}, we note that translational displacements u_i are antisymmetric to inversion. As a result, we obtain

$$\Phi_{ij}(lk, l'k') = \Phi_{ij}(lk, l''k')$$

and the sum of the two terms in the summation over l' arising from the interaction of the central molecule with the two molecules $(l'k')$, $(l''k')$ gives

$$2\Phi_{ij}(lk, l'k') \cos [\mathbf{q} \cdot \mathbf{R}(l'k'; lk)]$$

Since there exists a partner related by the center of inversion for every molecule, we conclude that this part of the dynamic matrix is real.

Next we consider libration-libration elements, $M_{\alpha\beta}$. We note that the librational displacements u_α are symmetric to inversion and we therefore reach again the same conclusion, that this part of the dynamical matrix is also real. However, in view of what has been noted, we find that the translation-libration interaction elements $M_{i\alpha}$ are pure imaginary. This results from the fact that we obtain here

$$\Phi_{i\alpha}(lk, l'k') = -\Phi_{i\alpha}(lk, l''k')$$

and therefore the sum of the corresponding two terms in the sum over l' [Eq. (2.9)] is

$$2i\Phi_{i\alpha}(lk, l'k') \sin [\mathbf{q} \cdot \mathbf{R}(l'k'; lk)]$$

It is also found that the following relations hold for the dynamical matrix elements:

$$M_{ij}(\mathbf{q}kk') = M_{ji}(\mathbf{q}k'k)$$
$$M_{\alpha\beta}(\mathbf{q}kk') = M_{\beta\alpha}(\mathbf{q}k'k)$$
$$M_{i\alpha}(\mathbf{q}kk') = -M_{\alpha i}(\mathbf{q}k'k)$$

and the matrix is therefore hermitian as expected and the roots are all real.

The dynamical matrix consists then of two square blocks composed of real elements (the translation-translation block and the libration-libration block) and two off-diagonal, generally rectangular blocks of pure imaginary elements (translation-libration interactions). Such a matrix can be transformed (Cochran and Pawley, 1964) into a real symmetric matrix of the same dimension by a transformation of the coordinates, which leaves all translational coordinates unchanged and transforms all librational coordinates $u_\alpha(lk)$ as follows:

$$iu_\alpha(lk) = u_\zeta(lk)$$

Solution of the resulting matrix gives real eigenvectors in terms of the u_i, u_ζ and resubstitution results in complex eigenvectors in terms of the u_i, u_α, such that the librational part is multiplied by i. The physical significance of this form is to be found in that the librational motions are out of phase with the translational motions, and the phase difference is $\pi/2$.

E. Symmetry Relations between Force Constants

The symmetry of the crystal structure imposes conditions on the elements of the dynamic matrix. As is evident from (2.9), the matrix elements are linear combinations of interaction force constants $\Phi_{ij}(lk, l'k')$ between a pair of molecules (lk), $(l'k')$ modulated by appropriate phase factors. The symmetry operations of the crystal map the lattice on itself.

It is evident that the force constants $\Phi_{ij}(lk, l'k')$ must be invariant to such transformations and, as a result, relations are obtained between different force constants. As a result, it frequently is possible to express the dynamic matrix in terms of a limited number of distinct parameters, the number depending on the crystal structure.

There exist several possible reasons for making use of the symmetry relations between force constants. In some instances, the independent coupling constants are treated as parameters and are determined by fitting the observed frequencies, much as is done in molecular vibrational spectroscopy (Wilson, Decius, and Cross, 1955). This procedure is useful only for simple structures, containing, e.g., one molecule per primitive cell and having a structure of very high symmetry. Such is the case for hexamine (Cochran and Pawley, 1964). In other cases, the symmetry may be useful for checking the calculation or reducing computer time. In addition to hexamine, the symmetry relations between the force constants have been investigated for the naphthalene structure $C_{2h}^5 - P2_{1/c}$ (Pawley, 1967) and for the α-N_2 or CO_2 structure $T_h^6 - Pa3$ (Schnepp and Ron, 1969). It must be remembered that symmetry requires relations only between force constants representing interactions between a central molecule and neighboring molecules in the same shell (at equal distances from the central molecule). As a result, as more shells of neighboring molecules are taken into account, the symmetry relations become progressively less useful.

We shall present a few examples of symmetry relations. Let us first consider an interaction force constant $\Phi_{ij}(lk, l'k')$ where the molecules (lk) and $(l'k')$ both lie on an element of symmetry. Then application of this symmetry operation (rotation axis or reflection plane) brings the pair of molecules into self-coincidence. This requires the $\Phi_{ij}(lk, l'k')$ to be identical before and after transformation. However, a change of sign may be required, and it can then be concluded that the particular force constant must vanish. For example, assume that the symmetry element is a twofold axis parallel to x. Then $\Phi_{yz}(lk, l'k')$ will not change sign but $\Phi_{xy}(lk, l'k')$ changes sign and must vanish. If the infinitesimal librational displacements about the molecular coordinate axes u_α, u_β, u_γ are used (Section IIC.6), then we can easily verify for a rotation about the molecular x-axis $u_\alpha \to u_\alpha$, $u_\beta \to -u_\beta$, $u_\gamma \to -u_\gamma$. We then conclude that $\Phi_{\alpha\beta}(lk, l'k')$, $\Phi_{x\beta}(lk, l'k')$, $\Phi_{z\alpha}(lk, l'k')$, etc., must all be identically zero. For a reflection in, e.g., the xy-plane, $u_\alpha \to -u_\alpha$, $u_\beta \to -u_\beta$, $u_\gamma \to u_\gamma$, and now $\Phi_{xz}(lk, l'k')$, $\Phi_{x\alpha}(lk, l'k')$, $\Phi_{\alpha\gamma}(lk, l'k')$, etc., must vanish. If, on the other hand, we use the Eulerian angle displacements u_θ, $u_\omega = \sin\theta_0 u_\phi$ (Schnepp and Ron, 1969), where the displacements of the linear molecules are measured from their equilibrium positions and the coordinate system is taken to be crystal-fixed, then we obtain for a twofold axis parallel to the crystal x-axis: $u_\theta \to -u_\theta$, $u_\omega \to -u_\omega$. In this case, $\Phi_{x\theta}(lk, l'k')$ and $\Phi_{x\omega}(lk, l'k')$ vanish. Similarly, u_θ, u_ω change sign under a twofold rotation about the y-axis, but they remain invariant under twofold rotation about the z-axis. Transformations of the type considered so far have been called Type A by Cochran and Pawley (1964). The transformation properties of u_θ, u_ω under some of the symmetry operations are summarized in Table II.

The u_α, u_β, u_γ can be thought of as polar vector components (as opposed to axial vector components u_x, u_y, u_z) and they transform accordingly. When the lattice dynamical problem is treated in terms of the dynamical variable $u_x u_y u_z u_\alpha u_\beta u_\gamma$, Cochran and Pawley have pointed out that the two-molecule interaction force constants $\Phi_{ij}(lk, l'k')$ can be treated as a two-dimensional tensor of dimension six. If $\mathbf{S}$ is the cartesian coordinate transformation matrix corresponding to a symmetry transformation, then the six-dimensional transformation matrix is

$$\mathbf{T} = \begin{pmatrix} \mathbf{S} & 0 \\ 0 & \mathbf{S} \det \mathbf{S} \end{pmatrix} \tag{2.70}$$

Here det $\mathbf{S}$ is the determinant of the matrix $\mathbf{S}$ and this is $+1$ if the operation is a proper rotation and -1 if it is an improper rotation. Then the transformed tensor is $\mathbf{T}\boldsymbol{\Phi}(lk, l'k')\tilde{\mathbf{T}}$ and for the case where the pair of molecules is transformed into themselves, we have the condition

$$\mathbf{T}\boldsymbol{\Phi}(lk, l'k')\tilde{\mathbf{T}} = \boldsymbol{\Phi}(lk, l'k') \tag{2.71}$$

The requirement for the vanishing of certain elements $\Phi_{ij}(lk, l'k')$ as discussed here can then be readily derived from (2.71).

For another type of symmetry operation, the origin molecule ($lk = 0$) is transformed into itself, but the interacting molecule ($l'k'$) is mapped on another molecule ($l''k'$) of the same type (site) but in another cell. The operation of inversion is an example and such an operation has been termed Type B by Cochran and Pawley (1964). Making use of the transformation matrix $\mathbf{T}$ [Eq. (70)], they obtain for inversion in the center of symmetry located at molecule ($lk = 0$):

$$\begin{aligned} \mathbf{T}\boldsymbol{\Phi}(0, l'k')\tilde{\mathbf{T}} &= \boldsymbol{\Phi}(0, l''k') = \Phi(0, -l'k') \\ &= \boldsymbol{\Phi}(l'k', 0) \text{ from translational invariance} \\ &= \tilde{\boldsymbol{\Phi}}(0, l'k') \text{ from the definition of } \boldsymbol{\Phi} \end{aligned}$$

Here we have designated the origin molecule (lk) as (0) and we note also that the position vectors of the molecules ($l'k'$), ($l''k'$) are related by

$$\mathbf{r}(l'k') = -\mathbf{r}(l''k')$$

On the other hand, in terms of the Eulerian librational displacement coordinates, using Table II, for a particular component of $\mathbf{\Phi}(0, l'k')$, we similarly obtain

$$\mathbf{P}_I\Phi_{xy}(0, l'k') = \Phi_{xy}(0, -l'k')$$

where $\mathbf{P}_I$ symbolizes the inversion operation. As a result, we conclude

$$\Phi_{xy}(0, l'k') = \Phi_{xy}(0, -l'k')$$

Application of this procedure gives the following relationships:

$$\Phi_{xy}(0, l'k') = \Phi_{xy}(0, -l'k') = \Phi_{xy}(l'k', 0) = \Phi_{yx}(0, l'k')$$
$$\Phi_{x\theta}(0, l'k') = -\Phi_{x\theta}(0, -l'k') = -\Phi_{x\theta}(l'k', 0) = -\Phi_{\theta x}(0, l'k')$$
$$\Phi_{\theta\omega}(0, l'k') = \Phi_{\omega\theta}(0, l'k')$$

The two types of symmetry transformation considered thus far are the only ones, aside from translations, that occur in a symmorphic space group (composed of rotation and reflection operations). Most molecular crystals, however, belong to nonsymmorphic space groups which contain screw axes and glide planes in addition to pure proper and improper rotations. The space groups C_{2h}^5 (naphthalene, anthracene) and T_h^6(α-N_2, CO_2) are both examples. For a twofold screw axis operation (e.g., axis parallel to x) the rotation is accompanied by a translation composed of half unit cell vectors, (e.g., $\frac{1}{2}x + \frac{1}{2}y$). Application of such an operation maps one pair of molecules on another pair, neither of them remaining the same:

$$\mathbf{P}_R\Phi_{ij}(lk, l'k') = \pm\Phi_{ij}(l''k'', l'''k''')$$

where $\mathbf{P}_R$ symbolizes the symmetry operation R. The upper or lower sign may apply in different cases. Again we conclude from this result that

$$\Phi_{ij}(lk, l'k') = \pm\Phi_{ij}(l''k'', l'''k''')$$

Invariance with respect to translations by lattice vectors in conjunction with the preceding considerations result in relationships among force constants of the interaction between the origin molecule and different neighbors in the same shell. Pawley (1967) refers to this class of operations as Type C and has discussed the resulting symmetry relations as applied to the lattice dynamics of naphthalene and anthracene (space group C_{2h}^5). Schnepp and Ron (1969) have made use of the symmetry restrictions caused by the three screw diads occurring in space group T_h^6 in their treatment of α-N_2. For this case, the following examples will illustrate the type of results obtained by application of the operation $C_2{}^x$. The molecular designations are listed in Table II together with the transformation properties of the librational coordinates:

$$\Phi_{xy}(1, 2_4) = -\Phi_{xy}(2_1, 1) = -\Phi_{yx}(1, 2_1)$$
$$\Phi_{x\theta}(1, 2_4) = -\Phi_{x\theta}(2_1, 1) = -\Phi_{\theta x}(1, 2_1)$$
$$\Phi_{y\omega}(1, 2_4) = \Phi_{y\omega}(2_1, 1) = \Phi_{\omega y}(1, 2_1)$$
$$\Phi_{\theta\omega}(1, 2_4) = \Phi_{\theta\omega}(2_1, 1) = \Phi_{\omega\theta}(1, 2_1)$$

The triad symmetry operations deserve special comment. Schnepp and Ron (1969) have listed the transformations of the Eulerian librational coordinates u_θ, u_ω for the structure T_h^6 (Table II). Here, we should recall, the coordinate axes used are space-fixed and the angles θ, ϕ (or θ, ω) are also measured in this crystal system. The triad axes are coincident with the figures axes of the linear molecules. Their directions are the four body diagonals of the cube, corresponding to the four molecules per primitive unit cell. The transformations of u_θ, u_ω or u_θ, u_ϕ were derived by following the cartesian coordinates of an extreme end of the linear molecule relative to the molecular center as origin and by using the exact transformations between x, y, z and θ, ϕ. It was found that the resulting transformations are nonlinear (Table II). The triad axis coincident with molecule 1 relates the interaction force constants for molecules $(1, 3_1)$ or $(1, 4_1)$ with those of $(1, 2_1)$. It was then further found that only first-order terms of the transformations appear in such relations. The triad transformations were also useful to establish relations between the "self-terms" of the form $\Phi_{\alpha\beta}(lk, lk)$, $\alpha, \beta = \theta, \omega$ (or ϕ). By applying the threefold rotation about the figure axis of molecule 1 to $\Phi_{\theta\theta}(1, 1)$, $\Phi_{\omega\omega}(1, 1)$ and $\Phi_{\theta\omega}(1, 1)$ the following relations were proved:

$$\Phi_{\theta\theta}(1, 1) = \Phi_{\omega\omega}(1, 1)$$
$$\Phi_{\theta\omega}(1, 1) = 0$$

Here again it turned out that only the linear part of the θ, ω transformations were of significance. However, second-order terms were found to be important for further simplification of the dynamic matrix elements as follows. The total potential Φ was expressed as a sum over pair potentials $v(lk, l'k')$:

$$\Phi = \frac{1}{2} \sum_{lk} \sum_{l'k'} v(lk, l'k')$$

It was then possible to express all nearest neighbor $\Phi_{ij}(lk, l'k')$ in terms of the derivatives of one specific pair potential $v(1, 2_1)$, where 1 designates the central molecule and 2_1 is a specific nearest neighbor molecule (Table II). As long as $lk \neq l'k'$, the contributions to $\Phi_{ij}(lk, l'k')$ will vanish for all pair potentials $v(\lambda\kappa, \lambda'\kappa')$ except the one $v(lk, l'k')$ since the others do not contain the coordinates of molecules lk, $l'k'$ [see definition, (2.4)]. Thus

$$\Phi_{ij}(lk, l'k') = \frac{\partial^2 v(lk, l'k')}{\partial u_i(lk)\, \partial u_j(l'k')}$$

But for the "self-terms" of the type $\Phi_{ij}(lk, lk)$ or $\Phi_{ij}(0, 0)$ all pair potentials $v(lk, \lambda\kappa)$ for all $\lambda\kappa$ must be included. It was then useful to transform terms like

$$\frac{\partial^2 v(1, \lambda\kappa)}{\partial u_\alpha(1)\, \partial u_\beta(1)}, \qquad \alpha, \beta = \theta, \omega$$

into expressions containing only

$$\frac{\partial^2 v(1, 2_1)}{\partial u_\alpha(1)\, \partial u_\beta(1)}.$$

In particular, for transformations of this kind to simplify the diagonal terms ($\alpha = \beta$, $\alpha = \theta$, ω or ϕ) of the dynamical matrix, the transformations of the θ, ω to second order proved to be important. The neglect of second-order contributions led to error of the order of 40%.

If the pair potentials used are functions of distances between atoms of the molecules concerned, then it can easily be shown that

$$\Phi_{ij}(lk, l'k') = -\Phi_{ij}(l'k', lk)$$

for $i, j = x, y, z$. No such simple relation exists for librational displacements ($i, j = \theta, \omega$). Using this relation in addition to all the symmetry properties discussed, Schnepp and Ron (1969) were able to express the 20×20 dynamical matrix for the lattice dynamical problem in terms of a limited number of independent derivatives of one specific pair potential. Here only interactions with the 12 nearest neighbor molecules at equal distances from the center were considered. Six parameters were sufficient to express all elements in the 12-dimensional translation block (the second derivatives of $v(1, 2_1)$ with respect to x_1x_1, y_1y_1, z_1z_1, x_1y_1, x_1z_1, y_1z_1). Ten independent derivatives had to be computed for the librational block, eight second derivatives of $v(1, 2_1)$ (with respect to $\theta_1\theta_1$, $\theta_{2_1}\theta_{2_1}$, $\theta_1\theta_{2_1}$, $\omega_1\omega_1$, $\omega_2\omega_{2_1}$, $\omega_1\omega_{2_1}$, $\theta_1\omega_{2_1}$, $\omega_1\theta_{2_1}$), and two first derivatives (with respect to θ_1 and θ_{2_1}). For the interaction blocks, 12 further second derivatives of $v(1, 2_1)$ had to be calculated (all combinations of x_1, y_1, z_1 with θ_1, ω_1, θ_{2_1}, ω_{2_1}).

For the hexamine problem (Cochran and Pawley, 1964) five independent parameters were needed to describe the interactions with the eight nearest neighbors and five more parameters for the interactions with the six next-nearest neighbors. The dynamical matrix is of dimension 6×6.

F. Applications and Comparison with Experiment

1. *Inorganic Molecular Solids*

Solid α-Nitrogen. Tables III and IV summarize the available information on the $\mathbf{q} = 0$ modes of α–N_2. The translational modes in this case are separable from the librations. The two infrared active translational modes frequencies were first reported by Anderson and Leroi (1966) and by Ron and Schnepp (1967). Both these groups studied solid films prepared by vapor deposition on a cold window. St. Louis and Schnepp (1969) investigated the far infrared absorption of solid samples prepared by cooling the liquid in a closed cell. It was found that the frequencies were not affected appreciably by the method of sample preparation, but the lines were narrower when the solid was formed under equilibrium conditions. In particular, the lower frequency line was found to be remarkably sharp (half-intensity width 0.3 cm^{-1} or less). The frequencies of these translational lattice modes were found to be well reproduced by a calculation (Ron and Schnepp, 1967) based on classical harmonic lattice dynamics and using a pair potential consisting of an intermolecular Lennard-Jones term (with gas phase determined parameters) and an angle-dependent quadrupole-quadrupole term (Column I in Table III, Walmsley and Pople, 1964). The former term is decisive for the translational frequencies, the latter contributing only on the order of 10%. Another potential model ("diatomic potential") using atom-atom interaction terms reproduces these frequencies equally well (Column II in Table III; Kuan, Warshel, and Schnepp, 1970). It can be concluded that the frequencies of the translational modes of symmetry T_u are not very sensitive to the potential model.

The Raman spectrum of α–N_2 was described by Cahill and Leroi (1969a) and by Brith, Ron, and Schnepp (1969). Both groups observed only two lines, at 31.5 cm^{-1} and 35.8 cm^{-1}, as compared to group theoretical prediction of three Raman active modes ($E_g + 2T_g$). From intensity ratio considerations, both groups were led to assume an accidental degeneracy and assigned the most intense line at 31.5 cm^{-1} as containing the E_g mode as well as one of the T_g vibrations. Kuan, Warshel, and Schnepp (1969) found that the quadrupole-quadrupole potential could not reproduce the librational frequencies as assigned. These authors used a parametrized diatomic potential and were successful in correlating the lattice mode frequencies with the crystal energy and nearest neighbor distance, a total of seven observables. Three potential parameters were necessary—two parameters of the 6–12 atom-atom interaction term and the distance between the two centers of interaction in the molecule

TABLE III

Lattice Energy, Lattice Vibrational Frequencies, and Intensity Ratios for α–N_2 [Frequencies are in units of cm^{-1} with half-intensity widths in parentheses. Calculated values are listed for two potential models: (I) Lennard-Jones potential plus quadrupole-quadrupole term (Ron and Schnepp, 1967; Walmsley and Pople, 1964). Nearest neighbor interactions only; (II) Diatomic Potential (Kuan, Warshel, and Schnepp, 1970) with 6–12 terms; Raman assignments as in Cahill and Leroi (1969) and Brith, Ron, and Schnepp (1969).]

		Calculated	
	Experimental	I	II
Lattice energy (k cal/mole)	−1.808		−1.810
Translational modes			
A_u	Inactive		47.5
E_u	Inactive		54.9
$T_u(Q_2)$	48.8(0.3)[a–c]	48[c]	51.3
$T_u(Q_1)$	70(6)[a–c]	71[c]	75.7
Infrared intensity ratio			
$I(Q_1)/I(Q_2)$	1.0 ± 0.2[b]	2.3	1.1
Librational modes			
E_g	31.5(1.4)[d–e]	27.8	26.5
$T_g(Q_2)$	31.5[d–e]	38.4	29.6
$T_g(Q_1)$	35.8(1.4)[d–e]	70.2	35.7
Raman intensity ratio			
$I(E_g) + I(T_g, Q_2)/I(T_g, Q_1)$	3.6	32.1	12.0

[a] Anderson and Leroi, 1966.
[b] St. Louis and Schnepp, 1969.
[c] Ron and Schnepp, 1967.
[d] Cahill and Leroi, 1969a.
[e] Brith, Ron, and Schnepp, 1969.

(effective interatomic distance). The results of this calculation are presented in Column II of Table III. It was also found that 6-exponential atom-atom potential terms were equally successful.

Recently Anderson, Sun, and Donkersloot (1970) reinvestigated the Raman spectrum of α–N_2. These workers observed three lines, as reported in Table IV. They observed a very weak line at 60 cm^{-1} of 4 cm^{-1} half-intensity width. The intensity ratio as observed here for the 32 cm^{-1} and 36 cm^{-1} lines was 2.8 as compared to 3.6 for the earlier investigations. Anderson and co-workers assigned the weak line at 60 cm^{-1} to the missing T_g mode and then found that the librational frequency pattern is well reproduced by a quadrupole-quadrupole potential. However, the molecular quadrupole moment as obtained from gas phase measurements does not give good absolute frequency values, i.e., they are too high by about

TABLE IV

Librational Frequencies and Intensity Ratios for α–N_2 [Frequencies are in units of cm^{-1}. Calculated values are as given by Anderson, Sun, and Donkersloot (1970) for the Lennard-Jones plus quadrupole-quadrupole potential. Raman experimental frequencies, intensity ratios and assignments are those of Anderson, Sun, and Donkersloot (1970).

Column I: Molecular quadrupole $\theta = 1.26 \times 10^{-26}$ esu, 4 neighbor shells. *Column II:* $\theta = 1.52 \times 10^{-26}$ esu or best gas phase value (Stogryn and Stogryn, 1966).]

		Calculated	
Lattice mode	Experimental	I	II
E_g	32(1.5)	29.0	35.0
$T_g(Q_2)$	36.5(1.5)	37.4	45.1
$T_g(Q_1)$	60(4)	61.5	74.3
Raman intensity ratios			
$I(E_g):I(T_g, Q_2):I(T_g, Q_1)$	2.8:1.0:0.1	2.8:1.0:0.12	

20% (Column II of Table IV). Therefore the molecular quadrupole moment is used as adjustable parameter and good agreement was obtained (Column I of Table IV) for a value of θ of 1.23×10^{-26} esu or 1.26×10^{-26} (depending on the number of neighbor shells included as compared to the best gas phase value of 1.52×10^{-26} esu. Anderson and co-workers also point out that the second-nearest neighbor shell contributes significantly to the librational frequencies calculated from the quadrupole-quadrupole potential. For example, the calculated E_g frequency is increased by 30% over the value obtained from nearest neighbor interaction only. Anderson and co-workers also showed that the observed intensity ratios of the Raman lines agreed very well with those calculated using the librational eigenvectors given by Walmsley and Pople (1964) with the results of intensity calculations of Brith, Ron, and Schnepp (1969).

It appears at this time that there is disagreement over the librational frequency assignment in α–N_2. From a purely experimental point of view, the discrepancy between the observations of Brith, Ron, and Schnepp (1969) and Cahill and Leroi (1969) on the one hand and those of Anderson, Sun, and Donkersloot (1970) on the other cannot be resolved. This is due to the fact that the method of sample preparation was different, which could have led to small differences in the observations. Anderson investigated a solid layer prepared by vapor deposition on a cold metal surface, whereas the other workers used solids prepared by cooling liquid in closed cells. Strains in solids prepared by nonequilibrium means could cause the appearance of weak Raman lines and the line in dispute is indeed very weak, i.e., about 3% of the intense line at 32 cm^{-1}. Furthermore, the difference in intensity ratio of the two low-frequency lines as observed with samples prepared by the two methods (3.6 as compared to 2.8) could be similarly accounted for. On the other hand, the line widths observed by Anderson and co-workers were no greater than those reported by the other groups. Anderson and co-workers also mention that the 60 cm^{-1} Raman line could represent an overtone of the intense line at 32 cm^{-1} in view of its greater width, as expected for a combination since only total $\mathbf{q}$ must be zero.

Anderson, Sun, and Donkersloot (1970), in the calculations used to support their assignment, were forced to use the molecular quadrupole moment as adjustable parameter rather than using the free molecule value. Further, the frequencies obtained with the molecular value were too high, whereas frequencies calculated by a similar method for CO_2 were too low (Walmsley and Pople, 1964). In the latter case, it has been argued that the quadrupole-quadrupole potential may be expected to give low frequencies since it does not sufficiently take into account the molecular shape and atom-atom repulsion. No physical reason for the discrepancy in the opposite direction has been offered. It must be concluded that the physical significance of the use of the molecular quadrupole moment as adjustable parameter is unclear. Therefore, such a calculation cannot be accepted as significant support for the assignment.

In addition, serious basic theoretical problems are encountered with the application of a harmonic expansion of the potential in terms of angular displacements to the calculation of the librational frequencies of α–N_2 (Section IIC.2). It is expected that quantum mechanical calculations now in progress (Jacobi and Schnepp, 1971) will clarify this problem. Further experimental work also is required to settle the question of the dependence of the intensity of the 60 cm^{-1} Raman line reported by Anderson and co-workers on sample preparation. Only such additional work can contribute further to the assignments of the $\mathbf{q} = 0$ librational frequencies.

Schnepp and Ron (1969) carried out a complete lattice dynamical calculation for α–N_2 throughout the Brillouin zone, using the potential model of Kuan, Warshel, and Schnepp (1969). This model contains three parameters which were calibrated for the optical modes [librational assignments of Brith, Ron, and Schnepp (1969)] and the equilibrium properties of the solid. The lattice dynamics was formulated in terms of Eulerian angle librational displacements. Dispersion curves for the symmetry directions of the Brillouin zone and density of states functions were reported with and without translation-libration interactions. The results clearly demonstrated the importance of these interactions. Schnepp and Ron also calculated the densities of two-phonon states in an effort to account for the large difference in widths of the two infrared absorption lines in terms of the usual optical phonon to two-phonon relaxation mechanism. However, the results could not reproduce the observed ratio of 20 or more. Again, the validity of the application of the harmonic lattice dynamics to the librations of α–N_2 must remain in doubt until further theoretical work (Section IIC.2).

Anderson, Sun, and Donkersloot (1970) and Cahill and Leroi (1969a) reported a 30 cm^{-1} broad absorption peaking at about 80 cm^{-1}. This feature is, presumably, due to two-phonon processes.

Table III includes experimental and calculated intensity ratios. The ratio for the infrared absorptions is better reproduced using the eigenvectors obtained from the diatomic potential calculation than from the quadrupole-quadrupole angle-dependent term; the eigenvectors were found to be sensitive to the anisotropic potential since the force constant mixing the corresponding symmetry coordinates is dependent only on this part of the potential. On the other hand, the Raman intensity ratio cannot be fitted with the assignment assumed in Table III.

Donkersloot and Walmsley (1970) discussed calculations for the $\mathbf{q} = 0$ lattice modes of α–N_2. They used two potential models. First these authors assumed an atom-atom potential with 6–12 terms similar to that of Kuan, Warshel, and Schnepp (1969), but they adopted a more restrictive procedure for the evaluation of the parameters. Second, Donkersloot and Walmsley assumed an explicit charge distribution (monopole model), which was compatible with the experimental value of the molecular quadrupole moment. Neither of these models reproduced the librational mode frequencies satisfactorily.

The crystal structure of solid CO_2 has been discussed as arising from minimization of the interactions between molecular quadrupole moments (Buckingham, 1959). Solid α–N_2 is isomorphous with CO_2 and therefore the same arguments hold here. Kuan, Warshel, and Schnepp (1969) showed that the diatomic potential used by these authors can equally well account for the observed structure. It was found that the minima in energy with molecular orientations are caused by the minima in the repulsion between molecular ends. This structure can therefore be accounted for by any anisotropic pair potential which contains end-to-end repulsions and relative end-to-center attraction.

Suzuki and Schnepp (1971) have shown that the specific heat of solid α–N_2 can be calculated in good agreement with experiment from the results of Schnepp and Ron (1969). These authors also have shown that the potential model of Kuan and others gives good results for second virial coefficients of N_2 gas. They used the statistical mechanics results for the diatomic model given by Sweet and Steele (1967).

The crystal structure of α–N_2 has been investigated using a single crystal (Jordan et al., 1964). The accurate structure was reported to be $P2_13$ (T^4) or slightly distorted $Pa3(T_h^6)$. However, none of the spectroscopic investigations have given any supportive evidence for deviation from $Pa3$. It must therefore be concluded that the distortion is too small to manifest itself spectroscopically. The α–N_2 form is stable below 35.6°K and at this temperature a transition takes place to a hexagonal structure (β–N_2), which is stable up to the melting point. No infrared absorptions are predicted for the β-structure and none have been observed (St. Louis and Schnepp, 1969). One Raman active translational mode and two active librational modes are predicted in the β-phase but no discrete Raman lines have been observed (Cahill and Leroi, 1969a). Presumably this is due to the orientationally disordered nature of the crystal structure in the β-phase.

Solid CO_2. Table V summarizes the available data for solid CO_2. The far infrared absorption spectrum in the lattice mode region was reported first by Anderson and Walmsley (1964) at 77°K and later by Ron and Schnepp (1967) at 20°K. Recently, helium temperature measurements were carried out by Kuan (1969) and 35°K measurements by Brown and King (1970). Of these workers, only Kuan prepared the solid sample from the vapor under equilibrium conditions. The results show that the frequencies of the two infrared active modes are not very sensitive to temperature or to sample preparation. On the other hand, the line widths observed by Kuan are about half those reported by Brown and King. Kuan's line widths are given in Table V. Moreover, the measured intensity ratios differ markedly (both values are listed in the table).

The low-temperature Raman spectrum has been studied by Ito and Suzuki (1968) and by Cahill and Leroi (1969a). The latter authors studied the temperature dependence of the frequencies and found them to be quite sensitive between 212 and 81°K. Only the liquid helium temperature values of Ito and Suzuki (1968) are given in Table V, and these agree very well with the frequencies reported by Cahill and Leroi (1969) for 15°K. In the case of solid CO_2, all three predicted Raman lines have been observed and reliably assigned.

As has been shown in Section IIC.2, the harmonic approximation is expected to be applicable to the treatment of the librational motions of CO_2. Additional support has been obtained for this recently (Suzuki and Schnepp, 1971). It is therefore possible to judge the validity of intermolecular pair potentials by comparison of calculated lattice frequencies with experiment. Table V summarizes available experimental and theoretical results for the $\mathbf{q} = 0$ modes. It is again seen here, as was the case for α–N_2, that the translational frequencies can be closely reproduced from a gas phase Lennard-Jones potential (Walmsley and

TABLE V

Lattice Energy, Lattice Vibrational Frequencies, and Intensity Ratios for Solid CO_2
[Frequencies are in units of cm^{-1} with half-intensity widths in parentheses. Experimental frequencies and intensity ratios are those for liquid helium temperature. Calculated values are listed for two potential models: (I) Lennard-Jones potential plus quadrupole-quadrupole term (Walmsley and Pople, 1964). Nearest neighbor interactions only; (II) Diatomic potential with 6–12 terms (Suzuki and Schnepp, 1971). Interactions to 10 neighbor shells included.]

	Experimental	Calculated	
		I	II
Lattice energy (kcal/mole)	−6.58		−6.38
Translational modes			
A_u	Inactive	94	109.4
E_u	Inactive	77	116.9
$T_u(Q_2)$	68.2[a,b](0.9)[a]	74	68.8
$T_u(Q_1)$	117.2[a,b](2.1)[a]	113	135.0
Infrared intensity ratio			
$I(Q_1)/I(Q_2)$	4.4[a]	5.4	1.9
	1.5[b]		
Librational modes			
E_g	72(2)[c,d]	35	69.0
$T_g(Q_2)$	92(3)[c,d]	48	88.2
$T_g(Q_1)$	136(6)[c,d]	88	138.5
Raman intensity ratios			
$I(E_g):I(T_g, Q_2):I(T_g, Q_1)$	10:2.4:1[d] (at 88°K)	20:7:1[e]	5.3:1:1.6

[a] Kuan, 1969.
[b] Brown and King, 1970.
[c] Ito and Suzuki, 1968.
[d] Cahill and Leroi, 1969a.
[e] Calculated using the eigenvectors of Walmsley and Pople (1964) and the intensity calculation of Brith, Ron, and Schnepp (1969). Note a misprint in Walmsley and Pople pointed out by Anderson, Sun, and Donkersloot (1970). The eigenvectors are: $0.42S_{15} + 0.91S_{18}$ and $0.91S_{15} - 0.42S_{18}$.

Pople, 1964). Furthermore, the diatomic potential model gives a good fit. However, the librational modes definitely cannot be fitted with the quadrupole-quadrupole potential term, even if we consider that interactions with additional shells contribute appreciably, as has been pointed out by Anderson, Sun, and Donkersloot (1970). Including these, we obtain frequencies of 48, 59, and 97 cm^{-1}, still far below the experimental values. On the other hand, the diatomic potential model of Suzuki and Schnepp (1971) gives a good overall fit. The calculated intensity ratios are, however, not in good agreement with experiment, although the general trends are reproduced.

Suzuki and Schnepp (1971) have carried out a complete lattice dynamical treatment throughout the Brillouin zone. Dispersion curves, density of state function, and specific heat have been calculated. The specific heat is in very good agreement with measurements. The group theory of the space group *Pa*3 was discussed for symmetry points and symmetry directions and it was shown that the calculations are in agreement with the predicted degrees of degeneracy and compatability relations (correlations). It may be concluded that CO_2 is the system which lends itself best for further work. In particular, neutron scattering investigations (both coherent and incoherent) would be of great value. Such additional experimental measurements will make possible further investigations into the intermolecular pair potential of CO_2.

Walmsley (1968) applied the lattice dynamical calculation of elastic constants to molecular solids. He calculated these constants for solid CO_2 and compared his results to the incomplete experimental data available. Agreement was found to be very good. Walmsley used an intermolecular pair potential consisting of Lennard-Jones and quadrupole-quadrupole terms (Walmsley and Pople, 1964). He pointed out that the contribution of the angle-dependent part of the potential to the absolute values of the elastic constants is relatively small. On the other hand, this part of the potential determines the relation between c_{12} and c_{44}, which are equal if angle dependence is ignored.

CO and N_2O. These solids, CO in its low temperature or α-form and N_2O within its entire solid range, have structures similar to α–N_2 and CO_2. However, since the molecules lack centers of symmetry, the space group is $P2_13(T^4)$. The translations are not separable from librations even at $\mathbf{q} = 0$ and consequently four lines are expected in the infrared and five lines in the Raman spectrum. Furthermore, the structures are disordered due to the similarity of the end atoms. The far infrared absorption spectrum of CO (Anderson and Leroi, 1966; Ron and Schnepp, 1967) is similar to that of α–N_2 and consists of a line at 50 cm^{-1} and a broader line centered at 86 cm^{-1}. Cahill and Leroi (1969a) and Anderson, Sun, and Donkersloot (1970) have described the Raman spectrum of CO, which consists of a single broad band centered at about 48 cm^{-1} at 12°K. The latter authors report five discernible peaks and shoulders between 38 and 95 cm^{-1}. They also observed an additional weak line at 90 cm^{-1}. Assignments were suggested. The infrared peaks do not appear in the Raman spectrum, contrary to expectation. Anderson and Walmsley (1964) have observed a single broad absorption in the infrared spectrum of solid N_2O. This band peaks at 113 cm^{-1} and has a width of 35 cm^{-1} at half intensity. The Raman spectrum, on the other hand, consists of three lines, similar to the CO_2 spectrum, but the lines are all 7 cm^{-1} wide and occur at somewhat lower frequencies (68, 82, and 124.5 cm^{-1}).

Comparison between the spectra of α–CO and N_2O reveals that for α–CO the infrared spectrum resembles that expected for an ordered symmetric molecule like CO_2, whereas the Raman spectrum bears evidence of disorder. On the other hand, the Raman spectrum of N_2O is that expected for the ordered crystal of a symmetric molecule and the infrared spectrum is that of a disordered solid. No satisfactory interpretation of these results are available at present. Shinoda and Enokido (1969) have carried out a lattice dynamical calculation for an ordered α–CO solid using a Lennard-Jones potential plus a quadrupole-quadrupole term and added correction terms arising from the anisotropy of the dispersion and repulsive forces. They found that their calculated heat capacity agreed well with experiment below about 15°K but deviated negatively above this temperature. These authors gave only two $\mathbf{q} = 0$ frequencies, which are in reasonable agreement with the observed infrared absorptions.

C_2N_2. Richardson and Nixon (1968) investigated the infrared and Raman spectra of solid cyanogen in the lattice mode region. The crystal is orthorhombic, belonging to the space group $Pbca(D_{2h}^{15})$ with four molecules at centrosymmetric sites. This structure can be thought of as derived from the CO_2-structure ($Pa3$) by reorienting the axes of the linear molecules away from the body diagonal directions such that the three screw axes parallel to the crystal axes are retained. In the present case, the axes of the orthorhombic primitive cell are not very different from each other, ranging from 6.19 to 7.08 Å. As a result of the lower symmetry structure, all degeneracies are removed and a much larger number of infrared and Raman active modes are expected for $\mathbf{q} = 0$. In fact, six infrared active translational modes are expected and all have been observed. In the Raman spectrum eight librational lines are expected and seven distinct lines have been observed. Calculations were carried out for the $\mathbf{q} = 0$ frequencies using a Lennard-Jones plus quadrupole-quadrupole potential, but only moderate agreement was achieved. However, a satisfactory fit was obtained for all frequencies using an atom-atom 6–12 potential between nitrogen atoms only. Only one adjustable parameter was used since the parameter σ was determined from the condition of uniform stress (equilibrium condition at the observed lattice parameter). The distance between interaction centers was taken as the observed distance between nitrogen atoms in the C_2N_2 molecule.

The Hydrogen Halide Solids. Anderson, Gebbie, and Walmsley (1964) first described the far infrared spectra of the hydrogen and deuterium chlorides and bromides at 77°K in the region 20–400 cm^{-1}. On the basis of the changes on isotopic substitution these authors assigned translational and librational modes. It seems that these are in fact separable to a large extent, although such a separation is not expected from symmetry. However, the moments of inertia of these molecules are relatively small, causing high librational frequencies with consequent near-separation from the appreciably lower translational modes. Anderson and co-workers also showed that their observations were consistent with a planar array structure with four molecules per unit cell. Some intermolecular force constants were derived. Arnold and Heastie (1967) investigated these solids in different phases. HCl solid undergoes a transition to a cubic phase at 98.4°K. HBr undergoes a transition to an orthorhombic structure at 89.7°K and to the cubic structure at 116.7°K. The low-temperature phases are all orthorhombic. Ito, Suzuki, and Yokoyama (1969) carried out an extensive investigation of the Raman spectra of HBr and HCl and their deuterium analogs at different temperatures. They conclude that their results are consistent with the planar zigzag structure determined by neutron diffraction for the low-temperature form of the DCl solid. Ito and co-workers succeeded in carrying out a normal coordinate analysis for the hydrogen halides by using a force field which took into account explicitly the hydrogen bonds; these authors used eight force constants including two "bending" constants (for the in-plane $X \cdots H\text{–}X$ angle). The $\mathbf{q} = 0$ vibrations fitted, including six lattice modes and two frequencies in the region of the internal molecular stretch. The force constants determined were judged to be of reasonable magnitudes. Trevino et al. (1968) also carried out a calculation of the lattice frequencies of solid HCl and DCl in the low-temperature phase. These authors used a force field of a type similar to that of Ito and co-workers but made some different assignments.

Ito, Suzuki, and Yokoyama (1969) also investigated the Raman spectrum of the cubic phase of HCl and found one broad band which peaks at about 200 cm^{-1}. Arnold and Heastie (1967) had observed a broad infrared band for this phase with a peak at 240 cm^{-1}. In view of the high frequency peak, this band is believed to represent mainly librational motion. Ito and co-workers discussed the spectroscopic results in terms of large amplitude oscillations of the molecules resulting in rotational disorder. Ito (1970) studied the Raman spectrum of HBr in the three phases between 77 and 210°K. These observations are again discussed in the light of models of orientational disorder.

Leech and Peachey (1968) investigated the phonon frequency distribution in the low-temperature phase of solid HCl. The force field used was of the atom-atom type with parameters fitted to reproduce the frequencies of the observed infrared-active modes at $\mathbf{q} = 0$.

It was pointed out in Section IIC.2 that the librational motions of the hydrogen halides are not expected to be satisfactorily described in terms of classical harmonic treatments. Additional theoretical work is therefore necessary to elucidate the character of these motions.

The infrared and Raman spectra of solid HF and DF were described by Kittelberger and Hornig (1967). A lattice vibrational band was observed near 200 cm^{-1} for both solids and the Raman spectra were found to contain the same frequency line and one additional optical mode each. The observed coincidence for one line in infrared and Raman spectra is relevant to crystal structure considerations.

The Halogen Solids. The far infrared spectra of solid Cl_2, Br_2, and I_2 were reported by Walmsley and Anderson (1964). Two translational lattice modes were observed in accordance with theoretical predictions for the known crystal structure. The Raman spectrum of the Cl_2 crystal was studied by Suzuki, Yokoyama, and Ito (1969). The librational lattice mode frequencies were assigned satisfactorily on the basis of the known layer structure. These authors carried out a normal-coordinate analysis and frequency calculation for the $\mathbf{q} = 0$ modes and found it necessary to include angle bending force constants. This observation is interpreted as indicating the importance of charge-transfer interaction in the solid. Cahill and Leroi (1969b) also studied the Raman spectra of solid Cl_2 and Br_2. They concluded that the intermolecular forces in solid Br_2 are stronger than those in solid Cl_2.

Miscellaneous Solids. Anderson and Walmsley (1965) described the far infrared spectra of ammonia, H_2S, and their deuterated analogs. The far infrared spectra of H_2S and D_2S have also been reported by Taimasalu and Robinson (1965). Giguere and Chapados (1966) reported the far infrared spectra of solid H_2O_2 and D_2O_2. All lattice modes predicted to be active were observed. Tubino and Zerbi (1970) studied the phonon curves and frequency distributions for a hydrogen-bonded system: the β-form of solid formic acid and of its deuterated isotopes. The low-energy phonons arising from the hydrogen bond were found to be strongly dispersed.

Bertie, Whalley, and co-workers have studied the far infrared spectra of ices. Bertie and Whalley (1964a, 1964b) discussed the spectra of two ices. Whalley and Bertie (1967) discussed the translational lattice vibrations of orientationally disordered solids and applied this theory to the interpretation of the infrared spectra of the ices Ih and Ic (Bertie and Whalley, 1967). Prask, Boutin, and Yip (1968) investigated the frequency spectrum of hexagonal H_2O ice by inelastic incoherent neutron scattering. Bertie, Labbé, and Whalley (1968a) described the far infrared spectra of the ices II, V, and IX. The spectra of II and IX were found to be sharp, as expected for an ordered solid, whereas that of V is broad, as expected for an orientationally disordered solid. These results are in agreement with previous information concerning the studied phases. Bertie, Labbé, and Whalley (1968b) investigated ice VI and concluded that it is orientationally disordered.

2. *Organic Molecular Solids*

Hexamethylenetetramine. This is the organic molecular solid for which the most extensive experimental information is available at this time. Dolling and Powell (1970) have described dispersion curves for several symmetry directions measured by coherent inelastic neutron scattering. These authors investigated the fully deuterated compound at 100 and 298°K, since hydrogen has a large cross section for incoherent scattering. This work followed earlier reports (Powell, 1968; Powell and Dolling, 1969). One very significant result of this work was the assignment of the $\mathbf{q} = 0$ librational mode at 63 cm^{-1} as compared to the value of 40 cm^{-1} which had been accepted since its assignment by Couture-Mathieu et al. (1951) and Cheutin and Mathieu (1956) on the basis of Raman results. The solid is body-centered-cubic with one molecule per primitive cell and therefore only one optic mode is expected at $\mathbf{q} = 0$. Dolling and Powell (1970) found that there is, indeed, a high peak in the density of states in the region of 40 cm^{-1} due to librational modes of $\mathbf{q} \neq 0$ and the Raman spectrum presumably reflects this peak; in any case; the $\mathbf{q} = 0$ librational mode is not predicted to be Raman active on group theoretical grounds.

Cochran and Pawley (1964) chose this solid, because of its simple structure, for the first complete lattice dynamical treatment of a molecular crystal. They did not use a specific intermolecular pair potential but instead used symmetry to reduce the number of independent parameters in the dynamic equation. They used four parameters for nearest-neighbor force constants and then attempted to determine part of these parameters from experimental measurements, i.e., elastic constants and librational optical mode. Unfortunately, at that time, the number of observables was very restricted. Dolling and Powell (1970) used an approach similar to that of Cochran and Pawley (1964) to interpret their large body of experimental results. They used six parameters to characterize each of the nearest neighbor and second-nearest neighbor force constants, a total of 12 parameters, and obtained a very good fit for the dispersion curves. Models using fewer parameters were somewhat less successful. Dolling and Powell also calculated the frequency distribution function for the hydrogen-compound from an eight-parameter model fitted to their measurements and compared this with the incoherent neutron scattering results of Bührer, Hälg, and Schneider (1967). Agreement was generally poor. However, Becka (1962) had observed peaks in such measurements which agreed well with prominent peaks in the distribution calculated by Dolling and Powell. These authors also calculated the heat capacity and found it to agree well with experiment up to about 60°K; above this temperature the internal modes make themselves felt. These modes were not taken into account in the calculation. Moderate agreement with experimental elastic constants was achieved.

Rafizadeh and Yip (1970) used their "extended point mass" method to reduce the number of independent parameters in the lattice dynamical treatment of hexamethylenetetramine. They successfully compared their result with the neutron scattering measurements of Powell and Dolling (1969) and then unpublished results by these authors (Dolling and Powell, Powell, 1970). They also calculated the frequency distribution function and succeeded to fit the experimental specific heat up to about 60°K. Rafizadeh and Yip showed that they could obtain good agreement above 60°K by including the internal modes. They also could fit the inelastic incoherent neutron scattering results of Becka (1962). These authors also made the interesting observation that they could achieve quite a good, although not quantitative fit for the dispersion curves by using a potential model with four parameters calibrated by four observables: elastic constants and the now corrected librational frequency at $\mathbf{q} = 0$.

Benzene. Table VI summarizes the status of our understanding of the lattice vibrations of solid benzene. Both Raman and infrared spectra have been measured. Harada and Shimanouchi (1966, 1967) carried out lattice dynamical calculations using the matrix method (see Section IIC.4). These authors first took into account H–H interactions only but the frequencies so calculated were too low. They then carried out calculations in which they included both H–H and C–H repulsive interactions and obtained much improved agreement (listings in Table VI). The atom-atom potential terms were exponential repulsion terms for which some parameters from previous authors were used and others were obtained by fitting to the Raman and infrared data. Oliver and Walmsley (1969) treated the $\mathbf{q} = 0$ librational lattice modes of benzene using direction cosine displacement coordinates (see Section IIC.5). These authors also calculated the translational modes. They used atom-atom potentials with three sets of parameters but concluded that those due to Williams (1967) are best suited. These are the frequencies listed in Table VI. It is seen from the table that very good agreement with experiment is achieved, and in particular for the translational modes the agreement is nearly perfect. The treatment of Oliver and Walmsley does not make the approximation inherent in the matrix method. On the other hand, the calculations of Harada and Shimanouchi (1967) take into account mixing with internal molecular modes. It is therefore not possible to compare the two sets of results in order to judge the effect of the approximation discussed in Section IIC.4.

TABLE VI
Lattice Vibrations of Solid Benzene

Species	Raman-active modes			
	Observed		Calculated	
	I and S[a]		H and S[b]	O and W[c]
D_{2h}	(4°K)	(138°K)	(138°K)	(138°K)
A_g	(100)	(90)	102	94
	86	79	80	76
	64	57	55	43
B_{1g}	136	126	131	134
	(86)	(79)	99	93
	69	61	65	48
B_{2g}	(100)	(90)	106	100
	100	90	102	87
	86	79	82	81
B_{3g}	136	126	128	131
	107	100	96	85
	(86)	(79)	81	61
	H and S[b] (140°K)		H and S[b] (138°K)	O and W[c] (138°K)
A_u			86	91
			66	61
			54	53
B_{1u}	85		79	82
	70		66	69
B_{2u}	94		86	96
			55	56
B_{3u}	94		87	93
			49	53

[a] Ito and Shigeoka, 1966a.
[b] Harada and Shimanouchi, 1967.
[c] Oliver and Walmsley, 1969; potential parameters of Williams, 1967.

Rush (1967) measured the inelastic incoherent neutron scattering from solid benzene and determined the phonon frequency distribution. Nakamura and Miyazawa (1969) carried out a complete Brillouin zone lattice dynamics calculation using the potential constants of Harada and Shimanouchi (1967), assuming the benzene molecule to be a nonvibrating rigid body. These authors first obtained agreement to better than 6% with Harada and Shimanouchi for the $\mathbf{q} = 0$ modes. They then calculated the frequency distribution and were able to fit the experimental specific heat to better than 2% by taking into account the internal modes as well and making the appropriate correction for the difference between the calculated C_v and the measured C_p. Nakamura and Miyazawa further calculated $G(\nu)$ (or frequency distribution weighted by the squared H-amplitudes) to allow comparison with the measurements of Rush (1967). The highest peak in $G(\nu)$ at 90 cm^{-1} was observed by Rush but the other calculated peaks at about 50 and 125 cm^{-1} appeared barely as shoulders in Rush's results.

Logan et al. (1970) measured the inelastic incoherent neutron scattering from a crystalline sample of benzene at temperatures in the range 77–270°K They found qualitative agreement at low temperatures with the one-phonon neutron scattering cross section calculated from $G(\nu)$ of Nakamura and Miyazawa (1969).

Bernstein (1970) carried out a matrix method lattice dynamical calculation for solid benzene. He treated both internal and external modes and described his results in terms of dispersion curves. This author discusses the potential and future profitable directions of approach which are expected to lead to further progress.

Tarina (1967) reported inelastic incoherent neutron scattering results for solid benzene. No sharp features were observed at 77°K and only a broad peak at about 100 cm^{-1} was found.

Naphthalene. Table VII summarizes the results for the lattice vibrations

TABLE VII
Lattice Vibrations of Solid Naphthalene

Species	Observed				Calculated			
	Raman			IR				
C_h	K and R[a]	W[b] (300°K)	I[c] (77°K)	H and S[d] (300°K)	H and S[d] (300°K)		P[e] (300°K)	
A_g	127	146	120		115	119	134	139
	76	90	88		85	86	94	92
	54	71	69		58	62	61	62
B_g	109	124	141		82	93	126	129
	74	86	81		70	77	70	77
	46	59	56		42	47	77	49
A_u				98	97	104		88
				53	54	58		47
B_u				66	74	76		58

[a] Kastler and Rousset, 1941.
[b] Wilkinson, 1966 (from Pawley, 1967).
[c] Ito et al., 1967.
[d] Harada and Shimanouchi, 1966.
[e] Pawley, 1967.

of solid naphthalene as of 1967. The table is reproduced from Schnepp (1969). Raman and infrared spectra in the lattice vibration region have been observed. Harada and Shimanouchi (1966) calculated the $\mathbf{q} = 0$ modes using the matrix method of lattice dynamics (Section IIC.4). These authors found here as in the case of benzene that they had to use C–H interaction terms in addition to H–H terms; both sets are given in the table. Pawley (1967) carried out a lattice dynamical calculation for the whole Brillouin zone using infinitesimal librational displacement coordinates (Section IIC.6). He used an atom-atom potential with six-exponential terms with parameters given by Kitaigorodskii (1966). The results for $\mathbf{q} = 0$ could be compared to available experimental results and were judged to be satisfactory. Two sets of calculated results of Palwey are listed in Table VII. The first set was calculated assuming librational motion about the principal molecular axes, whereas the second and more correct set was obtained without this assumption. It was then found that the normal modes in some cases represented librational motions about axes appreciably different from the molecular axes. Internal molecular vibrations were not considered and this assumed separation was believed to be the source of some error. Reasonably good agreement with experiment was obtained.

Suzuki, Yokoyama, and Ito (1967) studied the polarized Raman spectrum of single crystals and made definitive assignments on this basis. These authors also studied the temperature dependence of the spectrum between 300 and 400°K. Hadni and co-workers (1969) studied the far infrared absorptions of single naphthalene crystals in polarized light and confirmed thereby the assignments made for two infrared active lattice vibrations by Harada and Shimanouchi (1966) at 98 cm^{-1} (A_u) and at 66 cm^{-1} (B_u). Pawley and Cyvin (1970) have described a lattice dynamical treatment for solid naphthalene including the internal molecular vibrations. They obtained dispersion curves for both external and internal modes.

Anthracene. Pawley (1967) carried out a complete lattice dynamical treatment of anthracene by the same methods used in his treatment of naphthalene. He compared the results to the Raman measurements of Wilkinson (1966). Suzuki, Yokoyama, and Ito (1967) reported the Raman spectrum of anthracene in the lattice frequency region including single crystal measurements, and Colombo and Mathieu (1960) described the optical properties and Raman spectra of this solid. Bree and Kydd (1968) measured the infrared absorption spectra in polarized light and assigned the expected two a_u and one b_u lattice modes. Considerable interaction with internal modes is expected due to the low frequencies of the latter; in fact, the spectral regions overlap in this case. Hadni and co-workers (1969) also reported the far infrared spectra of single crystals in polarized light. These authors' assignment agrees with that of Bree and Kydd on the b_u mode (63 cm^{-1}) and on one of the a_u modes (126 cm^{-1}) but they disagree concerning the second a_u mode.

Acetylene. Acetylene has a phase transition in its solid range at 113°K. Above this temperature the structure is well established and has space group $Pa3(T_h^6)$, or identical with α–N_2 and CO_2. The low-temperature structure is not well known. It is believed to be orthorhombic and the factor group assumed is D_{2h}. Hatsenbuhler, Cahill, and Leroi (1969) reported the Raman spectrum and Ito, Yokoyama, and Suzuki (1970) have given a complete description of the Raman spectrum in both phases. In the low-temperature phase, three sharp lines were observed in the lattice mode region, of which the highest frequency line (171 cm^{-1}) is assigned as a two-phonon band. The other observed frequencies are assigned as $\mathbf{q} = 0$ librations. In the high-temperature phase the three expected librational modes were observed. Calculations using a quadrupole-quadrupole angle-dependent potential gave very accurate relative frequencies and the experimental frequencies could accordingly be fitted with one parameter, the molecular quadrupole moment. This moment is not well established but the value obtained from microwave line broadening gives frequencies which are too high. Ito, Yokoyama, and Suzuki (1970) suggested that the lower temperature solid has a layer-type structure of space group D_{2h}^{18}.

Anderson and Smith (1966) measured the far infrared spectrum of solid acetylene in the low-temperature phase at 77°K. They observed two lines each for C_2H_2 and for C_2D_2 and assigned them as translational lattice modes. Schwartz, Ron, and Kimel (1969) prepared their sample from the liquid in a closed sample cell and studied the far infrared absorption in both phases. In the low-temperature phase two absorptions were observed, similar to the observations of Anderson and Smith. However, in the high-temperature phase no sharp lines were observed. There was only a very broad and very weak absorption with a peak at 110 cm^{-1}. The spectrum change at the phase transition was reversible. The authors interpret their observations in terms of rotational disorder in the high-temperature phase. Schwartz, Ron, and Kimel (1971) investigated the temperature dependence of the infrared intensity and found it to decrease with increasing temperature. This result was compared to the parallel observation in α–N_2 by St. Louis and Schnepp (1969) and interpreted in the same way (Section IV); large-amplitude librational motion decreases the effective molecular quadrupole moment which is responsible for the absorption intensity. This interpretation correlates well with the large amplitude (26°) obtained from Raman studies by Cahill and Leroi (1969a) for the high-temperature phase of acetylene. As a result, no discrete lines are observed in the infrared absorption of this phase (Schwartz, Ron, and Kimel, 1969).

Ethylene. Brith and Ron (1969) reported the far infrared spectrum of solid ethylene at 8 and at 20°K. Two lines were observed and from the isotopic frequency ratio (from C_2H_4 and C_2D_4) they were assigned as translational modes. These authors calculated the translational modes by using a Lennard-Jones potential and an atom-atom repulsion potential. Reasonable agreement was obtained for space group D_{2h}^{12} as determined for the carbon skeleton but for the C_{2h}^{5} structure assumed by Brecher and Halford (1961) a third mode is expected at low frequency. Blumenfeld, Reddy, and Welsh (1970) investigated the Raman spectrum of solid ethylene at 79°K. They succeeded in preparing good and clear samples and assigned four observed lines as librational lattice modes. These were compared to the calculations by Dows (1962) using H–H repulsion terms.

The crystal structure of ethylene has been the subject of much discussion. Dows (1962) showed that the discussion of Brecher and Halford (1961) left open the possibility of two structures, $P2_1/n11$ (the a-axis case) and $P12_1/n1$ (the b-axis case), depending on which screw axis is preserved. The solid is monoclinic (C_{2h}) with the angle β accidentally 90°. Blumenfeld, Reddy, and Welsh carried out calculations for the splittings of the internal vibrational modes using atom-atom repulsion potential similar to that proposed by Dows (1962) and concluded from comparison of their results with experiment that the structure is most likely the b-axis case, $P2_1/n1$. Taddei and Giglio (1970) carried out calculations of the lattice frequencies ($\mathbf{q} = 0$ modes) of ethylene using an atom-atom potential with six-exponential terms. They tried six sets of parameters for this potential model. The authors found considerable sensitivity of their results to the range of interaction included. They also found that the attractive part of the potential contributed greatly to the frequencies. Neither of the two possible structures was favored over the other on the basis of comparison of their results with the measured infrared frequencies (Brith and Ron, 1969). Schwartz, Ron, and Kimel (1971) remeasured the far infrared spectrum of solid ethylene and again observed a decrease of intensity with increasing temperature. This behavior is analogous to that found by these authors for the low-temperature phase of acetylene and ethane. The interpretation is again similar to that offered for the parallel phenomenon in α–N_2 (Section IV). These authors also carried out calculations using a "diatomic potential" (Sweet and Steele, 1967). They found that they could reconcile the observation of only two translational modes in the infrared with the C_{2h}^{5} structure although three active modes are predicted; their results gave very close accidental degeneracy for two of the modes. Schwartz, Ron, and Kimel (1971) also calculated gas phase second virial coefficients from their crystal frequency fitted diatomic potential and obtained reasonable agreement with experimental values.

Miscellaneous Molecules. Ito, Suzuki, and Yokoyama (1967) reported the Raman spectra of *p*-dichloro and *p*-dibromobenzene and tabulated their lattice mode frequencies. Fleming, Turner, and Chantry (1970) described the far infrared spectra of the solid benzene monohalides and assigned their active lattice vibrations. Hadni and co-workers (1969) measured the far infrared spectrum of solid durene using single crystals and polarized light. Luty et al. (1970) investigated the inelastic incoherent neutron scattering of acenaphthene and carried out lattice dynamical calculations for this solid.

Ito and Shigeoka (1966) investigated the lattice vibrational Raman spectrum of solid pyrazine and compared their measurements to calculations which the same authors carried out. They correlated their work with the matrix method calculations of Harada and Shimanouchi (1966). Of particular interest was the N–H hydrogen bond force constant. Also Gerbaux and Hadni (1968) studied the lattice vibrations of solid pyrazine. Colombo (1967) reported the low-frequency Raman spectrum of *p*-toluidine solid.

Brot et al. (1968) studied the far infrared spectra of liquid and solid tertiary butyl chloride. The study is of interest in connection with rotations in solid phases. The lattice vibrations of thiourea and its deuterated analog were studied by Takahashi and Schrader (1967), who also carried out calculations of the optical mode frequencies and achieved good agreement with infrared and Raman measurements.

Janik and co-workers (1969) carried out incoherent inelastic neutron scattering studies on solid methane. They studied the lattice vibrations and molecular rotation in this solid near the melting point. Ito (1964) reported the lattice vibrational Raman spectrum of solid methyl iodide. Durig, Craven, and Bragin (1970) studied the low-frequency vibrations in the solid phases of two classes of molecules: $(CH_3)_3MCl$ and $(CH_3)_3MBr$ where M was C, Si, and Ge. The infrared spectra of solid CCl_4, benzene, and CS_2, pure and activated by the impurities I_2 and HCl, were studied by Munier and Hadni (1968). Colombo (1968) described the low-frequency Raman spectrum of single crystals of imidazole.

The far infrared spectrum of solid hydrazine was studied by Baglin, Bush, and Durig (1967). Solid methylamine was investigated by infrared and Raman spectroscopy by Durig, Bush, and Baglin (1968). Solid cyanamide was studied by Durig, Walker, and Baglin (1968). The far infrared and Raman spectra of dimethylsulfoxide were investigated by Durig, Player, and Bragin (1970). Shimanouchi and Harada (1964) applied their matrix method computations to solid cyanuric acid, uracil, and diketopiperazine. These authors also measured the far infrared spectra of the solids and characterized their hydrogen bond force constants.

Colombo and Moreau (1967) studied Raman line widths in solids as a function of temperature and Korshunov, Shabanov, and Rusetskii (1967) carried out similar studies for solid α-hydroxynaphthalene.

III. QUANTUM LATTICE DYNAMICS

A. General Remarks

Using classical lattice dynamics in the harmonic approximation, we expand the potential about the equilibrium positions and retain quadratic terms in the displacements only. In some cases, when the vibrating particles are very light and/or the binding energies are very weak, this approximation breaks down, yielding imaginary phonon frequencies (de Wette and Nijboer, 1965). In these cases the potential is so shallow that even in the ground and lowest excited states the observed intermolecular distances in the solid are appreciably larger than the minimum positions of the intermolecular potential, as determined from gas data. This "expansion" of the lattice constant is due to the pressure of the zero-point kinetic energy, which in very anharmonic solids is an appreciable portion of the total binding energy. In quantum solids this expansion of the lattice constant is beyond the inflection point of the potential, so that applying the classical harmonic approximation in this case gives negative force constants, and as a result imaginary lattice frequencies.

A different approach is required, allowing for quantum-mechanical effects due to anharmonicities and large zero-point energies, and avoiding ambiguous series expansions of the potential. This has been achieved by the self-consistent-field phonon scheme developed recently (Fredkin and Werthamer, 1965). This theory and its application to solid helium have been reviewed recently in detail (Guyer, 1969). In this section we present a simple derivation and discussion of the essentials of the method, with some applications to molecular solids. It should be noted that librations in low-barrier angular potential wells have to be treated quantum-mechanically too, as pointed out in a previous section. The formalism presented here is not directly applicable to librations, but it is now being extended to rotational degrees of freedom (Jacobi and Schnepp, 1971).

The main idea of SCF phonons is to calculate variationally the vibrational energy levels (and consequently phonon frequencies), the trial wavefunctions being harmonic oscillator wavefunctions or modifications of these. Those wavefunctions are exact wavefunctions of some harmonic potential, and the parameters characterizing the wavefunctions and the potential are its curvature, or effective force constant, and its minima, or equilibrium positions. Those parameters being determined from the variational principle, the calculations are done with the complete potential assumed, and no series expansion is carried out.

B. Single-Particle Model

Some important characteristics of SCF-phonons can be illustrated very simply in terms of a single particle moving in a one-dimensional potential (Koehler, 1968). The Hamiltonian of this simple system is

$$H = -\frac{\hbar^2}{2m}\frac{d^2}{dx^2} + V(x) \tag{3.1}$$

As variational wavefunctions of the ground and excited states we choose harmonic oscillator wavefunctions

$$\begin{aligned} \psi_0^{(h)} &= \left(\frac{\alpha}{\pi}\right)^{1/4} \exp\left[-\tfrac{1}{2}\alpha u^2\right] \\ \psi_1^{(h)} &= (2\alpha)^{1/2} u \psi_0^{(h)} \end{aligned} \tag{3.2}$$

where $u = x - x_0$, x_0 being some equilibrium position (not necessarily the minimum of the given potential). Those wavefunctions are exact solutions of a harmonic oscillator of frequency ω_h, related to α by

$$\alpha = \frac{m\omega_h}{\hbar} \tag{3.3}$$

The force constant of this harmonic oscillator is

$$\phi = m\omega_h^{\ 2} = \frac{\alpha^2\hbar^2}{m} \tag{3.4}$$

The variational energies of the true Hamiltonian in those approximate states are

$$\begin{aligned} E_0 &\equiv \langle 0|\, H\, |0\rangle = \frac{\hbar^2\alpha}{4m} + \langle 0|\, V\, |0\rangle \\ E_1 &\equiv \langle 1|\, H\, |1\rangle = \frac{3\hbar^2\alpha}{4m} + \langle 1|\, V\, |1\rangle \end{aligned} \tag{3.5}$$

The first terms are kinetic energies. They are obtained either directly or by noting that for a harmonic oscillator the zero-point kinetic energy is $\frac{1}{4}\hbar\omega_h$ and the first excited state kinetic energy is $\frac{3}{4}\hbar\omega_h$. The energy difference is

$$\Delta E \equiv \hbar\omega = \tfrac{1}{2}\hbar\omega_h + \frac{1}{2\alpha}\langle 0|\, \frac{d^2V}{dx^2}\, |0\rangle \tag{3.6}$$

where the last term was obtained by integrating by parts twice. The two terms represent the kinetic and potential energy excitations, respectively.

We now determine α (and x_0) by minimizing the ground-state energy. This gives

$$\frac{\hbar^2\alpha^2}{m} = \left(\frac{\alpha}{\pi}\right)^{1/2} \int \frac{d^2V(x_0 + u)}{du^2} \exp\,[-\alpha u^2]\, du$$

or

$$\phi = \langle \psi_0^{(h)}(u;\phi)|\, \frac{d^2V(x_0+u)}{du^2}\, |\psi_0^{(h)}(u;\phi)\rangle$$

$$= \frac{d^2}{dx_0^{\,2}} \langle \psi_0^{(h)}(u;\phi)|\, V(x_0+u)\, |\psi_0^{(h)}(u;\phi)\rangle \tag{3.7}$$

This is the fundamental result of the SCF-phonon method. It replaces the classical force constants which are the second derivatives of the potential at equilibrium by the quantum mechanical average of the second derivative of the given potential function, or, alternatively by the second derivative of the average potential with respect to the effective equilibrium position. In the classical limit of narrowly peaked wavefunctions the result reduces to the classical result. Since the wavefunctions depend on the force constant through (3.4), (3.7) must be solved self-consistently. In practice, α is found by minimizing the ground-state energy numerically.

Substituting this variationally determined value of α into (3.6), the potential energy excitation becomes

$$\frac{1}{2\alpha} \cdot \frac{\hbar^2\alpha^2}{m} = \tfrac{1}{2}\hbar\omega_h$$

so that

$$\omega = \omega_h \tag{3.8}$$

Thus in this simple model α is directly related to the vibrational frequency by (3.3).

C. Three-Dimensional Case

Assuming a pair potential, the Hamiltonian is

$$H = \sum_{lk} T(lk) + \frac{1}{2} \sum_{lk,l'k'} v[|\mathbf{r}(lk) - \mathbf{r}(l'k')|] \tag{3.9}$$

Here l denotes the unit cell ($l = 1, \ldots, N$) and k a site on the cell ($k = 1, \ldots, z$). For a system of bosons the ground-state vibrational wavefunction of the crystal is a Hartree product,

$$\Psi = \prod_{lki} \psi_i(lk) \tag{3.10}$$

where $i = x, y, z$ denotes the cartesian displacement component.

A numerical solution of the self-consistent Hartree equations (Nosanow and Shaw, 1962) showed that the exact one-particle wavefunction can be approximated very well by a gaussian. Therefore we use harmonic oscillator wavefunctions as trial wavefunctions, with parameters to be determined variationally:

$$\psi_i(lk) = \left(\frac{\alpha_{ikl}}{\pi}\right)^{1/4} \exp\left[-\tfrac{1}{2}\alpha_{ikl}u_i^2(lk)\right] \tag{3.11}$$

By the translational symmetry of the crystal, α cannot depend on the unit cell, but it may depend on the vibrational degree of freedom characterized by the indices k and i ($3Z$ degrees of freedom altogether). In any particular case the symmetry of the problem will reduce the number of independent α, and sometimes they may all be equal. We present the formalism in the general case, when

$$\alpha_{ikl} = \alpha_{ik}$$

The total wavefunction becomes

$$\Psi = \left(\frac{\prod_{ik} \alpha_{ik}}{\pi^{3Z}}\right)^{N/4} \exp\left[-\frac{1}{2}\sum_{ik} \alpha_{ik} \sum_{l} u_i^2(lk)\right] \tag{3.12}$$

As in (3.5), the average ground-state kinetic energy of degree of freedom ikl is

$$\langle T\rangle_{ikl} = \frac{\hbar^2\alpha_{ik}}{4m}$$

so that the ground-state average energy of the crystal is

$$E_0 = \frac{N\hbar^2}{4m}\sum_{i'k'} \alpha_{i'k'} + \left(\frac{\prod_{i'k'} \alpha_{i'k'}}{\pi^{3Z}}\right)^{N/2} \int V \exp\left[-\frac{1}{2}\sum_{i'k'} \alpha_{i'k'} \sum_{l} u_i^2(lk)\right] d\tau \tag{3.13}$$

The values of α_{ik} are obtained by minimizing this expression with respect to α_{ik}. This is done numerically, by first reducing the integrals to one-dimensional integrals. For a pair potential [Eq. (3.9)] the integrals are six-dimensional to start with. For a central potential, elimination of the center of mass further reduces them to three-dimensional integrals. The angular integration is trivial (it is less trivial for angle-dependent potentials, but in the most interesting cases it can still be performed analytically), so we are left with one-dimensional radial integrals (de Wette, Nosanow, and Werthamer, 1967) which are α-dependent and can be evaluated numerically.

The values of α also may be obtained formally by differentiating E_0

with respect to them:

$$0 = \frac{\partial E_0}{\partial \alpha_{ik}} = \frac{N\hbar^2}{4m} + \frac{N}{2\alpha_{ik}} \int \Psi^2 V \, d\tau - \sum_{l=1}^{N} \int u_i^2(lk) \Psi^2 V \, d\tau$$

The last integral is transformed by two partial integrations to

$$\int u_i^2(lk) \Psi^2 V \, d\tau = \frac{1}{2\alpha_{ik}} \int \Psi^2 V \, d\tau + \frac{1}{4\alpha_{ik}^2} \int \Psi^2 \frac{\partial^2 V}{\partial u_i^2(lk)} \, d\tau$$

The first term on the right-hand side cancels with the second term on the right-hand side of the previous equation, so that

$$\frac{\hbar^2 \alpha_{ik}^2}{m} = \frac{1}{N} \sum_{l=1}^{N} \int \Psi^2 \frac{\partial^2 V}{\partial u_i^2(lk)} \, d\tau \tag{3.14}$$

This is the obvious generalization of (3.7). It imposes a self-consistent condition as the wavefunction is α-dependent [Eq. (3.11)].

The usual procedure for calculating the lattice frequencies is to replace the classical equation (Born and Huang, 1954),

$$0 = \left| m\omega^2(\mathbf{q}) \, \delta_{kk'} \delta_{ii'} - \sum_{l} \left(\frac{\partial^2 V}{\partial u_i(lk) \, \partial u_{i'}(0k')} \right)_0 \exp i\mathbf{q} \cdot [\mathbf{R}(lk) - \mathbf{R}(0k')] \right| \tag{3.15}$$

by a similar equation in which the potential is replaced by its quantum-mechanical average with respect to the wavefunctions given by (3.12). Before doing that we rewrite the classical equation of motion for a system involving pair potentials only [Eq. (3.9)]:

$$V = \frac{1}{2} \sum_{l'k', l''k''} v[|\mathbf{R}(l'k') - \mathbf{R}(l''k'') + \mathbf{u}(l'k') - \mathbf{u}(l''k'')|]$$

Substitution into (3.15), after some algebra, gives

$$0 = \left| m\omega^2(\mathbf{q}) \, \delta_{kk'} \delta_{ii'} - \sum_{l''k''}' \{ \delta_{k''k} \exp i\mathbf{q} \cdot [\mathbf{R}(l''k'') - \mathbf{R}(0k')] - \delta_{k'k} \} \times \left[\frac{\partial^2 v_{l''k'',0k'}}{\partial u_i(l''k'') \, \partial u_{i'}(0k')} \right]_0 \right| \tag{3.16}$$

where the prime denotes that $l''k'' \neq 0k'$. In the corresponding quantum-mechanical equation the effective force constants are now replaced by

$$\left(\frac{\partial^2 v_{l''k'',0k'}}{\partial u_i(l''k'') \, \partial u_{i'}(0k')} \right)_0 \to \langle \psi(l''k'')\psi(0k') | \frac{\partial^2 v_{l''k'',0k'}}{\partial u_i(l''k'') \, \partial u_{i'}(0k')} | \psi(l''k'')\psi(0k') \rangle$$

$$= \frac{\partial^2}{\partial R_i(l''k'') \, \partial R_{i'}(0k')} \langle \psi(l''k'')\psi(0k') | \, v(l''k'', 0k') \, | \psi(l''k'')\psi(0k') \rangle \tag{3.17}$$

where the values of α found in minimizing the ground-state energy are used in the wavefunctions. The integrations are over the displacement; therefore for central forces we can replace the second derivatives with respect to displacements by second derivatives with respect to effective equilibrium positions and take those derivatives out of the integrals. This self-consistent Hartree description of the solid yields an Einstein oscillator phonon spectrum, the so-called RPA-phonons (Guyer, 1969).

This method has been used successfully for the treatment of quantum solids. We point out, however, that an alternative approach, providing an explicit quantum-mechanical treatment of excited states, is possible. In this approach we write exciton-like vibrationally excited wavefunctions, having the translational symmetry of the crystal. Using these wavefunctions as a basis, a secular equation for the excited states is obtained by applying the variational principle to the excited states. Work along these lines is now in progress and will be reported separately (Jacobi and Schnepp, 1971).

D. Correlation Corrections

1. *Introduction*

The description of the quantum solid by Hartree wavefunctions has several limitations. A Hartree treatment (Nasanow and Shaw, 1962) correlates only the average motion of different particles but does not correlate them directly. First, potentials with strong short-range repulsions require that the probability amplitude be sharply diminished when two particles approach one another. No such behavior is provided by one-particle wavefunctions and a modification involving explicit two-particle wavefunctions is required. Furthermore, the crystal structure imposes some long-range correlation between the displacements of various molecules. Thus far only the geometry of the crystal structure is reflected in the lattice sums [Eqs. (3.15) and (3.17)], and it is important to check the effect of including this effect directly on the wavefunctions.

2. *Long-Range Correlation*

Wavefunctions incorporating the correlation of displacements of different particles are the correlated gaussians (Koehler, 1966, 1968),

$$\Psi = \left(\frac{\det G}{\pi^{3ZN}}\right)^{1/4} \exp\left(-\frac{1}{2}\sum_{\substack{lki \\ l'k'i'}} G_{l'k'i'}^{lki} u_i(lk) u_{i'}(l'k')\right) \tag{3.18}$$

The nondiagonal terms of the matrix G describe the correlation between various vibrational degrees of freedom. If G is diagonal, the wavefunction reduces to the Hartree product previously discussed [Eq. (3.12)]. Now all the matrix elements of G are regarded as variational parameters to be determined self-consistently by minimizing the ground-state energy of the system. Using the correlated gaussians, the average ground-state energy is now*

$$E_0 = \frac{\hbar^2}{4m} \sum_{l'k'i'} G_{l'k'i'}^{l'k'i'} + \left(\frac{\det G}{\pi^{3ZN}}\right)^{1/2} \int V \exp\left(-\sum_{\substack{lki \\ l'k'i'}} G_{l'k'i'}^{lki} u_i(lk) u_{i'}(l'k')\right) d\tau \tag{3.19}$$

Minimizing the ground-state energy now gives

$$0 = \frac{\partial E_0}{\partial G_{l'k'i'}^{lki}} = \frac{\hbar^2}{4m} \delta_{l'k'i'}^{lki} + \tfrac{1}{2}(G^{-1})_{l'k'i'}^{lki} \langle\Psi|\, V \,|\Psi\rangle - \langle\Psi|\, u_i(lk) u_{i'}(l'k') V \,|\Psi\rangle$$

Successive partial integrations give the following result (again misprinted in Koehler, 1968) for the last term on the right-hand side:

$$\begin{aligned} \langle\Psi|\, u_i(lk) u_{i'}(l'k') V \,|\Psi\rangle &= \tfrac{1}{2}(G^{-1})_{l'k'i'}^{lki} \langle\Psi|\, V \,|\Psi\rangle \\ &\quad + \frac{1}{4} \sum_{\substack{l''k''i'' \\ l'''k'''i'''}} (G^{-1})_{l''k''i''}^{lki} \langle\Psi|\, \frac{\partial^2 V}{\partial u_{i''}(l''k'')\, \partial u_{i'''}(l'''k''')} \,|\Psi\rangle (G^{-1})_{l'k'i'}^{l'''k'''i'''} \end{aligned}$$

The minimization condition is now

$$\frac{\hbar^2}{m} \mathbf{I} = \mathbf{G}^{-1} \mathbf{\Phi} \mathbf{G}^{-1}$$

Matrix notation has been used, where

$$\Phi_{l'k'i'}^{lki} \equiv \langle\Psi|\, \frac{\partial^2 V}{\partial u_i(lk)\, \partial u_{i'}(l'k')} \,|\Psi\rangle \tag{3.20}$$

is seen to be an effective force constant in terms of correlated gaussians. The minimum condition, in matrix notation, is now

$$\frac{\hbar^2}{m} \mathbf{G}^2 = \mathbf{\Phi} \equiv \langle\Psi|\, \frac{\partial^2 V}{\partial \mathbf{u}\, \partial \mathbf{u}} \,|\Psi\rangle \tag{3.21}$$

* The following equation, which is misprinted in Koehler (1968), is very useful in evaluating kinetic energy matrix elements:

$$\begin{aligned} \langle\Psi|\, f \frac{\partial^2}{u_i^2(lk)} g \,|\Psi\rangle &= -\frac{G_{lki}^{lki}}{2} \langle\Psi|\, fg \,|\Psi\rangle \\ &\quad + \frac{1}{4} \langle\Psi|\, f \frac{\partial^2 g}{\partial u_i^2(lk)} - 2 \frac{\partial f}{\partial u_i(lk)} \frac{\partial g}{\partial u_i(lk)} + g \frac{\partial^2 f}{\partial u_i^2(lk)} \,|\Psi\rangle \end{aligned}$$

This is the generalization of (3.14) and (3.17), and it is a self-consistent equation for the effective force constants, as the wavefunction Ψ depends on G (or Φ). This equation has been used to obtain the force constants by iterations (Koehler, 1966), and once the force constants are known, (3.16) is used to calculate the lattice frequencies. In practice, (3.16), (3.17), and (3.21) are used successively in iteration until self-consistency is obtained (Morley and Kliewer, 1969).

With the force constants given by (3.21), the correlated gaussian wavefunction (3.18) is the exact solution of the harmonic Hamiltonian,

$$H_h = T + \frac{1}{2} \sum_{\substack{lki \\ l'k'i'}} \Phi_{l'k'i'}^{lki} u_i(lk) u_{i'}(l'k')$$

Thus the full SCF-phonon scheme finds the most general *harmonic* potential that fits the given potential and uses its exact wavefunctions as variational wavefunctions. The treatment in the previous section [Eqs. (3.10)–(3.17)] is not self-consistent, but the RPA-phonons it gives are a very good first step in the fully self-consistent scheme. Calculations in both schemes have been done for solid helium, although short-range correlations were treated differently in both cases. It is believed (Guyer, 1969) that there is no significant difference between the two schemes.

3. *Short-Range Correlation*

The usual method of including short-range correlations is to employ Jastrow-type functions:

$$\Psi = \Psi_h \prod_{lk,l'k'} f[|\mathbf{r}(lk) - \mathbf{r}(l'k')|] \tag{3.22}$$

where Ψ_h is the harmonic wavefunction previously used (correlated or uncorrelated). The two-particle correlation function $f(r)$ must have the following properties to describe short-range correlations:

1. The strong repulsion of the potential requires that $f(r) \to 0$ as $r \to 0$, so that the overlap between neighbors is greatly reduced.
2. At large distances there are no short-range correlation effects, so that $f(r) \to 1$ as $r \to \infty$.

In addition, the correlation functions most often used have a maximum very close or at the minimum of the potential.

There are several approaches to the choice of a correlation function. When the correlation effect is not drastic, it is sufficient to assume a relatively simple form for it, such as (Nosanow, 1966)

$$f(r) = \exp[-Kv(r)] \tag{3.23}$$

where $v(r)$ is the given potential and K is an additional variational parameter characterizing the correlation effects. A few attempts have been made to obtain better correlation functions by deriving differential equations for them from variational principles and trying to solve them (Brueckner and Frohberg, 1965; Mullin, Nosanow, and Steinback, 1969), or by constructing a fully self-consistent scheme of equations of motion for both one-particle and correlation functions (Guyer and Zaen, 1969). Here we are interested only in the modifications due to any correlation function $f(r)$. For simplicity we consider the RPA scheme, although the same treatment has been done in the SCF scheme as well (Koehler, 1967).

Using the Jastrow-type wavefunction, the ground-state energy is first written in a form similar to (3.13). The correlation function modifies seriously both the kinetic and potential energies. The contribution of correlation to the kinetic energy is usually regarded as an effective anharmonicity which is added to the potential to redefine an effective potential:

$$v'_{\text{eff}} = v - \frac{\hbar^2}{2m} \nabla^2 \ln f \tag{3.24}$$

To evaluate the average potential energy, it is necessary to employ cluster expansions (Brueckner and Frohberg, 1965; Nosanow, 1966). The first term in the expansion involves pair correlations only (which means the replacement of averages of products by products of averages), which gives for the ground state energy

$$E_0 = \frac{N\hbar^2}{4m} \sum_{l'k'} \alpha_{l'k'} + \frac{1}{2} \sum_{lk,l'k'} \langle \psi(lk)\psi(l'k')|\, v_{\text{eff}}(lk, l'k')\, |\psi(lk)\psi(l'k')\rangle \tag{3.25}$$

where

$$v_{\text{eff}}(lk, l'k') = \frac{f^2(lk, l'k')v'_{\text{eff}}(lk, l'k')}{\langle \psi(lk)\psi(l'k')|\, f^2(lk, l'k')\, |\psi(lk)\psi(l'k')\rangle} \tag{3.26}$$

It should be noted that while the integrals in (3.13) are divergent due to the strong repulsions, so that a cut-off procedure is required, the effective potential goes to zero with r and all the integrals become convergent.

This effective potential has been used to obtain lattice frequencies, using (3.15) and (3.17) (de Wette, Nosanow, and Werthamer, 1967; Klump, Schnepp, and Nosanow, 1970). In addition the ground-state energy must be minimized, thus obtaining an additional dependence of the effective force constants on the correlation. This minimization gives

(Gillis, Werthamer, and Koehler, 1968; Guyer, 1969)

$$\Phi_{l'k'i'}^{lki} = \langle\psi(lk)\psi(l'k')|\frac{\partial^2 v_{\text{eff}}(lk, l'k')}{\partial u_i(lk)\,\partial u_{i'}(l'k')}|\psi(lk)\psi(l'k')\rangle$$
$$- \langle\psi(lk)\psi(l'k')|\, v_{\text{eff}}(lk, l'k')\,|\psi(lk)\psi(l'k')\rangle$$
$$\times \frac{\langle\psi(lk)\psi(l'k')|\, \partial^2 f^2(lk, l'k')/\partial u_i(lk)\,\partial u_{i'}(l'k')\,|\psi(lk)\psi(l'k')\rangle}{\langle\psi(lk)\psi(l'k')|\, f^2(lk, l'k')\,|\psi(lk)\psi(l'k')\rangle} \tag{3.27}$$

As pointed out by Koehler (1968), even the solution obtained from these modified force constants is not the most meaningful one, because it is obtained from the harmonic part only of the wavefunction (3.22). He shows that the complete dispersion curves are not obtained from the force constant matrix Φ alone, but first it has to be transformed by a unitary matrix diagonalizing the matrix of second derivatives of the correlation function f^2. For $\mathbf{q} \to 0$, however, the frequencies coincide and (3.27) is sufficient for the calculation of lattice frequencies.

E. Temperature Dependence

Concluding the theoretical review, we note that the general formulation (Fredkin and Werthamer, 1965) and some later applications (Gillis, Werthamer, and Koehler, 1968) of quantum lattice dynamics include also a temperature dependence. In that scheme wavefunctions are replaced by density matrices, which are temperature-dependent through thermal Boltzmann factors, and instead of minimizing the energy we minimize the free energy. The normal modes of the system are then obtained by considering Liouville's equation of motion for the density matrix and looking for the poles of the linear response to an external perturbation which is finally made vanishingly small.

F. Applications and Comparison with Experiment

1. *Solid Helium*

Guyer (1969) has reviewed the field of quantum lattice dynamics and applications to solid helium in great detail. He also compared the theoretical with experimental results quite exhaustively. We shall here review the applications to this system only briefly.

de Wette, Nosanow, and Werthamer (1967) applied the self-consistent phonon method in the RPA approximation (Section IIIC) to body-centered-cubic (bcc) solid helium. These authors used a Jastrow-type function (3.22) and included short-range correlation by using an effective potential as in (3.24) and a pair correlation function as given in (3.23). de Wette and co-workers calculated dispersion curves and deduced sound velocities from these. The crystal energy was calculated to be 5°K per atom as compared to the experimental value of −1°K.

Koehler (1967) applied his full SCF lattice dynamics formalism to solid He^3. This formalism includes both long- and short-range correlation (Sections IIID.2 and IIID.3) and goes beyond the RPA approximation inasmuch as the relevant equations are solved by iteration until self-consistency is reached. The ground-state crystal energy and phonon frequencies were calculated. The crystal energy was found here to be lower than that calculated from the RPA or zeroth-order approximation (de Wette, Nosanow, and Werthamer, 1967) but it was still found to be positive, i.e., no binding energy was obtained. The sound velocities obtained from the slopes of the calculated dispersion curves again agreed very well with experimental measurements.

Gillis, Koehler, and Werthamer (1968) applied the full SCF phonon theory to hcp He^4. These authors had available results of inelastic coherent neutron diffraction studies carried out at Brookhaven (Minkiewicz et al., 1968) for the *a*- and *c*-axes at the molar volume of 21.1 cm^2. They found that the theoretical dispersion curves agreed well with the experimental results in the linear acoustic region but were too high for the acoustic branches near the zone boundary and even more so for the optical branches. Gillis, Koehler, and Werthamer (1968) also calculated the phonon frequencies for a smaller molar volume (16.0 cm^3) to compare with the inelastic neutron scattering experiments carried out at Iowa State University (Brun et al., 1968). Again, the agreement was fair in the acoustic region but the theoretical results for other regions of the Brillouin zone were now lower than the experimental values. The volume dependence of phonon frequencies and crystal energy was discussed over a wide volume range and it was concluded that the theoretical volume dependences are not satisfactory. The crystal energy obtained from the theory is again lower by 4–5 cal/mole than that predicted by the RPA method by Nosanow (1966). Sound velocities were calculated and compared to experimental results of Franck and Wanner (1968) for longitudinal velocities and obtained excellent agreement. Gillis, Koehler, and Werthamer (1968) concluded that improvement of the theoretical treatment is to be expected from use of a better short-range pair correlation function instead of that used in their work [see (3.23)]. They also found that their results were not very sensitive to the choice of interatomic potential since the large amplitude of the helium atoms requires averaging over a significant region of space. Conversely, it is concluded that the study of the phonon spectrum is not expected to give detailed information concerning this potential.

Morley and Kliewer (1969) treated the lattice dynamics of solid He^3 and He^4 at high densities, i.e., in the range of molar volumes of 10–16 cm^3. At these densities both isotopic solids have the hcp structure. These authors used the full SCF method of Keohler (1968) adapted for this structure. They argue, however, that short-range correlation is not of importance at high densities and therefore use correlated gaussina wavefunctions and omit the pair-correlation function factors. Dispersion curves, frequency distribution functions, sound velocities, elastic constants, and Debye temperatures at 0°K were calculated and were described as functions of molar volume within the considered range. In particular, the Debye temperatures were compared with experiment and it was found that agreement becomes worse as the molar volume increases. This trend is as expected. The theoretical values were much too high for molar volumes of 14.0 and 16.0 cm^3 but at high densities the theoretical values approach the experimental results. On the other hand, it can be concluded from the appropriate figure that at molar volumes of 18–20 cm^3 the RPA theory including short-range correlation (Nosanow 1966) gives very good results.

Werthamer (1970) has recently treated the cross section for neutron scattering from phonons in solid helium.

2. *Other Rare Gas Solids*

Nosanow and Shaw (1962) were the first to apply quantum lattice dynamical methods to the rare gas solids. Very good results were obtained for the crystal energies. Koehler (1966) concluded that solid neon cannot be treated by the traditional classical harmonic approximation. He used the SCF phonon method and obtained self-consistency after three cycles. His crystal energy is somewhat better (−219°K) than that of Nosanow and Shaw (−215°K). Gillis, Werthamer, and Koehler (1968) used a density matrix formulation of the SCF theory to include temperature dependence for solid neon and argon. They minimized the free energy of the solid at each temperature to obtain a theoretical nearest neighbor distance. The results were in very good agreement with experiment for argon in the temperature range 0–84°K. The calculated sound velocities were also very good, and for the specific heat agreement was good up to about 30°K. Above this temperature the theoretical results were too low and the discrepancy increased with increasing temperature. For solid neon, the nearest neighbor distances obtained in the range 0–24°K were consistently low by 4% for one set of potential parameters and 1.5% for another set. Good agreement was also obtained for the specific heat. In addition, the compressibilities and sound velocities were calculated as functions of temperature.

Recently, Glyde (1970) has investigated the importance of anharmonicities in the solidified rare gases. He did this by fitting the Lennard-Jones and Morse potential parameters to the 0°K sublimation energy, lattice spacing, and inverse compressibility. Performing this calculation both in the harmonic approximation and in the self-consistent scheme, Glyde concludes that the effect on the potential of neglecting anharmonicity is insignificant for xenon and krypton, small for argon, but appreciable for neon. He also finds an appreciable modification in the dispersion curves of neon.

Werthamer, Gray, and Koehler (1970) have recently calculated the Raman scattering cross sections in solid neon and argon.

3. *Solid Hydrogen*

Solid parahydrogen is hexagonal-close-packed at all temperatures below the melting point (13.9°K). If the sample contains at least 60% orthohydrogen (normal hydrogen contains 75% ortho) a λ-type heat capacity anomaly occurs at 1.5–2°K depending on the composition (Ahlers and Orttung, 1964). It is now known that the lower temperature phase is face-centered-cubic (Schuch and Mills, 1966; Schuch, Mills, and Depatie, 1968; Mucker et al., 1968) and, furthermore, the accurate structure is actually of lower symmetry and is believed to be $Pa3(T_h^6)$, analogous to α–N_2 and CO_2 (Mucker et al., 1968; Hardy, Silvera, and McTague, 1971; Hardy et al., 1968; Coll and Harris, 1970). A similar phase exists for paradeuterium.

McTague, Silvera, and Hardy (1970) investigated the Raman spectrum of hexagonal hydrogen and observed the *TO* mode at 37.4 cm^{-1} for p–H_2, at 38.4 cm^{-1} for o–H_2, at 36.9 cm^{-1} for o–D_2 and at 37.4 cm^{-1} for p–D_2. Nielsen, Moller, and Meyer (1970) have recently reported coherent inelastic neutron scattering results for hcp hydrogen. They determined the dispersion curves for the symmetry directions of the Brillouin zone. Klein and Koehler (1970) carried out complete SCF calculations for the phonon frequencies, lattice energies and nearest neighbor distances for hcp H_2 and D_2 to two different stages of refinement. Both these treatments gave very good results for the $\mathbf{q} = 0$ optical phonon mode as compared with experiment. Moreover, the crystal energies as obtained from the calculations, both for hydrogen and deuterium, were within 10% of experimental values. No detailed comparison with the neutron scattering results were made but the *TO* mode of p–D_2 from these measurements agrees very well with the calculations of Klein and Koehler and with the Raman measurements (McTague, Silvera, and Hardy, 1970; Silvera, Hardy, and McTague, 1969).

The optical phonons of cubic hydrogen and deuterium were investigated in the far infrared by Hardy et al. (1968). Although two active modes (belonging to representation T_u) are expected, three lines were in fact found. These results proved that long-range rotational ordering does occur in solid hydrogen as had been predicted by Raich and James (1966). If the $J = 1$ rotational state remained degenerate, no optical modes would be expected to exist since then the structure would be simple fcc, as in solid argon. The lowest two lines in the spectrum (at 62.2 and 80.0 cm^{-1} for o–H_2 and at 57.4 and 74.5 cm^{-1} for p–D_2) were assigned as the optical modes of symmetry T_u. The third line (at 93 cm^{-1} for o–H_2 and at 85 cm^{-1} for p–D_2) is much broader (7–10 cm^{-1}) and is believed to be a combination band, probably with librons (Mertens, Biem, and Hah, 1968). Klump, Schnepp, and Nosanow (1970) calculated the optical phonons of cubic hydrogen and deuterium using the RPA approximation including short-range correlation. Using the gas phase Lennard-Jones parameters and the observed crystal spacing, these authors obtained optical mode frequencies of 63.8 and 87.0 cm^{-1} for hydrogen and 53.1 and 74.4 cm^{-1} for deuterium, in close agreement with experiment. Furthermore, the calculated ground-state crystal energies are in good agreement with thermodynamically determined energies (for a recent summary of available data see Schnepp, 1970). A recent investigation (Jacobi, 1970) showed that short-range correlation is much more important in the ground state of hydrogen than in deuterium. In hydrogen it accounts for 30% of the crystal energy, but in deuterium for only 5%.

IV. INTENSITIES OF LATTICE VIBRATION SPECTRA

A. Introduction

Few papers have appeared which are concerned with the intensities of infrared absorption by lattice modes in molecular solids. For polar molecules, it is clear that librational motions will cause the absorption of light by a mechanism analogous to that which occurs for rotational spectra in the gas phase. Calculations have been made of such infrared absorption intensities by Schnepp (1967) and by Friedrich (1970). Friedrich used a theory which takes into account the polarizability of the medium in a specific way. For this case, the intensity is proportional to the square of the molecular dipole moment. Section IVB.2 is concerned with this topic.

A mechanism by which translational motions can acquire infrared absorption intensity has also been proposed (Schnepp, 1967). We first point out that an array of molecular dipoles cannot produce such intensity, unless they undergo angular displacements. However, polar molecules do not have a center of symmetry and therefore librational and translational motions are not separable by symmetry even at the Γ-point ($\mathbf{q} = 0$) and thus all normal modes of the solid represent mixtures of such degrees of freedom. As stated, only the librations of the dipole moments produce intensity (or a dipole derivative). It has, however, been shown (Schnepp, 1967) that the permanent molecular quadrupole moment causes infrared intensity of the correct order of magnitude (Ron and Schnepp, 1967) by an induction mechanism. This mechanism was proposed primarily for nonpolar molecular solids but may also make an appreciable contribution in the case of polar molecules. This subject is treated in Section IVB.3.

In addition to permanent and induced molecular dipole moments, transition moments caused by charge cloud distortion during collisions have been discussed for gas phase systems (Kiss, Gush, and Welsh, 1959; van Kranendonk, 1958). It was found that this "distortion moment" contribution is small and decreases rapidly with temperature. In the case of gas phase N_2, the observed absorption intensity (Colpa and Ketelaar, 1958; Ketelaar and Rettschnick, 1964) could be wholly accounted for on the basis of the molecular quadrupole-induced mechanism. We do not consider the distortion moment further here.

The Raman intensities of librational modes in molecular solids have also received attention. These intensities can be related to the molecular polarizabilities by a model which considers them to be additive in the solid (Fruehling, 1951; Brith, Ron, and Schnepp, 1969; Cahill and Leroi, 1969a). We discuss this treatment in Section IVC. In Section IVD we consider applications and comparison with experimental measurements.

B. Infrared Intensities

1. *General*

The following features are common to all treatments of infrared intensities. It is assumed that the dipole moment of the system can be written as a sum over molecular moments $\boldsymbol{\mu}(lk)$:

$$\boldsymbol{\mu} = \sum_{kl} \boldsymbol{\mu}(lk)(\mathbf{R}^n \boldsymbol{\Lambda}^n) \tag{4.1}$$

Here $\mathbf{R}^n$ symbolizes the positions of molecular centers of mass and $\boldsymbol{\Lambda}^n$ the orientations of all molecules in the crystal. The absorption intensity is determined by the transition dipole moment matrix element:

$$\langle \Psi'^0 | \boldsymbol{\mu} | \Psi' \rangle \tag{4.2}$$

The optic lattice modes are assumed to be harmonic and the wavefunctions of the system Ψ' are written as products over independent harmonic oscillator functions $\psi(Q_m)$ of normal coordinates Q_m:

$$\Psi' = \prod_m \psi(Q_m) \tag{4.3}$$

Only $\mathbf{q} = 0$ modes need to be considered since only those are optically active. In order to calculate the transition matrix element (4.2), we substitute from (4.1), after expanding the dipole moment in terms of the normal coordinates in the usual way:

$$\begin{aligned} \boldsymbol{\mu} &= \sum_{lk} \boldsymbol{\mu}(lk) \\ &= \sum_{lk} \boldsymbol{\mu}(lk)(0) + \sum_{lk} \sum_{m} \left[\frac{\partial \boldsymbol{\mu}(lk)}{\partial Q_m}\right]_0 Q_m \end{aligned} \tag{4.4}$$

and substituting (4.4) in (4.2) and making use of (4.3) we obtain

$$\langle \Psi^0 | \, \boldsymbol{\mu} \, | \Psi \rangle = \langle \psi^0(Q_n) | \, Q_n \, | \psi(Q_n) \rangle \sum_{lk} \left[\frac{\partial \boldsymbol{\mu}(lk)}{\partial Q_n}\right]_0 \tag{4.5}$$

The intensity problem then reduces to the evaluation of the sum

$$\sum_{lk} \left[\partial \boldsymbol{\mu}(lk)/\partial Q_n\right]_0$$

since the factor outside the sum, the harmonic oscillator transition matrix element, is well known (Wilson, Decius, and Cross, 1955).

Often it is more convenient to work with symmetry coordinates S_m and to calculate the dipole derivatives with respect to the S_m. We then use the orthogonal transformation $\mathbf{U}$, which transforms the symmetry coordinates to normal coordinates,

$$Q_n = \sum_m U_{nm} S_m$$

and obtain the relationship

$$\left[\frac{\partial \boldsymbol{\mu}(lk)}{\partial Q_n}\right]_0 = \sum_m U_{nm} \left[\frac{\partial \boldsymbol{\mu}(lk)}{\partial S_m}\right]_0 \tag{4.6}$$

and $\mathbf{U}$ is obtained from the solution of the lattice dynamics problem at $\mathbf{q} = 0$. The Q_m are the eigenvectors of this problem.

For comparison with experiment we use the integrated absorption strength (Mulliken, 1939; Seitz, 1940):

$$\int nk(\tilde{\nu}) \, d\tilde{\nu} = 2.61 \times 10^{18} G' \tilde{\nu}''_{,} D'_{,\prime\prime}$$

Here n is the index of refraction (real part) and $k(\tilde{\nu})$ is the absorption coefficient in units of cm^{-1} of gas path length measured at NTP:

$$k = \frac{1}{l} \ln \left(\frac{I_0}{I}\right)$$

The frequency $\tilde{\nu}$ is measured in cm^{-1} units. G' is the degeneracy factor for the upper state, $\tilde{\nu}'_{,}''$ the wavenumber of the transition. $D'_{''}$ is the square of the molecular transition matrix element, in units of cm^2. We assume that the absorption intensities we deal with are small enough so that n can be taken to be a constant over the absorption band. Then we obtain

$$\int k(\tilde{\nu})\, d\tilde{\nu} = 2.61 \times 10^{18} n^{-1} G' \tilde{\nu}'_{,}'' D'_{''} \tag{4.7}$$

If the integrated absorption is to be expressed in units of cm per millimole or darks (Overend, 1963; Friedrich, 1970), then the absorption coefficient ε is measured in units of cm^2 per millimole, and we have

$$\int \varepsilon(\tilde{\nu})\, d\tilde{\nu} = 5.85 \times 10^{19} n^{-1} G' \tilde{\nu}'_{,}'' D'_{''} \tag{4.8}$$

The square of the molecular transition matrix element $D'_{''}$ is related to (4.5) except for the fact that $\boldsymbol{\mu}$ represents the dipole moment of the whole crystal. Thus we have to normalize it by dividing by the total number of molecules. If N is the number of unit cells and Z the number of molecules per unit cell,

$$\begin{aligned} D'_{''} &= \frac{1}{2N} |\langle \Psi'^{0}|\, \boldsymbol{\mu}\, |\Psi'\rangle|^2 \\ &= \frac{1}{2N} \left| \langle \psi^0(Q_n)|\, Q_n\, |\psi(Q_n)\rangle \sum_{lk} \left[\frac{\partial \boldsymbol{\mu}(lk)}{\partial Q_n}\right]_0 \right|^2 \end{aligned} \tag{4.9}$$

2. *Intrinsic Transition Moment—Librations of Polar Molecules*

Let the permanent dipole moment of the molecules in the solid be m. We are concerned with the evaluation of the transition moment (4.5) or we have to calculate the sum:

$$\sum_{lk} \left[\frac{\partial \boldsymbol{\mu}(lk)}{\partial Q_n}\right]_0 = \sum_m U_{nm} \sum_{lk} \left[\frac{\partial \boldsymbol{\mu}(lk)}{\partial S_m}\right]_0 \tag{4.10}$$

If we use the direction cosine displacement coordinates to describe the librational modes (Walmsley and Pople, 1964; Schnepp, 1967), we write

$$\boldsymbol{\mu}(lk) = m\boldsymbol{\Lambda}(lk) \tag{4.11}$$

or for the ith component

$$\mu_i(lk) = m[\Lambda_i^0(lk) + \lambda_i(lk)], \qquad i = x, y, z \tag{4.12}$$

Here $\boldsymbol{\Lambda}(lk)$ is the unit vector of the molecular axis which is parallel to the permanent dipole moment, and $\boldsymbol{\lambda}(lk)$ is the vector of the displacement characteristic of the libration. $\boldsymbol{\Lambda}^0(lk)$ refers to the equilibrium orientation.

Then we obtain

$$\left[\frac{\partial \boldsymbol{\mu}(lk)}{\partial S_m}\right]_0 = \sum_i \sum_{l'k'} \left[\frac{\partial \boldsymbol{\mu}(lk)}{\partial \Lambda_i(l'k')}\right]_0 \frac{\partial \Lambda_i(l'k')}{\partial S_m}$$

$$= N^{-1/2} \sum_{ik'} T''_{m,ik'} \sum_{l'} \left[\frac{\partial \boldsymbol{\mu}(lk)}{\partial \Lambda_i(l'k')}\right]_0 \tag{4.13}$$

since the symmetry coordinates for $\mathbf{q} = 0$ are written

$$S_m = \sum_{ik} T''_{m,ik} \lambda_i(\mathbf{0}, k) \tag{4.14}$$

with

$$\lambda_i(\mathbf{q}, k) = N^{-1/2} \sum_l \exp\left[i\mathbf{q} \cdot \mathbf{R}(lk)\right] \lambda_i(lk) \tag{4.15}$$

The transformation $\mathbf{T}''$ is orthogonal, which is a condition for obtaining (4.13). We see from (4.12) that $\boldsymbol{\mu}(lk)$ is a function only of $\boldsymbol{\Lambda}(lk)$ and therefore the sum over $l'k'$ in (4.13) reduces to a single term:

$$\left[\frac{\partial \boldsymbol{\mu}(lk)}{\partial S_m}\right]_0 = N^{-1/2} \sum_i T''_{m,ik} \frac{\partial \boldsymbol{\mu}(lk)}{\partial \Lambda_i(lk)} \tag{4.16}$$

From (110) we also obtain

$$\left(\frac{\partial \mu_i(lk)}{\partial \Lambda_j(lk)}\right)_0 = m\delta_{ij} \tag{4.17}$$

If, on the other hand, we were to use Eulerian angle displacement coordinates (Schnepp and Ron, 1969), the equivalent of (4.11) would be

$$\boldsymbol{\mu}(lk) = m\boldsymbol{\Lambda}(lk) = m[\sin\theta \cos\phi \hat{\imath} + \sin\theta \sin\phi \hat{\jmath} + \cos\theta k] \tag{4.18}$$

from which the derivatives $[\partial \boldsymbol{\mu}(lk)/\partial u_\alpha(lk)]_0$ for $\alpha = \theta, \phi$ can be calculated for a known crystal structure. The coordinate frames (both cartesian and spherical) are here space-fixed. We therefore can obtain the derivatives which parallel (4.17) and, in analogy to (4.16) we can calculate

$$\left(\frac{\partial \boldsymbol{\mu}(lk)}{\partial S_m}\right)_0 = N^{-1/2} \sum_\alpha T''_{m,\alpha k} \frac{\partial \boldsymbol{\mu}(lk)}{\partial u_\alpha(lk)} \tag{4.19}$$

It is, of course, straightforward to write the relations which parallel (4.14) and (4.15) and the symmetry coordinates are assumed to be written in terms of the appropriate $u_\alpha(\mathbf{q}k)$, $\alpha = \theta, \phi$. It is clearly not necessary to go by way of the S_m and we could consider the eigenvectors Q_m expressed in terms of the u_α by way of the transformation $\mathbf{U}$. Use of symmetry coordinates S_m (4.14) is only a matter of convenience since symmetry can be used to factor the secular equation.

The transition matrix element is calculated from (4.9) by substituting the result of (4.13) and (4.10) or of (4.19) and (4.10) as the case may be. Further, the harmonic oscillator matrix element which appears as a factor in (4.9) is given by

$$\langle \psi^0(Q_n) | \, Q_n \, | \psi(Q_n) \rangle = \left(\frac{16.8}{I\tilde{\nu}} \right)^{1/2}$$

where I is the molecular moment of inertia in atomic mass units square angstroms and $\tilde{\nu}$ is the frequency of the transition in cm^{-1} for a one-phonon excitation. Comparison with experiment is carried out by using (4.7) or (4.8).

Friedrich (1970) has discussed the local field correction for the present problem. Following the work of Decius (1968) and Mandel and Mazur (1958), the total dipole moment $\mathbf{p}$ at site j in the crystal is given by

$$\mathbf{p}_j = \mathbf{m}_j + \boldsymbol{\alpha}_j \mathbf{F}_j \qquad (4.20)$$

where $\mathbf{m}_j$ is the permanent dipole moment, $\boldsymbol{\alpha}_j$ is the polarizability tensor, and $\mathbf{F}_j$ the local field. This field is composed of the external imposed field $\mathbf{E}_j$ shielded by the dipole moments $\mathbf{p}_i$ throughout the solid. The local field is composed of the external imposed field $\mathbf{E}_j$ shielded by the dipole moments $\mathbf{p}_i$ throughout the solid. The local field at site $\mathbf{F}_j$ can then be written as follows:

$$\mathbf{F}_j = \mathbf{E}_j - \sum_{i \neq j} \mathbf{T}_{ji} \mathbf{p}_i \qquad (4.21)$$

The operator $\mathbf{T}_{ji}$ converts the dipole moment $\mathbf{p}_i$ at site i (composed of permanent and induced moments as given by (4.20) into the electric field produced by it at site j. Elimination of $\mathbf{F}_i$ between (4.20) and (4.21) and retaining only first-order terms in $\mathbf{E}$ gives

$$\mathbf{p}_j = \sum_i [\mathbf{U} + \boldsymbol{\alpha} \mathbf{T}]_{ji}^{-1} (\mathbf{m} + \boldsymbol{\alpha} \mathbf{E})_i$$

where $\mathbf{U}$ is the unit tensor. As a result of these considerations, the transition matrix element (4.5) must be multiplied by the factor

$$(\mathbf{U} + \boldsymbol{\alpha} \mathbf{S})^{-1} \qquad (4.22)$$

where $\mathbf{S}$ represents a matrix of order equal to the number of molecules per unit cell and each element is a lattice sum (Decius, 1968).

3. *Induced Transition Moment—Translational Modes*

In this section we consider the dipole transition intensity induced by permanent multimoments. Schnepp (1967) has calculated the quadrupole-induced infrared absorption intensity of translational lattice modes in molecular solids. This author also found that the permanent dipole-induced intensity was negligible in the example considered, i.e., solid CO. This conclusion is valid for any crystal structure where the molecular sites are related by inversion in a center of symmetry.

In the crystal-fixed coordinate system we express the induced dipole moment $\boldsymbol{\mu}(lk)$ in terms of the local electric field $\mathbf{E}(lk)$ and the polarizability tensor $\boldsymbol{\alpha}(lk)$:

$$\boldsymbol{\mu}(lk) = \boldsymbol{\alpha}(lk)\mathbf{E}(lk) \tag{4.23}$$

It is simple to write $\boldsymbol{\alpha}'$ in the molecular coordinate frame. For example, for a linear molecule, we have

$$\boldsymbol{\alpha}' = \begin{pmatrix} \alpha_{\perp} & 0 & 0 \\ 0 & \alpha_{\perp} & 0 \\ 0 & 0 & \alpha_{\parallel} \end{pmatrix} \tag{4.24a}$$

If the transformation from the crystal system to the molecular axis system is $\mathbf{a}$, then

$$\boldsymbol{\alpha} = \mathbf{a}^{-1}\boldsymbol{\alpha}'\mathbf{a} \tag{4.24b}$$

We then obtain the dipole moment derivative:

$$\frac{\partial\boldsymbol{\mu}(lk)}{\partial S_m} = \boldsymbol{\alpha}(lk)\left(\frac{\partial\mathbf{E}(lk)}{\partial S_m}\right)_0 + \left(\frac{\partial\boldsymbol{\alpha}(lk)}{\partial S_m}\right)_0 \mathbf{E}_0(lk) \tag{4.25}$$

The second term in this equation vanishes if the crystal structure is centro-symmetric, since then $\mathbf{E}_0(lk)$ must vanish. Even if this is not so, the polarizability will change only with an orientational displacement and therefore $\partial\boldsymbol{\alpha}(lk)/\partial S_m$ is nonzero only for librational motions and depends on the anisotropy of the polarizability $(\alpha_{\parallel} - \alpha_{\perp})$. This term is therefore expected to be always small. It has, in fact, been investigated for solid CO and has been found to be negligible (Schnepp, 1967) compared to the intrinsic intensity of the librations (see preceding section). Here we consider only the first term in (4.25). We have, therefore, to calculate $\partial\mathbf{E}(lk)/\partial S_m$.

By analogy to (4.16), for a translational symmetry coordinate S_m we now obtain

$$\left(\frac{\partial\mathbf{E}(lk)}{\partial S_m}\right)_0 = \sum_{\substack{i \\ l'k'}} \left[\frac{\partial\mathbf{E}(lk)}{\partial R_i(l'k')}\right]_0 \frac{\partial R_i(l'k')}{\partial S_m}$$

$$= N^{-1/2} \sum_{ik'} T'_{m,ik'} \sum_{l} \left[\frac{\partial\mathbf{E}(lk)}{\partial R_i(l'k')}\right]_0 \tag{4.26}$$

Here $R_i(l'k')$ is the ith component of the instantaneous position vector of molecule $(l'k')$:

$$\mathbf{R}(l'k') = \mathbf{R}^0(l'k') + \mathbf{u}(l'k')$$

where $\mathbf{R}^0(l'k')$ designates the equilibrium position of the center of mass. The symmetry coordinate S_m is, in principle, a function of all translational displacements $\mathbf{u}(l'k')$ of all molecules in the solid. By analogy with (4.14) and (4.15) we have assumed the following relations:

$$S_m = \sum_{ik} T'_{m,ik} u_i(\mathbf{0}, k) \tag{4.27}$$

$$u_i(\mathbf{q}, k) = N^{-1/2} \sum_{l} \exp\,[i\mathbf{q} \cdot \mathbf{R}(lk)] u_i(lk) \tag{4.28}$$

We also have made use of the fact that the transformation $\mathbf{T}'$ is independent of unit cell for $\mathbf{q} = 0$. An equation can be written paralleling (4.26) for librational symmetry coordinates but has been found not to contribute in the cases investigated (Ron and Schnepp, 1967).

The electric field $\mathbf{E}(lk)$ at the site of molecule (lk) can be expressed as a sum of contributions by the neighboring molecules (rs):

$$\mathbf{E}(lk) = \sum_{rs} \mathbf{E}(lk, rs)[\mathbf{R}(rs), \mathbf{\Lambda}(rs)] \tag{4.29}$$

The contribution by molecule (rs) is a function of the position and orientation (symbolized by $\mathbf{R}$ and $\mathbf{\Lambda}$) of this molecule. Substituting (4.29) into (4.26) we note that we shall obtain nonzero contributions only when $(l'k')$ is either one of (lk), i.e., the central molecule or (rs), the contributing molecule. Therefore, the double sum over $l'k'$ and over rs reduces to a single sum with two terms:

$$\left[\frac{\partial \mathbf{E}(lk)}{\partial S_m}\right]_0 = N^{-1/2} \sum_{i} \sum_{l'k'} \left\{ T'_{m,ik'} \left[\frac{\partial \mathbf{E}(lk; l'k')}{\partial R_i(l'k')}\right]_0 + T'_{m,ik} \left[\frac{\partial \mathbf{E}(lk; l'k')}{\partial R_i(lk)}\right]_0 \right\} \tag{4.30}$$

It can be shown that for any pair of central molecule (lk) and contributing molecule $(l'k')$

$$\left[\frac{\partial \mathbf{E}(lk; l'k')}{\partial R_i(l'k')}\right]_0 = -\left[\frac{\partial \mathbf{E}(lk; l'k')}{\partial R_i(lk)}\right]_0$$

As a result, it is possible to simplify the summation in (4.30) for cases where the coefficients $T'_{m,ik}$ for a particular symmetry coordinate S_m all have the same absolute magnitude and different signs. This turns out to be a useful case for application to α-N_2 and CO_2 (Schnepp, 1967).

For nonpolar molecules, the major contribution to the induced infrared intensity of translational lattice modes may be expected to be due to molecular quadrupole moments. The quadrupole moment Θ is defined by:

$$\Theta = \sum_k e_k(z_k'^2 - x_k'^2)$$

where the summation is over all particles and e_k is the charge of particle k. The primes indicate the molecular coordinate frame. The field due to a quadrupole moment can be expressed in a number of ways, and the derivatives of this field as required for (4.30) are given in Table I of the paper by Schnepp (1967). The corresponding dipole moment derivative is then obtained from (4.25), while remembering that $\boldsymbol{\alpha}(lk)$ is a tensor. By using the result of (4.25) in (4.6) and finally (4.5), the transition moment matrix element is obtained. The summation over lk in (4.5) simply implies multiplication by ZN, the total number of molecules in the solid, since all molecules have the same environment as long as we neglect surface effects.

The resultant intensity in the case of a solid of diatomic molecules is proportional to $\Theta^2\alpha_{\mathrm{av}}{}^2/s^{10}$ where Θ is the molecular quadrupole moment α_{av} is the average polarizability given by

$$\alpha_{\mathrm{av}} = \tfrac{1}{3}(\alpha_{\parallel} + 2\alpha_{\perp})$$

and s is the nearest neighbor distance, or unit cell parameter.

It can be reasoned that a local field correction should be made for the present case as has been described in the preceding section. On the basis of simple considerations, this correction would consist of the multiplicative factor as given in (4.22). Such a correction for an induced transition moment has not been discussed in the literature.

For comparing calculation with experiment, it is often convenient to substitute (4.6) into (4.5) and to take out the sum over m, or over symmetry coordinates contributing to a given normal coordinate. It then becomes clear that it is useful to calculate the sum:

$$\sum_{lk} \left(\frac{\partial \boldsymbol{\mu}(lk)}{\partial S_m}\right)_0$$

Schnepp (1967) has given the values of these sums for two infrared active symmetry coordinates of a crystal of structure T_h^6, appropriate for the solids α-N_2 and CO_2. The symmetry coordinate designations are those of Walmsley and Pople (1964) and the coordinates S_4 and S_7 are components of triply degenerate sets. The results are

$$\begin{aligned}
\sum_{lk}\left[\frac{\partial \mu_z(lk)}{\partial S_4}\right]_0 &= \frac{4N^{1/2}\Theta\alpha_{\mathrm{av}}}{s^5}\cdot(-73.5 - 141\kappa) \\
\sum_{lk}\left[\frac{\partial \mu_y(lk)}{\partial S_7}\right]_0 &= \frac{4N^{1/2}\Theta\alpha_{\mathrm{av}}}{s^5}\cdot(-73.5)
\end{aligned} \tag{4.31}$$

where $\kappa = (\alpha_{\parallel} - \alpha_{\perp})/(\alpha_{\parallel} + 2\alpha_{\perp}) = \frac{1}{3}(\alpha_{\parallel} - \alpha_{\perp})/\alpha_{\mathrm{av}}$. All other components vanish. The other components of each of the degenerate sets give identical results for different cartesian components since the representation T_u of the factor group T_h is chosen in the basis x, y, z.

Returning now to the form of (4.5) referred to above, after substitution in it of (4.6) we note that the *sum* of the intensity contributions of all normal coordinates belonging to the same irreducible representation is independent of the force constant problem. This is so because the transformation U_{nm} is always orthogonal. Moreover, this sum of intensities is equal to the sum of intensity contributions by the symmetry coordinates belonging to this irreducible representation. For this reason, such sums should be compared to experimental measurements, if the intensity theory is to be tested independently of the force constant problem.

C. Raman Intensities

In this case we are interested in calculating the matrix element

$$\langle \Psi_0 | \boldsymbol{\alpha} | \Psi \rangle \tag{4.32}$$

We assume that the polarizability $\boldsymbol{\alpha}$ can be expressed as a sum of molecular polarizabilities $\boldsymbol{\alpha}(lk)$ which are taken to be in the crystal-fixed coordinate frame. We then expand the polarizability, as usual, in the normal coordinates:

$$\boldsymbol{\alpha} = \sum_{lk} \boldsymbol{\alpha}^0(lk) + \sum_n \sum_{lk} \frac{\partial \boldsymbol{\alpha}(lk)}{\partial Q_n} Q_n \tag{4.33}$$

As in the case of the dipole moment matrix element (4.5), we obtain on substitution of (4.33) into (4.32) a product of the well-known harmonic oscillator transition moment matrix element and the sum of polarizability derivatives:

$$\langle \Psi_0 | \boldsymbol{\alpha} | \Psi \rangle = \langle \Psi_0 | Q_n | \Psi \rangle \sum_{lk} \frac{\partial \boldsymbol{\alpha}(lk)}{\partial Q_n} \tag{4.34}$$

We use (4.24a) and (4.24b) and the direction cosine displacement coordinates of Walmsley and Pople (1964) to describe the application to a linear molecule as given by Brith, Ron, and Schnepp (1969). Equivalent treatments have been given by Cahill (1968) and by Richardson and Nixon (1968). The components $\Lambda_i(lk)$ of the unit vector $\boldsymbol{\Lambda}(lk)$ are the instantaneous direction cosine of the molecular axis. Part of the components of the matrix of the transformation from molecular (primed) to crystal (unprimed coordinates (4.24b) can be identified with the Λ_i, i.e.,

$\Lambda_x \equiv a_{31}$, $\Lambda_y \equiv a_{32}$, $\Lambda_z = a_{33}$. Then from (4.24b) we obtain for the ij-component of $\boldsymbol{\alpha}(lk)$

$$\begin{aligned}\alpha_{ij}(lk) &= \alpha_{\perp}\delta_{ij} + \Lambda_i(lk)\Lambda_j(lk)(\alpha_{\parallel} - \alpha_{\perp}) \\ &= \alpha_{\perp}\delta_{ij} + 3\alpha_{\text{av}}\kappa\Lambda_i(lk)\Lambda_j(lk)\end{aligned} \tag{4.35}$$

Here δ_{ij} is the Kronecker symbol and the polarizability anisotropy κ and α_{av} has been previously defined. Differentiation of (4.35) yields

$$\begin{aligned}\left(\frac{\partial\alpha_{ij}(lk)}{\partial S_m}\right)_0 &= \left(\frac{\partial\alpha_{ij}(lk)}{\partial\Lambda_i(lk)}\right)_0 \frac{\partial\Lambda_i(lk)}{\partial S_m} + \left(\frac{\partial\alpha_{ij}(lk)}{\partial\Lambda_j(lk)}\right)_0 \frac{\partial\Lambda_j(lk)}{\partial S_m} \\ &= 3N^{-1/2}\alpha_{\text{av}}\kappa[T''_{m,ik}\Lambda_j^0(lk) + T''_{m,jk}\Lambda_i^0(lk)]\end{aligned} \tag{4.36}$$

The transformation $\mathbf{T}''$ connects the symmetry coordinates with the direction cosine librational displacement coordinates $\lambda_i(lk)$ as before. We are now in a position to calculate the polarizability derivative tensor components if the symmetry coordinates S_m or the normal coordinates Q_n are known in terms of the librational displacement coordinates.

For single crystal samples, the crystal-fixed coordinate system is identical to the laboratory system. Therefore the intensities of the Raman scattered light in the various polarization components can be obtained directly in terms of the polarizability component derivatives from Table II of Brith, Ron, and Schnepp (1969) (reproduced from Wilson, Decius and Cross, 1955). If, however, a randomly oriented powder is investigated, the results must be averaged over orientations, a procedure also described in these two references. If the derivatives have been calculated with respect to the symmetry coordinates S_m, the derivatives with respect to the normal coordinates Q_n are given in analogy to (4.10). Finally, the observed intensity is obtained by substitution in the equation as follows (Woodward, 1967):

$$I_f = \frac{2\pi^3\nu^4}{c^3} \cdot N\alpha_{\text{av}}{}^2\kappa^2\left[1 - \exp\left(-\frac{h\nu}{kT}\right)\right]^{-1} \cdot \frac{h}{\Delta\nu \cdot 8\pi^2 I}\left(\frac{\partial\alpha_{FF'}}{\partial Q_n}\right)^2 \tag{4.37}$$

Here ν is the frequency of the exciting light and $\Delta\nu$ the Raman shift or the frequency of the vibration. The factor containing $\Delta\nu$ in the denominator is the harmonic oscillator matrix element squared.

D. Applications and Comparison with Experiment

1. *α-Nitrogen*

The absolute infrared intensities of the two translational lattice mode lines at 49 and 70 cm^{-1} were calculated by Ron and Schnepp (1967) using the quadrupole-induced theory of Schnepp (1967). These authors made crude estimates of the observed intensities and found them to be of the correct order of magnitude. The intensities were subsequently carefully measured by St. Louis and Schnepp (1969). They formed the solid sample by cooling a liquid sample in an enclosed cell to obtain first the higher temperature phase β-N_2. The lower temperature phase α-N_2 is stable below 36°K. The calculated and measured intensities are given in Table VII. The accuracy of the measurements is 25%. Since both active modes belong to the representation T_u of the factor group T_h^6, the sum of their intensities is to be compared to theoretical prediction to test the intensity theory. Their calculated ratio is strongly dependent on the eigenvectors which are calculated from the lattice dynamics problem. This has been discussed in Section IVB.3.

TABLE VIII
Infrared Intensities of the Translational Optic Modes of Solid α-N_2 at 12°K (St. Louis and Schnepp, 1969)

Lattice mode	Frequency (cm^{-1})	$\int k(\tilde{\nu})\, d\tilde{\nu}$ ($cm^{-2} \cdot atm^{-1}$)	$I(Q_1) + I(Q_2)$	
			Exptl.	Calc.
$Q_1(T_u)$	49	0.019 ± 0.005		
$Q_2(T_u)$	70	0.023 ± 0.005	0.040	0.16[a]

[a] The value given by Ron and Schnepp (1967) must be divided by the refractive index of the solid, taken as 1.25.

The intensity measurements as given in Table VIII were carried out at 12°K. Some uncertainty must be attached to the results in view of their temperature dependence. It is expected on the basis of group theoretical considerations that the infrared absorptions would disappear at the phase transition at 36°K. This prediction was verified by experiment (St. Louis and Schnepp, 1969). However, these authors also found that the integrated absorption decreased steeply with increasing temperature long before reaching the phase transition. In fact, the intensity of the 49 cm^{-1} line was found to decrease by 20% between 12 and 20°K and by a further 50% of the peak intensity between 20 and 30°K. The extrapolated intensity versus temperature curve crosses the zero line at 32°K. This temperature dependence could not be accounted for on the basis of thermal expansion of the lattice, even though the quadrupole-induced intensity depends on the inverse tenth power of the lattice parameter. It was suggested by St. Louis and Schnepp (1969) that the decrease in intensity occurs due to increasing population of excited librational states. The anisotropic part of the intermolecular potential is expected to be small (of the order of 15% of the crystal energy) and therefore the barrier to molecular rotation is expected to be about 100 cm^{-1}. Increased orientational freedom would cause a decrease of the effective molecular quadrupole moment resulting in lower induced intensities compared to those predicted by assuming fixed molecular axes (Schnepp, 1967).

As seen from Table VIII, the observed total intensity is lower than the calculated value by a factor of at least 4. It was not considered reasonable that neglect of the distortion contribution to the transition moment could account for this discrepancy; such neglect might be expected to predict a still higher intensity. It was suggested that the cause for the disagreement may be found at least partly in residual orientational disorder even at low temperature, similar to that discussed in connection with the temperature dependence of the intensity in the preceding paragraph. Another possible weakness of the theory is the assumption that the induced intensity is attributed wholly to the molecular quadrupole field. It is, however, not likely that this approximation would be better for solid CO_2 for which very good agreement between theory and experiment has been reported (Kuan, 1969; Brown and King, 1970).

The theoretical formulation of the Raman intensities of the librational modes of linear molecules in a crystal presented in Section IVC were applied to α-N_2 by Cahill and Leroi (1969a) and by Brith, Ron, and Schnepp (1969). Both these groups obtained the result that the scattering power of the E_g mode should be twice the sum of the scattering powers of the two T_g modes. The scattering power is the square of the polarizability derivative element and does not include the other factors in (4.37). Since only two of the three expected lines were observed, it was important to find an assignment that would, at least approximately, satisfy this condition. Brith, Ron, and Schnepp (1969) and also Cahill and Leroi (1969a) suggested on this basis that the Raman line at 31.5 cm^{-1} represents the E_g mode and in addition contains T_g intensity, i.e., that one of the T_g modes is accidentally closely degenerate with the E_g mode. More recently, Anderson, Sun, and Donkersloot (1970) have found a very weak Raman line at 60 cm^{-1} and these authors found that the intensity relations given by Brith, Ron, and Schnepp (1969) are very accurately satisfied by assigning this line as the second T_g mode. Anderson and co-workers used a quadrupole-quadrupole potential to calculate their eigenvectors. It should be noted that the Raman intensity theory cannot be tested independently of the lattice dynamics problem even if we consider sums of intensities of modes of the same symmetry. This is due to the fact that the theoretically predicted intensity contains the frequency of the Raman displacement as seen from (4.37). This is not so for infrared intensities since there the frequency which appears in the harmonic oscillator matrix element is cancelled by the frequency factor in the expression of the integrated absorption coefficient in terms of the dipole transition matrix element [(4.7) and (4.8)].

2. *Carbon Dioxide*

The absolute infrared absorption intensities of the two translational mode lines at 68 and 114 cm^{-1} were calculated by Ron and Schnepp (1967) using the quadrupole induction theory. These authors also found the predicted intensities to be in order-of-magnitude type agreement with rough measurements. Brown and King (1970) and Kuan (1969) made

TABLE IX
Infrared Intensities of the Translational Optic Modes of Solid CO_2

Lattice modes	Frequency (cm^{-1})	$\int k(\tilde{\nu})\, d\tilde{\nu}$ ($cm^{-2} \cdot atm^{-1}$)		$I(Q_1) + I(Q_2)$		
				Exptl.		
		B and K^a (33°K)	K^b (7°K)	B and K^a	K^b	Calc.[c]
$Q_1(T_u)$	68	1.16	0.75	2.93	4.07	2.81
$Q_2(T_u)$	114	1.77	3.32			

[a] Brown and King, 1970.
[b] Kuan, 1969.
[c] Ron and Schnepp, 1967. The value given there is divided by the refractive index of the solid, 1.44.

absolute intensity measurements using different sample preparation techniques. The former authors deposited CO_2 gas on a cold substrate at 35 or 77°K, without annealing. Kuan, on the other hand, grew his solid sample from the vapor in a closed cell under equilibrium conditions. Table IX lists the experimental and theoretical intensities. It is seen that the total intensities observed by both groups are in fairly good agreement with the theoretical prediction. Brown and King's "selected value" is a perfect fit within the experimental error limits, whereas Kuan's value is about 45% too high. (It should be noted that the calculated values given by Ron and Schnepp, 1967, must be divided by the refractive index or 1.44). However, the fluctuations in the data of Brown and King (their Table I) are such as to minimize the significance of this difference. On the other hand, the intensity ratios are significantly different (1.54 and 4.4, respectively). Furthermore, Brown and King reported no temperature dependence of the intensity, whereas Kuan observed a decrease by almost 50% between liquid helium temperature and 100°K. It should be noted that these apparently different findings may be reconciled since Brown and King made measurements only at 33 and 77°K, in which range Kuan found an intensity decrease by only 25%. Moreover, the line widths observed by these two groups differ by a factor of 2 (Kuan's being the smaller), testifying to the significance of the difference of sample preparation technique. In any case, we can conclude that the quadrupole induction theory is entirely successful for the translational lattice modes of CO_2.

The absolute Raman intensities of CO_2 have been measured by Cahill and Leroi (1969a) by observing the lattice modes and an intramolecular mode during the same experiment. This procedure produced relative measurements which were calibrated by using absolute gas phase measurements for the CO_2 internal mode available in the literature. It was found that the observed intensities agreed very well with those predicted by a theory similar to that described in Section IV.C.

3. *Nitrogen Dioxide*

Cahill and Leroi (1969a) have discussed the absolute Raman intensities of this solid in the lattice vibration region. They found here as in the case of solid CO_2 that agreement between experimental and calculated absolute intensities was excellent.

4. *Solid Hydrogen*

The infrared absorption intensities of the translational optic modes of cubic ortho-hydrogen and of cubic para-deuterium (below 1.5°K) were measured by Hardy et al. (1968). These authors also calculated the predicted quadrupole-induced intensities from the theory of Schnepp (1967).

These solids were assumed to have a structure analogous to α-N_2 and CO_2 except that the molecular axes are not fixed in space but undergo practically free rotation. The space diagonal of the cubic cell is here the principal symmetry axis and serves as axis of quantization for the rotational angular momentum in the $J = 1$ state. Since at low temperature all molecules are in the state $M_J = 0$, the expectation value of the molecular quadrupole moments in this state was used for the intensity calculation. This amounts to a reduction from the static value by a factor of $\frac{2}{5}$. The experimental difficulties precluded precise measurements, but it is seen from Table X that agreement between theory and experiment is within a factor of two for H_2 and very good for D_2. It is seen also that the predicted intensity is greater for H_2 than for D_2. D_2 has a smaller lattice constant (5.081 Å as compared with 5.312 Å for H_2), which leads to a larger induced dipole moment derivative, but the mass appears in the denominator of the harmonic oscillator matrix element and reduces the calculated intensity for D_2 below that of H_2.

TABLE X
Infrared Intensities of the Translational Optic Modes of Cubic Hydrogen at 1.3°K (Hardy et al., 1968)

Substance	Lattice mode	Frequency (cm^{-1})	$\int k(\tilde{\nu})\, d\tilde{\nu}$ ($cm^{-2} \cdot atm^{-1}$)	$I(Q_1) + I(Q_2)$ Exptl.	$I(Q_1) + I(Q_2)$ Calc.[a]
H_2	$Q_1(T_u)$	62.2	5.4	15	34
	$Q_2(T_u)$	80.0	10		
D_2	$Q_1(T_u)$	57.4	8.3		
	$Q_2(T_u)$	74.5	20	28.3	26

[a] No correction was made for the refractive index of the solid.

5. *Cyanogen*

Richardson and Nixon (1968) have measured the relative infrared and Raman intensities of the lattice modes of cyanogen (NCCN), a linear molecule. The six infrared active modes are all translational motions and the quadrupole-induced intensities were calculated using the theory of Schnepp (1967) adapted for a different crystal structure. It was found that the experimental intensities of individual lines could be fitted to the theory within factors of two to three by adjusting the sample thickness, i.e., one parameter. However, if sums of intensities for each symmetry type are compared, to separate the intensity problem from the lattice dynamics problem, much better agreement is obtained, well within a factor of two. The relative Raman intensities of the librational lattice modes were also found to give good agreement between theory and experiment.

6. *Halogen Solids*

David and Person (1968) measured the absolute infrared intensities of the translational lattice modes of the solid halogens. They did not carry out a complete calculation adapted to the crystal structure of these solids. These authors found, however, very good correlation between the measured absolute intensities and the molecular properties, i.e., quadrupole moments and polarizabilities, as expected from the theory.

7. *Carbon Monoxide*

Ron and Schnepp (1967) calculated the quadrupole-induced intensity of solid CO for the translational symmetry coordinates and the "intrinsic" intensity due to the molecular dipole moment for the librational symmetry coordinates. They found that even for such a small dipole moment (0.11 D) the intrinsic intensity outweighs the quadrupole-induced intensity by more than a factor of two. However, the total integrated intensity was in agreement with experiment. The experimental uncertainty was here considerable, amounting to 200%, since the sample thickness was uncertain.

8. *ClCN and BrCN*

Friedrich (1970) has measured the absolute infrared intensities of the two optic librational modes of the solids of ClCN and BrCN. He also calculated the intrinsic intensities of these modes using a dipolar coupling theory referred to in Section IVB.2. Good agreement was obtained between calculated and measured values. It was found that the local field correction [Eq. (4.22)] causes an increase of about 100% in the theoretical intensities for these substances, whereas the Lorentz effective field correction gave rise to an increase of less than 50%. The samples were prepared by deposition of gas on a cold window and the sample thicknesses were measured by an interference fringe technique.

9. *HCl and HBr*

Carlson and Friedrich (1971) have recently investigated the infrared intensities of the translational and librational modes of HCl and HBr. These authors found that the observed intensities were lower than predicted by about 70% when the gas phase dipole moment was used.

10. *OCS*

Friedrich (1967) measured the absolute infrared intensities of the librational mode of OCS crystal. In his calculation, he introduced an effective field correction factor, which had been determined by other authors by fitting to the observed infrared intensity of the intramolecular bending mode in the solid. Agreement between measured and calculated intensities was then excellent.

E. Discussion of Results

From the experimental results described here and their comparison to theory, we can say that generally the theory may serve as a useful guide. Calculations have been successful in predicitng the quadrupole-induced intensities of translational modes with fair accuracy. As a result, such calculations can be very useful in planning experimental measurements. It was originally thought that intensity measurements may be useful to determine molecular quadrupole moments. Such determinations would be very desirable since only few may be really accepted as well-established, namely, those for H_2, N_2, and CO_2. For solid H_2 and CO_2 we can conclude that agreement between theoretical calculation and experimental measurement of infrared intensities is good, for the latter excellent. In the case of solid α-N_2, an uncertainty still remains, by a factor of four or so. Since the intensity is proportional to the square of the molecular quadrupole moment, we may conclude at this stage that intensity measurements would allow the measurement of the quadrupole moment, even in problematic cases, within a factor of two and often to considerably greater accuracy.

Agreement between theory and experiment for the infrared intensity of librational modes of polar molecules in solids is generally good. In these cases, the local field correction is considerable and must be taken into account. This presents a serious problem and some uncertainties remain in the methods of dealing with it.

The calculated relative Raman intensities agree very well with experiment and for CO_2 and NO_2 the agreement has been shown to be excellent even for the absolute Raman intensities (Cahill and Leroi, 1969a).

APPENDIX. GROUP THEORY AND SYMMETRY IN THE BRILLOUIN ZONE AS APPLIED TO LATTICE DYNAMICS OF MOLECULAR SOLIDS

I. Introduction

Several good sources treat space groups and group theory of the Brillouin zone (Koster, 1957; Slater, 1965). In particular, applications of lattice dynamics have been treated by Maradudin and Vosko (1968) and by Montgomery (1969). Venkataraman and Sahni (1970) have included a discussion of symmetry in their review. In recent years, several books have appeared which contain comprehensive listings of the irreducible representations of space groups (Zak, 1969; Slater, 1965; Kovalev, 1965; Miller and Love, 1967).

It is our aim to give a simple and clear description of some of the principles with emphasis on application, rather than on rigorous and complete theory. In particular, we are concerned here with the determination of the symmetry species of lattice vibrations at symmetry points and along symmetry directions inside and on the surface of the Brillouin zone. Compatibility relations or correlations between irreducible representations at different symmetry points also are discussed. We emphasize the illustration of the principles by their application to two molecular crystals. The first example, naphthalene, is a monoclinic crystal of space group $C_{2h}^5(P2_1/c)$ and the second example, carbon dioxide is a cubic crystal belonging to $T_h^6(Pa3)$. These two space groups are representative of a considerable number of molecular solids. The reader who follows the principles described in this appendix should experience no difficulties in carrying out similar operations for other space groups with the aid of tables of the irreducible representations.

The space groups are listed and described in the *International Tables for X-Ray Crystallography* (1962). For our purposes, we are interested only in the "primitive unit cell," or the smallest unit cell that can be used to reproduce the crystal by means of translations only. The primitive unit cell often has lower symmetry than the conventional unit cell (Kittel, 1968). According to the international nomenclature for space groups, the letter P denotes a primitive lattice. If we deal with such a crystal, the conventional unit cell is the primitive cell. If, however, the group designation begins with another letter (e.g., I, F, C), then the primitive cell has to be determined and this is smaller and less symmetric than the conventional unit cell. The unit cell vectors of the primitive cell are called primitive translation vectors.

The symmetry operations of the space group, apart from primitive translations, may include only rotations and reflections (proper and improper rotations) including inversion in a center of symmetry. Such a group is called *symmorphic*. Space groups of interest include, in addition, screw axes and glide planes. These operations require nonprimitive translations to be combined with rotations and reflections, respectively. These translations are submultiples of the primitive translation vectors. It is important to choose one origin for all operations, since the nonprimitive translations are dependent on this choice, but none of the results discussed here depend on the particular choice of origin. Zak (1969) also showed that the characters of the irreducible representations are independent of the particular choice of origin.

In order to construct the reducible representation of the translations or of the librations, at a particular point in the Brillouin zone (for a particular wave vector $\mathbf{q}$), we must investigate the transformations of the basis functions under the pertinent symmetry operations. As basis functions we use the "translationally symmetrized" functions which already belong to irreducible representations of the group of pure translations by integral lattice vectors. The basis functions were given in Eq. (2.11). They are

$$u_j(\mathbf{q}k) = N^{-1/2} \sum_l \exp i[\mathbf{q} \cdot \mathbf{R}(lk)] u_j(lk) \tag{A.1}$$

Here $u_j(lk)$ designates the j-component of displacement of the molecule at the k-site in the l-cell, whose position vector is $\mathbf{R}(lk)$. The wave vector $\mathbf{q}$ is a vector in the inverse lattice.

The basis vectors of the inverse lattice $\mathbf{b}_1$, $\mathbf{b}_2$, $\mathbf{b}_3$ are defined in terms of the basis vectors of the lattice $\mathbf{a}_1$, $\mathbf{a}_2$, $\mathbf{a}_3$ as follows:

$$\mathbf{b}_1 = 2\pi \frac{\mathbf{a}_2 \times \mathbf{a}_3}{[\mathbf{a}_1 \mathbf{a}_2 \mathbf{a}_3]}$$

and similar relations for $\mathbf{b}_2$, $\mathbf{b}_3$. For an orthogonal lattice, the preceding equation reduces to $\mathbf{b}_1 = (2\pi/a_1)\hat{\imath}$, etc., where $\hat{\imath}$ is a unit vector in the direction of $\mathbf{a}$. We have then the important and general relationship

$$\mathbf{a}_i \cdot \mathbf{b}_j = 2\pi\delta_{ij}$$

A vector composed of integral multiples of the inverse lattice basis vectors is called an "inverse lattice vector."

II. Transformation of the Basis Functions

Consider a symmetry operation of the space group, designated by $\{\mathbf{S} \mid \mathbf{v}(S) + \mathbf{T}\}$. The operator $\mathbf{S}$ represents a point group operation executed at the chosen origin. The nonprimitive translation associated with $\mathbf{S}$ is designated $\mathbf{v}(S)$ and $\mathbf{T}$ is a lattice translation vector. The result of this operation on a vector $\mathbf{R}$ of the lattice is defined by

$$\{\mathbf{S} \mid \mathbf{v}(S) + \mathbf{T}\}\mathbf{R} = \mathbf{SR} + \mathbf{v}(S) + \mathbf{T} \tag{A.2}$$

The set of operators $\mathbf{S}$ form what is called the point group of the space group. We shall not be concerned further with the translations $\mathbf{T}$ since we are already dealing with properly symmetrized functions (A.1), which belong to the representation $\mathbf{q}$ of the translation group. We shall therefore consider the effect of applying the operation $\{\mathbf{S} \mid \mathbf{v}(S)\}$ to the $u_j(\mathbf{q}k)$.

Let us consider the result of the application of $\{\mathbf{S} \mid \mathbf{v}(S)\}$ to one of the terms of the sum over the unit cells l in $u_j(\mathbf{q}k)$, say $u_j(lk) \exp i[\mathbf{q} \cdot \mathbf{R}(lk)]$. We note that the operation will transform the molecule (lk) into another molecule $(l'k')$. Moreover, $u_j(lk)$ will generally be transformed into a linear combination of displacements of the new molecule $u_{j'}(l'k')$. However, the exponent is a scalar and remains unchanged. Thus

$$\{\mathbf{S} \mid \mathbf{v}(S)\}u_j(lk) \exp i[\mathbf{q} \cdot \mathbf{R}(lk)] = \sum_{j'} C_{jj'}(S)u_{j'}(l'k') \exp i[\mathbf{q} \cdot \mathbf{R}(lk)] \tag{A.3}$$

For example, if $\mathbf{S}$ is a twofold rotation about a coordinate axis, then $u_x(lk)$ will transform into $\pm u_x(l'k')$ and the sum over j' contains only a single term with $C_{jj'}(S) = \pm 1$ depending on the rotation axis. On the other hand, a three-fold rotation may transform $u_x(lk)$ into a linear combination of $u_x(l'k')$ and $u_y(l'k')$ with coefficients $C_{jj'}(S)$ forming the corresponding rotation matrix.

Now we note that all terms $u_j(lk)$ transform into terms $u_{j'}(l'k')$, or all terms occurring in the basis function $u_j(\mathbf{q}k)$ transform into combinations of terms occurring in the $u_{j'}(\mathbf{q}k')$. A particular coefficient $C_{jj'}$ is independent of l and therefore is common to all terms in $u_{j'}(\mathbf{q}k')$.

Next we investigate the effect of the transformation on the exponential factor. In $u_{j'}(\mathbf{q}k')$, the term $u_{j'}(l'k')$ is multiplied by the factor

$$\exp i[\mathbf{q} \cdot \mathbf{R}(l'k')]$$

We now determine the factor by which we must multiply the term resulting from the transformation to convert it to the term appearing in $u_{j'}(\mathbf{q}k')$. From (A.3) we find

$$\begin{aligned}\{\mathbf{S} \mid \mathbf{v}(S)\}u_j(lk) &\exp i[\mathbf{q} \cdot \mathbf{R}(lk)] \\ &= \sum_{j'} C_{jj'}(S)u_{j'}(l'k') \exp i[\mathbf{q} \cdot \mathbf{R}(lk)] \\ &= \sum_{j'} C_{jj'}(S)u_{j'}(l'k') \exp i[\mathbf{q} \cdot \mathbf{R}(l'k')] \exp i\mathbf{q} \cdot [\mathbf{R}(lk) - \mathbf{R}(l'k')] \quad \text{(A.4)}\end{aligned}$$

Therefore, apart from the factor $C_{jj'}(S)$, as a result of the transformation the particular term in $u_{j'}(\mathbf{q}k')$ has been effectively multiplied by the factor $\exp i\mathbf{q} \cdot [\mathbf{R}(lk) - \mathbf{R}(l'k')]$. But we know that

$$\mathbf{R}(l'k') = \{\mathbf{S} \mid \mathbf{v}(S)\}\mathbf{R}(lk)$$

and therefore

$$\begin{aligned}\exp i\mathbf{q} \cdot [\mathbf{R}(lk) - \mathbf{R}(l'k') &= \exp i\mathbf{q} \cdot [\mathbf{R}(lk) - \{\mathbf{S} \mid \mathbf{v}(S)\}\mathbf{R}(lk)] \\ &= \exp i\mathbf{q} \cdot \mathbf{R}(lk) \exp i\mathbf{q} \cdot [-\mathbf{S}\mathbf{R}(lk) - \mathbf{v}(\mathbf{S})]\end{aligned} \quad \text{(A.5)}$$

The last equality was obtained using the definition (A.2). It can further be shown that

$$\mathbf{q} \cdot [\mathbf{S}\mathbf{R}(lk)] = [\mathbf{S}^{-1}\mathbf{q}] \cdot \mathbf{R}(lk) \quad \text{(A.6)}$$

and then from (A.5) we obtain

$$\begin{aligned}\exp i\mathbf{q} \cdot [\mathbf{R}(lk) - \mathbf{R}(l'k')] &= \exp i[\mathbf{q} - S^{-1}\mathbf{q}] \cdot \mathbf{R}(lk) \exp [-i\mathbf{q}\mathbf{v}(S)] \\ &\equiv f_k(S, \mathbf{q})\end{aligned} \quad \text{(A.7)}$$

If $f_k(S, \mathbf{q})$, as we have named the expression given in (A.7), were independent of the unit cell l, we could describe the transformation of $u_j(\mathbf{q}k)$ as follows:

$$\{\mathbf{S} \mid \mathbf{v}(S)\}u_j(\mathbf{q}k) = f_k(S, \mathbf{q}) \sum_{j'} C_{jj'}(S)u_{j'}(\mathbf{q}k') \quad \text{(A.8)}$$

Such a transformation behavior would generate the representation spanned by the basis functions $u_j(\mathbf{q}k)$. It is therefore necessary to investigate the requirements for $f_k(S, \mathbf{q})$ to be independent of l.

For $\mathbf{q} = 0$, usually called point Γ in the Brillouin zone, it follows from (A.7) that $f_k(S, \mathbf{q})$ is unity for all space group operations $\{\mathbf{S} \mid \mathbf{v}(S)\}$. Therefore all operations transform the basis functions according to (A.8) and the symmetry group of point Γ is the full point group of the space group.

The reader interested only in symmetry properties at $\mathbf{q} = 0$, need not continue further in this section and can proceed to Section III.

We now consider values of $\mathbf{q}$ different from zero. For $\{\mathbf{S} \mid \mathbf{v}(S)\}$ being the identity operation $\mathbf{E}$ [of course for $\mathbf{S} = \mathbf{E}$, $\mathbf{v}(\mathbf{E}) = 0$] we obtain from (A.7) that $f_k(S, \mathbf{q})$ is unity for all $\mathbf{q}$ since $\mathbf{S}^{-1}\mathbf{q} = \mathbf{q}$ independently of the value of $\mathbf{q}$. For any point group operation $\mathbf{S}$ for which $\mathbf{q} = \mathbf{S}\mathbf{q}$ (it follows that also $\mathbf{S}^{-1}\mathbf{q} = \mathbf{q}$), we find that

$$f_k(S, \mathbf{q}) = \exp\,[-i\mathbf{q} \cdot \mathbf{v}(S)] \tag{A.9}$$

For such a case, therefore, $f_k(S, \mathbf{q})$ is again independent of l. It is now a root of unity, but not necessarily unity. In fact, the value of $f_k(S, \mathbf{q})$ depends on $\mathbf{q}$ and on the nonprimitive translation $\mathbf{v}(S)$ associated with $\mathbf{S}$. We also note that in this case $f_k(S, \mathbf{q})$ is independent of the site k. The point group symmetry operations which transform $\mathbf{q}$ into itself or which satisfy the requirement $\mathbf{q} = S\mathbf{q}$ are rotations about an axis coincident with $\mathbf{q}$ and reflections in planes containing $\mathbf{q}$. These are the valid symmetry operations only for values of $\mathbf{q}$ inside the Brillouin zone.

For $\mathbf{q}$ corresponding to points on the surface of the first Brillouin zone, additional symmetry operations can be found which satisfy (A.8) or for which $f_k(S, \mathbf{q})$ is independent of l. These are proper and improper rotations which change $\mathbf{q}$ by a vector of the reciprocal lattice. For such an operation we then have

$$\mathbf{q} - S^{-1}\mathbf{q} = n_1\mathbf{b}_1 + n_2\mathbf{b}_2 + n_3\mathbf{b}_3 \tag{A.10}$$

where n_1, n_2, n_3 are zero or unity and $\mathbf{b}_1$, $\mathbf{b}_2$, $\mathbf{b}_3$ are the basis vectors of the reciprocal lattice as previously defined. For example, consider the monoclinic crystal lattice of naphthalene, C_{2h}^5. Following Zak (1969), Kovalev (1965), and Miller and Love (1967), we choose the z-axis as the unique axis. Then the x- and y-axes define a plane perpendicular to z, although they are not perpendicular to each other. Then

$$\mathbf{b}_3 = \frac{2\pi}{a_3}\,\hat{k}$$

The Brillouin zone is, as usual, taken from $-\pi/a_3$ to π/a_3 in the z-direction (unit vector $\hat{k}$). The point $(0, 0, \pi/a_3)$ is called point Z, and the corresponding wave vector is

$$\mathbf{q} = \frac{\pi}{a_3}\,\hat{k}$$

Clearly, the point group operations $\mathbf{E}$ and $\mathbf{C}_2{}^z$ transform $\mathbf{q}$ into itself. However, the inversion $\mathbf{1}$ and the reflection in the xy-plane σ^z transforms $\mathbf{q}$ into $-(\pi/a_3)\hat{k}$, or

$$\mathbf{q} - S^{-1}\mathbf{q} = \frac{2\pi}{a_3}\,\hat{k} \tag{A.11}$$

Substituting this result in (A.7), we obtain

$$f_k(S, \mathbf{q}) = \exp 2\pi i\left[\left(\frac{\hat{k}}{a_3}\right) \cdot \mathbf{R}(lk)\right] \exp\left[-i\mathbf{q} \cdot \mathbf{v}(S)\right] \tag{A.12}$$

The molecular sites are chosen as (0 0 0) and ($\frac{1}{2}$ 0 $\frac{1}{2}$), corresponding to the sites $k = 1$ and 2, respectively. The value of the first factor in (A.12) is different for the two sites, since the position vectors are

$$\begin{aligned} \mathbf{R}(l1) &= l_1 a_1 \hat{\imath} + l_2 a_2 \hat{\jmath} + l_3 a_3 \hat{k} \\ \mathbf{R}(l2) &= (l_1 + \tfrac{1}{2}) a_1 \hat{\mathbf{i}} + l_2 a_2 \hat{\jmath} + (l_3 + \tfrac{1}{2}) a_3 \hat{k} \end{aligned} \tag{A.13}$$

Here l_1, l_2, l_3 are integers designating a specific cell l. Substituting in (A.12) we find that the first exponential in $f_k(S, \mathbf{q})$ as given in (A.12) has the value

$$\begin{aligned} \exp 2\pi i l_3 &= 1 && \text{for } k = 1 \\ \exp 2\pi i(l_3 + \tfrac{1}{2}) &= \exp(\pi i) = -1 && \text{for } k = 2 \end{aligned} \tag{A.14}$$

These values are independent of l and, as a result, we now can summarize the results as follows for point Z:

$\mathbf{S} = \mathbf{E}, \mathbf{C}_2^z$:

$$f_k(S, \mathbf{q}) = \exp\left[-i\mathbf{q} \cdot \mathbf{v}(S)\right], \qquad k = 1, 2$$

$\mathbf{S} = \mathbf{I}, \boldsymbol{\sigma}^z$:

$$\begin{aligned} f_1(S, \mathbf{q}) &= \exp\left[-i\mathbf{q} \cdot \mathbf{v}(S)\right] \\ f_2(S, \mathbf{q}) &= -\exp\left[-i\mathbf{q} \cdot \mathbf{v}(S)\right] \end{aligned} \tag{A.15}$$

We now need to consider the nonprimitive translations $\mathbf{v}(S)$ associated with $\{\mathbf{S} \mid \mathbf{v}(S)\}$. The $\mathbf{v}(S)$ must be consistent with the molecular sites and with the choice of origin.

Zak (1969) and Miller and Love (1967) choose as origin a center of symmetry and, as indicated, we have chosen a molecular site ($k = 1$) at the origin. Then the symmetry operations* are as follows:

$$\mathbf{E}, \mathbf{I}, \{\mathbf{C}_2^z \mid \tfrac{1}{2}\, 0\, \tfrac{1}{2}\}, \qquad \{\boldsymbol{\sigma}^z \mid \tfrac{1}{2}\, 0\, \tfrac{1}{2}\}$$

These designations of the operations define the nonprimitive translations $\mathbf{v}(S)$ in terms of the lattice basis vectors $\mathbf{a}_1$, $\mathbf{a}_2$, $\mathbf{a}_3$. For example, we have

$$\begin{aligned} \mathbf{S} &= \mathbf{E}, \mathbf{I}, && \mathbf{v}(S) = 0 \\ \mathbf{S} &= \mathbf{C}_2^z\ \sigma^z, && \mathbf{v}(S) = \frac{a_1}{2}\hat{\imath} + \frac{a_3}{2}\hat{k} \end{aligned} \tag{A.16}$$

Substitution in (A.15) gives the results listed in Table A.I.

* These are the operations listed by Miller and Love (1967). Zak (1969) seems to have a misprint. Kovalev (1965) chose as origin ($\frac{1}{4}$, 0, $\frac{1}{4}$) relative to the origin chosen here.

TABLE A.I
Values of the Factor $f_k(S, \mathbf{q})$ defined by Eq. (A.7) for the Point Z of the Brillouin Zone of the Naphthalene Crystal. Space Group $C_{2h}^5(P2_1/c)$

Symmetry operation	$f_1(S, \mathbf{q})$	$f_2(S, \mathbf{q})$
$\mathbf{E}$	1	1
$\mathbf{I}$	1	-1
$\{\mathbf{C}_2{}^z \mid \frac{1}{2}\,0\,\frac{1}{2}\}$	$-i$	$-i$
$\{\boldsymbol{\sigma}^z \mid \frac{1}{2}\,0\,\frac{1}{2}\}$	$-i$	i

III. Determination of Symmetry Species of Lattice Vibrations and Application to Solid Naphthalene

Before we can construct the reducible representation spanning the lattice vibrations of naphthalene, we still require the transformation coefficients $C_{jj'}(S)$ of (A.8). We begin with the Γ-point, or $\mathbf{q} = 0$.

We note that the operations $\mathbf{E}$, $\mathbf{I}$ transform molecules of like site into each other, whereas $\{\mathbf{C}_2 \mid \frac{1}{3}\,0\,\frac{1}{2}\}$ and $\{\boldsymbol{\sigma}^z \mid \frac{1}{2}\,0\,\frac{1}{2}\}$ interchange sites. The u_j for $j = x, y, z$ (translational displacements) transform like polar vector components, whereas for $j = \alpha, \beta, \gamma$ (librational displacements; see Section IIC.6) the u_j transform like axial vector components. We then obtain, for example, the following transformations in the form of (A.8) for $\mathbf{q} = 0$ [the factors $f_k(S, \mathbf{q})$ and $C_{jj'}$ are shown]:

$$\begin{aligned}
\{I\}u_x(\mathbf{0}k) &= (1)(-1)u_x(\mathbf{0}k) = -u_x(\mathbf{0}k), \qquad k = 1, 2 \\
\{\mathbf{C}_2{}^z \mid \tfrac{1}{2}\,0\,\tfrac{1}{2}\}u_x(\mathbf{0}1) &= (1)(-1)u_x(\mathbf{0}2) = -u_x(\mathbf{0}2) \\
\{I\}u_\alpha(\mathbf{0}k) &= (1)(1)u_\alpha(\mathbf{0}k) = u_\alpha(\mathbf{0}k), \qquad k = 1, 2 \\
\{\mathbf{C}_2{}^z \mid \tfrac{1}{2}\,0\,\tfrac{1}{2}\}u_\alpha(\mathbf{0}1) &= (1)(-1)u_\alpha(\mathbf{0}2) = -u_\alpha(\mathbf{0}2)
\end{aligned} \tag{A.17}$$

On the other hand, for the Z-point, discussed in the previous section, we obtain the following results using Table A.I:

$$\begin{aligned}
\{I\}u_x(\mathbf{q}1) &= (1)(-1)u_x(\mathbf{q}1) = -u_x(\mathbf{q}1) \\
\{I\}u_x(\mathbf{q}2) &= (-1)(-1)u_x(\mathbf{q}2) = u_x(\mathbf{q}2) \\
\{\mathbf{C}_2{}^z \mid \tfrac{1}{2}\,0\,\tfrac{1}{2}\}u_x(\mathbf{q}1) &= (-i)(-1)u_x(\mathbf{q}2) = iu_x(\mathbf{q}2) \\
\{\sigma^z \mid \tfrac{1}{2}\,0\,\tfrac{1}{2}\}u_x(\mathbf{q}1) &= (-i)(1)u_x(\mathbf{q}2) = -iu_x(\mathbf{q}2) \\
\{\sigma^z \mid \tfrac{1}{2}\,0\,\tfrac{1}{2}\}u_x(\mathbf{q}2) &= (i)(1)u_x(\mathbf{q}2) = iu_x(\mathbf{q}2)
\end{aligned} \tag{A.18}$$

The parallel results are easily obtained for $j = y, z$ and for $j = \alpha, \beta, \gamma$ since the $f_k(S, \mathbf{q})$ remain the same and the $C_{jj'}(S)$ depend on simple geometric considerations only.

TABLE A.II

Character Table for the Γ-Point of $C_{2h}^5(P2_1/c)$ (The character of the reducible representations spanning the translations and the librations Γ_{trans} and Γ_{libr} of the naphthalene crystal are included)

Γ-Point	E	$\{C_2^z \mid \frac{1}{2}\,0\,\frac{1}{2}\}$	I	$\{\sigma^z \mid \frac{1}{2}\,0\,\frac{1}{2}\}$
Γ_1^+	1	1	1	1
Γ_1^-	1	1	−1	−1
Γ_2^+	1	−1	1	−1
Γ_2^-	1	−1	−1	1
Γ_{trans}	6	0	−6	0
Γ_{libr}	6	0	6	0

We are now ready to construct the reducible representations spanned by the translations and by the librations separately at the Γ-point ($\mathbf{q} = 0$). Since there are two molecules in the unit cell, both representations are six-dimensional, and we require only the characters of the matrix traces. Clearly, $\{C_2^z \mid \frac{1}{2}\,0\,\frac{1}{2}\}$ and $\{\sigma^z \mid \frac{1}{2}\,0\,\frac{1}{2}\}$ have only zeroes on the diagonal since they exchange sites and therefore the corresponding traces are zero. The matrix of **E** has all 1 on the diagonal and therefore the trace is 6, for both Γ_{trans} and Γ_{libr}. The traces of I are −6 and 6 for Γ_{trans} and Γ_{libr}, respectively, as is easily concluded from the summary given in (A.17). The character table for the Γ-point is presented in Table A.II. This is simply the character table of the point group C_{2h}. The characters of the reducible representations Γ_{trans} and Γ_{libr} for the naphthalene crystal are included in the table.

Table A.III contains the character table for point Z of C_{2h}^5. This little group has only one two-dimensional representation and we therefore expect all eigenvalues at that point to be doubly degenerate. Using the transformations summarized in (A.18) and similar additional relations, we construct the characters of the reducible representations Z_{trans} and

TABLE A.III

Character Table for the Z-Point of $C_{2h}^5(P2_1/c)$ (The characters of the reducible representations spanning the translations and the librations Γ_{trans} and Γ_{libr} of the naphthalene crystal are included)

Z-Point	E	$\{C_2^z \mid \frac{1}{2}\,0\,\frac{1}{2}\}$	I	$\{\sigma^z \mid \frac{1}{2}\,0\frac{1}{2}\}$
Z	2	0	0	0
Z_{trans}	6	0	0	0
Z_{libr}	6	0	0	0

Z_{libr} for the naphthalene crystal. These have been included in Table A.III.

We now must determine the irreducible representations contained in each of Γ_{trans}, Γ_{libr}, Z_{trans}, and Z_{libr}. We use the following equation (Wigner, 1931, Slater, 1965):

$$n_r = \frac{1}{h} \sum_S \chi(S) \chi^{(r)}(S) \tag{A.19}$$

Here n_r is the number of times representation r is contained in the reducible representation. $\chi(S)$ is the trace corresponding to the group element **S** in the character of the reducible representation and $\chi^{(r)}(S)$ is the trace in representation r. The sum is over all group elements and h is the number of elements in the group.

Application of (A.19) to $\chi(\Gamma_{\text{trans}})$ in Table A.II gives the following results:

$$\begin{aligned} \Gamma_{\text{trans}} &= 3\Gamma_1^- + 3\Gamma_2^- \\ \Gamma_{\text{libr}} &= 3\Gamma_1^+ + 3\Gamma_2^+ \end{aligned} \tag{A.20}$$

We therefore conclude that at the Γ-point the translations are separable by symmetry from the librations since they belong to distinct representations. All eigenvalues at Γ are nondegenerate. The zero-frequency acoustic modes are here contained and span $\Gamma_1^- + 2\Gamma_2^-$. Thus there will be three nonzero frequency translational modes $(2\Gamma_1^- + \Gamma_2^-)$ and six librational modes at $\mathbf{q} = 0$. For the Z-point we obtain

$$\begin{aligned} Z_{\text{trans}} &= 3Z \\ Z_{\text{libr}} &= 3Z \end{aligned} \tag{A.21}$$

Here all modes of the system belong to the same representation and therefore translations and librations are mixed. As already stated, all eigenvalues are doubly degenerate.

When applying (A.19) to a problem concerning the Z-point, we note that it is necessary only to construct $\chi(E)$ in this case, since all other $\chi^{(z)}(S)$ are zeroes. Thus inspection of the applicable character table in advance may often save a great deal of work.

We now consider the symmetry direction Δ which joins the points Γ and Z. The wave vector for any point along this direction has the value

$$\mathbf{q} = \xi \frac{\pi}{a_3} \hat{k}, \qquad 0 < \xi < 1$$

Two point group symmetry operations, **E** and $\mathbf{C}_2{}^z$, transform the vector **q** into itself. As discussed in Section II, for any symmetry point within the zone, only these operations are valid and as a consequence $f_k(S, \mathbf{q})$ is obtained from (A.9) and is independent of the site k. We then have

$$
\begin{aligned}
f_k(S, \mathbf{q}) &= 1 && \text{for } \mathbf{S} = \mathbf{E}, && k = 1, 2 \\
f_k(S, \mathbf{q}) &= \exp\left[-i\xi\frac{\pi}{2}\right] && \text{for } \mathbf{S} = \{\mathbf{C}_2^z \mid \tfrac{1}{2}\, 0\, \tfrac{1}{2}\}, && k = 1, 2
\end{aligned}
\tag{A.22}
$$

The group applicable to all points along Δ is isomorphic with the point group C_2 and there are two one-dimensional irreducible representations. Functions belonging to Δ_1 are symmetric with respect to C_2^z and functions belonging to Δ_2 are antisymmetric. Comparison with Table A.II immediately leads to the conclusion that Γ_1^+ and Γ_1^- correlate with Δ_1 and Γ_2^+ and Γ_2^- correlate with Δ_2. On the other hand, we note that representation Z of the Z-point (Table A-III) correlates with $\Delta_1 + \Delta_2$. To determine this fact, we note that the traces for $\mathbf{E}$ and $\{C \mid \frac{1}{2}\, 0\, \frac{1}{2}\}$ in the character of Z are 2 and 0, respectively. This is a reducible representation in C_2 and reduces as indicated. Translations and librations interact along Δ.

As a result, we can predict the following behavior of the dispersion curves when $\mathbf{q}$ varies from 0 (Γ-point) to $(\pi/a_3)\hat{k}$ along Δ. At Γ, as already noted, there are four representations, translations belong to two of these (Γ_1^-, Γ_2^-) and librations belong to the other two (Γ_1^+, Γ_2^+). But along Δ, translations belonging to Γ_1^- mix with librations belonging to Γ_1^+ since both these representations are compatible with Δ_1. Similar considerations hold for Γ_2^-, Γ_2^+ and Δ_2. Then, as the Z-point is reached, one branch belonging to Δ_1 merges with one branch belonging to Δ_2 to form a doubly degenerate mode belonging to Z. These compatibility relations are summarized in Table A.IV.

It is often useful to construct "symmetry coordinates" for a given symmetry point in the Brillouin zone. This is done by making use of the transformations discussed and the character tables. The applicable formula

TABLE A.IV
Compatibility Relations along Direction Δ of C_{2h}^5

Γ-Point $\mathbf{q} = 0$	Δ-Direction $\mathbf{q} = \xi(\pi/a_3)\hat{k},\ 0 < \xi < 1$	Z-Point $\mathbf{q} = (\pi/a_3)\hat{k}$
Γ_1^+		
	Δ_1	
Γ_1^-		
		Z
Γ_2^+		
	Δ_2	
Γ_2^-		

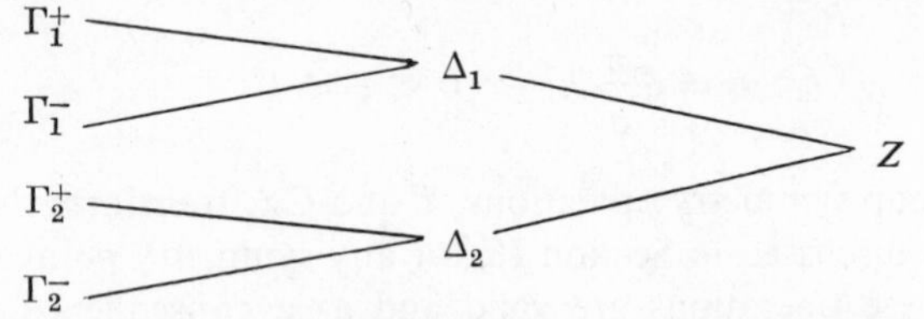

is given by Wigner (1931), Slater (1965), and others:

$$\Phi^{(r)} = \sum_{S} \chi^{(r)}(S)\{\mathbf{S} \mid v(\mathbf{S})\}F \tag{A.23}$$

where F is an arbitrary function which can be operated on by the transformation $\{\mathbf{S} \mid \mathbf{v}(S)\}$. $\Phi^{(r)}$ is then a function that belongs to the irreducible representation r. Further details concerning this procedure are described by Slater (1965). The symmetry coordinates so formed, using one of the basis functions (A.1) as F in (A.23) at a time, cause the factoring of the dynamical equation into diagonal blocks since the interaction terms between symmetry coordinates belonging to different representations vanish.

IV. Application to Solid CO_2

Solid CO_2 belongs to the space group $T_h^6 - Pa3$. The CO_2 molecules form a face-centered cubic lattice (see Table II for molecular positions). However, the molecular symmetry axes of the CO_2 molecules have four distinct directions in space, corresponding to the four body diagonals of the primitive, simple cubic unit cell. The correspondences between positions and axis orientations are also clarified in Table II. The symmetry operations of T_h^6 are listed in Table A.V. The notation is that of Zak (1969).

TABLE A.V
The Symmetry Operations of the Space Group $T_h^6 - Pa3$
(Notation is that of Zak, 1969).

$\mathbf{E}$

$\mathbf{C}_3^{2xyz} \left\{\mathbf{C}_3^{2x\bar{y}\bar{z}} \middle| 0\,\frac{a}{2}\,\frac{a}{2}\right\} \left\{\mathbf{C}_3^{2\bar{x}y\bar{z}} \middle| \frac{a}{2}\,0\,\frac{a}{2}\right\} \left\{\mathbf{C}_3^{2\bar{x}\bar{y}z} \middle| \frac{a}{2}\,\frac{a}{2}\,0\right\}$

$\mathbf{C}_3^{xyz} \left\{\mathbf{C}_3^{x\bar{y}\bar{z}} \middle| \frac{a}{2}\,0\,\frac{a}{2}\right\} \left\{\mathbf{C}_3^{\bar{x}y\bar{z}} \middle| \frac{a}{2}\,\frac{a}{2}\,0\right\} \left\{\mathbf{C}_3^{\bar{x}\bar{y}z} \middle| 0\,\frac{a}{2}\,\frac{a}{2}\right\}$

$\left\{\mathbf{U}^x \middle| \frac{a}{2}\,\frac{a}{2}\,0\right\} \left\{\mathbf{U}^y \middle| 0\,\frac{a}{2}\,\frac{a}{2}\right\} \left\{\mathbf{U}^z \middle| \frac{a}{2}\,0\,\frac{a}{2}\right\}$

and all products of these operations with $\mathbf{I}$

The twofold axes are designated $\mathbf{U}$ with the axis x, y, or z as superscripts. The threefold axes are designated $\mathbf{C}_3$ with the axis direction specified by the superscript. Thus $C^{\bar{x}\bar{y}z}$ specifies an axis parallel to $[-1\ -1\ 1]$ through the origin. The origin (000) is chosen at the site of a molecule of type 1 (Table II) whose orientation is [111]. We see that the four trigonal axes are parallel to the four body diagonals of the cube and therefore to the symmetry axes of the four types of molecule. A unit cell is made up of four molecules, each belonging to a different type. The character table for the Γ-point is given in Table A.VI. This is the character table of point group T_h. The character table for the R-point is given in Table A.VII.

We can now use the procedures already outlined to determine the transformation properties of the basis functions (A.1) in this crystal. As

TABLE A.VI

Character Table for the Γ-Point ($\mathbf{q} = 0$) of $T_h^6 - Pa3$ [The designations of the representations according to various authors are given. The characters for operations $\{\mathbf{S} \mid \mathbf{v}(S)\} \cdot \mathbf{I}$ are not given but are obtained by multiplying the characters of the $\{\mathbf{S} \mid \mathbf{v}(S)\}$ by ± 1. The full operation designations are given in Table A.V. $\omega = \exp(2\pi i/3)$. The reducible representations Γ_{trans} and Γ_{libr} for solid CO_2 are included.]

Slater (1965)	E	$3U$	C_3^{2xyz}	$3C_3^2$	C^{xyz}	$3C_3$	1	Miller and Love (1967)	Zak (1969)
Γ_1^+	1	1	1	1	1	1	1	Γ_1^+	—
Γ_2^+	1	1	ω	ω	ω^*	ω^*	1	Γ_3^+	—
Γ_3^+	1	1	ω^*	ω^*	ω	ω	1	Γ_2^+	—
Γ_4^+	3	-1	0	0	0	0	3	Γ_4^+	—
Γ_1^-	1	1	1	1	1	1	-1	Γ_1^-	—
Γ_2^-	1	1	ω	ω	ω^*	ω^*	-1	Γ_3^-	—
Γ_3^-	1	1	ω^*	ω^*	ω	ω	-1	Γ_2^-	—
Γ_4^-	3	-1	0	0	0	0	-3	Γ_4^-	—
Γ_{trans}	12	0	0	0	0	0	-12		
Γ_{libr}	8	0	-1	-1	-1	-1	8		

TABLE A.VII

Character Table for the R-Point [$\mathbf{q} = \pi/a(\hat{\imath} + \hat{\jmath} + \hat{k})$] of $T_h^6 - Pa3$ [The designations of the representations according to various authors are given. The characters for operations $\{\mathbf{S} \mid \mathbf{v}(S)\} \cdot \mathbf{I}$ are not given but are obtained by multiplying the characters of the $\{\mathbf{S} \mid \mathbf{v}(S)\}$ by ± 1. The full operation designations are given in Table A-V. $\omega = \exp(2\pi i/3)$. The reducible representations R_{trans} and R_{libr} for solid CO_2 are included.]

Slater (1965)	E	$3U$	C_3^{2xyz}	$3C_3^2$	C_3^{xyz}	$3C_3$	1	Miller and Love (1967)	Zak[a] (1969)
R_1^+	2	0	$-\omega$	ω	$-\omega^*$	$-\omega^*$	2	R_1^+	R_3
R_2^+	2	0	$-\omega^*$	ω^*	$-\omega$	$-\omega$	2	R_3^+	R_2
R_3^+	2	0	-1	1	-1	-1	2	R_2^+	R_1
R_1^-	2	0	$-\omega$	ω	$-\omega^*$	$-\omega^*$	-2	R_1^-	R_6
R_2^-	2	0	$-\omega^*$	ω^*	$-\omega$	$-\omega$	-2	R_3^-	R_5
R_3^-	2	0	-1	1	-1	-1	-2	R_2^-	R_4
R_{trans}	12	0	0	0	0	0	-12		
R_{libr}	8	0	-1	1	-1	-1	8		

[a] There is a misprint in Zak's table. The traces of elements $3C_3^2$ for R_2 and R_3 are interchanged.

before, the translational displacements correspond to $u_j(lk)$ with $j = x, y, z$ in the crystal coordinate frame. The two librational displacements correspond to the Eulerian angles θ, ϕ as described in Section IIC.7. We shall here restrict outselves to the construction of the reducible representations Γ_{trans} and Γ_{libr}.

For the Γ-point ($\mathbf{q} = 0$) we again see from (A.7) that all $f_k(S\mathbf{0})$ are unity. We therefore have to determine only the $C_{jj'}(S)$. We desire first to construct Γ_{trans}, the reducible representation of the translations for solid CO_2 included in Table A.VI. We require only diagonal terms of the matrix representation for this purpose. The identity element **E** is, of course, again diagonal and has trace 12 (four molecules per primitive unit cell). The inversion **I** has the effect of exchanging molecules of equal type among themselves and therefore its trace is -12. The twofold axes **U** always interchange sites and therefore their diagonal terms are all zeroes. A symmetry element $\mathbf{C}_3$ exchanges the molecules of one of the four types among themselves and interchanges the other sites. The sublattice transformed into itself is that whose molecular figure axis is parallel to the particular $\mathbf{C}_3$ axis. In principle, this sublattice will contribute to the diagonal. However, the threefold rotations interchange all cartesian axes x, y, z (e.g., $\mathbf{C}_3^{xyz}$ causes the interchanges $x \rightarrow y$, $y \rightarrow z$, $z \rightarrow x$), so that again we obtain no nonzero elements on the diagonal.

To construct the reducible representation of the librations we shall refer to the transformation properties of the θ- and ϕ- (or ω-) displacement coordinates given in Table II, Part B. The trace of **E** needs no further comment. The traces corresponding to the twofold axes **U** are again zero, as before, since all sites are interchanged. The trace of **I** is now eight as obtained from Table II, since all $u_\theta(lk)$ go into $u_\theta(l'k')$, and all $u_\phi(lk)$ go into $u_\phi(l'k')$. As already mentioned, each $\mathbf{C}_3$ operation leaves one site invariant (although interchanging unit cells l). Table II lists the transformations of the u_θ and u_ω (related to u_ϕ) to second order. For the purposes of symmetry transformations, only first-order terms need be considered, since the transformations must be valid for infinitesimally small displacements. We therefore see that we obtain a contribution of -1 to the diagonal for each site that remains unchanged. Therefore we obtain -1 for the trace of each operation $\mathbf{C}_3$. The resulting Γ_{libr} is also included in Table A.VI.

Reduction of Γ_{trans} and Γ_{libr} presents no further difficulty. We use (A.19) and Table A.VI and obtain the following results:

$$\begin{aligned} \Gamma_{\text{trans}} &= \Gamma_1^- + (\Gamma_2^- + \Gamma_3^-) + 3\Gamma_4^- \\ &= A_u + E_u + 3T_u \\ \Gamma_{\text{trans}} &= (\Gamma_2^+ + \Gamma_3^+) + 2\Gamma_4^+ \\ &= E_g + 2T_g \end{aligned} \tag{A.24}$$

Here the representation designations of Slater (1965) are given as well as the conventional designations for point group T_h. The representations Γ_2^+ and Γ_3^+ are degenerate since they have mutually complex conjugate representations. Again, Γ_{trans} includes the zero-frequency acoustic modes which span Γ_4^- (or T_u). As a result, there are four distinct nonzero translational modes $[\Gamma_1^- + (\Gamma_2^- + \Gamma_3^-) + 2\Gamma_4^-]$ at $\mathbf{q} = 0$, one of these doubly degenerate and two triply degenerate. Two of these are infrared active (Γ_4^-). On the other hand, there are three distinct librational modes, one doubly degenerate and two triply degenerate. All three modes are Raman active.

We now turn our attention to the construction of R_{trans}, the reducible representation of the translational displacements at the R-point, with $\mathbf{q} = (\pi/a)(\hat{\jmath} + \hat{\jmath} + \hat{k})$. R_{trans} is given in Table A-VII. The results for $\mathbf{E}$, $3\mathbf{U}$ and $\mathbf{C}_3$ follow immediately from the foregoing discussion. Only $\mathbf{I}$ needs some attention. We have already seen that the diagonal transformation coefficients $C_{jj}(S)$ contribute -1 to the diagonal for each $u_j(\mathbf{q}k)$. We must, however, here investigate the factor $f_k(S, \mathbf{q})$ [Eq. (A.7)]. For the R-point, we obtain for $\mathbf{I}$

$$\mathbf{q} - \mathbf{S}^{-1}\mathbf{q} = \frac{2\pi}{a}(\hat{\imath} + \hat{\jmath} + \hat{k})$$

and substituting a general $\mathbf{R}(lk)$ for each of $k = 1, 2, 3, 4$ (see Table II), we obtain the result that $f_k(S, \mathbf{q})$ is unity for all k. In obtaining this result, we have used the fact that $\mathbf{v}(S) = 0$ in the present case. We thus obtain the trace of -12 as before.

We now construct the representation R_{libr}, which is also given in Table A.VII. From the preceding remarks, the traces corresponding to $\mathbf{E}$, $3\mathbf{U}$ and $\mathbf{I}$ immediately follow. Also the transformation coefficients $C_{jj'}(S)$ have been discussed. We thus need only consider the factors $f_k(S, \mathbf{q})$. For the elements $\mathbf{C}_3^{2xyz}$ and $\mathbf{C}_3^{xyz}$, we have $S\mathbf{q} = \mathbf{q}$ and also $\mathbf{v}(S) = 0$. Therefore we obtain from (A.7) that for these elements the $f_k(S, \mathbf{q})$ are independent of site k and are unity. Hence we obtain the same result as for the Γ-point. For the element $\left\{C_3^{x\bar{y}\bar{z}}\, \frac{a}{2}\, 0\, \frac{a}{2}\right\}$, for example, $\mathbf{v}(S) = \frac{a}{2}(\hat{\imath} + \hat{k})$ and we obtain

$$\exp\left[-i\mathbf{q} \cdot \mathbf{v}(S)\right] = \exp\left[-i\pi\right] = -1$$

Also,

$$\mathbf{q} - \mathbf{S}^{-1}\mathbf{q} = \mathbf{q} - \mathbf{C}^{2x\bar{y}\bar{z}}\mathbf{q} = \frac{2\pi}{a}(\hat{\imath} + \hat{k})$$

The position vector need only be considered for $k = 2$ since only this site is left invariant. It is

$$\mathbf{R}(l2) = [(l_1 + \tfrac{1}{2})\hat{\imath} + (l_2 + \tfrac{1}{2})\hat{\jmath} + l_3\hat{k}]a$$

and substituting these values in (A.7), we obtain $f_2(S, \mathbf{q}) = (-1)(-1) = 1$. As a result, the trace is again as for the Γ-point, or -1. Similar results are obtained for the remaining $\mathbf{C}_3$ operations. However, the remaining $\mathbf{C}_3{}^2$ operations do not give the same result. For example, for $\left\{\mathbf{C}_3^{2x\bar{y}\bar{z}} \middle| 0 \frac{a}{2} \frac{a}{2}\right\}$, $\mathbf{v}(S) = \frac{a}{2}(\hat{\jmath} + \hat{k})$ and again we have

$$\exp[-i\mathbf{q} \cdot \mathbf{v}(S)] = \exp(-i\pi) = -1$$

But in this case

$$\mathbf{q} - \mathbf{S}^{-1}\mathbf{q} = \frac{2\pi}{a}(\hat{\imath} + \hat{\jmath})$$

and using this result and $\mathbf{R}(l2)$ as before, we obtain, from (A.7), $f_2(S, \mathbf{q}) = (-1)(1) = -1$. Therefore the trace for this element is 1 instead of -1. The same result is obtained for the two remaining $\mathbf{C}_3{}^2$ elements.

Reduction of R_{trans} and R_{libr} gives the following results (using the representation names of Slater, 1965):

$$\Gamma_{\text{trans}} = 2(R_1^- + R_2^-) + 2R_3^-$$
$$\Gamma_{\text{libr}} = (R_1^+ + R_2^+) + 2R_3^+$$

The representations R_1^+ and R_2^+ (or R_1^- and R_2^-) have mutually complex conjugate representations and are therefore degenerate. Each of these representation is two-dimensional to begin with and therefore fourfold degeneracy occurs due to each pair of $(R_1 + R_2)$. This type of degeneracy is commonly referred to as additional degeneracy due to time reversal symmetry (Maradudin and Vosko, 1968; Herring, 1937). These authors have developed criteria for such additional degeneracy and the present case corresponds to their third type. In addition, they classified another type of additional degeneracy as of the "second type." In this case, degeneracy occurs between two equivalent representations in pairs. The representations R_3^+ and R_3^- satisfy this condition, and as a result, $2R_3^+$ cause fourfold degeneracy and so do $2R_3^-$. We can now summarize the lattice modes of solid CO_2 at the R-point. The translations are here separable from the librations. There are three distinct translational frequencies, each mode being fourfold degenerate [two belonging to $(R_1^- + R_2^-)$ and 1 belonging to $2R_3^-$]. There are two distinct librational frequencies, each being again fourfold degenerate [one belonging to $(R_1^+ + R_2^+)$ and the second to $2R_3^+$]. Recent calculations of the dispersion curves of solid CO_2 (Suzuki and Schnepp, 1971) have indeed shown that the degeneracies occur as described here.*

* One of the authors (O. S.) is indebted to Dr. Gerald Dolling for pointing out the theoretical basis for the time-reversal degeneracies of the "second type."

TABLE A.VIII

Character Table for the Λ-Direction [$\mathbf{q} = \xi\pi/a(\hat{\imath} + \hat{\jmath} + \hat{k})$, $0 < \xi < 1$] of $T_h^6 - Pa3$ [The designations of the representations according to two sources are given. $\omega = \exp(2\pi i/3)$.]

Slater (1965)	E	C_3^{2xyz}	C^{xyz}	Miller and Love (1967)
Λ_1	1	1	1	Λ_1
Λ_2	1	ω	ω^*	Λ_2
Λ_3	1	ω^*	ω	Λ_3

The symmetry direction $\mathbf{q} = \xi \frac{\pi}{a} (\hat{\imath} + \hat{\jmath} + \hat{k})$, $0 < \xi < 1$ or the body diagonal of the cubic zone is designated Λ. For such a wave vector, the symmetry operations $\mathbf{E}$, $\mathbf{C}_3^{2xyz}$, $\mathbf{C}_3^{xyz}$ transform $\mathbf{q}$ into itself and therefore the appropriate little group is isomorphic with C_3. The character table is given in Table A.VIII. Since C_3 is a subgroup of T_h, the correlation (or compatibility) relations between the Γ-point and the Λ-direction are easily established. Further, the correlations between the representations at the R-point and the Λ-direction are found by considering the traces corresponding to $\mathbf{E}$, $\mathbf{C}_3^{2xyz}$ and $\mathbf{C}^{xyz}$ in the characters of the irreducible R-representations. These form reducible representations of Λ and reduction using (19) and Table A.VIII gives the result sought. The compatibility relations between Γ, Λ, and R are summarized in Table A.IX.

On the basis of this table, we can describe the dispersion curves for the lattice vibrations of solid CO_2 as we proceed from the Γ-point, along Λ to the R-point. At Γ, the translations and librations are separable as described, but along Λ they do not belong to distinct irreducible representations and are therefore allowed to interact. However, at R they are

TABLE A.IX

Compatibility Relations along Direction Λ of $T_h^6 - Pa3$

Γ-Point $\mathbf{q} = 0$	Λ-Direction $\mathbf{q} = \xi(\pi/a)(\hat{\imath} + \hat{\jmath} + \hat{k})$, $0 < \xi < 1$	R-Point $\mathbf{q} = (\pi/a)(\hat{\imath} + \hat{\jmath} + \hat{k})$
Γ_1^+, Γ_1^-		$(R_1^+ + R_2^+)$
	Λ_1	
$\Gamma_4^+\Gamma_4^-$		$(R_1^- + R_2^-)$
$(\Gamma_2^+ + \Gamma_3^-)$		
	$(\Lambda_2 + \Lambda_3)$	
$(\Gamma_2^- + \Gamma_3^-)$		R_3^+, R_3^-

again separable. The nondegenerate representations Γ_1 correlate uniquely with Λ_1 and contribute to the $(R_1 + R_2)$ fourfold degenerate modes. The twofold degenerate modes spanning $(\Gamma_2 + \Gamma_3)$ correlate with the twofold degenerate modes $(\Lambda_2 + \Lambda_3)$ and these correlate with the two-dimensional representations R_3 (which, as described, form fourfold degeneracies in pairs). Further, $(\Lambda_2 + \Lambda_3)$ also correlates with the fourfold degenerate $(R_1 + R_2)$. Beginning from the R-point, $(R_1 + R_2)$ reduces to $2\Lambda_1 + (\Lambda_2 + \Lambda_3)$.

The procedures for the symmetry classifications of the lattice vibrational modes at additional symmetry points of the Brillouin zone can be modeled on the preceding ideas. The necessary character tables are given in the sources cited.

References

Ahlers, G., and W. H. Orttung (1964). *Phys. Rev.*, **133A**, 1642.
Anderson, A., H. A. Gebbie, and S. H. Walmsley (1964). *Mol. Phys*, **7**, 401.
Anderson, A., and G. E. Leroi (1966). *J. Chem. Phys.*, **45**, 4359.
Anderson, A., and W. H. Smith (1966). *J. Chem. Phys* , **44**, 4216.
Anderson, A., T. S. Sun, and M. C. A. Donkersloot (1970). *Can. J. Phys.*, **48**, 2265.
Anderson, A., and S. H. Walmsley (1964). *Mol. Phys.*, **7**, 583.
Anderson, A., and S. H. Walmsley (1965). *Mol. Phys.*, **9**, 1.
Arnold, G. M., and R. Heastie (1967). *Chem. Phys. Letters*, **1**, 51.
Baglin, F. G., S. F. Bush, and J. R. Durig (1967). *J. Chem. Phys.*, **47**, 2104.
Becka, L. N. (1962). *J. Chem. Phys.*, **37**, 431.
Bernstein, E. R. (1970). *J. Chem. Phys.*, **52**, 4701.
Bertie, J. E., H. J. Labbé, and E. Whalley (1968a). *J. Chem. Phys.*, **48**, 775.
Bertie, J. E., H. J. Labbé, and E. Whalley (1968b). *J. Chem. Phys.*, **49**, 2141.
Bertie, J. E., and E. Whalley (1964a). *J. Chem. Phys.*, **31**, 575.
Bertie, J. E., and E. Whalley (1964b). *J. Chem. Phys.*, **40**, 1637.
Bertie, J. E., and E. Whalley (1967). *J. Chem. Phys.*, **47**, 1271.
Blumenfeld, S. M., S. P. Reddy, and H. L. Welsh (1970). *Can. J. Phys.*, **48**, 513.
Born, M., and K. Huang (1954). *Dynamical Theory of Crystal Lattices*, Clarendon Press, Oxford.
Brecher, C., and R. S. Halford (1961). *J. Chem. Phys.*, **35**, 1109.
Bree, A., and R. A. Kydd (1968). *J. Chem. Phys.*, **48**, 5319.
Brith, M., and A. Ron (1969). *J. Chem. Phys.*, **50**, 3053.
Brith, M., A. Ron, and O. Schnepp (1969). *J. Chem. Phys.*, **51**, 1318.
Brot, C., B. Lassier, G. W. Chantry, and H. A. Gebbie (1968). *Spectrochim. Acta*, **24**, 295.
Brown, K. G., and W. T. King (1970). *J. Chem. Phys.*, **52**, 4437.
Brueckner, K. A. and J. Frohberg (1965). *Prog. Theoret. Phys.* (Kyoto), Suppl. 383.
Brun, T. O., S. K. Sinha, L. A. Swenson, and C. R. Tilford (1968). *Proc. IAEA Symposium on Inelastic Neutron Scattering*, Copenhagen.
Buckingham, A. D. (1959). *Quart. Rev.*, **13**, 183.
Bührer, W., W. Hälg, and T. Schneider (1967). *Rep. Inst. Reaktorforsch.*, Wurenlingen, No. AF-SSP-9.
Cahill, J. E. (1968). Thesis, Princeton University.
Cahill, J. E. and G. E. Leroi (1969a). *J. Chem. Phys.*, **51**, 1324.

Cahill, J. E. and G. E. Leroi (1969b). *J. Chem. Phys.*, **51**, 4514.
Carlson, R. E., and H. B. Friedrich (1971). *J. Chem. Phys.*, **54**,
Cheutin, A., and J. P. Mathieu (1956). *J. Chem. Phys.*, **53**, 106.
Cochran, W., and R. A. Cowley (1967). *Handbuch der Physik*, Bd. XXV/2a, Springer, Berlin, Heidelberg, New York, p. 59.
Cochran, W., and G. S. Pawley (1964). *Proc. Roy. Soc.* (*London*), **A280**, 1.
Coll C. F., III, and A Brooks Harris (1970). *Phys. Rev.*, **B2**, 1176, App. A.
Colombo, L. (1967). *Spectrochim. Acta*, **23A**, 1561.
Colombo, L. (1968). *J. Chem. Phys.*, **49**, 4688.
Colombo, L., and J. P. Mathieu (1960). *Bull, Soc. Franc. Mineral. Crist.* **83**, 250.
Colombo, L. and D. Moreau (1967). *Compt. Rend.*, **B265**, 487.
Colpa, J. P., and J. A. A. Ketelaar (1958). *Mol. Phys.*, **1**, 343.
Coulson, C. A., and C. W. Haigh (1963). *Tetrahedron*, **19**, 527.
Couture-Mathieu, L., J. P. Mathieu, J. Cremer, and H. Poulet (1951). *J. Chem. Phys.*, **48**, 1.
Cruikshank, D. W. (1958). *Rev. Mod. Phys.*, **30**, 163.
David, J. C., and W. B. Person (1968). *J. Chem. Phys.*, **48**, 510.
de Boer, J. (1942). *Physica*, **9**, 363.
de Wette, F. W., and B. R. A. Nijboer (1965). *Phys. Letters*, **18**, 19.
de Wette, F. W., L. H. Nosanow, and N. R. Werthamer (1967). *Phys. Rev.*, **162**, 824.
Decius, J. C. (1968). *J. Chem. Phys.*, **49**, 1387.
Devonshire, A. F. (1935). *Proc. Roy. Soc.* (*London*), **A153**, 601.
Dolling, G., and B. M. Powell (1970). *Proc. Roy. Soc.* (*London*), **A319**, 209.
Donkersloot, M. C. A., and S. H. Walmsley (1970). *Mol. Phys.*, **19**, 183.
Dows, D. A. (1962). *J. Chem. Phys.*, **36**, 2836.
Dows, D. A. (1963, 1965). *Physics and Chemistry of the Organic Solid State*, Wiley-Interscience, New York, Chapter II.
Dows, D. A. (1966). *J. Chem. Phys.*, **63**, 168.
Durig, J. R., S. F. Bush, and F. G. Baglin (1968). *J. Chem. Phys.*, **49**, 2106.
Durig, J. R., S. M. Craven, and J. Bragin (1970). *J. Chem. Phys.*, **51**, 5663.
Durig, J. R., C. M. Player, and J. Bragin (1970). *J. Chem. Phys.*, **52**, 4224.
Durig, J. R., M. Walker, and F. G. Baglin (1968). *J. Chem. Phys.*, **48**, 4675.
Dymond, J. H., and B. J. Alder (1968). *Chem. Phys. Lett.*, **2**, 54.
Fleming, J. W., P. A. Turner, and G. W. Chantry (1970). *Mol. Phys.*, **19**, 853.
Franck, J. P., and R. Wanner (1968). Cited in Gillis, Koehler, and Werthamer.
Fredkin, D. R., and N. R. Werthamer (1965). *Phys. Rev.*, **138**, A1527.
Friedrich, H. B. (1967). *J. Chem. Phys.*, **47**, 4269.
Friedrich, H. B. (1970). *J. Chem. Phys.*, **52**, 3005.
Fruehling, A. (1950). *J. Chem. Phys.*, **18**, 1119.
Fruehling, A. (1951). *Annln. Phys.*, **6**, 401.
Gerbaux, X., and A. Hadni (1968). *J. Chem. Phys.*, **48**, 955.
Giguere, P. A., and C. Chapados (1966). *Spectrochim. Acta*, **22**, 1131.
Gillis, N. S., T. R. Koehler, and N. R. Werthamer (1968). *Phys. Rev.*, **175**, 1110.
Gillis, N. S., N. R. Werthamer, and T. R. Koehler (1968). *Phys. Rev.*, **165**, 951.
Glyde, H. R. (1970). *J. Phys.*, **C3**, 810.
Goldstein, H. (1950). *Classical Mechanics*, Addison-Wesley, Reading, Mass.
Guyer, R. A. (1969). *Solid State Physics*, Vol. 23, F. Seitz, D. Turnbull, and H. Ehrenreich, Ed., Academic Press, New York, p. 413.
Guyer, R. A., and L. I. Zane (1969). *Phys. Rev.*, **188**, 445.
Hadni, A., B. Wyncke, G. Morlet, and X. Gerbaux (1969). *J. Chem. Phys.*, **51**, 3514.

Harada, I., and T. Shimanouchi (1966). *J. Chem. Phys.*, **44,** 2016.
Harada, I., and T. Shimanouchi (1967). *J. Chem. Phys.*, **46,** 2708.
Hardy, W. N., I. F. Silvera, K. N. Klump, and O. Schnepp (1968). *Phys. Rev. Letters*, **21,** 291.
Hardy, W. N., I. F. Silvera, and J. P. McTague (1971). *Phys. Rev. Letters*, **26,** 127.
Hatsenbuhler, D. A., J. E. Cahill, and G. E. Leroi (1969). *Symposium on Spectroscopy and Molecular Structure*, Columbus, Ohio, Paper Q6.
Herring, C. (1937). *Phys. Rev.*, **52,** 361.
Horton, G. H. (1968). *Am. J. Phys.*, **36,** 93.
Internation Union of Crystallography (1952). *International Tables for X-Ray Crystallography*, Vol. 1, The Kynoch Press, Birmingham, England.
Ito, M. (1969). *J. Chem. Phys.*, **41,** 1255.
Ito, M. (1970). *Chem. Phys. Letters*, **7,** 439.
Ito, M., and T. Shigeoka (1966a). *Spectrochim. Acta*, **22,** 1029.
Ito, M., and T. Shigeoka (1966b). *J. Chem. Phys.*, **44,** 1001.
Ito, M., and M. Suzuki (1968). Unpublished work.
Ito, M., M. Suzuki, and T. Yokoyama (1967). *Bull. Chem. Soc. Japan*, **40,** 2461.
Ito, M., M. Suzuki, and T. Yokoyama (1969). *J. Chem. Phys.*, **50,** 2949.
Ito, M., T. Yokoyama, and M. Suzuki (1970). *Spectrochim. Acta*, **26A,** 695.
Jacobi, N. (1970). Unpublished work.
Jacobi, N., and O. Schnepp (1971). Unpublished work.
Janik, J. A., K. Otnes, G. Solt, and G. Kosaly (1969). *Disc. Faraday Soc.*, **48,** 87.
Jordan, T. H., H. W. Smith, W. E. Streib, and W. N. Lipscomb (1969), *J. Chem. Phys.* **41,** 756.
Kastler, A., and A. Rousset (1951). *J. Phys. Radium*, **2,** 49.
Ketelaar, J. A. A., and R. P. H. Rettschnick (1964). *Mol. Phys.*, **7,** 161.
Kiss, Z. J., H. P. Gush, and H. L. Welsh (1959). *Can. J. Phys.*, **37,** 362.
Kitaigorodskii, A. (1966). *J. Chim. Phys.*, **63,** 9.
Kittel, C. (1968). *Introduction to Solid State Physics*, 3rd ed., Wiley, New York.
Klein, M. L., and T. R. Koehler (1970). *J. Physics*, **C3,** L102.
Klump, K. N., O. Schnepp, and L. H. Nosanow (1970). *Phys. Rev.*, **B1,** 2496.
Koehler, T. R. (1966). *Phys. Rev. Letters*, **17,** 89.
Koehler, T. R. (1967). *Phys. Rev. Letters*, **18,** 654.
Koehler, T. R. (1968). *Phys. Rev.*, **165,** 942.
Korshunov, A. V., V. F. Shabanov, and Y. S. Rusetskii (1967). *Izvest. Sibirsk. Otd. Akad. Nauk SSSR*, **41.**
Koster, G. F. (1957). *Space Groups and their Representations*, Academic Press, New York.
Kovalev, O. V. (1965). *Irreducible Representations of the Space Groups*, Gordon and Breach, New York.
Kuan, T. S. (1969). Thesis, University of Southern California.
Kuan, T. S., A. Warshel, and O. Schnepp (1970). *J. Chem. Phys.*, **52,** 3012.
Landsberg P. T., Ed. (1969). *Solid State Theory: Methods and Applications*, Wiley-Interscience, New York.
Leech, J. W., and C. J. Peachey (1968). *Chem. Phys. Letters*, **1,** 643.
Lin, C. C., and J. D. Swalen (1959). *Rev. Mod. Phys.*, **31,** 841.
Logan, K. W., S. F. Trevino, H. J. Prask, and J. D. Gault (1970). *J. Chem. Phys.*, **53,** 3417.
Luty, T., J. A. Janik, A. Bajorek, I. Natkaniec, J. Domoslawski, and J. M. Janik (1970). *Report No. 724/PL*, Institute of Nuclear Physics, Cracow.

McTague, J. P., I. F. Silvera, and W. N. Hardy (1970). *Bull. Am. Phys. Soc.*, **15,** 296.
Mandel, M., and P. Mazur (1958). *Physica*, **24,** 116.
Maradudin, A. A., E. W. Montroll, and G. H. Weiss (1963). *Theory of Lattice Dynamics in the Harmonic Approximation*, (*Solid State Physics*, F. Seitz and D. Turnbull, Eds.), Academic Press, New York.
Maradudin, A. A., and S. H. Vosko (1968). *Rev. Mod. Phys.*, **40,** 1.
Martin, D. H. (1965). *Adv. Phys.*, **14,** 39.
Martin, P. H., and S. H. Walmsley (1969). *Disc. Faraday Soc.*, **48,** 49.
Mertens, F. G., W. Biem, and H. Hahn (1968). *Z. Physik*, **213,** 33.
Miller, S. C., and W. F. Love (1967). *Tables of Irreducible Representation of Space Groups*, Pruett Press, Boulder, Colo.
Minkiewicz, V. J., T. A. Kitchens, F. P. Lipschultz, R. Nathans, and G. Shirane (1968). *Phys. Rev.*, **174,** 267.
Mitra, S. S., and P. J. Gielisse (1964). *Progress in Infrared Spectroscopy*, Vol. II, Plenum Press, New York, p. 47.
Montgomery, H. (1969). *Proc. Roy. Soc.*, **A309,** 521 (1969).
Morley, G. L., and K. L. Kliewer (1969). *Phys. Rev.*, **180,** 245.
Mucker, F. K., P. M. Harris, D. White, and R. A. Erickson (1968). *J. Chem. Phys.*, **49,** 1922.
Mulliken, R. S. (1939). *J. Chem. Phys.*, **7,** 14.
Mullin, W. J., L. H. Nosanow, and P. M. Steinback (1969). *Phys. Rev.*, **188,** 410.
Munier, J. M., and A. Hadni (1968). *J. Mol. Structure*, **1,** 249.
Nakamura, M., and T. Miyazawa (169). *J. Chem. Phys.*, **51,** 3146.
Nielsen, M., H. B. Moller, and L. Meyer (1970). *Bull. Am. Phys. Soc.*, **15,** 383.
Nosanow, L. H. (1966). *Phys. Rev.*, **146,** 120.
Nosanow, L. H., and G. L. Shaw (1962). *Phys. Rev.*, **128,** 546.
Oliver, D. A., and S. H. Walmsley (1968). *Mol. Phys.*, **15,** 141.
Oliver, D. A., and S. H. Walmsley (1969). *Mol. Phys.*, **17,** 617.
Overend, J. (1963). *In Infrared Spectroscopy and Molecular Structure*, M. Davies, ed., Elsevier, New York, Chapter 10.
Pawley, G. S. (1967). *Physica Status Solidi*, **20,** 347.
Pawley, G. S., and S. J. Cyvin (1970). *J. Chem. Phys.*, **52,** 4073.
Piseri, L., and G. Zerbi (1968). *J. Mol. Spectr.*, **26,** 254.
Pitzer, K. S. (1953). *Quantum Chemistry*, Prentice-Hall, New York.
Powell, B. M. (1968). *Neutron Inelastic Scattering*, Vol. II, International Atomic Energy Agency, Vienna, p. 195.
Powell, B. M., and G. Dolling (1969). *Bull. Am. Phys. Soc.*, **14,** 301.
Prask, H., H. Boutin, and S. Yip (1968). *J. Chem. Phys.*, **48,** 3367.
Rafizadeh, H. A. and S. Yip (1970). *J. Chem. Phys.*, **53,** 315.
Raich, J. C., and H. M. James (1966). *Phys. Rev. Letters*, **16,** 173.
Richardson, P. M., and E. R. Nixon (1968). *J. Chem. Phys.*, **49,** 4276.
Ron, A., and O. Schnepp (1967). *J. Chem. Phys.*, **46,** 3991.
Rush, J. J. (1967). *J. Chem. Phys.*, **47,** 3936.
Sauer, P. (1966). *Z. Physik*, **194,** 360.
Schnepp, O. (1967). *J. Chem. Phys.*, **46,** 3983.
Schnepp, O. (1969). *Adv. At. Mol. Phys.*, **5,** 155.
Schnepp, O. (1970). *Phys. Rev.*, **2A,** 2574.
Schnepp, O., and A. Ron (1969). *Disc. Faraday Soc.*, **48,** 26.
Schomaker, V., and K. N. Trueblood (1968). *Acta Cryst.*, **B24,** 63.
Schuch, A. F., and R. L. Mills (1966). *Phys. Rev. Letters*, **16,** 616.

Schuch, A. F., R. L. Mills and D. A. Depatie (1968). *Phys. Rev.*, **165**, 1032.
Schwartz, Y. A., A. Ron, and S. Kimel (1969). *J. Chem. Phys.*, **51**, 1666.
Schwartz, Y. A., A. Ron, and S. Kimel (1971). *J. Chem. Phys.*, **54**, 99.
Seitz, F. (1940). *The Modern Theory of Solids*, McGraw-Hill, New York.
Shimanouchi, T. (1970). Private communication.
Shimanouchi, T., and I. Harada (1964). *J. Chem. Phys.*, **41**, 2651.
Shimanouchi, T., M. Tsuboi, and T. Miyazawa (1961). *J. Chem. Phys.*, **35**, 1597.
Shinoda, T., and H. Enokido (1969). *J. Phys. Soc. Japan*, **26**, 1353.
Silvera, I. F., W. N. Hardy, and J. P. McTague (1969). *Disc. Faraday Soc.*, **48**, 54.
Slater, J. C. (1965). *Quantum Theory of Molecules and Solids*, Vol. 2, McGraw-Hill, New York.
St. Louis, R. V., and O. Schnepp (1969). *J. Chem. Phys.*, **50**, 5177.
Suzuki, M. (1970). Unpublished work.
Suzuki, M., and O. Schnepp (1971). Unpublished work.
Suzuki, M., T. Yokoyama, and M. Ito (1967). *Spectrochim. Acta*, **24A**, 1091.
Suzuki, M., T. Yokoyama, and M. Ito (1969). *J. Chem. Phys.*, **50**, 3392.
Sweet, J. R., and W. A. Steele (1967). *J. Chem. Phys.*, **47**, 3029.
Taddei, G., and E. Giglio (1970). *J. Chem. Phys.*, **53**, 2768.
Taimasalu, P. and D. W. Robinson (1965). *Spectrochim. Acta*, **21**, 1921.
Takahashi, H., and B. Schrader (1967). *J. Chem. Phys.*, **47**, 3842.
Tarina, V. (1967). *J. Chem. Phys.*, **46**, 3273.
Trevino, S., H. Prask, T. Wall, and S. Yip (1968) *Neutron Inelastic Scattering*, Vol 1, International Atomic Energy Agency, Vienna, p. 345.
Tubino, R., and G. Zerbi (1970). *J. Chem. Phys.*, **53**, 1428.
van Kranendonk, J. (1958). *Physica*, **24**, 347.
Venkataraman, G., and V. L. Sahni (1970). *Rev. Mod. Phys.*, **42**, 409.
Walmsley, S. A. (1968). *J. Chem. Phys.*, **48**, 1438.
Walsmley, S. A., and A. Anderson (1964). *Moll Phys.*, **7**, 411.
Walmsley, S. H., and J. A. Pople (1964). *Mol. Phys.*, **8**, 345.
Werthamer, N. R. (1970). *Phys. Rev.*, **A2**, 2050.
Werthamer, N. R., R. L. Gray, and T. R. Koehler (1970). *Phys. Rev.*, **B2**, 951.
Whalley, E., and J. E. Bertie (1967). *J. Chem. Phys.*, **47**, 1264.
Wigner, E. (1931). *Gruppentheorie*. Friedr. Vieweg and Sohn Akt. Ges., Braunschweig.
Wilkinson, G. (1966). Unpublished; quoted by Pawley (1967).
Williams, D. E. (1967). *J. Chem. Phys.*, **47**, 4680.
Wilson, E. B., J. C. Decius, and P. C. Cross (1955). *Molecular Vibrations*, McGraw-Hill, New York.
Woodward, L. A. (1967). *Raman Spectroscopy*, H. A. Szymanski, Ed., Plenum Press, New York.
Zak, J., Ed. (1969). *Irreducible Representations of Space Groups*, W. A. Benjamin, New York.
Ziman, J. M. (1960). *Electrons and Phonons*, Clarendon Press, Oxford.

AN INTRODUCTION TO C^*-ALGEBRAIC METHODS IN PHYSICS†

GERARD G. EMCH ‡

The Theoretical Physics Institute, Nijmegen, The Netherlands

CONTENTS

† This chapter is based on a series of six lectures given at the Service de Chimie Physique II, Université Libre de Bruxelles, January–February 1971, under the auspices of the Solvay Seminars. The choice of the material presented reflects merely the particular bias of the lecturer, tempered by the interests of the audience.

‡ Permanent address: Departments of Mathematics and of Physics and Astronomy, The University of Rochester, Rochester, New York.

The nature of this article is that of a chapter in an abridged course in advanced quantum mechanics. It is geared specifically to the presentation of one method of mathematical physics, that of the C^*-algebraic approach.

The article is divided in three parts. The first part consists of an introduction in the form of a general motivation. The second briefly presents some of the tools and techniques involved, thus forming the material of the next two sections. The third part of the series, spread over the last three sections, discusses three illustrative applications chosen from the field of statistical mechanics: existence of certain phase transitions, symmetry breaking, and nonequilibrium.

Some bibliographical notes have been appended to each section. These have no pretense to completeness; they rather have to be considered as a possible introduction to the literature on the subject treated in the section.

Part I

GENERAL MOTIVATION

I. QUANTUM FIELD THEORY AND STATISTICAL MECHANICS: TWO FACETS OF THE SAME PROBLEM

A. Generalities

The elaboration of a physical theory often can be summarized by the following feedback process.

One first injects into the theory the following ingredients: (1) observable facts; (2) traditional physical wisdom; (3) mathematical techniques; and (4) simple theoretical models. One then "turns the crank" with the aim of getting two types of information: (5) physical predictions, and (6) insight into the first principles of the theory. The next step is to compare (5) with (1), and (6) with (2). Out of this comparison stems a new phase in which this scheme is repeated with an improved input (1′)–(4′).

The particular theory that is the subject of this article is that referred to as the *many-body problem*. Corresponding to the preceding list, we have (1) nontrivial scattering of quantum particles occurs, and phase transitions exist, (2) as summarized in the usual rules of quantum mechanics, (3) the functional analysis behind the use of Fock-space, and (4) the van Hove model for the interaction between a scalar meson field and punctual, recoilless, nucleons; and the BCS model for superconductivity. We obtain: (5) an asymptotic condition for scattering is satisfied for the van Hove model, and the BCS model leads to a self-consistent equation for the energy-gap, (6) primary importance of the algebraic structure of the set of observables. Whereas (5) makes plenty of physical sense, it still involves mathematical difficulties which we hope to dispel, in the next phase of the theory, by a systematic exploitation of (6). The object of the next five sections is then to demonstrate that such an approach is possible, that it provides a new approach to the many-body problem, and that it has interesting physical consequences.

B. The Usual Formalism of Quantum Mechanics

Let us first consider the case of a finite number of degrees of freedom, say $n = 1$. The "traditional wisdom" is that observables are (represented by) self-adjoint operators acting on a Hilbert space $\mathscr{H}$. They are "functions" of p and q with $[p, q] = -i\hbar I$. The natural realization is

$$\begin{cases} \mathscr{H} = \mathscr{L}^2(\mathbf{R}) \equiv \left\{\Psi : \mathbf{R} \to \mathbf{C} \,\middle|\, \int dx\, |\Psi(x)|^2 < \infty\right\} \\ (p\Psi)(x) = -i\hbar \dfrac{d\Psi}{dx}(x) \\ (q\Psi)(x) = x\Psi(x) \end{cases}$$

Remarks

1. We do not want to bother at this point with "domain questions" (see Stone, 1932).

2. We shall see, in the next section, in precisely which sense this realization is "natural."

3. The case of a finite number ($1 \leqslant n < \infty$) of degrees of freedom is described in a similar manner: $\mathscr{H} = \mathscr{L}^2(\mathbf{R}^n)$; $p_k = -i\hbar(\partial/\partial x_k)$; $q_k = x_k$ for $k = 1, 2, \ldots, n$.

We now want to pass to the case of an infinite number of degrees of freedom, and to a brief review of the so-called Fock-space formalism. The physical motivation behind this formalism is the need to have a space in which we can describe processes involving the creation and annihilation of particles.

The Hilbert space in which we usually work is

$$\mathscr{H}_F \equiv \bigoplus_{N=0}^{\infty} \mathscr{H}^{(N)}$$

where $\mathscr{H}^{(N)}$ are the N-particle subspaces

$$\mathscr{H}^{(N)} = (\otimes\, \mathscr{H})^s$$

with s denoting the fact that the N-fold tensor product of the 1-particle space $\mathscr{H}$ with itself has been properly symmetrized (or antisymmetrized). We consider the linear manifold $\mathscr{F} \subset \mathscr{H}_F$ characterized by imposing that every $\Psi \in \mathscr{F}$ has at most a finite number of nonvanishing N-particle components, i.e., $\Psi^{(N)} = 0$, except for at most a finite number of values of N.

For every $f \in \mathscr{H}$, we define on $\mathscr{F}$ the operator $a^*(f)$ by

$$[a^*(f)\Psi]^{(N)}(x_1, \ldots, x_N) = N^{-1/2} \sum_{i=1}^{N} f(x_i)\Psi^{(N-1)}(x_1, \ldots, \not{x}_i, \ldots, x_N)$$

for all $\Psi \in \mathscr{F}$. Clearly $a^*(f)\Psi$ differs from Ψ by the fact that it involves one more particle, namely one in wavefunction f. Hence $a^*(f)$ describes the creation operator for a particle in wavefunction f. We define the corresponding annihilation operator by

$$[a(f)\Psi]^{(N)}(x_1, \ldots, x_N) = (N+1)^{1/2} \int dx f(x)^* \Psi^{(N+1)}(x, x_1, \ldots, x_N)$$

It is then easy to check the following, on $\mathscr{F}$, and for all f and g in $\mathscr{H}$:

1. $a^*(f)$ coincides with the hermitian adjoint of $a(f)$.
2. $[a(f), a(g)] = 0 = [a^*(f), a^*(g)]$.
3. $[a(f), a^*(g)] = (f, g)I$.
4. There exists in $\mathscr{H}_F$ a vector Φ_0 (the "vacuum") such that

$$a(f)\Phi_0 = 0.$$

5. The number operator $N = \sum_{i=1}^{\infty} a^*(f_i)a(f_i)$, where $\{f_i\}$ is an orthonormal basis in $\mathscr{H}$, is well defined.

We should notice that:

6. The preceding description does not involve such objects as $a^*(x)$, $a(x)$, which should be viewed as operator-valued distributions.

7. The properties (4) and (5) above will *not* hold in general for the realizations which we will have to consider later on.

C. BCS Model

We first consider the case where the system is enclosed in a box of volume Ω. For simplicity, we take this box to be a cube of edge L and impose periodic boundary conditions. We define

$$f_p(x) = \Omega^{-1/2} e^{ipx} \qquad \text{with } p = 2\pi \frac{n}{L} \quad (n = 0, 1, \ldots)$$

$$a_{\pm}^*(p) = \text{the operator which creates an electron in wavefunction } f_p \text{ and with spin } s_z = \pm\tfrac{1}{2}$$

The Hamiltonian of the system is taken to be

$$H = \sum_p \varepsilon(p)\{a_+^*(p)a_+(p) + a_-^*(p)a_-(p)\} + \sum_{p,q} a_+^*(p)a_-^*(-p)\tilde{v}(p, q)a_-(-q)a_+(q)$$

where

$$\varepsilon(p) = \frac{p^2}{2m}$$

$$\tilde{v}(p, q) = \iint_{\Omega\times\Omega} dx\, dy f_p(x) v(x, y) f_q^*(y)$$

$$v \equiv \iint_{\mathbf{R}^6} dx\, dy\, |v(x, y)| < \infty$$

We then ask whether H can be diagonalized by a Bogoliubov transformation, i.e., whether there exist $\gamma_\pm^*(p)$ (fermi creation and annihilation operators) and $E(p)$ such that

$$\text{(i)} \quad H = \sum_p E(p)\{\gamma_+^*(p)\gamma_+(p) + \gamma_-^*(p)\gamma_-(p)\}$$

$$\begin{aligned} \text{(ii)} \quad \gamma_+^*(p) &= u(p)a_+^*(p) + v(p)a_-(-p) \\ \gamma_-^*(p) &= -v(-p)a_+(-p) + u(-p)a_-^*(p) \end{aligned}$$

Such a transformation exists if and only if $E(p)$, $u(p)$, $v(p)$ satisfy the following equation:

$$\begin{aligned} [H, \gamma_+^*(p)] &= E(p)\gamma_+^*(p) \\ &= \{\varepsilon(p)u(p) + \Delta(p)v(p)\}a_+^*(p) \\ &\quad + \{\Delta^*(p)u(p) - \varepsilon(p)v(p)\}a_-(-p) \end{aligned}$$

(and as above with $[H, \gamma_-^*(p)]$), where

$$\Delta(p) = \sum_q \tilde{v}(p, q)a_-(-q)a_+(q)$$

In general, $\Delta(p)$ will not be a c-number, and the preceding equation cannot be satisfied. Suppose, however, that we could actually treat (in some approximation) $\Delta(p)$ as a c-number. We would then have

$$\begin{aligned} E(p) &= \{\varepsilon(p)^2 + \Delta^*(p)\,\Delta(p)\}^{1/2} \\ \Delta(p) &= \left\langle \sum_q \tilde{v}(p, q)a_-(-q)a_+(q) \right\rangle_\beta \\ &= \sum_q \tilde{v}(p, q)\frac{\Delta(q)}{2E(q)} \tanh\left\{\beta\frac{E(q)}{2}\right\} \end{aligned}$$

where $\langle\cdots\rangle_\beta$ denotes the canonical-ensemble average. The agreement between the energy-gap $|\Delta(p)|^2$ computed from this self-consistency equation and its experimental value turns out to be remarkably good. We should therefore wonder in which sense the approximation that

$\Delta(p)$ is a c-number could be justified. The standard reason offered in the literature is that $\Delta(p)$ indeed becomes a c-number in the thermodynamical limit. This claim is based on the remark that

$$\begin{cases} [\Delta(p), a_{\pm}(q)] = 0 \\ \|[\Delta(p), a_{+}^{*}(q)]\| = \|\tilde{v}(p, q)a_{-}(-q)\| \leqslant \dfrac{v}{\Omega} \to 0 \qquad \text{as } \Omega \to \infty \\ \|[\Delta(p), a_{-}^{*}(q)]\| = \|\tilde{v}(p, q)a_{+}(q)\| \leqslant \dfrac{v}{\Omega} \to 0 \qquad \text{as } \Omega \to \infty \end{cases}$$

Hence in the thermodynamical limit $\Delta(p)$ commutes with all $a_{\pm}(q)$, $a_{\pm}^{*}(q)$ and seems therefore to become a c-number.

We could stop the analysis there and merely accept this "explanation." This would, however, lead us to miss two interesting problems. First, $\Delta(p)$ depends on β through the self-consistency equation, and then so does $E(p)$. How is it that the spectrum of H becomes temperature-dependent? Second, H is originally invariant under the gauge transformation $a \to ae^{i\alpha}$, whereas this is no longer the case for the diagonalized form of H. Again, how is that possible? The proper mathematical answer is that neither of these facts is possible within Fock-space formalism. The reader who is not yet convinced that this is a serious problem should try to follow, in the thermodynamical limit, what happens to the "dressed" vacuum of the $\gamma_{\pm}(p)$.

We shall come back to this point later in this section, and leave these problems open for a moment. The solution of an essentially similar problem will be presented in Section IVG.

D. van Hove Model

We want to consider the interaction of a quantum scalar meson field with recoilless nucleons, which we consider as classical point sources.

Let us first treat the case of a smooth source distribution $\rho(x)$ (say, ρ is μ-square integrable and continuous). Classically the equation of motion of the field F is then

$$[\nabla^2 - m^2 - \partial_t^{\,2}]F_t(x) = \rho(x)$$

Quantum mechanically we then have

$$H = H_0 + F(\rho)$$

where H_0 is the free-field Hamiltonian, and

$$F(\rho) = \Omega^{-1/2} \sum_k [2\omega_k]^{-1/2}\tilde{\rho}(k)(a_k + a_{-k}^{*})$$

It is then easy to show that there exists a unitary operator V such that

$$H = V^{-1}(H_0 + W \cdot I)V$$

with

$$W = \frac{1}{2}\int d^3k\omega_k^{-2}\tilde{\rho}^*(k)\tilde{\rho}(k)$$
$$= \frac{1}{2}\iint dx\, dy\rho^*(x)Y(x - y)\rho(y)$$

where

$$Y(x) = \frac{e^{-m|x|}}{4\pi\,|x|}$$

is the usual Yukawa potential.

A field $\hat{F}$ is then defined from the free field F by

$$\hat{F}(f) = V^{-1}F(f)V$$
$$= F(f) + c(f, \rho)$$

with

$$c(f, \rho) = \int d^3k\omega_k^{-2}\tilde{f}(k)^*\tilde{\rho}(k) + c.c.$$

If Φ_0 is the vacuum for the free-field, $V^{-1}\Phi_0 = \hat{\Phi}_0$ is then the vacuum for the field $\hat{F}$.

The time-evolution of the free field F is

$$F_t(f) = F(f_t) \qquad \text{with} \quad (f_t)^{\tilde{}}(k) = e^{-i\omega_k t}\tilde{f}(k)$$

whereas the time-evolution of $\hat{F}$, generated by the full Hamiltonian H, is given by

$$\hat{F}_t(f) = F_t(f) + c(f_t, \rho)$$

We then check easily that the usual asymptotic condition is satisfied for F taken as asymptotic field and $\hat{F}$ taken as interpolating field. There is therefore no scattering (as expected): $S = I$.

Interesting things now happen when we pass to the limit of a point source $[\rho(x) \to \delta(x)$, i.e., $\tilde{\rho}(k) \to 1]$. We check that the interpolating field $\hat{F}$ can still be defined (albeit with a slightly more restricted test function space) and still satisfies the asymptotic condition with the free field as asymptotic field. We therefore still have $S = I$. Up to here, everything seems in order. However, we should notice that in this limit:

1. $W \to \infty$ (i.e., infinite energy renormalization).
2. $\|H\Psi\| < \infty \Rightarrow \Psi = 0$ (i.e., the total Hamiltonian cannot be properly defined).
3. $\hat{\Phi}_0 \notin \mathscr{H}_F$ (since $\hat{\Phi}_0 \in \mathscr{H}_F \Leftrightarrow \rho \in \mathscr{L}_\mu{}^2$).

4. V cannot exist (for the same reason), and the Møller matrices of usual scattering theory cannot be defined.

5. The time-evolution $\hat{F} \rightarrow \hat{F}_t$ is not unitarily implemented.

Consequently, strict adherence to the usual postulates of quantum mechanics is not possible in this limit.

E. Summary

Both of the models just discussed present features which we want to remark upon. First, both are reported to be exactly solvable in some limit (i.e., thermodynamic limit for the BCS model, and point-source limit for the van Hove model). Second, in both cases the physics seems to come out right (i.e., energy-gap in BCS, and $S = I$ in van Hove). Third, both models show a strange mathematical behavior. Finally, both involve an infinite number of degrees of freedom.

We now take the attitude that the idealizations involved in these two models are *not* pathological, and that it is therefore relevant to understand these simple models in order to understand more sophisticated models. Furthermore, since we do not want to consider these models as pathological, we are confronted with the fact that their strange mathematical behavior must be due to some shortcoming of the formalism. More specifically, we are led to conclude that the traditional wisdom of quantum mechanics, as expressed in the Schrödinger-Fock formalism (Section IB), contains some mathematical assumption(s) which is (are) physically unwarranted.

We therefore decide to try to check whether the Fock-space formalism is as "natural" and as "universal" as it looks when viewed as a generalization of what happens when only a finite number of degrees of freedom are involved. We shall therefore try to formulate the theory without recourse to a Hilbert space, invoking only the algebraic structure of the set of observables on the systems to be considered. Moreover, in view of the great successes of ordinary quantum mechanics, we should proceed in such a way as to recover the usual theory whenever it is correct. In particular this means that we should modify the theory only in those of its aspects which are intimately connected with the occurrence of infinitely many degrees of freedom.

F. The C^*-Algebraic Approach

The fundamental assumption ("axiom," "postulate," or even "conjecture" would be acceptable words as well at this point) of the C^*-algebraic approach is that a physical system Σ can be described by a triple $\{\mathscr{A}, \mathscr{S}, \alpha\}$ where (1) $\mathscr{A}$ is a C^*-algebra, the self-adjoint elements of which can be interpreted as the *observables* on Σ, (2) $\mathscr{S}$ is the set of all positive, linear functionals on $\mathscr{A}$ which are normalized to 1. The elements $\phi \in \mathscr{S}$ are interpreted physically as *states* of Σ. The number $\langle\phi; A\rangle$ is interpreted as the expectation value of the observable A when the system is in the state ϕ. In addition, (3) α is a continuous homomorphism α: $G \to \text{Aut}(\mathscr{A})$ describing the action of the symmetry group G on the system. We now make the definitions involved more explicit and give some examples.

A C^*-*algebra* $\mathscr{A}$ is by definition a complex involutive Banach algebra such that $\|A^*A\| = \|A\|^2$ for all A in $\mathscr{A}$. This means:

1. $\mathscr{A}$ is a complex vector space, i.e., for every A and B in $\mathscr{A}$ and every λ and μ in $\mathbf{C}$ (complex numbers) there exists an element $(\lambda A + \mu B)$ in $\mathscr{A}$ and the usual associative and distributive laws hold.

2. A product AB is defined on $\mathscr{A}$, which equips it with the structure of an algebra; in particular the usual associative law, and distributive law with respect to (1) are satisfied.

3. A mapping $A \to A^*$ of $\mathscr{A}$ onto itself is given satisfying the following properties:

a. $(A^*)^* = A$.

b. $(\lambda A + \mu B)^* = \lambda^* A^* + \mu^* B^*$ (where λ^*, μ^* are the complex conjugate of λ, μ).

c. $(AB)^* = B^* A^*$.

[With the properties (1), (2) and (3) $\mathscr{A}$ is an involutive algebra. An element A in $\mathscr{A}$, such that $A^* = A$, is said to be *self-adjoint*.]

4. A mapping $A \to \|A\|$ from $\mathscr{A}$ to $\mathbf{R}^+$ (= positive numbers) is given satisfying the usual axioms of a *norm*, i.e.,

a. $\|A + B\| \leqslant \|A\| + \|B\|$.

b. $\|\lambda A\| = |\lambda| \cdot \|A\|$, $\quad [|\lambda| = +(\lambda^*\lambda)^{1/2}]$.

c. $\|AB\| \leqslant \|A\| \cdot \|B\|$.

d. $\|A^*\| = \|A\|$.

5. $\mathscr{A}$ is complete with respect to this norm, i.e., if $\{A_n\}$ is a Cauchy sequence in $\mathscr{A}$ (which means that for every $\varepsilon > 0$ there exists N_ε such that $\|A_n - A_m\| < \varepsilon$ for all $n, m > N_\varepsilon$), then there exists A in $\mathscr{A}$ such that for every $\varepsilon > 0$ there exists N_ε such that $\|A_n - A\| < \varepsilon$ for all $n > N_\varepsilon$.
[With the properties (1) to (5), $\mathscr{A}$ is a complex involutive Banach algebra.]
The property

6. $\|A^*A\| = \|A\|^2$

makes it a C^*-algebra.

Examples

1. $\mathbf{C}$ equipped with its usual structure (addition, multiplication, complex conjugation, and module) is a C^*-algebra.

2. $\mathscr{B}(\mathscr{H})$ the set of all bounded linear operator on a complex Hilbert space $\mathscr{H}$, equipped with the usual addition, multiplication, hermitian conjugation, and norm $\|A\| = \sup_{\substack{\Psi\in\mathscr{H} \\ \|\Psi\|=1}} \|A\Psi\|$ is a C^*-algebra.
3. $\mathscr{C}(\Gamma)$ the set of all continuous, complex valued functions on a compact Hausdorf space Γ, equipped with the usual addition, multiplication, and with the involution $f^*(x) = f(x)^*$, and the norm $\|f\| = \sup_{x\in\Gamma} |f(x)|$ is a C^*-algebra, etc . . . .

Now $\phi\colon \mathscr{A} \to\to \mathbf{C}$ is said to be a *state* on $\mathscr{A}$ if

a. $\langle\phi; \lambda A + \mu B\rangle = \lambda\langle\phi; A\rangle + \mu\langle\phi; B\rangle$.
b. $\langle\phi; A^*A\rangle \geqslant 0$.
c. $\sup_{\substack{A\in\mathscr{A} \\ \|A\|=1}} |\langle\phi; A\rangle| = 1$.

(Rem. If $\mathscr{A}$ has a unit I, which we will assume in the sequel for mathematical convenience, then (c) can be replaced simply by $\langle\phi; I\rangle = 1$).

A mapping α_g of $\mathscr{A}$ onto itself is said to be an automorphism of $\mathscr{A}$ [we then write $\alpha_g \in \mathrm{Aut}\,(\mathscr{A})$] if α_g respects the full C^*-algebraic structure of $\mathscr{A}$. The continuity of the mapping $\alpha\colon G \to \mathrm{Aut}\,(\mathscr{A})$ is taken here to mean that $\langle\phi; \alpha_g[A]\rangle$ is a continuous function of $g \in G$ for all ϕ in $\mathscr{S}$ and all A in $\mathscr{A}$ (it is actually sufficient to assume it for all ϕ in $\mathscr{S}$ and all A in $\mathscr{A}$ such that $A^* = A$, which then makes physical sense).

G. Historical Background

Whereas quantum mechanics followed a pragmatic path in its development, some physicists thought, even at an early stage, that difficulties might appear if the theory were not put on a more axiomatic basis as well. In particular, Jordan, von Neumann, and Wigner, considering the difficulties already accumulated by quantum field theory in the early 1930s, published about that time a series of papers in which they undertook an investigation which can properly be considered as the forerunner of the C^*-algebraic approach. However the paucity of mathematical tools available then limited their approach to the extent that this line of thought was not picked up again until the publication, in 1947, of Segal's pioneering paper. This paper, in turn, remained almost unnoticed in the physics community until 1963–1964 when, on the one hand, Haag and Kastler undertook to apply these ideas to local quantum field theories and, on the other hand, Araki and Woods demonstrated that these ideas are also relevant for a proper discussion of equilibrium statistical mechanics. Since then, the field has exhibited a properly exponential growth, its domain of applicability being extended today to quantum field theory, equilibrium and nonequilibrium statistical mechanics.

H. Bibliography

P. A. M. Dirac, *Principles of Quantum Mechanics*, Clarendon Press, Oxford, 1930.
M. H. Stone, *Linear Transformations in Hilbert Space*, Amer. Math. Soc. Colloq. Publ. XV, 1932.
V. Fock, *Z. Physik*, **75,** 622 (1932).
J. M. Cook, *Trans. Amer. Math. Soc.*, **74,** 222 (1953).
R. Haag, *Nuovo Cim.*, **25,** 287 (1962).
G. G. Emch and M. Guenin, *J. Math. Phys.*, **7,** 915 (1966).
L. van Hove, *Physica*, **18,** 145 (1952).
G. Barton, *Introduction to Advanced Field Theory*, Interscience, New York, 1963.
J. Dixmier, *Les C*-algèbres et leurs représentations*, Gauthier-Villars, 1964.
P. Jordan, J. von Neumann, and E. P. Wigner, *Ann. Math.*, **35,** 29 (1934).
I. E. Segal, *Ann. Math.*, **48,** 930 (1947).
R. Haag and D. Kastler, *J. Math. Phys.*, **5,** 848 (1964).
H. Araki and E. J. Woods, *J. Math. Phys.*, **4,** 637 (1963).

Part II

TOOLS AND TECHNIQUES

II. REPRESENTATION THEORY, THE RICHNESS OF THE C^*-ALGEBRAIC APPROACH

A. Generalities

We recall from the first section that we assumed that there exists, for the description of a physical system Σ, a C^*-algebra $\mathscr{A}$, the self-adjoint elements of which are observables on Σ. In specific problems, our first task will therefore be to determine the "proper" $\mathscr{A}$ (see, for instance, the case of the quasi-local algebra defined in Section IV). For the moment, we simply assume that we somehow got $\mathscr{A}$.

We further recall that we characterized a state ϕ on $\mathscr{A}$ by the collection $\{\langle\phi; A\rangle \mid A \in \mathscr{A}\}$ of the expectation-values of all "observables" in $\mathscr{A}$, when the system is in state ϕ. We assumed that ϕ is a positive linear functional on $\mathscr{A}$, normalized to 1.

The aim of this section is to show the following:

1. A Hilbert space formalism can be recovered.
2. For the case of systems with a finite number of degrees of freedom, he "old" quantum mechanics is recovered too.
3. For systems with an infinite number of degrees of freedom, an essentially new situation occurs which escapes the usual Fock-space formalism.

B. The GNS Construction

Definition. A mapping $\pi: \mathscr{A} \to \mathscr{B}(\mathscr{H})$ is said to be a *representation* of $\mathscr{A}$ in the Hilbert space $\mathscr{H}$ if

$$\text{(i)} \qquad \pi(\lambda A + \mu B) = \lambda\pi(A) + \mu\pi(B)$$

$$\text{(ii)} \qquad \pi(AB) = \pi(A)\pi(B)$$

$$\text{(iii)} \qquad \pi(A^*) = \pi(A)^*$$

for all A, B in $\mathscr{A}$ and all λ, μ in $\mathbf{C}$.

Note that we did not require $\|\pi(A)\| \leqslant \|A\|$. This property actually comes "free of charge" from the three conditions and the fact that $\mathscr{A}$ is an involutive Banach algebra and $\mathscr{B}(\mathscr{H})$ is a C^*-algebra. Moreover, if π is faithful (i.e., $\pi(A) = 0 \Rightarrow A = 0$), we have $\|\pi(A)\| = \|A\|$. Finally, every C^*-algebra admits a faithful representation in some Hilbert space. This comes as a corollary of the following theorem.

Theorem (GNS construction). To every state ϕ on a C*-algebra $\mathscr{A}$ correspond:

(1) a Hilbert space $\mathscr{H}_\phi$,
(2) a representation $\pi_\phi : \mathscr{A} \to \mathscr{B}(\mathscr{H}_\phi)$,
(3) a vector $\Phi \in \mathscr{H}_\phi$, such that

(a) $\overline{\pi_\phi(\mathscr{A})\Phi} = \mathscr{H}_\phi$
(b) $\langle \phi; A \rangle = (\Phi, \pi_\phi(A)\Phi) \;\forall\; A \in \mathscr{A}$.

This theorem plays such a central role in applications that it is indispensable to give at least a sketch of its proof. To this effect, we first notice that $\langle \phi; A^*B \rangle$ resembles very much to a scalar product. Indeed, we have

$$\langle \phi; A^*(\lambda B + \mu C) \rangle = \lambda \langle \phi; A^*B \rangle + \mu \langle \phi; A^*C \rangle$$

and

$$\langle \phi; A^*B \rangle^* = \langle \phi; B^*A \rangle$$

However, $\langle \phi; A^*A \rangle = 0$ does *not* imply, in general $A = 0$. We now compensate for this.

We define

$$\Phi_0 \equiv \{K \in \mathscr{A} \mid \langle \phi; K^*K \rangle = 0\}$$

and for every A in $\mathscr{A}$

$$\Phi_A \equiv \{B \in \mathscr{A} \mid B = A + K, K \in \Phi_0\}$$

We notice now that the set of all Φ_A obtained by letting A run over $\mathscr{A}$ can be equipped with the structure of a pre-Hilbert space. Specifically, we first define

$$\lambda \Phi_A + \mu \Phi_B \equiv \Phi_{\lambda A + \mu B}$$

and verify that this definition is consistent, i.e., that the right-hand side of this expression does *not* depend on the choice of A in Φ_A or on the choice of B in Φ_B. We then check easily that we get in this way a complex vector space.

The next step is to define

$$(\Phi_A, \Phi_B) \equiv \langle \phi; A^*B \rangle$$

and verify again that this definition is consistent, i.e., that the RHS of this expression does *not* depend on the choice of A in Φ_A or on the choice of B in Φ_B. We then check easily that $(\ldots, \ldots)$ is indeed a scalar product, and in particular that $\|\Phi_A\| \equiv (\Phi_A, \Phi_A)^{1/2} = 0$ implies $\Phi_A = \Phi_0$, which is indeed the "zero-vector" of our space.

With this structure

$$\mathscr{K}_\phi \equiv \{\Phi_A \mid A \in \mathscr{A}\}$$

is, by definition, a pre-Hilbert space. The Hilbert space $\mathscr{H}_\phi$ is then defined as the norm-completion of $\mathscr{K}_\phi$.

Our next task is to define π_ϕ. To this effect, we define, for each A in $\mathscr{A}$, $\pi_\phi(A)$ by

$$\pi_\phi(A)\Phi_B \equiv \Phi_{AB}$$

and verify again that this definition is consistent, i.e., that the RHS of this expression does *not* depend on the choice of B in Φ_B. Furthermore, we check easily that $\pi_\phi(A)$ is bounded (namely by $\|A\|$) on $\mathscr{K}_\phi$, so that it can be extended uniquely, by continuity, to a bounded operator on $\mathscr{H}_\phi$. It is then easy to check that π_ϕ is indeed a representation of $\mathscr{A}$ in $\mathscr{H}_\phi$. The proof is finally completed by noticing that the vector $\Phi \equiv \Phi_I$ satisfies trivially the conclusion of the theorem.

Q.E.D.

Remarks

The property (a) mentioned in the statement of the theorem is referred to by saying that Φ is a *cyclic vector* for $\pi_\phi(\mathscr{A})$ in $\mathscr{H}_\phi$, or, less specifically, by saying that π_ϕ is a *cyclic representation.*

Now if ν is another cyclic representation of $\mathscr{A}$ in some Hilbert space $\mathscr{H}$, with cyclic vector Ψ such that $\langle\phi; A\rangle = (\Psi, \nu(A)\Psi)$ for all A in $\mathscr{A}$, it is easy to check that the mapping $V: \mathscr{H}_\phi \to \mathscr{H}$ generated by $V\Phi_A = \nu(A)\Psi$ establishes a unitary equivalence between $\pi_\phi(\mathscr{A})$ and $\nu(\mathscr{A})$, i.e., (1) V is unitary, and (2) $V\pi_\phi(A)V^{-1} = \nu(A)\ \forall\ A \in \mathscr{A}$. This is often referred to by saying that ϕ determines the cyclic representation π_ϕ "up to a unitary equivalence."

We have tacitly assumed that $\mathscr{A}$ has a unit. This is, however, not an actual limitation on the validity of the theorem.

Example 1. Let $\mathscr{A}$ be the C^*-algebra at all 2×2 matrices with complex entries, and usual composition laws (discussed earlier). We then have

$$\mathscr{A} \equiv \{A = aI + b\sigma^x + c\sigma^y + d\sigma^z \mid a, b, c, d \in \mathbf{C}\}$$

where σ^x, σ^y, σ^z are the Pauli matrices. Let

$$\langle\phi; A\rangle = a + d$$

Then

$$\pi_\phi(A) = \begin{pmatrix} a+d & b-ic \\ b+ic & a-d \end{pmatrix} = A; \qquad \Phi = \begin{pmatrix} 1 \\ 0 \end{pmatrix}$$

Example 2. Let $\mathscr{A}$ be the same as in Example 1, and

$$\langle\phi; A\rangle = \mathrm{Tr}\,\rho A \qquad \text{with} \quad \rho = \frac{e^{-\beta H}}{\mathrm{Tr}\, e^{-\beta H}}$$

and $H = -B\sigma^z$. We have then

$$\langle\phi; A\rangle = a + td \qquad \text{with } 0 < t \equiv \tanh \beta B < 1$$

Then

$$\pi_\phi(A) = \begin{pmatrix} A & 0 \\ 0 & A \end{pmatrix}$$

and

$$\Phi = \begin{pmatrix} u \\ 0 \\ 0 \\ v \end{pmatrix} \qquad \text{with } \begin{cases} u = [\frac{1}{2}(1+t)]^{1/2} \\ v = [\frac{1}{2}(1-t)]^{1/2} \end{cases}$$

C. Irreducible Representations

The two examples just given seem to indicate that there is a relation between "ϕ pure state" and "π_ϕ irreducible." This is indeed the case.

Definitions. Let $\pi: \mathscr{A} \to \mathscr{B}(\mathscr{H})$ be a representation of $\mathscr{A}$ and $\mathscr{M}$ be a (closed) subspace of $\mathscr{H}$. $\mathscr{M}$ is said to be *stable* with respect to $\pi(\mathscr{A})$ if $\Psi \in \mathscr{M}$ implies $\pi(A)\Psi \in \mathscr{M}$ for all $A \in \mathscr{A}$. The representation π is said to be *irreducible* if the only stable subspaces which it admits are $\{0\}$ and $\mathscr{H}$.

Definitions. A state ϕ on the C^*-algebra $\mathscr{A}$ is said to be a *mixture* of the states ϕ_1 and ϕ_2 on $\mathscr{A}$ if there exists λ with $0 < \lambda < 1$ such that

$$\langle\phi; A\rangle = \lambda\langle\phi_1; A\rangle + (1-\lambda)\langle\phi_2; A\rangle$$

for all A in $\mathscr{A}$. A state ϕ is said to be *pure* if it cannot be written as a mixture of other states.

Definition. If $\mathscr{N}$ is a subset of $\mathscr{B}(\mathscr{H})$, let

$$\mathscr{N}' \equiv \{A \in \mathscr{B}(\mathscr{H}) \mid AN = NA \ \forall\ N \in \mathscr{N}\}$$

A self-adjoint subalgebra of $\mathscr{B}(\mathscr{H})$ is said to be a *von Neumann* algebra if

$$\mathscr{N} = \mathscr{N}'' \equiv (\mathscr{N}')'$$

Theorem. Let $\mathscr{A}$ be a C^*-algebra, and ϕ be a state on $\mathscr{A}$. Then the following conditions are equivalent:

1. ϕ is pure.
2. π_ϕ is irreducible.
3. $\pi_\phi(\mathscr{A})' = \{\lambda I \mid \lambda \in \mathbf{C}\}$.
4. $\pi_\phi(\mathscr{A})'' = \mathscr{B}(\mathscr{H}_\phi)$.
5. Every $\Psi \in \mathscr{H}_\phi$, $\Psi \neq 0$ is cyclic for $\pi_\phi(\mathscr{A})$.
6. Every $\Psi \in \mathscr{H}_\phi$, $\|\Psi\| = 1$ generates, via $\langle\psi; A\rangle \equiv (\Psi, \pi_\phi(A)\Psi)$, a pure state on $\mathscr{A}$.

D. Application to the Canonical Commutation Relations

The ingenuity shown by certain mathematicians in finding all kinds of fancy conditions under which the uniqueness of the Schrödinger representation can be proved is amazing. To avoid losing oneself in mathematical details one has a definite advantage in thinking first of the primary physical meaning of P and Q. In words: Q should be an operator describing the position of a particle, and P should be the generator of space translations. The mathematical translation of these requirements is that we want the following:

1. A Hilbert space $\mathscr{H}$.
2. A self-adjoint operator Q acting on $\mathscr{H}$.
3. A continuous, one-parameter group $\{U(a) \mid a \in \mathbf{R}\}$ of unitary operators acting on $\mathscr{H}$ in such a manner that (4) is true.
4. $U(a)QU(-a) = Q - aI$ for all a in $\mathbf{R}$.

We can then prove (without further assumptions) the following:

5. The spectrum of Q extends from $-\infty$ to $+\infty$.
6. The continuous, one-parameter group $\{V(b) \mid b \in \mathbf{R}\}$ of unitary operators acting on $\mathscr{H}$ and generated by Q satisfies $U(a)V(b) = e^{iab} \times V(b)U(a)$.
7. If $\{U(a), V(b) \mid a, b, \in \mathbf{R}\}$ is irreducible, $\mathscr{H}$ is separable.

The relations

$$(*)\begin{cases} U(a_1)U(a_2) = U(a_1 + a_2) \\ V(b_1)V(b_2) = V(b_1 + b_2) \\ U(a)V(b) = e^{iab}V(b)U(a) \end{cases}$$

required for all a_1, a_2, b_1, b_2, a, b in $\mathbf{R}$, *together* with the continuity requirements given above, *and* the fact that $U(a)$, $V(b)$ are unitary operators, are the so-called *CCR in Weyl's form*, for one degree of freedom. We can compound the three relations (*) into one, upon introducing for every $z = x + iy \in \mathbf{C}$, x and $y \in \mathbf{R}$:

$$W(z) \equiv U(x)V(y)e^{-ixy/2}$$

with this notation, (*) becomes:

$$W(z_1)W(z_2) = e^{i\,Im\,(z_1 * z_2)/2}W(z_1 + z_2)$$

and the unitary requirement becomes

$$W(z)^* = W(-z)$$

The passage to $n \leqslant \infty$ degrees of freedom is then obtained from the following substitution:

$\mathbf{C} \to \mathscr{T}_{\mathbf{C}}$ complex (pre-)Hilbert space with $n = \dim \mathscr{T}_{\mathbf{C}}$

$z_1^* z_2 \to (f_1, f_2)$ scalar product in $\mathscr{T}_{\mathbf{C}}$

$$W(z) \to W(f) \quad \text{with} \begin{cases} W(f_1)W(f_2) = e^{i\,Im\,(f_1, f_2)/2}W(f_1 + f_2) \\ W(f)^* = W(-f). \end{cases}$$

We pass here over the details (in particular the precise formulation of the appropriate continuity conditions) and assert that the following result can be proved as a corollary of the GNS construction theorem.

Result. To every $\hat{\phi}: \mathscr{T}_{\mathbf{C}} \to \mathbf{C}$ such that

(i) $\hat{\phi}(0) = 1$

(ii) $\hat{\phi}(f + \lambda g)$ is continuous in $\lambda \in \mathbf{R}$ for all $f, g \in \mathscr{T}_{\mathbf{C}}$.

(iii) For *every* finite sequence $\{\lambda_k, f_k \mid k = 1, 2, \ldots, m\}$ of elements λ_k in $\mathbf{C}$ and f_k in $\mathscr{T}_{\mathbf{C}}$,

$$\sum_{k,l=1}^{m} \lambda_l^* \lambda_k \hat{\phi}(f_k - f_l) \exp\left\{\frac{-i\,Im\,(f_l, f_k)}{2}\right\} \geqslant 0$$

corresponds a unique (up to unitary equivalence) continuous representation $\{W(f) \mid f \in \mathscr{T}_{\mathbf{C}}\}$ of the CCR, with cyclic vector Φ such that

$$\hat{\phi}(f) = (\Phi, W(f)\Phi)$$

for all f in $\mathscr{T}_{\mathbf{C}}$; and conversely.

E. von Neumann Theorem

We are now ready to reach our second aim stated at the beginning of this section.

We first notice that in the "natural" representation of the CCR in the case $n = 1$, mentioned in the first section, we have

$$\mathscr{H} = \mathscr{L}^2(\mathbf{R})$$

and

$$(Q\psi)(\xi) = \xi\psi(\xi), \qquad \text{i.e., } [V(b)\Psi](\xi) = e^{-ib\xi}\Psi(\xi)$$

$$(P\psi)(\xi) = -i\frac{d\psi}{d\xi}(\xi), \qquad \text{i.e., } [U(a)\Psi](\xi) = \Psi(\xi - a)$$

In particular, for

$$\Phi(\xi) = \pi^{-1/4}e^{-\xi^2}$$

we get

$$\hat{\phi}(z) = (\Phi, W(z)\Phi) = e^{-|z|^2/4}$$

We next make the preliminary remark that

$$d\mu(z) = (2\pi)^{-1/2} e^{-|z|^2/4}\, dz$$

is the unique normalized measure which is invariant with respect to all transformations $z \to z'$ such that $|z| = |z'|$. Suppose now that $W(z)$ is an arbitrary, irreducible representation of the CCR in Weyl form. Then

$$A \equiv \int d\mu(z) W(z) = A^*$$

We then verify that

$$AW(z)A = e^{-|z|^2/4} A$$

and also that $A \neq 0$. In particular (write $z = 0$ above), $A^2 = A$. Hence A is a nonzero orthogonal projector. Let Φ be such that $A\Phi = \Phi$ and $\|\Phi\| = 1$. Since $W(z)$ is assumed to be irreducible, Φ is cyclic. Again from the preceding expression, now for arbitrary z, we get

$$(\Phi, W(z)\Phi) = (\Phi, AW(z)A\Phi) = e^{-|z|^2/4}(\Phi, A\Phi)$$

i.e.,

$$\hat{\phi}(z) = e^{-|z|^2/4}$$

Hence this representation is unitarily equivalent to the usual representation in $\mathscr{L}^2(\mathbf{R})$ by the foregoing result; the case $n < \infty$ is obtained similarly with

$$d\mu(f) = (2\pi)^{-n/2} e^{-\|f\|^2/4}\, df$$

To conclude, for all $n < \infty$, there exists only one (up to unitary equivalence) representation of the CCR which is irreducible, and this representation is obtained explicitly in $\mathscr{L}^2(\mathbf{R}^n)$ by the usual correspondence principle of quantum mechanics.

F. Fock Space

The space is characterized by the existence of a cyclic vector Φ such that $a(f)\Phi = 0$ for all f in $\mathscr{T}_{\mathbf{C}}$.

It is not difficult to show (although it requires some ingenuity to give a short proof of it) that this condition implies that

$$\hat{\phi}_F(f) = (\Phi, W(f)\Phi) = e^{-\|f\|^2/4}$$

whatever $\mathscr{T}_{\mathbf{C}}$ is. Hence we know how to characterize the Fock representation. The proof given of the von Neumann theorem in the previous section collapses if we extend n to ∞; this reflects the absence of an invariant measure in the infinite dimensional $\mathscr{T}_{\mathbf{C}}$. In itself, this is not yet a proof that there exist other irreducible representations of the CCR which are not equivalent to the Fock representation. The proof is, however, achieved if we exhibit irreducible representations which do not admit a Φ leading to the above $\hat{\phi}_F$. This is exactly what occurs in the case of the van Hove model with the state corresponding to the "dressed vacuum," and there the nonexistence of the Møller matrices is just a reflection of the existence of inequivalent representation at the CCR, of physical relevance!

G. The Free Bose Gas

The didactic value of this model is that we can produce nontrivial $\hat{\phi}$ with elementary analytical tools. Consider first the case of the *ground state*. We enclose the system in a box of volume Ω. At finite density ρ we have $N = \rho\Omega$ particles in wavefunction $f_\Omega(x) = \Omega^{-1/2}$. Since $N < \infty$, we can use Fock space without loss of generality. The state of the system is obtained from the vector

$$\Psi = (N!)^{-1/2} a^*(f_\Omega)^N \Phi_0$$

We form then

$$\hat{\phi}_{\Omega,N}(f) = (\Psi, W(f)\Psi)$$

An elementary calculation then shows that

$$\hat{\phi}_{\Omega,N}(f) = \hat{\phi}_F(f) L_N\{\tfrac{1}{2}[(f_\Omega, f_1)^2 + (f_\Omega, f_2)^2]\}$$

where $L_n(x)$ denotes the nth Laguerre polynomial of argument x, and f_1 and f_2 are respectively the real and imaginary parts of f.

The passage to the thermodynamic limit is then done in close agreement with the traditional rule,

$$\hat{\phi}(f) = \lim_{\substack{N\to\infty \\ N/\Omega=\rho}} \hat{\phi}_{\Omega,N}(f)$$

and involves a nice little piece of classical analysis to show that

$$\hat{\phi}(f) = \hat{\phi}_F(f) J_0\{2\rho[\tilde{f}_1(0)^2 + \tilde{f}_2(0)^2]\}$$

where $J_0(x)$ denotes the zeroth Bessel function. We can then in principle construct or, with some luck, guess directly the corresponding representation of the CCR. This is left to the reader (if he should turn impatient with this "exercise," he can find comfort in the paper of Araki and Woods).

The case of the canonical equilibrium for chemical potential μ and natural temperature β is worked out in much the same way. We start in a finite box, define

$$\rho_\Omega = \frac{\exp\{-\beta(H - \mu N)\}}{\mathrm{Tr}\,[\exp\{-\beta(H - \mu N)\}]}$$

$$\hat{\phi}_\Omega(f) = \mathrm{Tr}\,\{\rho_\Omega W(f)\}$$

and then pass to the thermodynamic limit to get

$$\hat{\phi}_{\beta,\mu}(f) = \hat{\phi}_F(f) \exp\left\{-\tfrac{1}{2}\int dk\tilde{\rho}(k)[|\tilde{f}_1(k)|^2 + |\tilde{f}_2(k)|^2]\right\}$$

with

$$\tilde{\rho}(k) = (2\pi)^{-3}[\exp \beta(E_k - \mu) - 1]^{-1}$$

and

$$E_k = \frac{k^2}{2m}$$

The representation corresponding to this is easily checked to be

$$\begin{cases} \mathscr{H} = \mathscr{H}_F \otimes \mathscr{H}_F \\ U(f) = U_F([1 + \rho]^{1/2}f) \otimes U_F(-\rho^{1/2}f) \\ V(f) = V_F([1 + \rho]^{1/2}f) \otimes V_F(\rho^{1/2}f) \\ \Phi = \Phi_F \otimes \Phi_F \end{cases}$$

We shall come to the remarkable properties of this representation toward the end of the third section in connection with the Kubo-Martin-Schwinger condition.

H. Bibliography

J. Dixmer, *loc. cit.*

J. Manuceau, *Ann. Inst. Henri Poincaré*, **8,** 139 (1968).

J. v. Neumann, *Math. Ann.*, **104,** 570 (1931).

G. W. Mackey, *Induced Representations and Quantum Mechanics*, Benjamin, New York, 1969.

K. O. Friedrichs, *Mathematical Aspects of Q.F. Theory*, Interscience, New York, 1953. (Part I–V of this book, which is out of print, appeared in *Comm. Pure Appl. Math.*, **4, 5, 6** (1951–1953).

H. Araki and J. Woods, *loc. cit.*

III. SYMMETRIES, CLUSTER PROPERTIES, AND THE KMS CONDITION

A. Generalities

It seems physically reasonable to say that a symmetry of a physical theory is a transformation of this theory which preserves its structure. In the formalism studied in this article, this idea is embodied in the following definition. Let $\mathscr{A}$ be the C^*-algebra associated to the system Σ under consideration. Then a symmetry of Σ is defined as an automorphism of $\mathscr{A}$, i.e., as a mapping α of $\mathscr{A}$ onto itself, which satisfies the following properties:

$$\text{(i)} \qquad \alpha[\lambda A + \mu B] = \lambda\alpha[A] + \mu\alpha[B]$$
$$\text{(ii)} \qquad \alpha[AB] = \alpha[A]\alpha[B]$$
$$\text{(iii)} \qquad \alpha[A^*] = \alpha[A]^*$$

for all A, B in $\mathscr{A}$ and λ, μ in $\mathbf{C}$.

We define accordingly the action of a symmetry group G on $\mathscr{A}$ as an homomorphism $\alpha: G \to \text{Aut}(\mathscr{A})$, i.e., to each g in G, α attributes an automorphism α_g of $\mathscr{A}$ in such a manner that

(i) $\quad \alpha_{g_1}[\alpha_{g_2}[A]] = \alpha_{g_1 g_2}[A] \qquad$ for all g_1, g_2 in G and A in $\mathscr{A}$.

(ii) $\quad \langle\phi; \alpha_g[A]\rangle \qquad$ is a continuous function of g for all ϕ and A (resp. state and observable)

We finally say that a state ϕ is invariant under the action of G (or simply that ϕ is G-inv) if

$$\langle\phi; \alpha_g[A]\rangle = \langle\phi; A\rangle \qquad \text{for all } g \text{ in } G \text{ and } A \text{ in } \mathscr{A}$$

Some brief technical remarks may be in order here. It follows from the preceding definition that

$$\|\alpha_g[A]\| = \|A\| \qquad \text{for all } g \text{ in } G \text{ and } A \text{ in } \mathscr{A}$$

The definition of symmetry just given is not the most general we can think of. Actually a real definition of symmetry should involve the observables themselves, i.e., the self-adjoint elements of $\mathscr{A}$. Mathematically we can equip the self-adjoint part of $\mathscr{A}$ with a structure of Jordan algebra (with the product $A \cdot B = AB + BA$) and then impose that a symmetry be a Jordan automorphism, thus preserving only that structure. Physically, such a generalization of our initial definition is required if we want to give a proper account of symmetries such as time-reversal. Since we are mainly concerned in this article with connected symmetry groups, this generalization will be unnecessary in the immediate sequel.

When we say "group," we mean "topological group." Whereas the foregoing definitions make sense for any topological group of symmetries, we restrict ourselves here (and especially in Section IIIC) to locally compact groups. From the physicist's point of view this is not a very severe restriction.

We chose to define a symmetry in the "Heisenberg picture." The passage to the "Schrödinger picture" is obtained by defining $\nu_g: \mathscr{S} \to \mathscr{S}$ (space of states) by

$$\langle\nu_g[\phi]; A\rangle = \langle\phi; \alpha_g[A]\rangle$$

i.e., $\nu_g = \alpha_g^*$.

Whereas the preceding definitions work very nicely for space symmetries, gauge symmetries, etc., some minor adaptations might turn out to be necessary for dealing with the time-evolution. We shall ignore this problem in this section, and postpone it to the next.

B. Covariant Representations

Upon using the GNS construction it is easily checked that if ϕ is a G-inv state on $\mathscr{A}$ there exists in $\mathscr{H}_\phi$ a (strongly) continuous unitary representation $\{U_\phi(g) \mid g \in G\}$ of G such that (1) $U_\phi(g)\Phi = \Phi$ (where Φ is the GNS cyclic vector associated to ϕ), and (2) $\pi_\phi(a_g[A]) = U_\phi(g) \times \pi_\phi(g)U_\phi(g^{-1})$ for all g in G and A in $\mathscr{A}$. [The proof is trivially obtained by defining $U_\phi(g)\Phi_A = \Phi_{\alpha_g[A]}$ (Φ_A defined as in Section IIB).] The reader familiar with classical ergodic theory will recognize in this result an extension to quantum situations of the Koopman formalism for classical mechanics.

C. Averages

In several physical problems we are interested in quantities of the form

$$\overline{\langle \phi; \alpha_g[A] \rangle}^G$$

where ϕ is a state, A is an observable, and $^{-G}$ denotes some "invariant averaging process" with respect to G. Traditional examples are embodied in most formulations of statistical mechanics. One instance is the usual "ergodic average"

$$\lim_{T \to \infty} (2T)^{-1} \int_{-T}^{T} dt \langle \phi; \alpha_t[A] \rangle$$

another is the definition of some of the macroscopic observables as space averages (e.g., magnetization).

Since many averaging processes might be thought of, we first give a general definition of an "invariant mean" (or "invariant averaging process") on G. Let $\mathscr{C}(G)$ be the C^*-algebra of all complex valued, continuous, bounded functions on G, equipped with the usual operations:

$$\begin{aligned} f^*(g) &\equiv f(g)^* \\ (\lambda_1 f_1 + \lambda_2 f_2)(g) &\equiv \lambda_1 f_1(g) + \lambda_2 f_2(g) \\ (f_1 f_2)(g) &\equiv f_1(g) f_2(g) \\ \|f\| &\equiv \sup_{g \in G} |f(g)| \end{aligned}$$

We notice that for every A in $\mathscr{A}$ and every ϕ, state on $\mathscr{A}$, and for every $\alpha: G \to \text{Aut}(\mathscr{A})$, $\langle \phi; \alpha_g[A] \rangle$ considered as a function of g (at fixed A and ϕ) indeed belongs to $\mathscr{C}(G)$.

We now define a *mean* on G as a mapping $\langle \eta; \cdot \rangle : \mathscr{C}(G) \to \mathbf{C}$ satisfying the following properties:

1. η is linear.
2. $\langle \eta; f \rangle \geqslant 0$ for all positive f in $\mathscr{C}(G)$.
3. $\langle \eta; u \rangle = 1$ where $u(g) \equiv 1$ for all g in G.

These are indeed reasonable statements to make to define a mean. Mathematically, this definition amounts to saying that η is a *state* over the C^*-algebra $\mathscr{C}(G)$.

However, we do not want to be satisfied with "just" means. We want "invariant means." To define these, we must first define the action of G on $\mathscr{C}(G)$. There are essentially two ways to do this:

1. $$(h[f])(g) \equiv f(h^{-1}g)$$
2. $$([f]h)(g) \equiv f(gh)$$

We say that a mean is an *invariant mean* when

$$\langle \eta; h[f] \rangle = \langle \eta; f \rangle = \langle \eta; [f]h \rangle$$

for all f in $\mathscr{C}(G)$ and all h in G.

The existence of at least one invariant mean characterizes a class of groups, the *amenable groups*, which include *translation* groups $\mathbf{R}^n$, *rotation* groups O^n, and *Euclidian* groups $\mathbf{E}^n$. (*Rem.* the reason is $\mathbf{R}^n$ abelian, O^n compact, and $\mathbf{E}^n$ semi-direct product of $\mathbf{R}^n$ and O^n.) However, the Lorentz group is *not* amenable.

Returning to our central theme, we mention the following result, which gives the definition, in certain situations, of the invariant average of any observable. If

$$\phi \text{ is a } G\text{-inv state on } \mathscr{A}$$

$$\eta \text{ is an inv. mean on } G$$

then there exists a mapping

$$\eta_\phi : \mathscr{A} \to \pi_\phi(\mathscr{A})'' \cap U_\phi(G)'$$

such that

(i) $$\eta(\Psi_1, U_\phi(g)\pi_\phi(A)U_\phi(g^{-1})\Psi_2) = (\Psi_1, \eta_\phi[A]\Psi_2)$$

for all Ψ_1, Ψ_2 in $\mathscr{H}_\phi$ and all A in $\mathscr{A}$.

(ii) $$\eta_\phi[A]E_\phi = E_\phi \eta_\phi[A] = E_\phi \pi_\phi(A) E_\phi \;\forall\; A \in \mathscr{A}$$

where

$$E_\phi = \{\Psi \in \mathscr{H}_\phi \mid U_\phi(g)\Psi = \Psi \;\forall\; g \in G\}$$

(with E_ϕ we denote as well the projector on this subspace).

D. Extremal G-Invariant States, η-Clustering, and Ergodicity

1. *Preliminary Remarks*

We say that a G-inv state ϕ on $\mathscr{A}$ is *extremal G-invariant* if it cannot be decomposed (on a nontrivial manner) into a mixture of G-inv states. The extremal G-inv states form in some sense the "building blocks" of a covariant theory, and therefore we try to investigate their properties here.

One of the most exciting problems of equilibrium statistical mechanics is the problem of phase transitions. These are often characterized by the setting of a "long-range order" which will mean that (in some sense to be made precise later) we have

$$\text{"}\lim_{g\to\infty}\text{"}\ |\langle\phi;\alpha_g[A]B\rangle - \langle\phi;A\rangle\langle\phi;B\rangle| \neq 0$$

In classical mechanics, an alternate definition of ergodicity is to say that (in the Koopman formalism, see in particular Arnold-Avez) a system is ergodic if the projector

$$E_0 \equiv s - \lim_{T\to\infty}(2T)^{-1}\int_{-T}^{T} dtU_t$$

is one-dimensional. The following theorem shows how these notions (as well as some others) are linked.

2. *The Principal Theorem*

Let ϕ be a G-inv state on $\mathscr{A}$

η be an invariant mean on G

Then the following conditions are in the logical relation:

$$\begin{array}{ccccccc} & & (v) \Leftrightarrow (iv) & \Leftarrow & (vii) \Rightarrow & (viii) \\ & & \Downarrow & & & \Downarrow \\ (vi) \Leftarrow (iii) & \Leftarrow & (ii) & \Leftrightarrow & (i) & \Leftarrow (ix) \end{array}$$

Where

(i) ϕ is extremal G-inv.

(ii) $\pi_\phi(\mathscr{A})' \cap U_\phi(G)' = \{\lambda I\}$

(iii) $\{\pi_\phi(\mathscr{A}), U_\phi(G)\}'' \cap \pi_\phi(\mathscr{A})' \cap U_\phi(G)' = \{\lambda I\}$

(iv) $\eta\ \langle\phi;\alpha_g[A]B\rangle = \langle\phi;A\rangle\langle\phi;B\rangle\ \forall\ A, B \in \mathscr{A}$ (called η-clustering)

(v) E_ϕ is one-dimensional (called ergodicity for ϕ)

(vi) $\pi_\phi(\mathscr{A})'' \cap \pi_\phi(\mathscr{A})' \cap U_\phi(G)' = \{\lambda I\}$

(vii) $\eta_\phi[A] = \langle\phi;A\rangle I\ \forall\ A \in \mathscr{A}$ (i.e., averaged observables are "C-numbers")

(viii) $\tilde{\phi}$ is the only normal G-inv state on $\pi_\phi(\mathscr{A})''$ (see Dixmier for thorough definition)

(ix) ϕ is the only G-inv vector state on $\pi_\phi(\mathscr{A})$ [i.e., if $\langle\psi; A\rangle$ is G-inv for all A in $\mathscr{A}$ and $\langle\psi; A\rangle = (\Psi, \pi_\phi(A)\Psi)$ for some Ψ in $\mathscr{H}_\phi$ and all A in $\mathscr{A}$, then $\psi = \phi$].

If, furthermore,

$$\eta\langle\psi; \alpha_g[A]B - B\alpha_g[A]\rangle = 0$$

for all A, B in $\mathscr{A}$ and all ψ of the form $\langle\psi; A\rangle = (\Psi, \pi_\phi(A)\Psi)$ with Ψ in $E_\phi\mathscr{H}_\phi$ (this condition is called G-abelianess on ϕ), *then* the first five conditions above are equivalent.

If, moreover,

$$\eta\langle\phi; D^*(\alpha_g[A]B - B\alpha_g[A])D\rangle = 0$$

for all A, B, D in $\mathscr{A}$ (this condition is called η-abelianess on ϕ), *then* all nine conditions are equivalent.

As a corollary, let ϕ be extr. G-inv on $\mathscr{A}$. If in addition Φ (the cyclic GNS vector in $\mathscr{H}_\phi$) is also cyclic with respect to $\pi_\phi(\mathscr{A})'$, then:

1. The group G acts in a η-abelian manner on ϕ (whatever invariant mean on G is chosen).
2. The nine conditions of the theorem are satisfied.

We leave all proofs to the reader who might find useful hints in the literature mentioned in the bibliographical notes.

Note that in most physical applications, $\mathscr{A}$ is taken as the algebra of quasi-local observables on Σ (see, e.g., the next section). In this case, the translation group (and hence the euclidian group) *does* act in an η-abelian manner on all ϕ on $\mathscr{A}$ as a consequence of locality (i.e., the fact that two observables relative to disjoint regions of space commute). The question of whether the time-evolution acts in the same manner is much more delicate, and on the basis of explicit solutions of known models, we know that this property is by no means guaranteed in general.

E. Formulation of the KMS Condition

This condition was first recognized as a property of thermal Green functions by Kubo, Martin, and Schwinger (hence its name KMS). We begin by formulating it in its most naïve form. Let $\mathscr{A} = \mathscr{B}(\mathscr{H})$ and H be the Hamiltonian supposed to be defined in such a way that the density matrix $\rho = e^{-\beta H}/\mathrm{Tr}\, e^{-\beta H}$ exists. Consider then the state ϕ defined as

$$\langle\phi; A\rangle = \frac{\mathrm{Tr}\, e^{-\beta H} A}{\mathrm{Tr}\, e^{-\beta H}}$$

which is nothing but the state describing the canonical equilibrium ensemble. For all A and B in $\mathscr{A}$ we have

$$\langle\phi; \alpha_t[A]B\rangle = (\mathrm{Tr}\, e^{-\beta H})^{-1}\, \mathrm{Tr}\, e^{-\beta H} e^{iHt} A e^{-iHt} B$$

Upon using the cyclic invariance of the trace, we find

$$\langle\phi; \alpha_t[A]B\rangle = \langle\phi; B\alpha_{t+i\beta}[A]\rangle$$

for all A, B in $\mathscr{A}$. We can rewrite this identity in the form

$$\int dt f_0(t)\langle\phi; \alpha_t[A]B\rangle = \int dt f_{-\beta}(t)\langle\phi; B\alpha_t[A]\rangle$$

for all A, B in $\mathscr{A}$ and all $\tilde{f} \in \mathscr{D}$ ($\equiv$ space of all continuously differentiable functions to all order, which have compact support), with

$$f_\gamma(t) \equiv \int d\omega \tilde{f}(\omega) e^{i\omega(t+i\gamma)}$$

Before leaving this trivial case, we should notice that this condition determines ρ uniquely, and conversely H generates the only time-evolution for which $\langle\phi; A\rangle = \mathrm{Tr}\, \rho A$ satisfies this condition.

Now, on the basis of the results of Araki and Woods (*loc. cit.*), Haag, Hugenholtz, and Winnink recognized that this condition could be extended to the case of the *infinite*, free Bose gas, treated in the thermodynamic limit. Since then, several systems have been studied and proved to be such that, in the thermodynamic limit, the Gibbs state ϕ does indeed satisfy the KMS condition.

This condition was further proved (see already Haag, Hugenholtz, and Winnink) to have far-reaching consequences. Many speculations have been built from there, some of which we shall present in the next two sections. Let it suffice for the moment to mention that for classical systems, such as a lattice gas, the connection between the KMS condition and the usual definition of equilibrium states (in the thermodynamic limit) has also been established.

F. Elementary* Consequences of the KMS Condition

First, we should mention that ϕ KMS implies that ϕ is invariant under the time-evolution (which is certainly an indispensable property to have if we want at all to interpret the KMS condition as an equilibrium condition). As far as the structure of the representations goes (and the reader

* "Elementary" here does not refer so much to the fact that the following consequences can be proved easily, but rather to the fact that they (at least some of them) can be proved only under rather technical conditions which we cannot enter into here, but which will be satisfied for all the elementary models of the next section.

will check the veracity of the assertion below, in the case of the Bose gas already described) it could be mentioned that if ϕ is a KMS state on $\mathscr{A}$, then there exists an antiunitary operator C on $\mathscr{H}_\phi$ (with $C^2 = I$) such that (1) $C\Phi = \Phi$ (where Φ is the GNS cyclic vector), (2) $C\pi_\phi(\mathscr{A})''C = \pi_\phi(\mathscr{A})'$ (so that $\pi_\phi(\mathscr{A})'\Phi$ is dense in $\mathscr{H}_\phi$, i.e., Φ cyclic for $\pi_\phi(\mathscr{A})'$; see then III.D.2), and (3) $CU_\phi(t)C = U_\phi(t)$ (so that the spectrum of H_ϕ, the generator of $U_\phi(t)$, has to be symmetrical around the origin).

If ϕ is KMS, then

$$\mathscr{Z}_\phi(\mathscr{A}) \equiv \pi_\phi(\mathscr{A})'' \cap \pi_\phi(\mathscr{A})' \subseteq \pi_\phi(\mathscr{A})' \cap U_\phi(G)'$$

and hence the elements of the center of $\pi_\phi(\mathscr{A})''$ are left invariant by the time-evolution. This has an immediate physical consequence. Indeed, due to locality, the space average $\eta_\phi[A]$ of any local observable does belong to $\mathscr{Z}_\phi(\mathscr{A})$ and hence is time-invariant.

We now pass to the next step, the review of the properties of extremal KMS states (i.e., KMS states which cannot be decomposed as mixtures of other KMS states). As we shall understand later, we will want to conjecture that pure thermodynamical phases are extremal KMS states.

We should first notice that a KMS state ϕ is extremal KMS if and only if $\mathscr{Z}_\phi(\mathscr{A}) = \{\lambda I\}$ (i.e., if and only if $\pi_\phi(\mathscr{A})''$ is a factor, or if and only if ϕ is "primary"); this property will be crucial in establishing the uniqueness of the decomposition of an arbitrary KMS state into its extremal KMS components; i.e., if our conjecture is correct, the uniqueness of the decomposition of an equilibrium state as a mixture of pure thermodynamic phases.

We now want to mention that if ϕ is an *extr. KMS state* (of course with respect to the time-evolution) and if G is a symmetry of the theory (such as translation in space, or euclidian group) such that ϕ is G-inv and G acts in a η-abelian manner on ϕ, then all nine conditions of the principal theorem of III.D.2 are satisfied with respect to G. Actually, in the cases where $\mathscr{A}$ is the algebra of quasi-local observables on a physical system, and where G is the translation group (or for that matter the euclidian group itself), the η-clustering property can be considerably strengthened. We say that a state ϕ is *uniformly clustering* if given any $\epsilon > 0$ and any A in $\mathscr{A}$ there exist a (closed) finite region Ω in $\mathbf{R}^3$ such that

$$|\langle\phi; AB\rangle - \langle\phi; A\rangle\langle\phi; B\rangle| \leqslant \epsilon \, \|B\|$$

for all B "outside" Ω (we shall make this expression precise in the next section). Then we can show that ϕ KMS on $\mathscr{A}$ is extremal KMS if and only if ϕ is uniformly clustering.

G. Bibliography

V. I. Arnold and A. Avez, *Ergodic Problems of Classical Mechanics*, Benjamin, New York, 1968.

J. Diximier, *loc. cit.*

R. V. Kadison, *Topology*, **3,** Suppl. 2, 177 (1965).

S. Doplicher, R. V. Kadison, D. Kastler, and D. W. Robinson, *Comm. math. Phys.*, **6,** 101 (1967).

O. E. Lanford and D. Ruelle, *J. Math. Phys.*, **8,** 1460 (1967).

E. P. Greenleaf, *Invariant Means on Topological Groups*, Van Nostrand, Princeton, N.J., 1969.

R. Kubo, *J. Phys. Soc. Japan*, **12,** 570 (1957).

P. C. Martin and J. Schwinger, *Phys. Rev.*, **115,** 1342 (1959).

R. Haag, N. Hugenholtz, and M. Winnink, *Comm. math. Phys.*, **5,** 215 (1967).

H. J. Brascamp, *Comm. math. Phys.*, **18,** 82 (1970).

O. E. Lanford and D. Ruelle, *Comm. math. Phys.*, **13,** 194 (1969).

G. G. Emch, *Algebraic Methods in Statistical Mechanics and Quantum Field Theory*, Wiley, New York, 1972.

Part III
APPLICATIONS

IV. SPIN SYSTEMS AND LATTICE-GASES

A. Generalities

This section aims at presenting, on the basis of exactly soluble models, some support for the following conjectures:

1. A canonical equilibrium state (for an infinite system as well as for a finite system) should be a KMS state (see Section IIIE).
2. A pure thermodynamic phase should be an extremal KMS state (see Section IIIF).

As we saw, this tentative dynamical characterization of canonical equilibrium states and pure thermodynamic phases can be made appealing, from an abstract point of view, by the following two remarks. For conjecture (1), the works of Brascamp and of Lanford have established a connection between the KMS condition itself and other more direct definitions of equilibrium conditions (e.g; "variational principles," etc.). For conjecture (2), the consequences of the extremal KMS character of a state ϕ, as expounded at the end of Section IIIF are quite reminiscent of the properties we would expect a pure thermodynamic phase to satisfy.

The approach we follow in this section is more concrete in the sense that we consider some specific models and use them to check whether the preceding conjectures are, or are not, justified by the facts.

B. The Quasi-Local Algebra of a Spin System

We consider here, for sake of simplicity, a "cubic" lattice $\mathbf{Z}^n$, although the considerations proposed in this section can easily be transposed to any regular lattice. We suppose that to any site $i \in \mathbf{Z}^n$ is attached a finite-dimensional Hilbert space $\mathscr{H}_i$, and the algebra $\mathscr{A}(i) = \mathscr{B}(\mathscr{H}_i)$. For sake of simplicity again we assume that $\mathscr{A}(i)$ is (for all $i \in \mathbf{Z}^n$) the algebra of 2×2 matrices with complex entries. Physically, this means that we restrict our attention to those systems where a spin $\frac{1}{2}$ sits at each site of the lattice. The regularity of the lattice is important for several of the considerations to come, whereas the case of more general spins is easily obtained from the formalism exposed below. So, incidentally, is the case of a lattice-gas with m $(< \infty)$ components.

Under the foregoing assumptions, we now define our C^*-algebra of quasi-local observables. We first remark that to every finite collection Ω

of points in $\mathbf{Z}^n$ we can associate the C^*-algebra

$$\mathscr{A}(\Omega) = \bigotimes_{i \in \Omega} \mathscr{A}(i)$$

which, here, is just the algebra of $2^p \times 2^p$ matrices with complex entries, where p is the number of sites in Ω. We have hence obtained the C^*-algebra of all local observables relative to the finite region $\Omega \subset \mathbf{Z}^n$.

We now notice that the collection $\{\mathscr{A}(\Omega)\}$ of all local algebras obtained in this manner, where Ω runs over all *finite* regions of $\mathbf{Z}^n$, enjoys two remarkable properties, *isotony* and *locality*. To formulate these properties, we first notice that if $\Omega_1 \subseteq \Omega_2$ and if we write $\Omega_2 \backslash \Omega_1 \equiv \{i \in \Omega_2 \mid i \notin \Omega_1\}$ we have

$$\mathscr{A}(\Omega_2) = \mathscr{A}(\Omega_1) \otimes \mathscr{A}(\Omega_2 \backslash \Omega_1)$$

This trivial remark provides us with a natural (both "mathematically natural," and "physically natural") embedding of $\mathscr{A}(\Omega_1)$ into $\mathscr{A}(\Omega_2)$, which we denote by $j_{2,1}$ and which we define by

$$j_{2,1}[A] \equiv A \otimes I_{\Omega_2 \backslash \Omega_1} \qquad \text{for all } A \text{ in } \mathscr{A}(\Omega_1)$$

[where $I_{\Omega_2 \backslash \Omega_1}$ is evidently the identity in $\mathscr{A}(\Omega_2 \backslash \Omega_1)$.] This mapping clearly is injective, linear, and preserves products, adjoint conjugation (and hence norms); it further maps the identity in $\mathscr{A}(\Omega_1)$ onto the identity in $\mathscr{A}(\Omega_2)$. The two properties we previously referred to are then:

Isotony:

$$j_{3,1} = j_{3,2} j_{2,1}$$

whenever

$$\Omega_1 \subseteq \Omega_2 \subseteq \Omega_3$$

Locality

$$[j_{3,1}[A_1], j_{3,2}[A_2]] = 0 \qquad \text{for all } A_1 \text{ in } \mathscr{A}(\Omega_1) \text{ and } A_2 \text{ in } \mathscr{A}(\Omega_2)$$

whenever

$$\Omega_1 \cap \Omega_2 = \phi \qquad \text{and}$$
$$\Omega_3 \supseteq \Omega_1 \qquad \text{and} \qquad \Omega_3 \supseteq \Omega_2$$

We can then clearly define

$$\mathscr{A}_0 = \bigcup_{\Omega \subset \mathbf{Z}^n} \mathscr{A}(\Omega)$$

where the union is taken over all *finite* regions in $\mathbf{Z}^n$.

This is the algebra of *local* observables on our system. A norm is clearly defined on it (as well as all the other *-algebra operations), and we then define

$$\mathscr{A} = \overline{\bigcup_{\Omega \subset \mathbf{Z}^n} \mathscr{A}(\Omega)}$$

which is the completion, with respect to this norm, of $\mathscr{A}_0$. This algebra is referred to as the algebra of *quasi-local* observables on our system. It is a C^*-algebra, and actually it is the smallest C^*-algebra containing all $\mathscr{A}(\Omega)$ as sub-C^*-algebras. [$\mathscr{A}$ is the "C^*-inductive limit" of $\{\mathscr{A}(\Omega) \mid \Omega \subseteq \mathbf{Z}^n\}$].

We end with two remarks on the scope of the method. First, for sake of simplicity and definiteness, we considered only the case of quantum spin lattices. This restriction is by no means essential. For a continuous system $\mathscr{A}(\Omega)$ would have been the C^*-algebra of all bounded observables relative to a finite region $\Omega \subset \mathbf{R}^n$. For quantum systems, $\mathscr{A}(\Omega)$ would be, for instance, $\mathscr{B}(\mathscr{H}_\Omega)$; for classical systems, $\mathscr{A}(\Omega)$ would be $\mathscr{C}(\mathbf{K}_\Omega)$ where $\mathbf{K}_\Omega$ is the appropriate configuration space for Ω. Second, as long as attention is focused on *finite* systems, it is at best of epistemological value to emphasize the algebraic properties of $\mathscr{A}(\Omega)$. For infinite systems, however, the C^*-algebraic nature of $\mathscr{A}$ is all we have in general.

C. Time-Evolution

We now return to our lattice systems, and suppose that we are given, whenever the system is enclosed in a finite "box" Ω, a Hamiltonian H_Ω and hence a time-evolution,

$$\alpha_\Omega(t)[A] = e^{iH_\Omega t} A e^{-iH_\Omega t}$$

which for each time t maps $\mathscr{A}(\Omega)$ onto itself [and actually is an automorphism of $\mathscr{A}(\Omega)$].

The question then is to define the time-evolution for the infinite system described by $\mathscr{A}$. We usually say that we have this time-evolution if we have a (unique) homomorphism $\alpha: \mathbf{R} \to \mathrm{Aut}\,(\mathscr{A})$ such that, for every local observable A,

$$\lim_{\Omega \to \mathbf{Z}^n} \|\alpha_t[A] - \alpha_\Omega(t)[A]\| = 0$$

(where the collection of Ω over which the limit is taken is chosen in such a manner as to contain ultimately all finite regions in $\mathbf{Z}^n$).

That the time-evolution for an infinite, quantum spin system can be obtained in this manner has been proven for all finite-range interactions (i.e., $J \in \mathscr{B}_0$), and actually for all interactions $J \in \mathscr{B}_1$, the completion of $\mathscr{B}_0$ with respect to the norm

$$\|J\|_1 = \sum_{\Omega \ni 0} \|J(\Omega)\| \exp\{p(\Omega) - 1\}$$

D. Some Variations on the Theme of the Transfer Matrix Formalism

What we want to do in this subsection is just to present the transfer matrix formalism in such a way as to make possible extensions beyond its usual scope. For this kind of review the simplest model will be the best. Let us thus consider the case of the one-dimensional Ising model with nearest neighbor interactions.

A configuration of the system is prescribed by attributing to each site i in $\mathbf{Z}$ a value $\sigma_i = \pm 1$. In other words, a configuration is a function σ, defined on $\mathbf{Z}$ and taking its values in $\{+1, -1\}$. This is to say that the configuration space for the infinite system is $\mathbf{K} = \{+1, -1\}^{\mathbf{Z}}$, and similarly the configuration space for a finite subsystem $[a, b]$ is $\mathbf{K}_{[a,b]} = \{+1, -1\}^{[a,b]}$. We have then

$$\mathscr{A}([a, b]) = \mathscr{C}(\mathbf{K}_{[a,b]})$$

We define, as usual, the transfer matrix V as

$$V(\sigma_0, \sigma_1) = \exp\{-\beta H(1)(\sigma_0, \sigma_1)\}$$

with

$$H(1)(\sigma_0, \sigma_1) = -m\tilde{H}\tfrac{1}{2}(\sigma_0 + \sigma_1) - J_1\sigma_0\sigma_1$$

The trick now is to introduce $\mathscr{L}$ defined by

$$(\mathscr{L}A)(\sigma_1, \sigma_2, \ldots) = \tfrac{1}{2}\sum_{\sigma_0} V(\sigma_0, \sigma_1)A(\sigma_0, \sigma_1, \sigma_2, \ldots)$$

This operator, defined on $\mathscr{C}(\mathbf{K}_+)$, has quite remarkable properties. There exist indeed a state ν on $\mathscr{C}(\mathbf{K}_+)$, an element h in $\mathscr{C}(\mathbf{K}_+)$, and a positive number λ such that

(i) $\mathscr{L}^*\nu = \lambda\nu$ [i.e., $\langle\nu; \mathscr{L}A\rangle = \lambda\langle\nu; A\rangle\ \forall\ A \in \mathscr{C}(\mathbf{K}_+)$]
(ii) $\mathscr{L}h = \lambda h$
(iii) $\langle\nu; h\rangle = 1$
(iv) $\operatorname{norm-lim}_{n\to\infty} \lambda^{-n}\mathscr{L}^n A = \langle\nu; A\rangle h$
(v) $\lim_{b\to\infty} \langle\phi^{[1,b]}; A\rangle = \langle\nu; A\rangle\ \forall\ A \in \mathscr{C}(\mathbf{K}_{[1,c]})$, $c < b$ (where $\phi^{[1,b]}$ is the equilibrium state for $[1, b]$)
(vi) λ, h, ν are continuous in β, $\tilde{H}$, and J_1.

(Actually λ, h, and ν have very simple expressions in term of the usual transfer matrix; the reader might find it wise to look for these expressions by himself.)

It is to be emphasised that property (v), which can be obtained similarly for the semi-infinite system $(-\infty, 0)$, does establish the existence of the Gibbs state (i.e., the thermodynamic limit of the finite-volume canonical equilibrium state). Property (vi) establishes, on the other hand, the absence of phase transitions (as is well-known).

E. Generalizations to Classical Systems

The generalization to potentials with finite-range is evidently absolutely trivial and does not bring us out of the recognized territory where transfer matrix methods are known to be operative. However, the usual formalism of the transfer matrix collapses when the interactions, although they ultimately die out at infinity., extend nevertheless over an infinite range. The interest of the reformulation of this formalism in term of the operator $\mathscr{L}$, as presented in Section IVD, is that it can be extended to a rather wide class of interactions which die out rapidly enough at infinity. Specifically, if we denote by $J(\Omega)$ the many-body interaction between the σ_i with i in Ω, we impose

$$\sum_{k\geqslant 1} \sum_{0<i_1<\cdots<i_k} i_1 \cdots i_k \, |J(0, i_1, \ldots, i_k)| < \infty$$

In the particular case of a two-body interaction, this condition reduces to

$$\sum_{i>0} i \, |J(0, i)| < \infty$$

so that all interactions which die out at infinity more rapidly than i^{-2} are allowed.

We then define

$$(\mathscr{L}A)(\sigma_1, \sigma_2, \ldots) \equiv \sum_{\sigma_0} V(\sigma_0, \sigma_1, \sigma_2, \ldots) A(\sigma_0, \sigma_1, \sigma_2, \ldots)$$

with

$$V(\sigma_1, \sigma_2, \ldots) \equiv \exp\{-\beta H(1)(\sigma_1, \sigma_2, \ldots)\}$$

and

$$H(1)(\sigma_1, \sigma_2, \ldots) = -B\sigma_1 - \sum_{k\geqslant 1} \sum_{1<i_1<i_2<\cdots<i_k} J(1, i_1, i_2, \ldots, i_k)\sigma_1\sigma_{i_1} \cdots \sigma_{i_k}$$

which simply reduces to

$$-B\sigma_1 - \sum_{m>1} J(1, m)\sigma_1\sigma_m$$

in the case of a two-body interaction. Thus $H(1)$ is the sum of the energy due to the magnetic field at the site 1 and of the energy due to the interaction between this site and all sites to its right.

With these definitions, and under the condition mentioned earlier on the interactions, we can prove that $\mathscr{L}$ satisfies all the fundamental properties already encountered in the particular case of the preceding subsection. In particular, the Gibbs state, defined as the thermodynamic limit of the canonical equilibrium states for finite volumes, does exist and is smooth in the parameters of the models (i.e., magnetic field, temperature, interactions), so that these models do not exhibit phase transitions.

The proof proceeds in principle as a natural generalization of that established for the finite-range case. It presents an ingenious combination of "soft analysis" (repeated use of Schauder-Tychonov theorem) and "hard analysis" (tricky majorizations, etc.), which we do not want to reproduce here. It should, however, be noticed that we can show explicitly for these (one-dimensional) models that the effect of the boundary conditions adopted for the computation of the Gibbs state *does* disappear in the thermodynamic limit. This is a reflection of the fact that these models do not exhibit phase transitions.

The translation of these results from a chain of "classical spins" to a one-dimensional lattice-gas goes through trivially. The passage from there to a one-dimensional gas of hard-rods (interacting in addition via many-body potentials which satisfy conditions similar to those written for the discrete case at the beginning of this subsection) can be carried out as a natural generalization, thus providing a genuine extension of van Hove's original results.

Whereas it is essential that these systems be one-dimensional, their classical character can be dispensed with, as will be seen in Section IVF, thus providing another generalization of the remarks of Section IVD.

F. General Heisenberg Models in One Dimension

We now consider the case of an infinite, one-dimensional, regular lattice $\mathbf{Z}$, each site of which is occupied by a *quantum* spin σ_i (see Section IVB). We suppose that these spins interact via potentials which are (1) lattice-translation invariant, (2) finite-range (say r), and (3) arbitrarily anisotropic.

If $J(\Omega)$ denotes the potential relative to a finite collection Ω of sites of $\mathbf{Z}$, we define

$$H(n) \equiv \sum_{\Omega \subseteq [n,n+r]} \nu_r(\Omega)^{-1} J(\Omega)$$

where $\nu_r(\Omega)$ is the number of translates $[\Omega + m]$ of Ω which are still in $[n, n + r]$;

$$H(\Omega) \equiv \sum_{\substack{n \\ [n,n+r] \subseteq \Omega}} H(n)$$

The proper generalization of the transfer matrix is obtained by considering for $b > 1$ the operator $V_{[1,b]}$:

$$V_{[1,b]} \exp\{-\tfrac{1}{2}\beta H[2, b]\} \equiv \exp\{-\tfrac{1}{2}\beta H[1, b]\}$$

We notice that in the classical case, where $[H(n), H(m)] = 0$ for all n, m in $\mathbf{Z}$, this reduces simply to

$$V_{[1,b]} = \exp\{-\tfrac{1}{2}\beta H(1)\}$$

thus making obvious contact with the usual formalism of the transfer matrix.

Here enters in an essential manner the fact that we work within the C^*-algebra $\mathscr{A}$ (see Section IVB): we know how to define the limit of $V_{[1,b]}$ as $b \to \infty$. Specifically, we can prove that there exists $V \in \mathscr{A}$ such that

$$\lim_{b \to \infty} \| V_{[1,b]} - V \| = 0$$

With the element V of $\mathscr{A}$ so defined, we now introduce

$$\mathscr{L} A = \tfrac{1}{2} \tau_s(-1) \operatorname{Tr}_{[1]} V^* A V$$

where $\tau_s(-1)$ denotes the translation to the left by one lattice unit and $\operatorname{Tr}_{[1]}$ denotes the partial trace over the first site. We should emphasize here that, contrary to the classical case with finite-range interaction, the noncommutativity of $H(n)$, $H(m)$ leads in the general quantum case to a V which depends on the interactions in the whole semilattice $\mathbf{Z}^+$, so that V can turn out to be quite involved. However, its explicit form is not needed in the sequel; from its definition as the limit of $V_{[1,b]}$ in the norm, and from a proper choice of majorizations (here again we cannot enter into the details of that), we see that Schauder-Tychonov theorem may again be used to assert the existence of a positive constant λ, of a state ν_+ on $\mathscr{A}_+ = \overline{\bigcup_{\Omega \subset \mathbf{Z}^+} \mathscr{A}(\Omega)}$ and of a positive element h in $\mathscr{A}_+$ such that

$$\mathscr{L}^* \nu_+ = \lambda \nu_+; \qquad \mathscr{L} h = \lambda h$$
$$\langle \nu_+; h \rangle = 1$$

and

$$\lim_{N \to \infty} \| \lambda^{-N} \mathscr{L}^N A - \langle \nu_+; A \rangle h \| = 0$$

the latter limit holding for all A in $\mathscr{A}_+$. From this we obtain rather easily that

$$\lim_{b \to +\infty} \langle \phi^{[1,b]}; A \rangle = \langle \nu_+; A \rangle$$

where $\phi^{[1,b]}$ is the "canonical state" on $[1, b]$. We now proceed analogously for $\mathbf{Z}^-$ and reconstruct from these two steps the Gibbs state ϕ^G for the entire lattice.

Again, we prove that these systems do *not* exhibit phase transitions by showing that $\langle \phi^G; A \rangle$ are continuous (actually holomorphic) in the parameters of the theory. Hence ϕ^G is a pure thermodynamic phase in the usual sense.

In connection with the two conjectures that we started with, note that the construction of ϕ^G along the lines indicated above allows to prove the following facts:

1. ϕ^G satisfies the KMS condition.
2. ϕ^G is lattice-invariant.
3. ϕ^G satisfies uniform clustering properties.

From these facts follow then:

4. ϕ^G is extremal KMS.
5. ϕ^G is extremal lattice-invariant.

Our conjectures IVA are thus confirmed for a class of models which do not show phase transitions.

The next step in supporting these conjectures is to see how they check against models which do exhibit phase transitions. Such models are discussed in the next subsection.

G. Weiss Model for Antiferromagnetism

The considerations to be presented here seem to be readily extendible to any model that can be exactly soluble by molecular field methods (e.g., BCS). The physical drawbacks of these models are well-known. We retain here the Weiss model for antiferromagnetism for its didactic value, and consider simply the interactions of the form

$$H(\Omega) = -B \sum_{i\in\Omega} \sigma_i^z - \tfrac{1}{2} \sum_{i,j\in\Omega} J_{i,j}(\Omega)\sigma_i^z\sigma_j^z$$

with:

	$J_{i,j}(\Omega) = J_{j,i}(\Omega)$	real
	$J_{i,j}(\Omega) = J_{\lvert i-j\rvert}(\Omega)$	(transl. inv.)
	$J_n(\Omega) = J_{n+p}(\Omega)$	(periodic potential of period p)
	$J_n(\Omega) = pf(n)/N(\Omega)$	where $f(n) \neq f_\Omega(n)$ and $N(\Omega)$ = number of sites in Ω (this guarantees the "stability")

More general conditions would do as well, but there is obviously no point complicating things.

The first difficulty encountered with this kind of model is that the interaction does *not* satisfy the conditions under which α_t can be defined according to the rule given in Section IVC. We can nevertheless bypass this difficulty by defining α_t for a class of states which are sufficiently regular. (It turns out that this class is large enough for the purpose of equilibrium statistical mechanics.)

It is remarkable that the time-evolution obtained in this manner (remember that even for equilibrium states we want to be able to use time-correlation functions, such as for the formulation of the KMS condition) is in general not an automorphism of $\mathscr{A}$ but only of $\pi_\phi(\mathscr{A})''$. This leads to a slight, actually trivial, adaptation of the formulation of the KMS condition, which changes none of its main consequences.

We next consider the extremal KMS states at fixed temperature. These states can be computed directly from the KMS condition itself. They turn out to coincide exactly with the solutions of the "usual" self-consistency equations (i.e., those equations which are obtained for states on $\mathscr{A}$ by a straightforward extension of the usual arguments on the partition function).

The *main point* is that we can thus support the second of the initial conjectures by the analysis of these specific models: pure thermodynamic phases are indeed extremal KMS states. Again, one could object, as with all models, that this might just be an accident due to the peculiarities of the models. However, no counterexample is known (yet).

The consideration of these models, beside this main point, produces some more by-products which are worth mentioning here. For extremal KMS state, the time evolution can be shown to be an automorphism $\alpha_\phi(t)$ of $\pi_\phi(\mathscr{A})$, and hence of $\mathscr{A}$ (since all its representations are faithful). But even in this most simple case $\alpha_\phi(t)$ depends on the natural temperature β and does not act in an η-abelian manner on $\mathscr{A}$.

The pure thermodynamic phases obtained for these models are in general *not* invariant for the whole lattice-translation group **Z** but only for that subgroup of **Z** generated by the period of $J_{i,j}$, and this in spite of the fact that $J_{i,j}$ is supposed to be lattice-translation invariant. We thus have there an explicit example of a symmetry breaking, the physical origin of which is clear: existence of different sublattice magnetizations for models exhibiting antiferromagnetism [e.g., $f(n) = (-1)^n$].

Finally, a word of caution concerning the converse of the second conjecture. Whereas these models show that pure thermodynamic phases are extremal KMS states, there exist extremal KMS states which are not (stable) pure thermodynamic phases.

H. Bibliography

M. Kac, in *Fundamental Problems in Statistical Mechanics*, Vol. II, E. G. D. Cohen, Ed., North-Holland, Amsterdam, 1968, p. 71.

H. J. Brascamp, *Equilibrium States for a Classical Lattice Gas*, and *Equilibrium States for a One-Dimensional Lattice Gas*, Thesis, Gröningen, 1970. See also *Commun. math. Phys.*, **18**, 82 (1970).

O. E. Lanford, *High Temperature Expansions*, Cargèse Summer School, D. Kastler, Ed., Gordon and Breach, New York, 1969.

A. Guichardet, *Ann. Ec. Norm. Sup.*, **83**, 1 (1966).

D. Ruelle, *Statistical Mechanics*, Benjamin, New York, 1969.

D. W. Robinson, *Commun. math. Phys.*, **6**, 151 (1967); **7**, 337 (1968).

D. Ruelle, *Commun. Math. Phys.*, **9**, 267 (1968).

G. Gallavotti and S. Miracle-Sole, *J. Math. Phys.*, **11**, 147 (1970).

H. Araki, *Commun. math. Phys.*, **14**, 120 (1969).

G. G. Emch and H. J. F. Knops, *J. Math. Phys.*, **10**, 3008 (1970).

V. CRYSTALLIZATION AS A SYMMETRY-BREAKING PROCESS

A. Generalities

We say that a phase transition does occur and is accompanied by a symmetry breaking when the Gibbs state ϕ, computed at fixed values of the external parameters of the theory (such as temperature, magnetic field, chemical potential) is a KMS state without being extremal KMS, and moreover can be decomposed into pure thermodynamic phases (hence extremal KMS states) which have a lower symmetry than that of the interactions one starts from; it is understood that in the course of the computation of the thermodynamic limit one chooses the boundary conditions in such a way that ϕ itself inherits the full symmetry of the interactions.

For example, we would expect the following to occur. Take the Ising model in two dimensions, with $B = 0$ (the only case we know how to solve exactly); the interactions are obviously invariant against a change of *all* σ_i^z simultaneously into $(-\sigma_i^z)$. We refer to this as "flip-flop" symmetry. Periodic boundary conditions do respect this symmetry, so that at finite volume the canonical equilibrium state is flip-flop invariant. Thus the thermodynamic limit should provide a state which is flip-flop invariant as well. We then would expect that this state ϕ can be decomposed into two thermodynamic phases:

$$\phi = \tfrac{1}{2}(\phi_+ + \phi_-)$$

such that

$$\langle \phi_+ ; \sigma_i^z \rangle = -\langle \phi_- ; \sigma_i^z \rangle = \pm \mathscr{M}_0 \neq 0$$

(if one is under T_c). Obviously $\phi_\pm$ are not flip-flop invariant any more: they are actually flip-flop images of one another.

The opening statement of this section is obviously in line with the material developed in the preceding sections. It could be rephrased into a more general or more precise form. Although this rephrasing would not be proper now, we do want to emphasise two points on which no claim will be made. First, we shall not touch upon the problem of whether the pure thermodynamic phases, obtained in the decomposition of an invariant Gibbs state, could themselves be directly obtained as Gibbs states computed with boundary conditions that would break the symmetry. For Ising-type models we know something about this point. In the case of crystallization this question has not been settled yet. Second, we insist on the essentially equilibrium character of the description we propose. We do not attempt to give a description of the process by which crystals grow from a solution.

We are thus mainly concerned with the possibility of a consistent description of crystals within a theory which is basically Euclidian covariant. We only suggest a scheme. No specific Hamiltonian models are available. Still, within that scheme one can understand, on a more rigorous basis, Landau's argument on the absence of critical point in the liquid-solid phase transition.

B. Decomposition Theory

The first step in the theory is to provide a canonical decomposition of a KMS state into its extremal KMS components. Rather than stating the main theorem first, let us try to gain some feeling on how things should (and do!) go in a simplified case.

We assume thus that ϕ is a KMS state and that the representation space $\mathscr{H}_\phi$ of the representation $\pi_\phi(\mathscr{A})$ of $\mathscr{A}$ associated to ϕ by the GNS construction is separable. This can actually be justified on general terms. We now make our simplifying assumption. We remark that

$$\mathscr{Z}(\mathscr{A}) = \pi_\phi(\mathscr{A})'' \cap \pi_\phi(\mathscr{A})'$$

is an abelian von Neumann-algebra. We assume that it is generated by a single self-adjoint operator Z, which has a discrete spectrum. (This assumption will be removed pretty soon; it is now made only for illustrative purpose). With this assumption we can write

$$Z = \sum a_i P_i \qquad (\text{with } a_i \text{ in } \mathbf{R} \text{ and } P_i = P_i^* = P_i^2)$$

and

$$\mathscr{Z}(\mathscr{A}) = \{\sum z_i P_i \mid z_i \in \mathbf{C}\}$$

We then notice that Φ cyclic for $\pi_\phi(\mathscr{A})$ (and thus for $\pi_\phi(\mathscr{A})''$) implies Φ separating for $\pi_\phi(\mathscr{A})'$, i.e., $X \in \pi_\phi(\mathscr{A})'$ and $X\Phi = 0$ implies $X = 0$. Since

$$P_i \in \mathscr{Z}_\phi(\mathscr{A}) \subseteq \pi_\phi(\mathscr{A})' \qquad \text{and} \quad P_i \neq 0$$

We have $P_i\Phi \neq 0$ and hence

$$\lambda_i \equiv \langle\tilde{\phi}; P_i\rangle = (\Phi, P_i\Phi) \neq 0$$

We then form for every X in $\pi_\phi(\mathscr{A})''$:

$$\langle\psi_i; X\rangle \equiv \frac{\langle\tilde{\phi}; P_i X\rangle}{\langle\tilde{\phi}; P_i\rangle}$$

We then notice the following:

1. ψ_i is KMS [which follows from ϕ KMS and $P_i \in \mathscr{Z}_\phi(\mathscr{A})$].
2. $\mathscr{Z}_{\psi_i}(\mathscr{A})$ (which is "defined" on $P_i\mathscr{H}_\phi$) is $\{\lambda I_i \mid \lambda \in \mathbf{C}\}$.

Thus ψ_i is extremal KMS.

Furthermore, for any Z in $\mathscr{Z}_\phi(\mathscr{A})$ $(Z = \sum z_i P_i)$, we have

$$\langle\tilde{\phi}; ZA\rangle = \sum z_i \langle\psi_i; A\rangle \lambda_i \qquad (*)$$

i.e., in particular,

$$\langle\phi; A\rangle = \sum \langle\psi_i; A\rangle \lambda_i$$

so that we get a decomposition of ϕ into extremal KMS states. It is easy to see that this decomposition is unique.

We now rewrite (*) in a more "fancy" form. We define

$$\mu_\phi \equiv \sum \lambda_i \delta_{\psi_i}$$

where δ_{ψ_i} is the normalized measure on $\mathscr{S}$, which is concentrated on ψ_i. Notice that μ_ϕ is thus concentrated on the extremal KMS states, which decompose ϕ. We form

$$\varphi_z(\psi_i) \equiv z_i$$

Notice that

$$\varphi : Z\ (\in \mathscr{Z}_\phi(\mathscr{A})) \to \varphi_z[\in \mathscr{L}^\infty(\mathscr{S}, \mu_\phi)]$$

is an isomorphism. Then (*) can be rewritten as

$$\langle\tilde{\phi}; ZA\rangle = \int \varphi_z(\psi)\langle\psi; A\rangle\, d\mu_\phi(\psi)$$

The advantage of this notation is that it carries through (up to technical details which have no place here) when we suppress the restrictions imposed at the beginning of this section on $\mathscr{Z}_\phi(\mathscr{A})$.

We should remark that the "continuous" mixtures suggested by this integral form are not mathematical luxury. The reader will convince himself that it is a physical necessity if he wants to translate to an Heisenberg model the considerations of Section VA. The same will, of course, occur in the crystallization problem.

C. Restrictions Imposed on the Theory

The degree to which the following restrictions are necessary (or even mutually independent) for the elaboration of the theory varies from one to another. Some of them can actually be dispensed with entirely. Our attitude in this respect will be that (1) they permit us to describe the theory with a minimal expenditure of technicalities, (2) the theory so obtained seems to capture the essentials of the physical problems to be tackled, and (3) they have some immediate consequences, which we will list, that allow us to judge the value and strength of these restrictions.

We assume that we are given the following:

1. An algebra $\mathscr{A}$.
2. A symmetry $\alpha : \mathbf{R} \to \mathrm{Aut}\,(\mathscr{A})$ describing the time-evolution.

3. A symmetry $\tau: \mathbf{E}^3 \to \text{Aut}(\mathscr{A})$ describing the effect of space-transformations on the observables (Euclidian covariance of the theory), and such that $\alpha_t \tau_g = \tau_g \alpha_t$ for all $t \in \mathbf{R}$ and all $g \in \mathbf{E}^3$.

4. A state ϕ on $\mathscr{A}$ with the following properties:

a. ϕ is KMS with respect to $\{\alpha_t \mid t \in \mathbf{R}\}$.

b. ϕ is $\mathbf{E}^3$-invariant.

c. ϕ is locally normal (which means essentially that only a finite number of particles are allowed to coexist simultaneously in a finite subvolume of our infinite system).

d. ϕ is strongly transitive, which is to say that we assume:

d_1. There exists a state ϕ on $\mathscr{A}$ such that $\mu_\phi\,(0_\psi^E) = 1$ where $0_\psi^E \equiv \{\tau_g^*[\psi] \mid g \in E^3\}$.

d_2. There exists $\psi \in$ support of μ_ϕ such that $0_\psi^R \equiv \{\tau_g^*[\psi] \mid g \in \mathbf{R}^3\}$ is closed (with respect to w^*-topology) in $\mathscr{S}$).

These restrictions readily imply the following:

1. ϕ is extremal $\mathbf{E}^3$-invariant.

2. If ψ and ψ_0 are two states occurring in the decomposition of ϕ into extremal KMS states, then there exists $g \in \mathbf{E}^3$ such that $\tau_g^*[\psi_0] = \psi$ and if G_ψ (resp. G_{ψ_0}) is the space-symmetry inherited by ψ (resp. ψ_0) from the Euclidian symmetry of ϕ, then $G_\psi = gG_{\psi_0}g^{-1}$, which means that the symmetries of the pure phases occurring in the decomposition of ϕ belong to the same equivalence class of $\mathbf{E}^3$. In other words, the symmetry of ψ is determined, up to an isomorphism, by its belonging to the decomposition of ϕ. Hence this equivalence class is a characteristic of ϕ. We shall call it the *intrinsic symmetry* of ϕ.

3. G_ψ contains at least three noncoplanar translations (and irrelevant pathologies are avoided).

The program now is to classify the possible decompositions of ϕ according to its intrinsic symmetry and to look for alternate characterizations of the classes so obtained.

D. Fluid Phase

The first case that may occur is that ϕ itself is extremal KMS, i.e., $G_\psi = \mathbf{E}^3$. We want to interpret this case by saying that ϕ is a fluid phase. This interpretation is based on the fact that this case happens if and only if any one (and thus all) of the following properties is realized:

1. ϕ is extremal $\mathbf{R}^3$-invariant.

2. $\{U_\phi(a) \equiv \exp\{-iP_\phi a\} \mid a \in \mathbf{R}^3\}$ is such that P_ϕ admits only one discrete eigenvalue, $k = 0$, *and* (and this is the condition) this value is nondegenerate.

3. ϕ is uniformly clustering.

E. Nonfluid Phases

Suppose now that ϕ is not extremal KMS and let G_ψ be the intrinsic symmetry of ϕ. Define $H_\psi \equiv G_\psi \cap \mathbf{R}^3$ and form $\chi = \eta^R[\psi]$ where η^R is an invariant mean over $\mathbf{R}^3$. χ is then $\mathbf{R}^3$-invariant. Still it "remembers" quite precisely the symmetry H_ψ. Specifically, with

$$\{U_\chi(a) = \exp\{-iP_\chi a\} \mid a \in \mathbf{R}^3\}$$

we have that the discrete part of the spectrum of the momentum operator P_χ is

$$Sp_d(P_\chi) = H_\psi^* \equiv \{k \in \mathbf{R}^3 \mid k \cdot a \equiv 0(\text{mod } 2\pi)\forall\, a \in H_\psi\}$$

i.e., $Sp_d(P_\chi)$ is isomorphic to the reciprocal group of H_ψ, and in addition is nondegenerate. Consequently χ is not only $\mathbf{R}^3$-invariant but also extremal $\mathbf{R}^3$-invariant.

Only the following three cases can then occur.

Case 1. $H_\psi = \mathbf{R}^3$. Here only the rotation part of $\mathbf{E}^3$ is broken, and $\chi = \psi$. We evidently have

$$Sp_d(P_\psi) = Sp_d(P_\chi) = \{0\}$$

We can show that χ satisfies a clustering property which is weaker than uniform clustering but stronger than η-clustering, namely, weak mixing:

$$\eta^R\, |\langle\chi; \tau_a[A]B\rangle - \langle\chi; A\rangle\langle\chi; B\rangle| = 0$$

Case 2. H_ψ is continuous in one or two directions, but discrete in the other(s). Certainly O^3-invariance is broken, and part of $\mathbf{R}^3$-invariance also is. Then $Sp_d(P_\chi)$ contains points other than $k = 0$ and these lie in a plane (resp. on a line).

The clustering property characteristic of this case is "partial weak mixing," which is weaker than weak mixing but stronger than η-clustering, and is defined by

$$\eta^x\, |\eta^{y,z}\langle\chi; \tau_a[A]B\rangle - \langle\chi; A\rangle\langle\chi; B\rangle| = 0$$

or

$$\eta^{x,y}\, |\eta^z\langle\chi; \tau_a[A]B\rangle - \langle\chi; A\rangle\langle\chi; B\rangle| = 0$$

Case 3. H_ψ is generated by three noncoplanar translations. Here G_ψ is a crystalline group and $Sp_d(P_\chi)$ contains points which lie in three noncoplanar directions and is actually isomorphic with the reciprocal lattice of our crystal (H_ψ^*).

χ is then η-clustering (we recall that this means

$$\eta^R\langle\chi; \tau_a[A]B\rangle - \langle\chi; A\rangle\langle\chi; B\rangle = 0$$

and that it is the weakest clustering property compatible with the fact that χ is extremal $\mathbf{R}^3$-invariant).

F. Concluding Remarks

As we have seen, *either* ϕ is a (pure) fluid phase, is $\mathbf{E}^3$-invariant, extr. $\mathbf{R}^3$-inv, uniform clustering, *or* its intrinsic symmetry $G_\psi \subsetneq \mathbf{E}^3$. Within the second alternative, only three cases can occur which can be characterized *equivalently* by: (1) how much the translation symmetry is broken, i.e., a classification according to H_ψ, (2) spectrum properties, and (3) clustering properties.

As a by-product we see that the intuitive distinction between fluid phase (Section VD) and a crystalline phase (Case 3 of Section VE), on the basis of their symmetry, corresponds exactly to a classification on the basis of clustering properties. This allows us to place Landau's argument on the nonexistence of a critical point in the phase-solid phase transition, on a safer footing than the usual heuristic arguments.

Some interpretations of Cases 1 and 2 of Section VE have been attempted, but they are still not quite convincing.

Moreover, the problems that we said we did not want to consider (Section VA) remain open. Still, in view of the paucity of models for crystallisation, the theory presented here might be of value to prospective model builders.

G. Bibliography

G. G. Emch, H. J. F. Knops, and E. J. Verboven, *Commun. math. Phys.*, **8,** 300 (1968).

R. L. Dobrushin, *Funct. Anal. Appl.*, **2,** 31; 44 (1968).

D. Kastler, R. Haag, and L. Michel, *Central Decomposition of Ergodic States*. Sem. Phys. Th. Marseille 1967–8.

G. E. Uhlenbeck, in *Fundamental Problems in Statistical Mechanics*, Vol. II, E. G. D. Cohen, Ed., North-Holland, Amsterdam, 1968 (see in particular p. 17 for Landau's argument).

G. G. Emch, H. J. F. Knops, and E. J. Verboven, *J. Math. Phys.*, **11,** 1655 (1970).

M. Sirugue and M. Winnink, *Commun. math. Phys.*, **19,** 161 (1970).

VI. NONEQUILIBRIUM AND ERGODICITY

A. Generalities

In equilibrium statistical mechanics, and more particularly in the theory of phase transitions, the thermodynamic limit is considered to eliminate size effects (such as the "rounding-off" of isotherms) which tend to blur the essential features we are trying to understand. The same situation occurs in nonequilibrium statistical mechanics; here we mainly want to get rid of "recurrences," which are not only physically undesirable but also present hopelessly entangled situations (e.g., the mean recurrence time of the dog-flea model becomes a meaningless concept in the limit of large systems).

Traditional wisdom in the field of nonequilibrium statistical mechanics has it that one gets rid of recurrences, in the thermodynamic limit, because the spectrum of the Hamiltonian becomes "dense" or "continuous." This kind of statement must be taken *cum grano salis*. First, it is next to impossible to exhibit, in the usual formalism, an operator that would be mathematically well defined and that would be obtained from a finite system Hamiltonian by the limiting procedures that Hilbert space techniques (as commonly used) have to offer. Second, if this problem were to be ignored for a while, it would still be of interest to know how "dense" or "continuous" things become in the thermodynamic limit. And third, it would be useful to link these "how much" questions and answers with the *rate* of approach to equilibrium.

It would be quite unreasonable to deny that something is known in the direction of a solution of these fundamental problems. It is, however, the feeling of this author that it would be equally unreasonable to assert that the solutions known today are entirely satisfactory and leave no room for improvements, and even fundamental improvements. Furthermore, there is a hope that the C^*-algebraic approach might be able to offer some clues. This hope is based on evidence, some of which, the most elementary, is presented in this section as a germ for more sophisticated elaborations.

This section, based on the author's own line of thought, may be remiss in not presenting other advances in the field. However unfortunate this might be, it has to be taken as one of the consequences imposed by limitations in the available time. Furthermore, it must be admitted that many of the problems of bridging the gaps between various approaches still remain to be solved.

B. Description of One Experiment

A CaF_2 crystal is placed in a strong magnetic field B, say in the z-direction. We recall that the F nuclei carry a nuclear spin $I = \frac{1}{2}$. Thermal equilibrium is established: $\mu = (\mu^z, 0, 0)$. An RF pulse is then been applied, with the effect $\mu^z \to \mu^x$. The time development of the latter is then observed: $\mu^x(t)$ displays an oscillatory damping to the value $\mu^x(\infty) = 0$. This phenomenon has been explained on the basis of the dipolar interactions between the fluorine nuclei. No spin-lattice relaxation need be considered (so that the spin system as such can actually be considered as being isolated).

Whether this system is "dissipative" in an orthodox sense was apparently of no concern to the experimenters, since nobody explained to them (at the time) what had to be measured in this connection.

C. Elementary Theoretical Analysis

Hamiltonian

$$H = -B \sum_i \sigma_i^z - \sum_{i,j} \epsilon_{i,j} \sigma_i^z \sigma_j^z$$

(One-dimensional finite ring, periodic boundary conditions.) i.e.,

$$H = H_0 + V$$

with

$$H_0 = -B \sum_i \sigma_i^z, \qquad V = \sum_{n>0} H(n), \qquad \text{and} \quad H(n) = \epsilon(n) \sum_i \sigma_i^z \sigma_{i+n}^z$$

Initial State

$$\rho(0) = \frac{e^{-\beta H_0'}}{\operatorname{Tr} e^{-\beta H_0'}} \equiv \frac{e^{-\zeta S^x}}{\operatorname{Tr} e^{-\zeta S^x}}$$

To Be Computed

$$\langle S^x \rangle(t) = \operatorname{Tr} S^x \rho(t)$$

where $\rho(0) \to \rho(t)$ given by H above.

Results (no other approximation than the assumptions above):

$$\langle S^x \rangle(t) = \langle S^x \rangle(0) f_N(t) \cos 2Bt$$

$$\langle S^y \rangle(t) = -\langle S^x \rangle(0) f_N(t) \sin 2Bt$$

$$\langle S^z \rangle(t) = \langle S^z \rangle(0) = 0$$

with

$$f_N(t) = \prod_{n=1}^{N} \cos^2 \{2\epsilon(n)t\}$$

Take then:

$$\epsilon(n) = \epsilon_0 2^{-n}$$

so that

$$f(t) = \lim_{N \to \infty} f_N(t) = \left(\frac{\sin \epsilon_0 t}{\epsilon_0 t} \right)^2$$

$$f_N(t) = f(t) W_N(t)^{-1}$$

$$W_N(t) = \left(\frac{\sin \epsilon(N)t}{\epsilon(N)t} \right)^2$$

D. Elementary Exploitation of this Analysis

1. Zermelo's Paradox. There is indeed, for the finite system, a recurrence time $T_N \sim 2^N \pi / \epsilon_0$ (no probabilistic argument involved). These recurrences disappear in the "thermodynamic limit" ($N \to \infty$) and the experimental situation is explained, at least qualitatively; we find an oscillatory approach to equilibrium.

2. Loschmidt Paradox. $f(t) = f(-t)$ mirrors the reversibility, and this is evidently compatible with an approach to equilibrium.

3. Coarse-graining and **naive ergodic theory** can be illustrated on this model. Take $A = \sigma_0{}^x$ as the observable of interest. Two "macrocells" E_+ and E_- are occupied with probability $p_\pm(t)$. Then

$$\lim_{T\to\infty} \frac{1}{T}\int_0^T dt p_\pm(t) = \tfrac{1}{2}\{1 \pm (\tfrac{1}{2})^N[p_+(0) - p_-(0)]\}$$

which indeed tends to $\frac{1}{2}$ as N tends to infinity.

4. Due to the particular (!) choice of the space dependence of $\epsilon(n)$ we get a **non-Markovian approach to equilibrium**, which cannot be made Markovian by any amount of course-graining or time-smoothing.

5. Other concepts can be discussed and illustrated with this model and some of them have already been treated with some detail.

E. Algebraic Formulation

Condition on two-body potential

$$\epsilon(0) = 0; \qquad \sum_{j\in\mathbf{Z}^n} |\epsilon(j)| < \infty$$

Then (see Section IVC) α_t exists, giving proper time-evolution on the algebra $\mathscr{A} = \overline{\cup\mathscr{A}(\Omega)}$ of quasi-local observables on our system.

Moreover, for every finite $\Omega \subset \mathbf{Z}^n$ there exists $\tilde{H}_\Omega \in \mathscr{A}$ [*not* $\mathscr{A}(\Omega)$] such that

$$\alpha_t[A] = e^{i\tilde{H}_\Omega t}Ae^{-i\tilde{H}_\Omega t}$$

for all A in $\mathscr{A}(\Omega)$. For instance,

$$\tilde{H}_j = \tfrac{1}{2}\sum_{k\in\mathbf{Z}^n} \epsilon(|j-k|)\sigma_j{}^z\sigma_k{}^z$$

This can be considered as an "effective Hamiltonian," which describes the evolution of the observables that are relative to the region Ω *of the infinite system* $\mathbf{Z}^n$. Notice we do not have, and do not need, any such thing for $\Omega \to \mathbf{Z}^n$; the thermodynamic limit (or rather the infinite volume limit) is already taken into account in $\tilde{H}_\Omega$. With these preliminaries, the following results of interest are immediately proven:

1. $\eta\langle\phi; \alpha_t[A]\rangle$ depends only on ϕ and A but not on the invariant mean η chosen to compute it.

2.

$$\langle\phi; \alpha_t[\sigma_0{}^x]\rangle = \langle\phi; \sigma_0{}^x\rangle\hat{\mu}(t)$$

$$\hat{\mu}(t) = \prod_{j\in\mathbf{Z}^n} \cos\{2\epsilon(j)t\}$$

valid for a wider class of initial conditions than those mentioned in the "description of the experiment."

3. $\hat{\mu}(t)$ can be considered as the Fourier transform of a measure μ, and μ itself is easily linked to the spectral measure of $\pi(\tilde{H}_0)$ in the appropriate representation. We can thus establish a link between (a) spatial dependence of the interaction, (b) spectral measure of effective Hamiltonian, and (c) rate of approach to equilibrium. This link is explicitly known for a wide class of interaction (finite range, exponentials, Dyson type), and thus throws some light on the questions formulated at the beginning of this section.

We should also remark that the first of the preceding results is a mild kind of ergodic theorem. It suggests that an investigation along these lines should be pursued, with the principal aim to generalize the results of this section to more general potentials, and thus open the door to the study of a general class of models, the dissipative character of which could be decided upon and could eventually be linked to some ergodic behavior. A first step in this direction is presented in the next subsection.

F. Noncommutative Ergodic Theory

We have already seen two kinds of "ergodic theorems," the principal theorem (Section IIID.2) and the first result in Section VIE. Here is another such result, which also generalizes to a quantum situation, a well-known classical theorem.

Let $\{\mathscr{A}, \mathfrak{S}, \alpha_t\}$ be an abstract dynamical system where $\mathscr{A}$ is a C^*-algebra, $\mathfrak{S}$ is the set of states on $\mathscr{A}$, and α_t is a continuous group of automorphism of $\mathscr{A}$, representing the time-evolution of the system. For any ϕ in $\mathfrak{S}$, let

$$\mathscr{S}(\phi) = \{\psi \in \mathfrak{S} \mid \psi \leqslant a\phi \text{ for some } a \in \mathbf{R}\}$$

As is well-known, to every $\psi \in \mathscr{S}(\phi)$ corresponds a (positive) operator C in $\pi_\phi(\mathscr{A})'$ such that

$$\langle \psi; A \rangle = (C\Phi, \pi_\phi(A)C\Phi)$$

which we can rewrite as

$$\langle \psi; A \rangle = (\Psi, \pi_\phi(A)\Psi)$$

with

$$\Psi = C\Phi \in \mathscr{H}_\phi$$

Furthermore, let us denote by $\overline{\mathscr{S}(\phi)^n}$ the norm-closure of $\mathscr{S}(\phi)$. We then verify that for every ψ in $\overline{\mathscr{S}(\phi)^n}$, we can find an operator C (in general unbounded) with domain $\mathscr{D}(C)$ such that

$$\mathscr{D}(C) \supseteq \pi_\phi(\mathscr{A})\Phi$$
$$[C, \pi_\phi(A)]\pi_\phi(B)\Phi = 0 \ \forall\, A, B \in \mathscr{A}$$
$$\langle \psi; A \rangle = (C\Phi, \pi_\phi(A)C\Phi) \ \forall\, A \in \mathscr{A}$$

Now if ϕ is time-invariant, i.e.,

$$\phi \in \mathfrak{S}_T = \{\chi \in \mathfrak{S} \mid \alpha_t^*[\chi] = \chi \ \forall\, t \in \mathbf{R}\}$$

we can prove that for all $\psi \in \overline{\mathscr{S}(\phi)}^n$ there *exists* a *unique*

$$\tilde{\psi} \in \mathfrak{S}_T \cap \overline{\mathrm{co}\,\{\alpha_t^*[\psi]\}}^n$$

and, moreover,

$$\tilde{\psi} \in \overline{\mathscr{S}(\phi)}^n$$

(Note that if $\mathscr{T} \subseteq \mathfrak{S}$, we denote by co $(\mathscr{T})$ the set of all convex combinations of elements in $\mathscr{T}$ and by $\overline{\mathrm{co}(\mathscr{T})}^n$ the norm-closure of $\mathrm{co}(\mathscr{T})$).

The preceding result is well-known in the case where $\mathscr{A}$ is abelian (classical ergodic theory). The point here is that it is valid as well for quantum systems. It represents the proper adaptations of the first result of Section VIE to a much more general situation.

As we have noticed, $\psi \in \overline{\mathscr{S}(\phi)}^n$ is a vector state for $\pi_\phi(\mathscr{A})$, i.e.,

$$\overline{\mathscr{S}(\phi)^n} \subseteq \mathscr{V}_\phi$$

The question now is whether the equality sign might be realized. This indeed occurs in at least one case of interest, namely, when ϕ is a KMS state (and such that $\langle \phi; A^*A \rangle = 0$ implies $A = 0$). It follows then, from the cyclicity of Φ with respect (now) to $\pi(\mathscr{A})'$ that, in this case,

$$\overline{\mathscr{S}(\phi)^n} = \mathscr{V}_\phi$$

We notice further that $\overline{\mathscr{S}(\phi)}^n$ is a convex set. We have then in this case

$$\overline{\mathscr{S}(\phi)^n} = \mathscr{V}_\phi = \overline{\mathrm{co}\,\mathscr{V}_\phi}^n = \mathscr{D}_\phi$$

where $\mathscr{D}_\phi$ is the set of all density matrix states with respect to the representation $\pi_\phi(\mathscr{A})$.

Hence what the theorem says in this case is that, in the proper representation, the ergodic average of density matrices is still a density matrix, thus making possible the use of Liouville-space techniques. But this is *not* true for any representation, and we can easily produce counterexamples ("first ergodic paradox") if we stay in the usual formalism, that in which $\mathscr{H}$ is fixed once and for all.

Furthermore, we can show that if the system is submitted to a "gentle" perturbation (such as a local perturbation), then the equilibrium (i.e., KMS) state ϕ becomes $\psi \in \overline{\mathscr{S}(\phi)^n}$ and we can use the theorem to assert the existence and uniqueness of the average $\tilde{\psi}$ of ψ. Moreover, since $\tilde{\psi}$ is in $\mathscr{V}_\phi$, we can use theorem III.D.2 to assert that if ϕ is extremal in $\mathfrak{S}_T$ (or if ϕ is extremal KMS and ifα_t acts η-ab on ϕ), then $\tilde{\psi} = \phi$.

This result is thus clearly relevant when we want to apply ergodic theory to physical situations. It should, however, be stressed that this is but a first step in the problem of physical ergodic theory. Several problems remain, among them the question of Poincaré's invariants. The point we wanted to make is that in this kind of problem too the C^*-algebraic approach can be of definite use.

G. Bibliography

M. Kac, *Probability and Related Topics in Physical Sciences*, Interscience, New York 1959.

I. J. Lowe and R. E. Norberg, *Phys. Rev.*, **107**, 46 (1957).

G. G. Emch, *J. Math Phys.*, 7, 1198 (1966); 7, 1413 (1966); *Rigorous Results in Non Equilibrium Statistical Mechanics*, Lectures in Theoretical Physics, Boulder, Colo., 1965.

G. G. Emch and C. Favre, *Course-Graining in Liouville Space and Ergodicity*, Preprint, Geneva, 1965.

C. Radin, *Approach to Equilibrium in a Simple Model*, Ph.D. Thesis, Rochester, 1970; *J. Math. Phys.*, **11**, 2945 (1970); *Commun. math. Phys.*, **21**, 291 (1971); *Commun. math. Phys.* (to appear).

G. G. Emch and C. Radin, *J. Math. Phys.*, **12**, 2043 (1971).

R. Herman and T. Takesaki, *Commun. math. Phys.*, **19**, 142 (1970).

I. Prigogine and co-workers, Current papers in *Physica, Proc. NAS, Bull. Acad. Sc. Belgium*, etc.

Acknowledgments

These notes were written while the author was visiting the Service de Chimie Physique II, Université Libre de Bruxelles. The author would like to thank Professor I. Prigogine for his kind invitation and hospitality, as well as for many interesting discussions, and his suggestion that these lecture notes be published.

The assistance of Mr. J. W. Turner in the production of these notes is heartily acknowledged.

RADIATIVE AND NONRADIATIVE PROCESSES IN BENZENE*

C. S. PARMENTER

Department of Chemistry, Indiana University,
Bloomington, Indiana

CONTENTS

I. INTRODUCTION

Properties of excited electronic states of polyatomic molecules may be conveniently classified as "stationary" or "nonstationary." A comparison of the present knowledge about properties in these two classifications is interesting. Theoreticians and spectroscopists have been rather successful in developing good descriptions of energy levels and geometries of excited states, and hence the stationary type is in a comparatively advanced state of development. For example, the spectroscopic work of

* Contribution 1934 from the Chemical Laboratories of Indiana University.

D. A. Ramsay and co-workers has provided many constants for one of the excited singlet states (^1A_u) of glyoxal, and the rotational structure within many of the vibrational levels of this state can be described with exceedingly high precision.[1] Benzene, the subject of the present discussion, is another example where a fairly detailed description of numerous stationary aspects of both the ground and upper electronic states are now available. As a result of Callomon, Dunn, and Mills' beautiful analysis[2] of rotational band contours in the ultraviolet absorption spectrum, much is known about even the Coriolis constants for vibronic levels in the excited singlet state. Numerous other examples could be cited in which the geometries and many upper state constants in polyatomics are well known.

In contrast, knowledge of the most distinguishing feature of an excited state—its nonstationary character—is rather primitive. Photochemists and to a lesser extent spectroscopists have of course expended great efforts on excited state relaxation and have developed a general understanding of the types of relaxation channels accessible to a large polyatomic system. In addition, the nature of excited-state relaxation is understood for a given energy region, or for specific environmental conditions in many molecules. But data yielding even a modestly complete description of relaxation from a wide range of vibrational energies in a given electronic state of a given molecule are sparse. For example, little detail is known about relaxation of an electronic state under "isolated molecule" conditions where properties intrinsic to the molecule can be unambiguously separated from environmental effects. Even the basic association of internal conversion and intersystem crossing with isolated molecule behavior has occurred only since about 1965. Thus the disparity between characterization of stationary and nonstationary properties of excited electronic states is striking.

Aside from the fact that excited state decay is so important and so diffuse in characterization, another motive now exists for greater experimental efforts, and that is due to the resurgence of theoretical interest in the problem. As a result of early work by I. G. Ross and by Robinson and Frosch, and of the later efforts of Siebrand, Lin, Rice, and Jortner among many others, a theory of nonradiative transitions now exists that is capable of some quantitative success.[3] Its principal application has been to the problem of internal conversion among singlets or among triplets and intersystem crossing between singlet and triplet states. Recently this theory has been extended to discuss chemical relaxation (predissociation and isomerization) in polyatomics. Thus a general approach capable of detailed descriptions of nonradiative processes is beginning to emerge from theory. It is in need of equally detailed experimental data to guide its construction.

It is in this spirit that benzene has long held a prominent position among those benchmark systems most suitable for relating theory and experiment. With 30 vibrational degrees of freedom, benzene has properties typical of large and complex molecular systems, yet its high symmetry allows them to be examined in rather fine experimental detail. In turn its high symmetry also gives the molecule unusual theoretical accessibility. Thus the anticipation arises that a careful examination of excited state relaxation in this molecule will prove particularly rewarding in subsequent efforts to develop effective theoretical descriptions of relaxation phenomena in large molecules.

In the present discussion, a description of the current factual knowledge of excited state relaxation in benzene will be given. Emphasis will be placed on relaxation in the first excited singlet ($^1B_{2u}$) state in the gas phase where both the vibrational excitation and the environmental interaction of the relaxing molecules can be most precisely controlled. Contributions from spectroscopy and photochemistry have been more effectively intertwined in these studies than in most, and a uniquely detailed picture of the crucial role of vibrational excitation on both radiative and non-radiative decay is slowly emerging. In turn, these data are now proving useful in support of theoretical efforts to describe vibronically induced radiative relaxation and intersystem crossing to triplet states. It seems likely that continued study of relaxation from the $^1B_{2u}$ state will supply equally useful information about chemical relaxation and vibrational relaxation in large molecules.

The discourse will begin with brief discussion of the evidence for the positions and assignments of the lower $\pi\pi^*$ states in benzene, including a summary of the vibrational constants for the lowest $\pi\pi^*$ singlet state ($^1B_{2u}$). These data are still in evolution, and since they are essential to theoretical assessment of relaxation data, some care has been taken to sort out those constants which are securely observed and to indicate their source.

Attention will be then directed to some aspects of $^1B_{2u}$–$^1A_{1g}$ spectroscopy, which, perhaps surprisingly, is still an active area in both absorption and fluorescence. For example, numerous novel resonance fluorescence spectra have been obtained recently from single vibronic levels of isolated $^1B_{2u}$ molecules. These in turn have helped to extend the absorption assignments.

Discussion of the relaxation parameters of the $^1B_{2u}$ state in the vapor begins in Section IV. Specific effort has been made to limit discussion to the experimental characterization of the excited state decay. Since benzene occupies such a fundamental position among large molecules and since the experimental literature is so extensive, it seems appropriate merely to bring together in one place the descriptive facts about excited state decay and to present the evidence for them unencumbered with less secure interpretations. Accordingly, the theoretical discussion of $^1B_{2u}$ benzene relaxation, which is the subject of an increasing literature, will not be treated in detail.

Experiments that are now yielding the detailed picture of the vibrational structure in excited state relaxation are described in only an introductory manner in the last section of this chapter. The new experimental approaches and the extent to which they can define the vibrational structure can be seen in the preliminary experiments that are now in print. The bulk of the data, however, has not yet been published.

The work that has so far come forward on single vibronic level relaxation shows that these experiments are explicitly dependent on facts about relaxation from a Boltzmann distribution of levels for their interpretation. Without this, it is impossible to use the single-level data in support of theory. The new experiments thus place the Boltzmann studies in an extremely crucial role. For that reason, the derivation of a well supported and elementary set of facts about decay from the thermal vibrational levels of $^1B_{2u}$ benzene vapor is given in the first sections dealing with excited state relaxation.

The focus of this article on relaxation from $^1B_{2u}$ molecules in the vapor ignores important work on the radiative and nonradiative properties of other excited states. For example, the extensive studies of the $^3B_{1u}$ state in condensed phase comprise some of the foundations for the theory of intersystem crossing and have generated an interesting story about the triplet-state geometry. The growing literature on photochemical activity from the second and third excited singlets of benzene and from the first excited singlet in condensed media has also been omitted. Finally, there is an increasing interest in vibrational relaxation in the $^1B_{2u}$ state of benzene vapor, which has not been described in detail. It promises to provide the most complete picture of vibrational energy transfer yet available for a large molecule.

A compendium has recently appeared with much information about excited states in benzene and other aromatic molecules. It is a useful source of data.[4]

II. THE $\pi\pi^*$ EXCITED STATES

Interest generated by the central position of benzene among aromatic systems has yielded an extensive literature about its excited electronic states. In spite of this long history, fundamental statements about the lower excited states are still just emerging. For example, the assignment of the second excited singlet seems settled only within the past year. The positions as well as the assignments of singlets immediately above the third now comprise a continuing story. A similar situation exists for the triplets in that the second triplet state was not observed until 1965 and the positions of the third and higher triplet are still a matter of some debate.

A. The Singlet States

We briefly describe in this section the current status of characterization of the lower three excited singlet states. Each of these can be reached by absorption from the ground state, and these are the only states for which significant relaxation data exist.

Energies of the lower $\pi\pi^*$ states of benzene are shown in Figure 1. These singlet states are reached by absorption systems beginning near 2600 Å ($\epsilon \approx 60$), 2100 Å ($\epsilon \approx 6000$), and 1850 Å ($\epsilon = 60{,}000$), respectively. No clear evidence has arisen to indicate that additional singlet states lie hidden within the energy region encompassed by the 2600 and 2100 Å transitions, although the issue may not be completely settled for the latter region. A Rydberg transition lies within the 1850 Å $\pi\pi^*$ transition.

The symmetries of those singlet states and the calculation of their energies are discussed in many standard works. For example, a recent description is given by McGlynn, Azumi, and Kinoshita.[5] The π electron M.O.s of benzene are a_{2u}, e_{1g}, e_{2u}, and b_{2g} in order of increasing energy. Single-electron excitation out of the e_{1g} orbital of the ground-state configuration $(a_{2u})^2(e_{1g})^4$ gives the configurations $(a_{2u})^2(e_{1g})^3(e_{2u})^1$ and $(a_{2u})^2(e_{1g})^3(b_{2g})^1$. The former corresponds to B_{2u}, B_{1u}, and E_{1u} excited states and the latter yields an excited state of symmetry E_{2g}. These comprise the lowest energy $\pi\pi^*$ excited states and many calculations[6–15] of their positions have been offered.

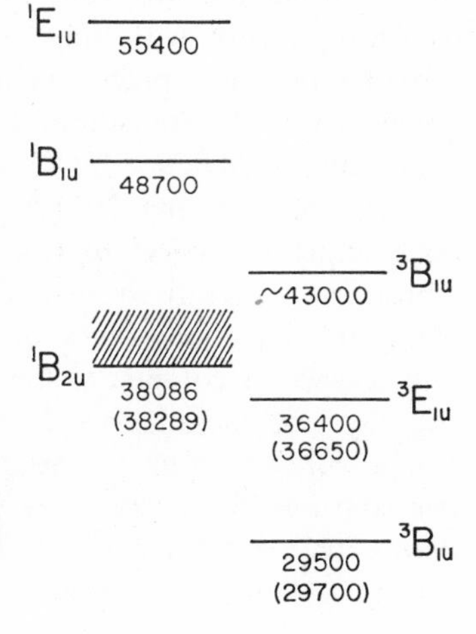

Fig. 1. The observed zero-point energies of the lower $\pi\pi^*$ states of benzene vapor. These are the states S_1, S_2, S_3 and T_1, T_2, T_3. The energies are given in cm^{-1}, and the values in parentheses are for C_6D_6. Evidence for the assignments and energies is discussed in the text. The hatched area extending about 3000 cm^{-1} above the $^1B_{2u}$ zero-point energy represents the approximate energy region where excitation results in fluorescence in the vapor. Fluorescence is not observed following excitation of higher vibrational levels in S_1 or of any levels in S_2 or S_3.

Two points of significance for this discussion arise from inspection of those calculations. We first see fairly consistent agreement in the calculation of the $^1B_{2u}$ energy which puts it near 5 eV, corresponding well with the position of the first absorption $S_1 \leftarrow S_0$ near 2600 Å. Although we must be careful in taking confidence from the energy calculations (some of them fit the B_{2u} energy to this spectroscopic position), the general consistency of ordering and the general match of energies of the $^1B_{2u}$ state with the 2600 Å transition give confidence based on theoretical calculation alone that the lowest absorption transition is indeed $^1B_{2u} \leftarrow {}^1A_{1g}$. Further confirmation apart from spectroscopic analysis comes from the fact that the $^1B_{2u} \leftarrow {}^1A_{1g}$ is electric-dipole forbidden and this is consistent with the low extinction coefficient of the 2600 Å transition.

The second point is that energy calculations cannot be used with confidence to assign the higher singlet states. The calculated positions and even the ordering are too sensitive to details of the calculation to allow this. We can say only that the intensity of absorption at 1850 Å is sufficiently large so that it may be the allowed transition $^1E_{1u} \leftarrow {}^1A_{1g}$, and thus the absorption near 2100 Å would be one of the forbidden transitions $^1B_{1u} \leftarrow {}^1A_{1g}$ or $^1E_{2g} \leftarrow {}^1A_{1g}$. It is quite apparent, however, that we must search among the tools of spectroscopy for criteria beyond intensity and energy of transitions for proper assignment of these (and higher) singlet states.

The vibrational analysis of the highly structured 2600 Å absorption system is of course one of the classic tales of spectroscopy, and it is told with superb clarity in two reviews.[16,17] Suffice it to say that the initial analysis of Sponer, Nordheim, Sklar, and Teller,[18] the further extensive work of Ingold and co-workers,[19] and most recently the rotational analysis of the electronic spectrum by Callomon, Dunn, and Mills[2] link this transition firmly with the $^1B_{2u}$ state. The last-mentioned work now provides the most precise value of the forbidden electronic origin (the 0,0 band) of the transition. The origins are at 38,086.1 cm^{-1} for C_6H_6 vapor and 38,289 cm^{-1} for C_6D_6 vapor.

Assignment of the 2100 Å transition has been difficult because the upper state is subject to fast nonradiative decay which broadens the vibrational structure to the point of diffuseness. This plus the normal hot band and sequence congestion characteristic of room temperature polyatomic spectra complicate the identification of the upper state ($^1B_{1u}$ or $^1E_{2g}$) by vibronic analysis. In turn, complexity due to site splittings has clouded the analysis of the spectrum obtained from benzene in low-temperature solid hosts where the removal of hot bands otherwise makes the spectrum more discrete. Recently, however, two studies have appeared, which together seem to point in favor of the assignment $^1B_{1u} \leftarrow {}^1A_{1g}$.

The first of these is based on two-photon absorption spectroscopy. In 1967, Honig, Jortner, and Szöke[20] suggested that two-photon absorption could discriminate between the $^1B_{1u} \leftarrow {}^1A_{1g}$ and $^1E_{2g} \leftarrow {}^1A_{1g}$ transitions on the basis of the absorption cross section alone. Following this proposal, Monson and McClain[21] have examined the 2100 Å region for two-photon absorption in pure liquid benzene. Absorption was not observed, and this establishes an upper bound on the cross section of about 1×10^{-51} cm^{4}sec/photon-molecule. The predicted cross section for the $^1B_{1u} \leftarrow {}^1A_{1g}$ transition is in the range 0.4 to 20 times 10^{-51}, whereas that for the $^1E_{2g} \leftarrow {}^1A_{1g}$ transition is between 10^4 and 10^5 times 10^{-51}. The experiment thus seems to rule out the $^1E_{2g} \leftarrow {}^1A_{1g}$ possibility, leaving by default an assignment of the 2100 Å transition as $^1B_{1u} \leftarrow {}^1A_{1g}$.

Recent work with single-photon absorption by C_6H_6 and deuterated benzenes in N_2 or inert gas matrices has supported this.[22,23] Using these matrices, Katz, Brith, Sharf, and Jortner have been able to distinguish site splittings from the intrinsic vibronic structure in the 2100 Å transition, and the activity of two vibrations (~520 and ~1480 cm^{-1}) becomes apparent. The 520 cm^{-1} frequency suggests the e_{2g} vibration ν_6, and the 1480 cm^{-1} frequency suggests the e_{2g} vibration ν_8 whose ground state frequency is 1596 cm^{-1} (the frequency of this vibration in an excited state is not securely known). These modes are consistent with the vibronic requirement for inducing the $^1B_{1u} \leftarrow {}^1A_{1g}$ transition and are distinct from the vibronic activity that would be associated with the $^1E_{1g} \leftarrow {}^1A_{1g}$ transition.

The position of the forbidden 0,0 band of the $^1B_{1u} \leftarrow {}^1A_{1g}$ transition in the vapor phase follows from this analysis. Dunn and Ingold[24] observe the lowest energy bands of the 2100 Å system to be a "doublet" at 49,005 and 49,215 cm^{-1}. This becomes a single vibronic feature at low temperatures,[23] so that one of its members in the vapor is a hot band. The vapor spacing is consistent with the 1–1 hot band sequence structure in the e_{2u} vibration ν_{16} (399 cm^{-1} in the ground state and 237 cm^{-1} in the $^1B_{2u}$ state) seen throughout the 2600 Å transition. On this basis, the lower energy component is the sequence band and the higher energy component is $0,0 + \nu e_{2g}$ (520 cm^{-1}). Thus the electronic origin in the vapor is probably within about 50 cm^{-1} of $49{,}215 - 520 = 48{,}695$ cm^{-1}.

The 1850 Å transition has been assigned to the $^1E_{1u}$ state on the basis that its large extinction coefficient would correspond with the allowed $^1E_{1u} \leftarrow {}^1A_{1g}$ transition calculated to be in that energy region. Support from vibrational analysis is difficult to secure in gas phase because the $\pi\pi^*$ component of that absorption (as opposed to the Rydberg transition also in that region) is without appreciable structure. However, the suppression of hot and sequence band congestion by using low-temperature solid hosts enables vibronic structure to be seen. Katz et al.[23] have been able to separate the Rydberg structure from the $\pi\pi^*$ structure in inert gas matrices since the former is blue shifted and the latter red shifted. The $\pi\pi^*$ structure appears as a progression in a vibration ~920 cm^{-1} in C_6H_6 and ~850 cm^{-1} in C_6D_6 corresponding to the a_{1g} vibration ν_1. This structure is consistent with a progression in ν_1 starting with the 0,0 band for an allowed transition such as $^1E_{1u} \leftarrow {}^1A_{1g}$. However, it is also consistent with the expected structure from a forbidden electronic transition that is vibronically induced. In that case, the ν_1 progression would start with a false origin displaced to the blue of the forbidden 0,0 band by one quantum of the active vibration. There seems no way to choose between these alternatives. However, it appears certain that two vibronic false origins do not exist as would be the case for the forbidden $^1E_{2g} \leftarrow {}^1A_{1g}$ transition. The 0,0 band can be taken from the first band in the vapor spectrum of Wilkinson,[25] and it is about 55,400 cm^{-1}.

Higher ($\pi\pi^*$?) singlets have been observed by $S_n \leftarrow S_1$ absorption in nanosecond spectroscopy.[26,27] Some assignments are proposed by Birks.[28]

B. The Triplet States

In one aspect, calculation of triplet energies has been more successful than that of singlets. Calculation has provided a fairly consistent energy ordering $^3B_{1u}$, $^3E_{1u}$, and $^3B_{2u}$ for the states T_1, T_2, and T_3. On the other hand, as with the singlets, the calculations do not provide firm statements concerning where these triplets lie.

Experimental resolution of the problem has proven a challenge, and the spectroscopic history of triplets is laden with false leads. The 1962 discussions of Kearns[29] and of Platt[30] describe this well. A recent review of both the calculations and spectroscopy has also appeared,[5] so that the detailed evidence for the assignments and positions of these triplets will not be repeated here. Only brief comments concerning these efforts will be offered.

Evidence for assignment of the 3400 Å transition in benzene as the $^3B_{1u} - {}^1A_{1g}$ transition is convincing.[29–31] The electronic origin of the transition in the vapor must be close (± 50 cm^{-1}) to 29,500 cm^{-1} (29,700 cm^{-1} for C_6D_6), but a more precise determination for the vapor phase is not yet available. These positions are derived from the broad 0,0 band in the O_2-perturbed absorption of benzene vapor.[32–34] The exceedingly low oscillator strength ($f \approx 10^{-10}$) of the spin- and symmetry-forbidden $^3B_{1u} - {}^1A_{1g}$ transition has so far precluded its observation in pure benzene vapor.

In contrast, the condensed phase spectroscopy is well developed. Very recent photoexcitation experiments[35] have revealed the absorption in a pure benzene crystal at 4.2°K, and direct ${}^3B_{1u} \leftarrow {}^1A_{1g}$ absorption has been observed in O_2-perturbed solid benzene.[36] The ${}^3B_{1u} - {}^1A_{1g}$ phosphorescence in condensed phase at low temperatures has been the object of numerous studies.[31,37–41]

The position of T_2 now seems securely identified from Colson and Bernstein's study[36] of O_2-perturbed absorption of solid benzene at 4.2°K and of absorption in a pure benzene crystal at 4.2°K. Their results place the ${}^3E_{1u}$ state within 50 cm^{-1} of 36,560 cm^{-1} and 36,724 cm^{-1} for O_2-perturbed solid C_6H_6 and C_6D_6, respectively. By comparison of their data on T_1 with data of T_1 in O_2-perturbed vapor we can estimate that the condensed phase T_2 values may be blue shifted by about 150 cm^{-1} from the vapor position. Thus the position of the ${}^3E_{1u}(T_2)$ state in the vapor is approximately ($\pm$200 cm^{-1}) 36,400 and 36,650 cm^{-1} for C_6H_6 and C_6D_6, respectively.

The position of the ${}^3B_{2u}$ state (T_3) is less certain. Theoretical estimates are diverse[29] and suggest only that T_3 lies somewhere in the region 40,000–60,000 cm^{-1} with a value toward the lower end more likely. McGlynn, Azumi, and Kinoshita[5] have suggested that the best estimate probably derives from Pariser's relationship,[42]

$$E({}^3E_{1u}) = \tfrac{1}{2}[E({}^3B_{1u}) + E({}^3B_{2u})]$$

The energies of the ${}^3B_{1u}$ and ${}^3E_{1u}$ states given here would then place the ${}^3B_{2u}$ state near 43,300 cm^{-1} in C_6H_6 vapor.

Comment on this proposition from optical spectroscopy is entirely lacking. The ${}^3B_{2u}$–${}^1A_{1g}$ transition is orbitally as well as spin-forbidden, so that the problems of observing it parallel those for the ${}^3B_{1u}$–${}^1A_{1g}$ transition discussed above. Furthermore, the difficulties are severely compounded by the fact that the transition certainly lies within the spectral region of the much stronger $S_1 \leftarrow S_0$ absorption. The ${}^3B_{2u}$ state might perhaps be seen by $T_3 \leftarrow T_1$ absorption, but this has not yet been reported.[43,44] That transition is itself orbitally forbidden.

We may turn to electron scattering[45–47] for help with this problem. Two sets of data[46,47] reveal transitions that correlate with the known positions of T_1 and T_2, and in each study a third transition is seen which may plausibly be assigned as $T_3 \leftarrow S_0$. There are certainly no known singlet states to which it could be assigned, and its intensity is consistent with a singlet-triplet combination. Further evidence is given in detail by Doering.[47]

The data given in Table I show the close correspondence between the studies. Unfortunately, correspondence between the $T_3 \leftarrow S_0$ transitions is poorest because the transition is a shoulder of the very strong $S_1 \leftarrow S_0$ transition.

TABLE I
Singlet–Triplet Transitions Observed with Electron-Scattering from C_6H_6 Vapor

Transition	Energy of Franck-Condon maximum (cm^{-1})	
	Doering[47]	Compton et al.[46]
$T_1 \leftarrow S_0$	31,800	31,500
$T_2 \leftarrow S_0$	38,400	38,000
$T_3 \leftarrow S_0$	45,200	43,600

Two points of importance for the present purpose arise from the data.

1. The maxima in the transitions $T_1 \leftarrow S_0$ and $T_2 \leftarrow S_0$ are in each case roughly 2000 cm^{-1} above the known transition origins, and a similar relationship is also observed between the Franck-Condon maxima and known origins in the $S_n \leftarrow S_0$ transitions. Thus the origin of the $T_3 \leftarrow S_0$ transition would be in the range 41,600 to 43,200 cm^{-1}.
2. The relative spacings of these triplets correspond well with the prediction of Pariser.[42] T_2 is approximately midway between T_1 and T_3.

Guided by these points, the position of T_3 in C_6H_6 vapor is most plausibly placed near 43,000 cm^{-1}.

No data are available for C_6D_6.

Still higher triplets have been seen recently by $T_n \leftarrow T_1$ absorption using nanosecond flash spectroscopy.[28,43,44]

C. Vibrational Frequencies

The modern efforts on this problem began with Wilson's[48] normal mode analysis in 1934 and have subsequently led to determination of every frequency in the ground electronic state, a smaller set of frequencies in the $^1B_{2u}$ state (S_1), and an approximate value for a few frequencies in other electronic states. The data for the electronic states S_0, S_1, and T_1 are collected in Table II. Some frequencies in other states are summarized by Herzberg.[16]

Two numbering schemes for vibrational modes of benzene are in current use. That of Wilson[48] is adopted in the present discussion because it has become firmly entrenched in the literature on electronic and vibrational spectroscopy of benzene. A majority of papers use this system rather than the more modern convention used in Herzberg's discussions.[16,49] The numbering of the two systems is related in Table II.

Ground-state frequencies are taken from the compilation of Callomon, Dunn, and Mills[2] (CDM), which itself is taken largely from the vapor phase data of Brodersen and Langseth,[50] supplemented by the former authors' data from high-resolution electronic spectroscopy. The value for ν_1 in Table II has been changed to 993.1 cm^{-1} to correct a misprint in the compilation of CDM. The original elucidation of these frequencies by Raman and infrared spectroscopy comes in great measure from the classic work of Ingold and co-workers on benzene and its isotopic modifications.[51,52] A review of this and related work is given by Dunn.[17]

Frequencies for the $^1B_{2u}$ state in Table II include only those that have been identified *by actual observation*. A complete list of all 20 $^1B_{2u}$

TABLE II

Observed Fundamental Vibrational Frequencies in the $^1A_{1g}$, $^1B_{2u}$, and $^3B_{1u}$ state of C_6H_6 and C_6D_6. (Those for C_6D_6 are in parentheses; the $^1A_{1g}$ and $^1B_{2u}$ frequencies are for vapor except where noted)

Vibrational mode[a]	Vibrational symmetry	Fundamentals (cm^{-1}) $^1A_{1g}$[b]	$^1B_{2u}$	$^3B_{1u}$[i]
1 (2)	a_{1g}	993.1[c] (945.6[d])	923.0[g] (879[h])	925 (881)
2 (1)	a_{1g}	3,073 (2,303)	3,130[h]	
3 (3)	a_{2g}	1,350 (1,059)		
4 (8)	b_{2g}	707 (599)		
5 (7)	b_{2g}	990 (829)		
6 (18)	e_{2g}	608.0[d] (580.2[d])	522.4[d] (498.0[d])	628 (575)
7 (15)	e_{2g}	3,056 (2,274)	3,077.2[d](2,349.6[d])	
8 (16)	e_{2g}	1,596[e] (1,558)		239, 252 (227, 237)
9 (17)	e_{2g}	1,178 (869)	~1,154[g]	1,188? (961?)
10 (11)	e_{1g}	846 (660)	585[h] (454[h])	
11 (4)	a_{2u}	674.0 (496.2[f])	514.8[d] (382[h])	
12 (6)	b_{1u}	1,010 (970)		
13 (5)	b_{1u}	3,057 (2,285)		
14 (9)	b_{2u}	1,309 (1,282)		
15 (10)	b_{2u}	1,146 (824)		
16 (20)	e_{2u}	398.6[d] (347.4[d])	237.3[d] (207.1[d])	244? (215?)
17 (19)	e_{2u}	967 (787)	719[g] (590[h])	
18 (14)	e_{1u}	1,037 (814)		
19 (13)	e_{1u}	1,482 (1,333)		
20 (12)	e_{1u}	3,064 (2,288)		

[a] Wilson's numbering.[48] Herzberg's numbering[49] is in parentheses.
[b] Vapor phase. Ref. 50 except where noted.
[c] Correction to Ref. 2 misprint.
[d] Ref. 2.
[e] Corrected for Fermi resonance with $(\nu_6 + \nu_1)$.
[f] A. Danti and R. C. Lord, *Spectrochim. Acta*, **13**, 180 (1958).
[g] Ref. 55.
[h] Ref. 54.
[i] Solid phase. Ref. 35. The degeneracy of ν_8 is split in the $^3B_{1u}$ state.

frequencies has been given in the work of Garforth, Ingold, and Poole,[53] but over half of these frequencies derive only from calculation. They should, however, be satisfactory for use as a set in approximate calculations of densities of states, and so forth.

The $^1B_{2u}$ frequencies come from analysis of the ultraviolet $^1B_{2u} \leftarrow {}^1A_{1g}$ vapor absorption spectrum. The first extensive compilation derived from the 1948 analysis of Garforth and Ingold[54] (GI), which yielded 11 frequencies. The high-resolution analysis of CDM in 1966 modified and refined that list. Perhaps the most significant change was deletion of the e_{2g} mode ν_8 following the demonstration that the rotational contours of vibronic bands induced by this mode would be too broad and featureless to be observed in the spectrum. The fact that ν_8 cannot be included among the observed frequencies is not always appreciated by others taking comfort in GI's "observed" value of $\nu_8 = 1470\ \text{cm}^{-1}$ for support of assignments in other electronic states.

The most recent examination of the $^1B_{2u}$–$^1A_{1g}$ transition in the vapor has been by Atkinson and Parmenter.[55] These efforts have yielded still further modification of the $^1B_{2u}$ vibrational frequencies. The b_{2g} modes ν_4 and ν_5 have been deleted from the "observed" classification, and ν_9 has been changed to about $1154\ \text{cm}^{-1}$. Presumably the assignments in ν_4 and ν_5 are also in error in the C_6D_6 analysis,[54] and these frequencies have been omitted in the C_6D_6 tabulation.

The only other $\pi\pi^*$ state for which fairly accurate vibrational frequencies are known is the lowest triplet, $^3B_{1u}$. The most precise values are probably those from the recent and elegant photoexcitation absorption study by Burland et al.[35] of the $^3B_{1u} \leftarrow {}^1A_{1g}$ transition in benzene crystals. A most significant result of this work is observation of a very low value of $\nu_8 \sim 250\ \text{cm}^{-1}$ in the $^3B_{1u}$ state (versus $1596\ \text{cm}^{-1}$ in the ground state). This value and the splitting of the degeneracy of ν_8 is in confirmation of earlier predictions by van der Waals, Berghuis, and de Groot.[56] The identification of ν_4 and ν_5 in the photoexcitation spectrum seems unlikely in view of the inactivity of these modes in the $^1B_{2u}$–$^1A_{1g}$ spectrum, and therefore we might prefer the alternate assignments suggested by Burland et al.

III. $^1B_{2u}$–$^1A_{1g}$ SPECTROSCOPY

Notation. The exceedingly convenient scheme introduced by CDM[2] is used in this and later discussions of benzene spectra. Assignments of vibrational structure in the spectra are given by notation such as $6_0^1 10_2^0 1_0^1$. This indicates that three vibrational modes ν_6, ν_{10}, and ν_1 are involved in the transition with $v = 0$ in both electronic states for all others. The superscript indicates the quanta of each active mode in the upper state ($v_6' = 1$, $v_{10}' = 0$, and $v_1' = 1$) and subscripts are for the lower electronic state. This notation is also used to discuss individual levels in a given electronic state. For example, the notation 6_1 would describe a lower electronic state with $v_6'' = 1$ and $v = 0$ for all other modes. The notation $6^2 1^1$ describes the excited electronic state with $v_6' = 2$ and $v_1' = 1$.

The ${}^1B_{2u}$–${}^1A_{1g}$ $(S_1 - S_0)$ transition in benzene has been a pioneer model for a number of aspects of electronic spectroscopy in larger polyatomics. Sklar's analysis[57] of the absorption spectrum in 1937 and its extension by Sponer et al.[18] provided the first extensive analysis of a vibronically induced transition as described theoretically by Herzberg and Teller in 1933.[58] The further extension of this analysis by Ingold and co-workers[19] has now assumed classic stature as *the* description of a Herzberg-Teller vibronic transition. The high-resolution band contour analysis of CDM[2] provides a unique elucidation of Coriolis coupling of degenerate vibrations in both the ground and excited states of a large molecule. Fluorescence excited in low-pressure benzene with the 2536 Å line of an Hg arc was among the earliest characterizations of large-molecule resonance fluorescence.[59–61] Finally, one of the best pictures of Franck-Condon factors in a polyatomic molecule comes from the ν_1 progression in the ${}^1B_{2u}$–${}^1A_{1g}$ absorption spectrum.[62,63]

From this extensive activity we might suspect that further examination of the transition offers little opportunity for additional information. However, recent activity in ${}^1B_{2u}$–${}^1A_{1g}$ spectroscopy belies this and illustrates once again the virtues of benzene as a model for large molecules. This activity has occurred in both fluorescence and absorption. Benzene has recently yielded the first extensive set of resonance fluorescence spectra from each of a variety of known single vibronic levels in an excited state of a large molecule.[64] In turn, information from these spectra has helped to provide a considerable extension of assignments in the ${}^1B_{2u}$–${}^1A_{1g}$ absorption spectrum. Because the fluorescence is both spectroscopically novel and subsequently basic to relaxation studies of the ${}^1B_{2u}$ state, it will be discussed briefly in the next section. Following that, some comments will be offered about absorption assignments that play a particularly important role in discussing both radiative and nonradiative decay from the ${}^1B_{2u}$ state.

A. Single Vibronic Level Fluorescence

One can see even in the low-resolution spectrum of Fig. 2 that the 2600 Å absorption spectrum of benzene is unusually discrete for a molecule of such size. Further examination at high resolution reveals many regions in the spectrum where vibronic bands are not seriously overlapped by others. This sets the stage for obtaining a unique set of resonance fluorescence spectra. By pumping one of these bands with tuned monochromatic light, ${}^1B_{2u}$ molecules can be prepared in the specific vibronic level reached

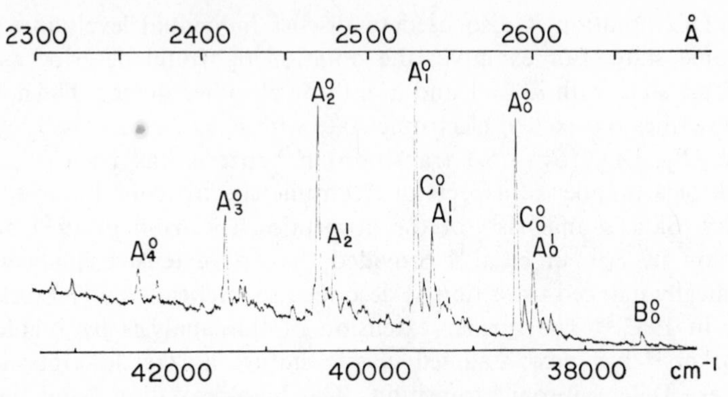

Fig. 2. The $^1B_{2u} \leftarrow {}^1A_{1g}$ absorption spectrum of benzene vapor at low resolution. The rising background at short wave lengths is an experimental artifact. The band notation is that of Garforth and Ingold.[54] The bands correspond to the following transitions: $B_0^0 = 6_1^0$; progression $A_n^0 = 6_0^1 1_0^n$; progression $A_n^1 = 6_0^1 16_1^1 1_0^n$; progression $C_n^0 = 6_1^2 1_0^n$.

by the absorption band. When pressures are sufficiently low, electronic relaxation of the $^1B_{2u}$ state will occur before collisions, and if one of the relaxation channels is radiative decay, resonance fluorescence will occur from the vibronic level optically selected in the absorption act. By this technique, resonance fluorescence has been observed from over 20 levels in the $^1B_{2u}$ state.[64–66]

The term single vibronic level (SVL) fluorescence has been used to distinguish this emission from the more general case of resonance fluorescence where optical selection of a *single* level does not necessarily occur either because monochromatic pumping was not used or because the excitation fell in a region of overlapping absorption bands.

The experimental problems associated with SVL fluorescence are presently severe,[66] and obtaining exciting radiation that is simultaneously tunable, monochromatic, and intense provides the principal challenge. Short-arc xenon arcs coupled with a monochromator have been used with benzene where an exciting bandpass of 15–45 cm^{-1} has been sufficiently monochromatic to yield SVL fluorescence from a number of states. Narrower bandpasses are needed to extend the list of available levels in the $^1B_{2u}$ state and to extend this type of experiment to other systems where spectral congestion in absorption is more severe. It is certain that this will be among the important applications of tunable lasers in the near future.

Three SVL fluorescence spectra from 0.1 torr of benzene are compared with fluorescence from a Boltzmann vibronic distribution (300°K) in Fig. 3. SVL fluorescence is generally observed to the red of excitation, and each spectrum from different vibronic levels is distinctive in appearance.

The open and simple structure of SVL spectra relative to the Boltzmann fluorescence in Fig. 3 is a consequence of the success of the experiment. The SVL spectra are truly limited to a single vibronic origin, whereas the Boltzmann fluorescence is not. (The density of vibrational states below about 3000 cm^{-1} in benzene is too low to allow appreciable vibrational mixing among those states.)

The simplicity of SVL fluorescence is unusual even among resonance spectra. This is emphasized in Fig. 4, where the SVL fluorescence spectrum from the level $6^1 1^1$ is compared with the resonance fluorescence excited by absorption from the 2537 Å Hg line. This originates from about five excited state vibrational levels reached by the several absorption bands that overlap the position of the Hg line.[67,68] Sharp structure exists with greater abundance in the Hg-excited spectrum, but, more significantly, an underlying "continuum" can be observed in this spectrum that is absent in the SVL fluorescence. This "continuum" is in fact a continuum of congestion. It arises from the multitude of discrete structure from those several emitting states that is too closely spaced to be resolved.

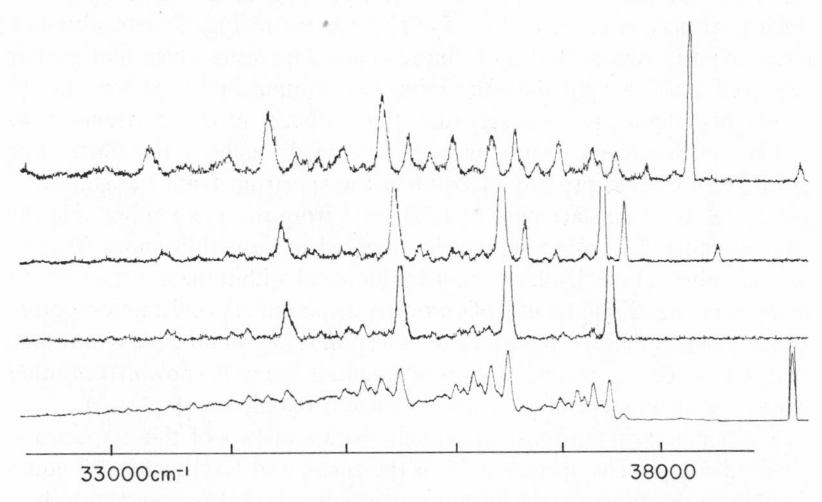

Fig. 3. Fluorescence spectra from the $^1B_{2u}$ state of benzene vapor. The upper three are SVL fluorescence spectra from the vibrational levels $6^1 1^1$ ($\epsilon_{vib} = 1446$ cm^{-1}), 6^1 ($\epsilon_{vib} = 523$ cm^{-1}), and the zero-point level, respectively, from the top. The bottom spectrum is from a thermal distribution of levels in the $^1B_{2u}$ state. In every case the exciting line is at the position of the structure at highest energy. Strong reabsorption attenuates fluorescence coincident with the exciting line in the upper two spectra. The exciting line for the bottom spectrum was the 2536 Å Hg line, and high total pressure (*ca.* 50 torr of added gas) was used for thermal equilibration of vibrational levels before emission.

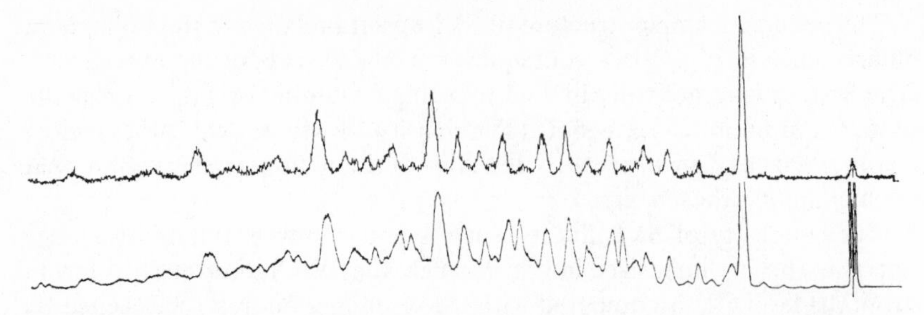

Fig. 4. A comparison of resonance fluorescence from the excited vibrational level $6^1 1^1$ (top spectrum) reached by excitation in the absorption band $6_0^1 1_0^1$ with resonance fluorescence from the several excited vibrational levels populated by absorption of light from the 2536 Å Hg line. The exciting lines (structure at far right) have been placed above one another to facilitate comparison of the fluorescence structure. Fluorescence is in each case to lower energy from excitation.

The SVL fluorescence spectra in Fig. 3 have been displayed with accurate wave length relationships. Comparison is facilitated, however, when they are shown with the exciting lines aligned above one another. Such a display is given in Fig. 5. (The spectra in Fig. 5 were obtained from an early version[64] of SVL fluorescence apparatus which had greater scattered exciting light than the improved equipment[65] used for Fig. 3.) With this alignment, we see that fluorescence structure occurs with similar displacements from the exciting line throughout the series. For example, the most prominent band in the spectrum from the state $6^1 1^1$ has a measured displacement of 1203 cm^{-1} from the exciting line and the analogous band in other spectra is observed to lie within about 20 cm^{-1} of that value. These displacements are identical within the accuracy of the measurements. Similar cases of common displacements exist among other bands. Each of these displacements corresponds to ground-state vibrational energies. Since the ground-state fundamentals are well known from other work, the assignment of SVL fluorescence is readily accomplished.

A schematic of the most prominent bands in two of those spectra is given in Fig. 6. The spectrum from the zero-point level in Figs. 3 and 5 contains a prominent series of fluorescence bands that is observed to be a progression in ν_1''. The principal intensity in the $^1B_{2u}$–$^1A_{1g}$ combination is induced by transitions with $\Delta v_6 = \pm 1$ and $\Delta v = 0$ in other nonsymmetric vibrations. From the zero-point level, $\Delta v_6 = +1$ only, and we see from the schematic in Fig. 6 that the progression origin is thus coincident with the position of the 6_1^0 absorption band used for excitation. The remaining bands in the progression are then observed at displacements of 992, 1984, 2976 cm^{-1}, etc., which are the fundamental and overtones of ν_1''.

The right-hand schematic in Fig. 6 shows the emission spectrum from the level 6^1 reached by excitation into the absorption band 6_0^1. The $\Delta v_6 = \pm 1$ selection rules allow in this case two progressions from the excited state 6^1. The $\Delta v_6 = -1$ progression origin is coincident with the exciting line and leads to members again at displacements corresponding to the fundamental and overtones of ν_6''. This progression yields bands matching precisely the displacements observed in the zero-point spectrum, even though the initial and final levels in each case are different. The $\Delta v_6 = +1$ transition yields the progression origin 6_2^1, and it is thus displaced by $2\nu_6'' = 1216\ \text{cm}^{-1}$ from the exciting line. Other members of this progression are displaced by $1216\ \text{cm}^{-1}$ plus the fundamental or overtones of ν_1''.

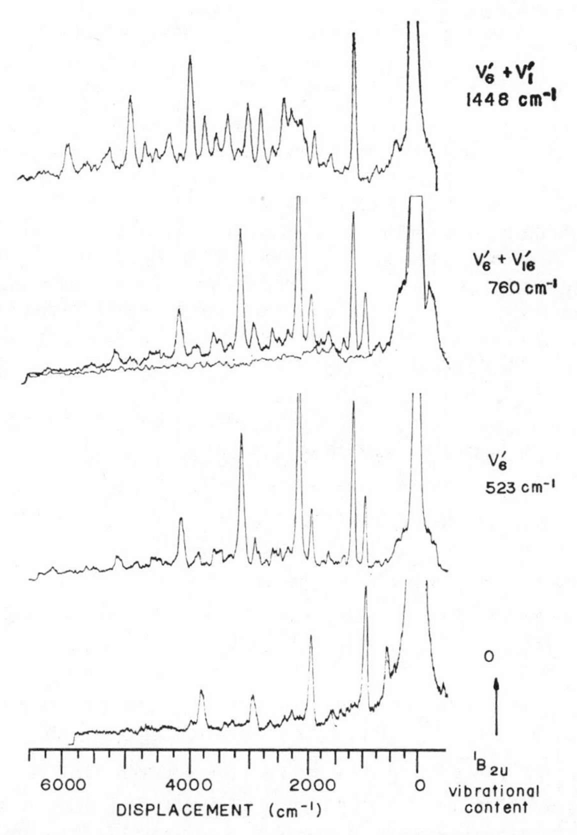

Fig. 5. SVL fluorescence from four levels in the $^1B_{2u}$ state of benzene vapor. The vibrational levels and vibrational energies of those levels are given to the right of each spectrum. The exciting lines are the structure to the right in each spectrum, and they have been aligned one above the other to show the common displacements of fluorescence structure from those lines.

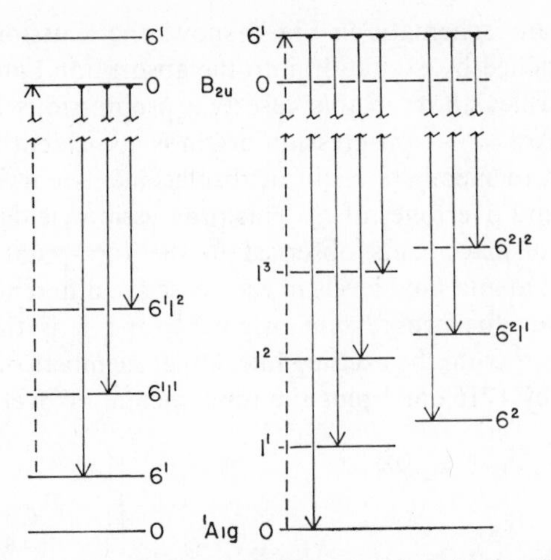

Fig. 6. A schematic of the most prominent progressions in two SVL fluorescence spectra from benzene. To the left is the spectrum from the zero-point level excited by pumping the 6^0_1 absorption band. Only one strong ν_1 progression induced by a $\Delta v_6 = \pm 1$ transition can occur in fluorescence ($\Delta v_6 = \pm 1$ only) and the progression origin is coincident with the exciting line. Fluorescence band displacements are then observed at energies equivalent to ν''_1, $2\nu''_1$, $3\nu''_1$, etc. To the right is the spectrum from 6^1 excited by the absorption 6^1_0. In this case, ν_1 progressions induced by both $\Delta v_6 = +1$ and $\Delta v_6 = -1$ transitions can occur in fluorescence. The second yields fluorescence band displacements identical to those in the zero-point spectrum. The $\Delta v_6 = +1$ progression yields a new series of displacements beginning with a band at $2\nu''_6 = 1212$ cm^{-1}.

Thus the spectrum from 6^1 has some displacements identical to those observed in zero-point fluorescence plus others that cannot occur in the zero-point spectrum. Extension of this type of analysis to the minor structure in these spectra allowed by the selection rules shows additional features that are common to both transitions and still others not mutually accessible. These are typical of the types of transition that give SVL fluorescence spectra so many common features in spite of different upper and lower states, but still introduce some unique qualities into each spectrum.

Detailed analyses of many SVL fluorescence spectra have been completed and are discussed in several reports.[64–66,68] As an example the spectrum from the level $6^1 1^1$ is shown with its complete assignment in Fig. 7.

Two general points of significance arise from analyses of SVL fluorescence. First, assignments are in every case observed to be consistent with the symmetry selection rules appropriate for a Herzberg-Teller vibronic transition (these rules are set forth in great detail for benzene by Garforth, Ingold, and Poole.[69]) No transitions have been found that violate these rules. With the recent availability of SVL fluorescence, there is now a large body of analyses of both fluorescence and absorption in the vapor phase at resolutions ranging from low to very high. In none of these has vibronic structure been found other than that allowed by the Herzberg-Teller selection rules for the vibronically induced ${}^1B_{2u}$–${}^1A_{1g}$ transition. The different structure that is characteristic of a symmetry-allowed transition occurring in spectra from benzene in low-temperature solid media[41,70,71] must not be intrinsic to the isolated molecule. It must arise from environmental perturbation.[65]

A second and exceedingly important consequence of the SVL fluorescence analysis has been its help in elucidation of the set of vibrational modes that carry the principle activity in the ${}^1B_{2u}$–${}^1A_{1g}$ optical transition. As described earlier, this can be accomplished fairly easily with fluorescence because the structure is established by ground-state vibrations whose frequencies are all known from other work.

The modes observed in SVL fluorescence form a consistent set from spectrum to spectrum. Only eight of the twenty vibrational frequencies in benzene are required to describe the more prominent structure in SVL fluorescence. Principal vibrational activity in fluorescence is limited to the

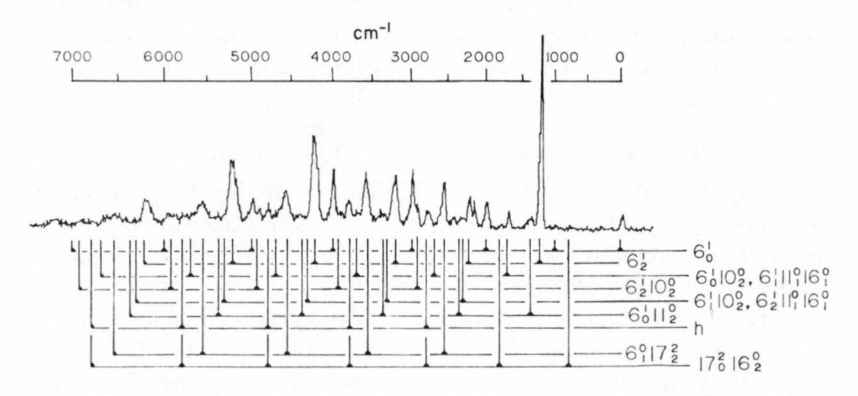

Fig. 7. SVL fluorescence principally from the excited state $6^1 1^1$ produced by pumping the $6^1_0 1^1_0$ absorption band at 38,600 cm^{-1} in C_6H_6 with radiation having a 50 cm^{-1} band width. The exciting position is marked zero on the cm^{-1} scale and displacements are to lower energy. The band width of the fluorescence spectrometer is 25 cm^{-1}. The experimental details are similar to those described elsewhere.[65] The assignments indicate the vibrations active in the transition. To each assignment should be added the component 1^1_n where n indicates the activity of $v_1'' = 0, 1, 2, \ldots$, with $v_1'' = 0$ being the first band in each progression. Note that in this spectrum, structure is identified from two upper states. Principal structure arises from the state $6^1 1^1$, but two progressions can be seen from the state 17^2. The strongest of these is the only intense transition expected from 17^2, and the other is visible only because it falls in a region vacant of other structure. The state 17^2 is excited by absorption in a very weak band 17^2_0 that lies approximately 5 cm^{-1} away from the maximum of the stronger $6^1_0 1^1_0$ absorption band.

modes ν_1, ν_6, ν_7, ν_9, ν_{10}, ν_{11}, ν_{16}, and ν_{17}. This set also matches the set derived from analysis of the absorption spectrum. The mode ν_2 should be added to the list of active vibrations, for it can be seen clearly in absorption. It would be difficult to observe ν_2 at the lower resolution of the SVL fluorescence, because its high frequency (3073 cm^{-1}) causes it to appear weakly in a region of fluorescence badly congested with progression members of other transitions.

The resulting list of nine active modes span five of the ten vibrational symmetries represented in benzene. Furthermore, every mode in those five symmetry groups is active with the sole exception of ν_8 in the species e_{2g}, and in fact even this exception may be an artifact. CDM have shown that ν_8, if active, would most probably appear in bands whose rotational structure is distributed without distinguishing maxima so that they would be hard to detect. Thus the interesting proposition occurs that vibronic activity in the $^1B_{2u}$–$^1A_{1g}$ transition is controlled to a good approximation by vibrational symmetry. Entire symmetry groups of vibrations rather than selected vibrations from within a group are either active or inactive. It should be remarked that classification of vibrations as active or inactive is of course a matter of degree. Principal vibronic structure involves those modes classified as active. Structure associated with nonzero quanta of other vibrations is probably also present in the spectrum, but those transitions comprise features that are very weak indeed.

B. Franck-Condon Factors in SVL Fluorescence

Among the interesting aspects of SVL fluorescence lies the opportunity to obtain a more extensive view of Franck-Condon factors in polyatomic spectra than heretofore has been possible. In all spectra, Franck-Condon factors appear most prominently as governing the relative intensities of members of a progression. In absorption, the progression in some vibration ν_x can generally be observed only from $v_x'' = 0$ because the Boltzmann factors are often too severe to see transitions from $v_x'' = 1$. In SVL fluorescence, however, this restriction is entirely removed. Emission can be generated from a long series of upper states with $v_x' = 0, 1, 2, \ldots$, by simply tuning to successive progression members X_0^n in excitation. Thus we can readily observe the various progressions $X_n^0, X_n^1, X_n^2, \ldots$, in fluorescence.

Numerous SVL fluorescence spectra of this type have been obtained from benzene, and they display the progression in ν_1 very clearly. For example, in the spectrum from the zero-point level of C_6H_6 alone, 1_n^0 progressions built on nine different progression origins can be observed and measured.[65] Similar opportunities exist for progressions of the type 1_n^1 and 1_n^2 in many other spectra.

Craig[62] has used the Franck-Condon factors measured from 1_0^n progressions in absorption to estimate the increase in the C—C bond distance in the $^1B_{2u}$ state of benzene. (His estimate of the increase as 0.037 Å is extremely close to estimates derived from other sources (see summary in CDM[2], p. 531). The problem now may be turned about to calculate the intensities expected in 1_n^0, 1_n^1 and 1_n^2 fluorescence progressions from the known upper and lower state geometries and vibrational frequencies.

Such intensities have been calculated from the Franck-Condon overlap tables of Henderson[72] and compared with intensities of progression members in a number of spectra.[66,73] Examples of the results for the fluorescence progressions 1_n^0 and 1_n^1 are given in Fig. 8, where the calculated intensities and the spectra containing prominent ν_1 progressions are displayed together. Although the calculations are only approximate due to the necessity of interpolating between increments in the tables of overlap integrals, a close match with the observed intensities is obtained. The marked difference between the progression from $v_1' = 0$ and $v_1' = 1$ is reproduced well. These progressions could hardly be more disparate in their intensity distributions, with the most intense member ($v_1'' = 1$) of the 1_n^0 progression corresponding to the member of the 1_n^1 progression with almost zero intensity. A fine discussion of the cancellation of overlap integrals leading to a small intensity in the 1_1^1 progression member is given by Smith.[63]

C. SVL Fluorescence and the $^1B_{2u} \leftarrow {^1A_{1g}}$ Absorption Spectrum

Interpretation of experiments using optical selection of single vibronic levels in the $^1B_{2u}$ state of benzene is dependent on correct assignment of the absorption bands used for selection.[74,75] Close examination of existing absorption analyses shows that much structure is still unassigned and that the assignment of other structure is uncertain. For this reason the $^1B_{2u} \leftarrow {^1A_{1g}}$ absorption spectrum has been reexamined.[55,74]

The most extensive analysis of the spectrum has been the 1948 report of GI.[54] Subsequently, two important developments have appeared. The first was the 1966 report of CDM,[2] which offered numerous refinements of band assignments and, most important, which offered analysis of the rotational contours of the vibronic bands. This now allows more precise calculation of the position of band maxima from vibrational constants. It also provides essential guidance in choosing between alternative assignments on the basis of vibronic band contours, and it reveals which and how many maxima in a given region can be associated with a single transition. The second development has been the assistance provided by analysis of SVL fluorescence. These two advances now provide tools

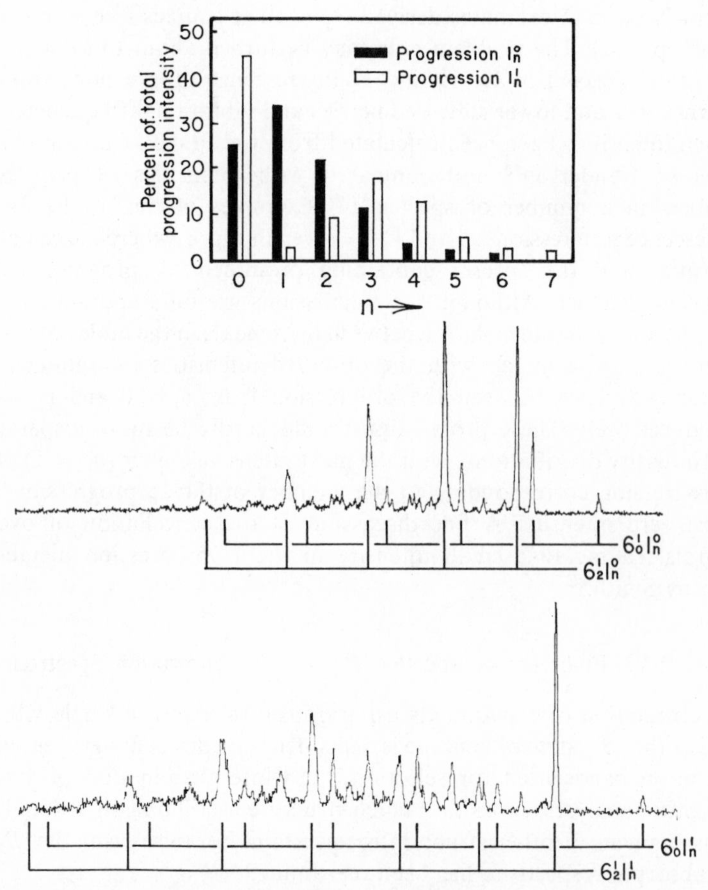

Fig. 8. Calculated relative intensities for the members of the progression $v_1' = 0 \rightarrow v_1'' = 0, 1, 2, \ldots$, and $v_1' = 1 \rightarrow v_1'' = 0, 1, 2, \ldots$, in the ${}^1B_{2u}$–${}^1A_{1g}$ transition of benzene. These were calculated with Henderson's tables[72] using parameters matching the known constants of benzene closely but not exactly. Comparison of these approximate intensities with those actually seen in SVL fluorescence can be made from the two spectra. The upper spectrum is SVL fluorescence from the excited level 6^1 and contains two prominent progressions of the type 1^0_n. One is built on the progression origin $6^1_0 1^0_0$ and the other on the progression origin $6^1_2 1^0_0$. The lower spectrum is from the excited level $6^1 1^1$ and contains 1^1_n progressions built on progression origins $6^1_0 1^1_0$ and $6^1_2 1^1_0$. The value of $n = 0, 1, 2, \ldots$, increases in every case to the left of these spectra. The transitions $6^1_0 1^0_0$ and $6^1_0 1^1_0$ are each reduced from their actual fluorescence intensities by reabsorption. The relative intensities of other transitions are fairly accurately displayed.

leading to considerable extension and refinement of vibronic assignments in the absorption spectrum.

The extended assignment[74] has been based on first identifying the progression origins in the spectrum after a careful analysis of the anharmonicity of ν_1' in known 1_0^n progressions built on various progression origins that vibronically induce the transitions. This immediately identifies most of the structure as progression members so that only the residue of progression origins need to be assigned to set the analysis.

The progression origins in turn are assigned with the same set of active modes used in SVL fluorescence analysis. By using only these modes, and by using them in only the *simplest possible combinations* allowed by the selection rules,[53] almost all progression origins of at least weak intensity are plausibly assigned. Due to a still inadequate knowledge of anharmonic constants, some ambiguity remains with respect to specific assignments of some weak neighboring bands near the calculated positions of several possible choices. But in the main the assignments form a consistent picture of vibronic activity and are in accord with a variety of other checks that can be made for their validity.

One of these checks derives directly from SVL fluorescence itself. In a number of areas, we can merely inquire if the structure in SVL fluorescence excited by monochromatic pumping of an absorption band in question is consistent with the emitting state predicted by the absorption analysis. An inquiry of this sort, for example, shows that the absorption bands assigned by GI[54] to activity in ν_8 do not in fact involve ν_8. This confirms the similar conclusion reached by CDM[2] through their rotational band contour analysis.

Another example is illustrated in the SVL fluorescence from the state $6^1 1^1$ displayed in Fig. 7. This is excited by pumping the strongest absorption band in the entire ${}^1B_{2u} \leftarrow {}^1A_{1g}$ transition, the so-called A_1^0 band representing the absorption $6_0^1 1_0^1$. Close inspection of this absorption band reveals structure not previously reported lying within its rotational contour that originates from a separate and nearly coincident vibronic transition. It lies close to the position calculated for the transition 17_0^2 which would be expected to occur weakly in absorption. Structure consistent with emission from the excited state 17^2 can be observed in SVL fluorescence when the A_1^0 band is pumped. A progression of significant intensity, consistent with the assignment $6_1^0 17_2^2 1_n^0$, is seen. The progression origin would be displaced by $\nu_6'' + 2\nu_{17}'' = 608 + 2 \times 967 = 2542\ \text{cm}^{-1}$ from the 17_0^2 absorption position. This progression is expected to exceed the intensity of all other progressions from 17^2 by at least an order of magnitude and, in the presence of another emitting state, it would be the only identifiable structure from 17^2. Furthermore, it can be seen that the

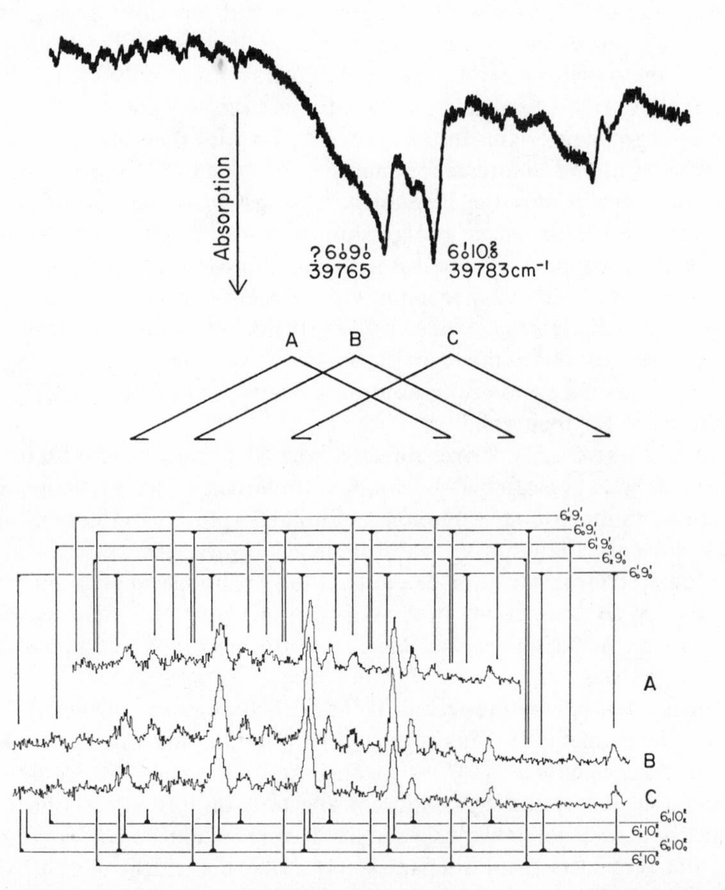

Fig. 9. *Top*. The O_0^0 absorption bands assigned as the transition $6_0^1 10_0^2$ by Garforth and Ingold. Revised assignments from recent analyses are indicated. The excitation profiles for the resonance fluorescence spectra below are indicated by the triangular brackets. The excitation bandpass is approximately 50 cm^{-1}. *Bottom*. Resonance emission from excitation into the O_0^0 band at each of the three positions indicated in top figure. Fluorescence bandpass is 75 cm^{-1}, and fluorescence is at lower energies from excitation. The positions of members of the most intense progressions from each of the excited states $6^1 10^2$ and $6^1 9^1$ are indicated in the assignments. Two very strong progressions are expected from each origin ($6_2^1 10_2^2 1_n^0$ and $6_0^1 10_2^2 1_n^0$; $6_2^1 9_1^1 1_n^0$ and $6_0^1 9_1^1 1_n^0$). Other structure from each excited state is expected to be much weaker. As excitation progresses from *C* to *A*, we observe diminution of structure from $6^1 10^2$ and growth of structure from $6^1 9^1$.

progression assigned as $16^0_1 17^2_2 1^0_n$ has a Franck-Condon intensity envelope that clearly identifies it as a 1^0_n progression. The other structure in that spectrum has an envelope characteristic of a 1^1_n progression (the 1^1_1 bands have very low intensity). In this case, SVL fluorescence has provided convincing confirmation of the 17^2_0 assignment in absorption.

There are numerous other examples of the utility of SVL fluorescence in probing absorption assignments. One of particular interest concerns the so-called O^0_0 absorption band of GI. This band actually consists of several maxima which have been the subject of considerable discussion by CDM in connection with assignment of some of that structure to the transition $6^1_0 10^2_0$. The bands are insufficiently separated to selectively excite one at a time in an SVL experiment leading to a resolved fluorescence spectrum, but they can be examined by attempting to produce resonance fluorescence with dominant rather than exclusive contributions from one terminating state or the other. Figure 9 shows the positions of excitation bandpasses that attempt to pump one absorption and then the other. The resulting fluorescence is also shown in Fig. 9. Excitation at position *C* produces a spectrum whose most prominent structure is completely consistent with that expected from the state $6^1 10^2$ populated by excitation into an absorption band $6^1_0 10^2_0$. This securely identifies the assignment of the 39,783 cm^{-1} component of the O^0_0 absorption bands as the latter transition. As excitation moves to position *B* and then to *A* to emphasize emission from the terminating state of the other component (at 39,765 cm^{-1}), the fluorescence changes. New structure grows in, and it is consistent with the most intense structure expected if that absorption band is $6^1_0 9^1_0$. This structure is the progression $6^1_2 9^1_1 10^0_n$, which falls in a region relatively free of other transitions so that its growth can be observed as excitation moves from *C* to *A*. Thus upon this tenuous evidence, ν_9 is now identified in absorption via the band $6^1_0 9^1_0$, and an approximate frequency of 1154 cm^{-1} is assigned to ν'_9. A previous and very tentative assignment[2] of ν'_9 to the Y^0_0 absorption band near 39,039 cm^{-1} is not consistent with the SVL fluorescence observed following excitation into that band.

IV. RELAXATION OF THE $^1B_{2u}$ STATE IN THE VAPOR—GENERAL CONSIDERATIONS

The electronic relaxation of the excited singlet states of benzene vapor is rather typical of polyatomic molecules. Relaxation is predominantly nonradiative, and by paths that are sensitive to the degree of vibrational excitation in the decaying electronic state. Indeed, vibrational excitation appears as perhaps the most crucial parameter in setting the course of excited state decay.

The general nature of excited singlet relaxation in benzene vapor can be quickly mapped out by reference to the energy level diagram in Fig. 1. Fluorescence is observed only when excitation is limited to the low vibrational regions of S_1 shown by the hatched area. Relaxation from higher levels of S_1 and from everywhere in S_2 and S_3 in the vapor is dominated by nonradiative paths. In fact, nonradiative decay is also observed in the low energy regions of S_1 where it competes with fluorescence. Nowhere does fluorescence become the principal relaxation channel.

As the competition between radiative and nonradiative decay is sensitive to vibrational excitation, so is the competition between various types of nonradiative decay. At very low vibrational energies in S_1, one of the nonradiative channels is known to be intersystem crossing to a triplet state, and this may be the only channel of consequence. At high energies of S_1, and in S_2 and S_3, photochemical activity is observed and indications of triplet formation no longer exist. Thus as we climb the energy ladder from the zero-point level of the lowest excited singlet state, the dominant nonradiative channel seems to transform from intersystem crossing to chemical relaxation. The following discussions of vapor phase ${}^1B_{2u}$ relaxation provide details of this picture and the methods used for its development.

The data derived from measurements of a variety of observables collected over an extended range of experimental parameters such as exciting wave length and band spread, temperature, gas pressures, added diluents, and, of course, phase. Rather than ordering these studies according to specific conditions of excitation, they are organized according to the vibrational distributions in the ${}^1B_{2u}$ state from which electronic relaxation actually proceeds. For example, many experiments probe relaxation from a Boltzmann distribution of vibrational states in S_1, and they are considered first. Then the discussion turns to experiments revealing the nature of relaxation from other vibrational domains in S_1, and finally from individual vibrational levels in S_1.

Work on the last area mentioned is just beginning to unfold at the time of this writing, and it is a most exciting development. Experiments from several laboratories are finally realizing the goal of mapping excited state decay in isolated molecules as we climb the vibrational manifold level-by-level. The present discussion deals with only those studies in this area that have appeared in print, and they provide an intriguing introduction to what ultimately will be learned. However, the interpretation of all experiments probing relaxation from single vibronic levels must depend on what has come from more conventional studies. These earlier studies will continue to play a crucial role in our understanding of the relaxation channels for the excited states of benzene.

An attempt will be made to separate primary facts about relaxation parameters from the less secure (and sometimes transitory) inferences derived by mechanistic interpretation of the data. Accordingly, for each vibrational domain, attention will be directed first to the elementary separation of excited state decay into the channels of radiative and nonradiative relaxation without effort to further identify the nature of the nonradiative decay. This question will then be discussed separately, for it is an involved and incompletely resolved issue.

Partitioning of relaxation into radiative and nonradiative channels requires two measurements. A measurement of ϕ_f, the quantum yield of fluorescence (defined as the fraction of excited molecules that decays by fluorescence), immediately gives the ratio of the radiative and nonradiative rates of decay. A subsequent measurement of τ_{obs}, the lifetime of the excited state observed in a time modulated experiment (a lifetime established by the sum of *all* relaxation processes) then allows assessment of the absolute rate of each type of relaxation.

In kinetic nomenclature, this "zero order" picture of excited state relaxation is described with only two rate constants:

k_r for radiative decay $\quad S_1 \rightarrow S_0 + h\nu$

k_{nr} for nonradiative decay $\quad S_1 \rightarrow$ all other channels

The quantum yield of fluorescence, ϕ_f, is thus defined as

$$\phi_f = \frac{k_r}{k_r + k_{nr}}$$

The observed lifetime of the decaying state, τ_{obs}, is defined as

$$\tau_{\text{obs}} = (k_r + k_{nr})^{-1},$$

so that

$$k_r = \frac{\phi_f}{\tau_{\text{obs}}}$$

and

$$k_{nr} = \frac{(1 - \phi_f)}{\tau_{\text{obs}}}$$

In this kinetic scheme, k_{nr} has been represented as a simple first-order constant. It may of course contain many constants, both first and second order, representing competing nonradiative decay channels. But a single pair of measurements, ϕ_f and τ_{obs}, in a single experiment will describe only an apparent first-order constant, k_{nr}, and this is the proper starting point for characterization of the data. Then subsequent examination of k_{nr} under a wide variety of experimental conditions can reveal the extent of its fine structure.

V. RELAXATION FROM THE THERMAL VIBRATIONAL LEVELS

Many relaxation studies have used the 2536 Å Hg resonance line to excite benzene in systems with total gas pressure above about 10 torr. Although this line excites vibrational levels about 2000 cm^{-1} above the zero-point level of the first excited singlet state (see Fig. 10), the fluorescence structure shows that radiative decay comes from a Boltzmann distribution of vibrational states.[59] The proposition thus arises that *all* electronic relaxation in these "high"-pressure systems occurs after thermal equilibration of the vibrational levels.

Two obvious criteria must be fulfilled to realize such a situation. First, collisions must not induce significant electronic decay during vibrational

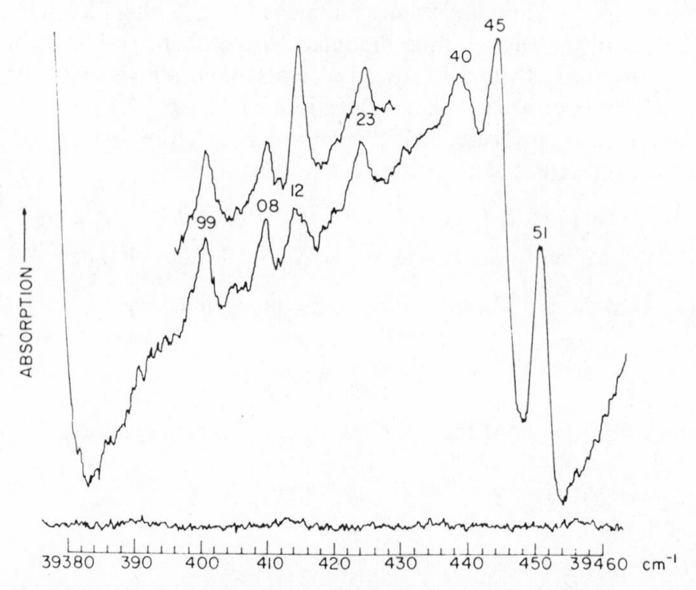

Fig. 10. Absorption in benzene near the region of the 2536.52 Å (39,412 cm^{-1} vac.) resonance line of Hg. The upper spectrum is that of benzene with mercury vapor to mark the spot of the resonance line. It can be seen by the sharp absorption near 39,412. The lower spectrum is that of benzene from which mercury vapor has carefully been excluded, and we can see the persistence of an absorption band not previously reported which is nearly coincident with the position of the resonance line. The rotational contour analysis of CDM[2] shows that those bands at energies higher than the Hg line (up to 39,451) have rotational tails that overlap the position of the resonance line. The transition with a maximum at 39,408 cm^{-1} may not have overlapping structure and those at lower energy certainly do not. Some of the absorption band have been assigned and involve $^1B_{2u}$ levels with about 1700–2000 cm^{-1} of vibrational energy in excess of zero-point energy. The trace at the bottom is emission from the Xe continuum used to obtain the absorption spectrum. The lamp was carefully selected to be free of mercury structure.

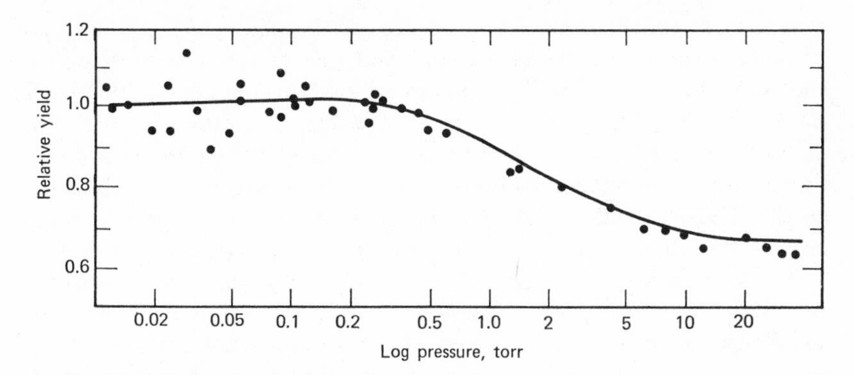

Fig. 11. Relative quantum yields of fluorescence from C_6H_6 as a function of total pressure following excitation with the 2536 Å Hg resonance line. Cyclohexane is used as a diluent in these experiments, but the relative yields are independent of its mole fraction. Data from Anderson and Kistiakowsky.[81]

equilibration. Second, electronic decay from *every* level reached in absorption plus *every* level populated during vibrational relaxation must be slow enough to insure the dominance of vibrational relaxation. The evidence in the paragraphs that follow indicates that both constraints are at least approximately satisfied when the pressure is at least 10 torr and when the exciting light is at 2520 Å or longer (but not shorter!) wave lengths.

The first criterion is most easily examined. In pure benzene vapor excited at 2536 Å or longer wave lengths, both the quantum yield[76,77] and the observed singlet state lifetime[78,79] show only a very shallow dependence on benzene pressure above 10 torr. Addition of foreign gases such as hydrocarbons to these experiments also has little (but sometimes finite) effect on fluorescence yields[80–82] or lifetimes.[83,84] Even in the extreme collisional limit of condensed phase at 77°K, the fluorescence yield of 0.2 matches that of the vapor.[85,86] (Condensed phase yields drop to about 0.05[86–89] at 300°K, but this is probably a special thermal effect somewhat apart from a collision-induced electronic decay. See Section IVC.)

Another clear indication of the role of collisions in ${}^1B_{2u}$ relaxation comes from the fluorescence yield experiments of Kistiakowsky and co-workers.[80,81] They have measured fluorescence yields as the system is taken from very low to high pressures during excitation with the 2536 Å Hg line. The data in Fig. 11 show that the yield remains constant at low pressures but begins to decrease when the pressure climbs above a few tenths of a torr. It subsequently reaches another constant limit in the region above 10 torr total pressure. *These fluorescence yields parallel exactly the pressure dependence of the fluorescence spectrum itself.* A change from true resonance fluorescence first becomes significant at total pressures of a few tenths of a torr,[90] and the structure reaches that of emission from the thermally equilibrated levels in the region 10–20 torr total pressure.[91] The fluorescence yield (and hence the ratio of radiative to nonradiative relaxation) in benzene with 2536 Å excitation seems established principally by the vibration levels from which electronic relaxation occurs. The principal role of collisions is to alter the distribution of those levels rather than to induce electronic decay in the collision act.

It should be noted that detailed studies of vibrational relaxation in the $^1B_{2u}$ state show that typically many hundred torr of added atomic or small molecule gases and even over 100 torr of larger polyatomics such as cyclohexane are required to *completely* equilibrate the spectrum.[82,92–94] However, we can define approximately the region of Boltzmann fluorescence as being that above about 10 torr of gas. Only meticulous examination of the emission spectrum reveals further change at higher pressures.

Having established that radiative relaxation is essentially from a thermal vibrational distribution above 10 torr, it is now necessary to discover whether *all* nonradiative relaxation occurs from this distribution. The most sensitive test comes from examination of high-pressure fluorescence yields as a function of exciting wave length. If nonradiative electronic decay from all levels reached in absorption and during vibrational relaxation is truly slower than the vibrational equilibration, the quantum yields will be independent of exciting wave length.

Noyes and Harter[95] have investigated this essential point. Their high-pressure yields are invariant within experimental error when excitation ranges from 2610 Å (near the long wave length limit of absorption) to about 2520 Å (see Fig. 12). At shorter wave lengths, the fluorescence yield declines even though the emission spectrum remains invariant,[66] and a fast electronic decay obviously competes with vibrational relaxation. Most important, however, electronic relaxation appears to proceed from only the thermal vibrational levels at excitation wave lengths longer than about 2520 Å.[96]

An additional check derives from other quantum yield measurements made at extremely low pressures. Relative fluorescence yields from three high-vibrational levels reached with excitation at wave lengths longer than 2520 Å have been measured by two independent methods.[66,97,98] One of these depends explicitly on the assumption of wave length-independent high-pressure yields, whereas the other does not. Yields by the two methods match closely and indicate that the assumption is valid.

As a further check, experiments in condensed phase are once again helpful. The fluorescence yield in solution appears invariant following excitation from the long wave length absorption edge of the 2600 Å system to below 2400 Å, and perhaps throughout the entire $S_1 \leftarrow S_0$ absorption

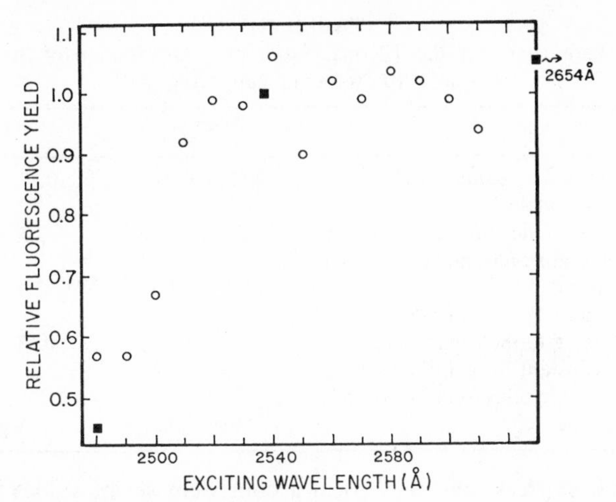

Fig. 12. Relative fluorescence quantum yields of benzene vapor at 300°K as a function of exciting wave. Those data indicated by squares are normalized to the point at 2536 Å and are from excitation with lines from a mercury arc. Those data indicated by circles are from another set obtained with light from a Xe arc coupled with a monochromator whose bandpass is about 6-15 Å. They are normalized to the average in that set between 2530 and 2590 Å. The two sets are independent and not normalized to each other. All quantum yields are obtained from mixtures of benzene and cyclohexane with benzene pressures in the range 0.5 to 3 torr and cyclohexane between 35 and 50 torr. Data from Noyes and Harter. [95]

spectrum.[99–101] Vibrational relaxation in solution intercepts even those high-vibrational levels in S_1 whose lifetime is too short to be affected by collisions at modest gas pressures. (Callomon[102] has recently shown that the lifetimes of some of those higher levels approach 10^{-12} sec. The quantum yield studies thus indicate rather securely that loss of several thousand wave numbers vibrational energy occurs within this time in solution.)

In the midst of all the data cited indicating "complete" vibrational relaxation prior to electronic decay at pressures above 10 torr, a single study suggests otherwise. Poole[77] reports that the high-pressure fluorescence yields with 2654 Å excitation (he quotes 2638 Å but an Hg medium pressure arc was used) exceed those obtained at 2536 Å by a factor of 1.6. However, the data of Noyes and Harter[95] in Fig. 12 indicates that the ratio of yields at these wave lengths (and using the same lamp) to be 1.05. The cause of the discrepancy is unknown.

A. Partitioning of Radiative and Nonradiative Decay

The yields, lifetimes, and rate constants for relaxation from the thermal vibrational levels in 10–20 torr of benzene vapor near 300°K are given in Table III. Both fluorescence yields and lifetimes are available from several

TABLE III
Relaxation Parameters for the Thermal Vibrational Distribution of States of $^1B_{2u}$ Benzene in 10–20 torr of Vapor near 300°K

		C_6H_6	C_6D_6
ϕ_f	Fluorescence quantum yield	0.18 ± 0.04	$0.32 \pm$?
τ_{obs}	Observed lifetime	75 ± 10 nsec	92 ± 10 nsec
k_r	Radiative relaxation rate constant	2.5×10^6 sec^{-1}	3.5×10^6 sec^{-1}
k_{nr}	Nonradiative relaxation rate constant	11.2×10^6 sec^{-1}	7.4×10^6 sec^{-1}
ϕ_T	Quantum yield of intersystem crossing to the triplet state	0.7 ($\pm$0.07?)	$0.63 \pm$?
k_2	Rate constant for collision-induced electronic relaxation by ground state benzene	10^{-2} to 10^{-3} torr^{-1} sec^{-1}	as C_6H_6

independent sources, and they form a consistent set of values. The data (particularly quantum yields) on deuterated benzene are perhaps less reliable because of limited measurements, and we must have due regard for their uncertainty when using them.

The absolute fluorescence yields for C_6H_6 derive with one exception from excitation at 2536 Å. Two of these are based on calibration against the phosphorescence yield of biacetyl,[103] which has been a gas phase standard for many years. The first reports a yield of about 0.23 ± 0.05 at 11 torr,[76] but this was subsequently remeasured by Poole as 0.17 in further evolution of that technique in the same laboratory.[77] A later measurement independent of the biacetyl standard is reported by Noyes, Mulac, and Harter.[96] They observe (in their Method B) that $\phi_f = 0.18 \pm 0.04$ at 11.5 torr. Further support for this value comes from its use in calibration of singlet relaxation quantum yields in fluorobenzene[104] and toluene[105] in which the sum of observed radiative and nonradiative yields is unity within experimental error. A higher benzene fluorescence yield would push that sum paradoxically above unity.

Figure 12 indicates that the fluorescence yield obtained with 2536 Å excitation is representative of the yield at longer exciting wave lengths as well.

The fluorescence yield of C_6D_6 derives only from Anderson and Kistiakowsky's measurement of the ratio of high-pressure yields from C_6D_6 and C_6H_6.[81] The C_6D_6 yield is greater by a factor of 1.8 with 2536 Å excitation, and thus $\phi_f = 0.32$. Poole observed yields from C_6D_6 in the region 1.2–5.2 torr that are less than Anderson and Kistiakowsky's value would suggest.[77]

Five lifetime measurements in high-pressure benzene vapor using various wave lengths and bandpasses of exciting radiation are available.

TABLE IV
Observed Lifetime of the $^1B_{2u}(s_1)$ State of C_6H_6 Vapor with a 300°K Distribution of Vibrational Levels

Pressure (torr)	Observed lifetime (nsec)	Authors
5–40	60 ($\pm^{20}_{10}$)	Brewer and Lee[78]
>30	80	Chen and Schlag[83]
20	74	Nishikawa and Ludwig[79]
50	72	Selinger and Ware[84]
760	80	Burton and Hunsiker[106]

Two of these tuned the exciting wave length to pump various absorption bands[79,84] and another measurement used the 2536 Å Hg resonance line for excitation.[106] In none of these experiments can there be found a significant dependence of the high-pressure lifetime on the initial vibrational distribution in the excited electronic state. The results collected in Table IV indicate that $\tau_{obs} = 75 \pm 10$ nsec for C_6H_6.

Two measurements[84,107] of the Boltzmann $^1B_{2u}$ lifetime of C_6D_6 are in essential agreement, and indicate a value of about 92 ± 10 nsec.

Several of the fluorescence yield[76,77] and lifetime studies[79,84] observe a modest pressure sensitivity. However, since the dependence is small relative to the uncertainties in the value themselves, it seems for most purposes quite appropriate to describe yields and lifetimes with single "high-pressure" values.

An approximate rate constant for collision-induced electronic relaxation of $^1B_{2u}$ benzene is given in Table III[76,77,79] Collision-induced relaxation accounts for not more than 10% of electronic decay of $^1B_{2u}$ benzene at 10–20 torr.

B. Identification of the Nonradiative Channel

The quantum yield of fluorescence is about 0.18 so that about 80% of the relaxation from the thermal vibrational levels is by nonradiative channels. Evidence in this section indicates that at least most of that nonradiative relaxation is intersystem crossing to the triplet state. The quantum yield of triplet formation is about 0.7, and within the uncertainties of the data, this combined with the small but finite collision-induced decay may account for all nonradiative decay. Triplet formation also occurs in C_6D_6, and the data again suggest that it may account for essentially all of the nonradiative decay.

Although identification of benzene triplet formation in solid media is a classic story,[37] its vapor phase history is rather recent. The triplet state does not emit in the vapor and several early efforts to detect it in absorption were not successful. Since 1960, however, abundant evidence has arisen to attest to its presence in the vapor following excitation of the lower vibrational levels of the $^1B_{2u}$ state. The first indication arose from Cundall and Palmer's observation[108] of sensitized butene-2 isomerization after optical excitation of benzene vapor. Many studies of that system are now consistent with their original hypothesis that the sensitizing species is $^3B_{1u}$ benzene. Shortly afterward, Ishikawa and Noyes[76] showed that irradiation of the 2600 Å singlet absorption system of benzene vapor could sensitize triplet but not singlet emission from biacetyl. Subsequently, time-dependent studies of this system showed the sensitizing species to have a lifetime several orders of magnitude longer than the excited singlet of benzene.[109] In other experiments, excitation of benzene at 2536 Å in the presence of O_2 produces $O_2(^1\Delta_g)$ at O_2 pressures so low that the energy transfer must come from a long-lived excited state of benzene.[110] Moreover, benzene-sensitized decomposition of a variety of small molecules give product distributions characteristic of triplet-triplet energy transfer.[111,112]

These and other studies[113] give evidence for triplet formation from the thermal levels of the $^1B_{2u}$ state in both C_6H_6 and C_6D_6 vapor. The problem now is to determine whether this triplet formation accounts for all or even a significant fraction of the nonradiative relaxation from those levels.

Quantitative assessment about triplet formation comes principally from two methods. The first is the sensitized *cis-trans* isomerization of butene-2 introduced by Cundall and Palmer in 1960.[108] This method has been used by many workers and the essential elements of the mechanism are well established.[114–121] The second method, based on sensitized emission from biacetyl, was introduced by Ishikawa and Noyes,[76] and is discussed in its most recent stage of evolution by Noyes and Harter.[121]

Both methods have been used with a variety of aromatic hydrocarbons in addition to benzene.[122] In some cases there are indications that one or both do not work well.[121] Unresolved complications of the mechanisms seem to be present. However, none of these problems are known to exist when the methods are used with benzene. They give results that are mutually consistent, and the methods survive various other mechanistic checks.

The butene method is the most straightforward and it has been the more thoroughly explored of the two methods. In essence, triplet benzene sensitizes *cis-* or *trans*-butene-2, and in the subsequent relaxation of triplet butene, isomerization occurs. The quantum yield of this isomerization can then be traced back to the quantum yield of triplet benzene formation.

The scheme can be made extremely simple because it is possible to use large amounts of butene-2, which pick up *all* the benzene triplets while leaving the excited singlet unperturbed. This point has been examined in detail by Lee, Denschlag, and Haninger,[119] who show that approximately

97% of the triplet benzene decay is by energy transfer to butene-2 when the butene is present at 1 torr. This estimate is consistent with the large cross sections for the energy transfer[119] and the long lifetime of triplet benzene.[109,123] At the same time, the fluorescence yield of benzene is unaffected by the addition of even up to 50 torr of *cis*-butene-2.[95,115] Thus butene does not influence electronic decay from the thermal levels of S_1 in benzene. The quantity we would like to measure, ϕ_T in benzene, is unaffected by the probe.

With complete scavenging of triplet benzene, a measurement of the yield of triplet butene provides a direct measure of ϕ_T in benzene. Triplet butene yields are assessed by monitoring the resulting cis-trans isomerization, and it is at this point that uncertainties in the mechanism arise. Little is known about the triplet state of butene-2. Various kinetic schemes have been proposed that distinguish but yet allow interconversion between "different" butene triplets formed from the cis and trans ground state. However, no data are available on this proposition, and the more detailed schemes always reduce to that of a common triplet in actual treatment of the data. Accordingly, the relaxation of the triplet butene (B^*) is characterized by the simple kinetic steps

$$\mathrm{B}^* \rightarrow \textit{cis}\text{-butene-2}$$

$$\mathrm{B}^* \rightarrow \textit{trans}\text{-butene-2}$$

with the fractions of B* branching to cis and trans ground-state butene given as b_c and b_t, respectively.

If the mechanism is correct, two measurements then define the triplet yield of benzene. When *cis*-butene is added at a pressure sufficient to scavenge all benzene triplets, the yield $\phi_{c\rightarrow t}$ for conversion of *cis* to *trans*-butene at *low conversions* is given by

$$\phi_{c\rightarrow t} = (\phi_T)(b_t) \tag{1}$$

where ϕ_T is the yield of triplet benzene. The caveat "low conversions" is essential, for this relationship is of course destroyed when enough *trans*-butene-2 accumulates to cause measureable back-reaction. Similarly, an experiment starting with only a *trans*-butene-2 adduct gives an initial yield

$$\phi_{t\rightarrow c} = (\phi_T)(b_c) \tag{2}$$

Since $(b_t + b_c) = 1$,

$$\phi_T = \phi_{c\rightarrow t} + \phi_{t\rightarrow c} \tag{3}$$

Thus measurements of two initial isomerization yields when butene pressures are high enough (a few torr) to scavenge all benzene triplets gives ϕ_T with a minimum of mechanistic assumptions.

Noyes and Harter[121] have very recently reported the triplet yield measurements at 2520 and 2590 Å using the method of Eq. (3). They observe the yield to be independent of exciting wave length and equal to 0.73.

Only three of the many other studies of this method report isomerization yields. The first[115] used exciting radiation spanning the region 2400–2600 Å where it is probable that at least some of the electronic relaxation does not go through the thermal vibrational levels. The second[116] uses 2536 Å excitation and the third[95] uses various longer exciting wave lengths in addition to the 2536 Å line. Thus the latter reports provide information about triplet formation from the thermal levels in $^1B_{2u}$ benzene vapor. But those authors measured only $\phi_{c\to t}$. Additional kinetic information or assumption is required to deduce ϕ_T from their data. We must obtain the branching fraction b_t in Eq. (1).

This fraction may be obtained with fewest mechanistic assumptions by measuring the *relative* yields $\phi_{c\to t}$ and $\phi_{t\to c}$. From (1) and (2),

$$\frac{\phi_{c\to t}}{\phi_{t\to c}} = \frac{b_t}{b_c} = \left(\frac{1}{b_t} - 1\right)^{-1}$$

Such data are given by Lee et al. for 2536 Å excitation. The ratio is 1.02 ± 0.02, so that $b_t \approx b_c \approx 0.5$. The ratio b_t/b_c has also been checked by other methods[119] and by other experimenters. All find it close to unity. The higher value given in the early work by Cundall et al.[115] also has been revised to a value near unity.[119]

With this value of b_t, the average yields of triplet benzene can be calculated from the data of Cundall and Davies[116] and of Noyes and Harter[95] and they are given in Table V. (The data of Noyes and Harter[95] at 2536 Å in Table V are nonsense, for they imply that the fraction of $^1B_{2u}$ molecules decaying by intersystem crossing to the triplet state is about 100%, which leaves little room to accommodate the 18% fluorescence yield. Abundant evidence suggests that the problem lies with the data. Checks of the method using similar excitation reveal no flaws in the assumed mechanism.[117,118] The method also seems to work well at longer exciting wave lengths in that the yields are independent of wave length as expected. Finally, the 2536 Å triplet yield of Cundall and Davies shows no such anomaly.[116])

The first five entries plus the 2536 Å data of Cundall and Davies[116] in Table V probe relaxation from the thermal vibrational levels of S_1. Within the scatter of the data, no convincing sensitivity to exciting wave length appears. The data rather consistently indicate that the triplet yield is about 0.7.

TABLE V
The Absolute Quantum Yields of Triplet Formation in Benzene Vapor at Total Pressures above Ten Torr as Derived from Butene-2 Sensitization Experiments (The yields of Noyes and Harter[95] and of Cundall and Davies[116] have been recalculated from their original data with a branching fraction $b_t = 0.50$)

Exciting wave length	Region of "principal" absorption	Total pressure (torr)	ϕ_T[b]	Reference
2690–2540 Å		18–71	$0.78\binom{0.86}{0.70}$	95
2620–2560 Å	A_0^0 band 2589 Å	19–24	$0.69\binom{0.69}{0.68}$	95
2590 Å[a]	A_0^0 band 2589 Å	15–20	0.73	121
2530–2490 Å	A_1^0 band 2529 Å	36–60	$0.67\binom{0.74}{0.61}$	95
2520 Å[a]	A_1^0 band 2529 Å	15–20	0.73	121
2500–2440 Å (approx.)	A_2^0 band 2471 Å	40–55	$0.64\binom{0.71}{0.61}$	95
2420 Å (approx.)	A_3^0 band 2417 Å		Very small	95
2536 Å Hg line		32–61	$0.99\binom{1.01}{0.95}$	95
2536 Å Hg line		2–20	0.72	116

[a] Bandpass not reported.
[b] High and low extremes are given in parentheses.

An independent estimate of the triplet yield from thermal vibrational levels derives from a different chemical method involving sensitized ketene dissociation in the presence of propane.[112] It gives a value of 0.71 for ϕ_T. Since this method is relatively untested, it is uncertain whether much comfort should be taken from the consistency of this value with the others.

Finally, the values of $\phi_T = 0.7$ compare well with that obtained by examination of sensitized phosphorescence in biacetyl. This technique is mechanistically more complicated than the butene-2 method. Its principal complication arises from the fact that energy transfer occurs from the singlet state as well as from the triplet state of benzene, and a rather subtle correction is required. Such corrections have not been applied to the original determination of ϕ_T by this method[76] and $\phi_T = 0.63$ derived from that work must be considered approximate and somewhat on the low side. (Ishikawa and Noyes'[76] first value of $\phi_T = 0.78$ was later revised[96] to 0.63 when a better fluorescence yield became available.)

Noyes and Harter[121] have repeated the biacetyl experiments with excitation of the strong A_0^0 absorption band at 2590 Å and also the A_1^0

band at 2530 Å. Their results for each wave length are as follows:

$$\text{excitation at } 2590 \text{ Å} \qquad \phi_T = 0.72 \pm 0.04$$

$$\text{excitation at } 2530 \text{ Å} \qquad \phi_T = 0.73 \pm 0.03$$

The errors represent uncertainty in making the corrections for singlet sensitization rather than scatter in the data, but uncertainty deriving from the absolute biacetyl emission yield basic to the calculation is even greater. It is at least 10%.

In summary, all of the estimates of ϕ_T from the thermal vibrational levels of $^1B_{2u}$ benzene provide a consistent picture, even though they derive from different techniques in different laboratories using different initial excitation conditions. In every case $\phi_T \approx 0.7$.

Only a single estimate of ϕ_T for C_6D_6 can be obtained from the literature. Cundall and Davies[116] have measured $\phi_{c\to t} = 0.31$ when *cis*-butene isomerization is sensitized by triplet C_6D_6 formed after excitation of benzene with the 2536 Å line. Equation (1) gives the yield of triplet C_6D_6 as

$$\phi_T = \frac{0.31}{0.50} = 0.63$$

C. Photochemistry from the Thermal Vibrational Levels

The sum of fluorescence and intersystem crossing quantum yields from the Boltzmann vibrational levels of S_1 appears to be slightly less than unity in both C_6H_6 and C_6D_6. Although the uncertainty is sufficient to mask a real difference from unity, it seems reasonable to inquire further for other relaxation channels, and to identify if possible the nature of the very inefficient collision-induced electronic relaxation that seems to be present.

An extensive literature concerning the photochemical activity of benzene[124] exists. Among this, we find only two studies of such activity under conditions in which benzene vapor undergoes electronic relaxation from the thermal levels of S_1. Wilzbach, Harkness, and Kaplan[125] provided the first glimpse of the subtle vapor phase photochemistry induced by 2536 Å absorption by demonstrating that benzene-1,3,5-d_3 rearranges to the 1,2,4 isomer. They postulated that this rearrangement derived from a benzvalene intermediate produced in the photochemical decay the excited singlet state. The quantum yield of that chemical decay would be at least 0.03.

Kaplan and Wilzbach[126] further explored benzvalene formation following 2536 Å excitation of benzene at high pressures, and their communication is an elegant description of the experimental intricasies associated with establishing quantum yields of photochemical relaxation channels in benzene (and even in merely identifying them). The photochemical products are unstable with respect to further thermal and photochemical transformations, and some isomerization of the products leads back to benzene itself. Thus identification of the primary photochemical quantum yields is difficult. Their work merely suggests that the quantum yield of benzvalene formation from the thermal levels is at least 0.02. It is not possible to assess the effect of increasing gas pressure on this primary yield because higher pressures also tend to stabilize benzvalene against further rearrangements from the hot vibrational levels in which it is initially formed.

Thus evidence exists for at least one chemical relaxation channel (leading to benzvalene) from the thermally accessible vibrational levels in S_1. However, not more than about 10% of S_1 relaxation can be via this channel. There are no data that can directly comment on the role of collisions in this relaxation, nor are there photochemical data that can indicate which vibrational levels, if not all, are effectively coupled to this channel.

On the other hand, there is a large body of data from other sources that collectively support but by no means prove the idea that this chemical relaxation occurs principally from vibrational levels more than 2500 cm^{-1} above the zero-point level. This is discussed in the next section.

VI. RELAXATION FROM THE "HIGHER" VIBRATIONAL LEVELS

The high-pressure fluorescence quantum yields in Fig. 12 fall off from their constant values when the exciting wave lengths become shorter than about 2500 Å. This corresponds to excitation of vibrational levels 2500–3000 cm^{-1} above the zero-point level, and it shows that nonradiative relaxation of those high levels is sufficiently fast so that gas collisions cannot relax them to the lower emitting levels even at collision rates of 10^9 sec^{-1}.

Some data are collected here that bear on relaxation from those levels. The data show that fast relaxation from high levels is accompanied by disappearance of all indications of triplet formation and by an increase in photochemical activity.

We begin by comparing the fall-off in fluorescence yields in Fig. 12 with an exploration of yields obtained from low-pressure experiments where collisions do not change the vibrational levels before their decay. A qualitative examination[64] of fluorescence yields from a variety of single levels is shown in Fig. 13. Those yields appear to have a rather sharp dependence on vibrational energy of the decaying state. The relative rate

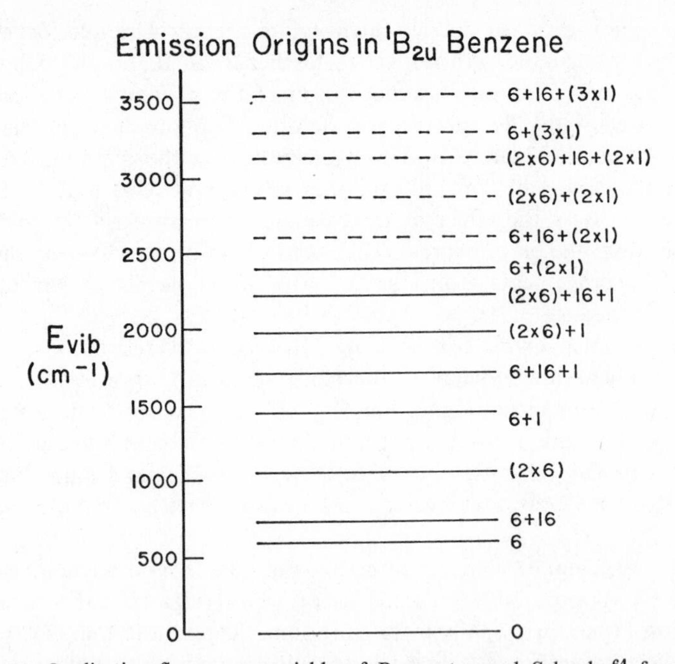

Fig. 13. Qualitative fluorescence yields of Parmenter and Schuyler[64] from single vibronic levels of benzene vapor. The solid lines indicate the levels from which moderate to strong fluorescence has been observed when they are individually excited with pressures low enough (*ca.* 0.2 torr) to preclude significant collisional deactivation prior to electronic decay. The dashed lines indicate levels from which emission was too weak to be observed. The notation on the right identifies the vibrational level. For example, $(2 \times 6) + 1$ indicates emission from the vibrational level $(2\nu_6 + \nu_1)$ of the $^1B_{2u}$ state.

of nonradiative decay appears to increase steeply when the vibrational energy exceeds about 2500 cm^{-1}. This roughly corresponds to the energy region where the fall-off in yields is first seen in Fig. 12.

More quantitative yield measurements have subsequently confirmed the observations in Fig. 13.[66,98] The drop in yield is also accompanied by a decrease in the lifetime of the excited states.[84,98] Callomon has pointed out that the lifetime of many levels above about 3300 cm^{-1} in fact becomes so short that diffuseness in the high-resolution absorption spectrum is apparent.[102,127] This is consistent with the fact that the quantum yield of fluorescence at shorter exciting wave lengths becomes too small to measure even at high pressures.[95]

There is also abundant evidence from condensed phase studies for the onset of a fast nonradiative decay from high levels. In condensed phase, we can see efficient thermal activation to those levels. The quantum yield of fluorescence in condensed phase at 77°K (where those levels cannot be activated) is about 0.2 with 2600–2400 Å excitation.[128–130] This yield closely reproduces that of the thermal levels of the gas phase. However, the condensed phase yield drops as temperature rises from 77°K, and Eastman[129] provides good arguments that the effect is due to efficient thermal activation to higher vibrational levels from which nonradiative decay then occurs. The yield declines to about 0.05 in many solvents[131] at room temperature. The activation energy for the nonradiative relaxation causing this decline is calculated to be on the order of 2000–2500 cm^{-1}.[129]

Lifetime measurements of fluorescence in condensed phase reproduce the temperature dependence. Greenleaf and King[132] find an activation energy of about 2400 cm^{-1} associated with the temperature-sensitive component of the fluorescence lifetime, and Helman[133] independently observes by a similar technique an apparent activation energy of about 2300 cm^{-1}. It thus seems certain that the drop in fluorescence yield from about 0.2 for the thermal levels of S_1 in the gas phase to about 0.05 for the condensed phase thermal levels at 300°K is simply due to the increased rate of collisional activation to high vibrational levels. If that process is stopped by cooling, then the gas and condensed phase yields become essentially identical.

The fast nonradiative decay may be attributed to one of three channels:[64]

1. A sudden increase in the S_1–S_0 internal conversion rate.
2. A sudden increase in the S_1–T intersystem crossing rate.
3. The onset of a chemical relaxation channel.

No evidence or apparent reason exists to support the first possibility. Indeed, that channel seems essentially noncompetitive in the gas phase decay from thermal S_1 levels,[134] and it is difficult to imagine how an S_1–S_0 internal conversion suddenly could become so efficient. Various authors have proposed that the internal conversion S_1–S_0 proceeds through an intermediate isomeric state,[124,129] but this author would prefer to classify this as chemical relaxation.

The second possibility seems to be ruled out on the basis of experimental evidence, although some might argue its plausibility from theory. Intersystem crossing would necessarily involve the second triplet[135] rather than the third. The third lies too high to produce the effect. The second triplet will begin to have appreciable densities of states isoenergetic with the 2500 cm^{-1} levels in the $^1B_{2u}$ state where the acceleration in nonradiative decay occurs, but a mechanism whereby this state can *suddenly* accelerate the rate of intersystem crossing by a factor approaching three orders of magnitude remains to be put forward. (Diffuseness in the absorption spectrum indicate some levels near 3300 cm^{-1} have lifetimes on the order of 10^{-11} sec.[102,127])

Experimental evidence uniformly suggests that triplet yields from higher levels fall off as does fluorescence. Efforts to detect triplet benzene in the vapor by the butene-2 isomerization method with excitation of high levels using 2420 Å radiation reveal none. There is negligible isomerization,[95] even though nonradiative decay in benzene is fast enough to reduce the emission yield nearly to zero with that excitation. Poole[77] implies in discussion of his gas phase experiments (although he does not present the data) that the biacetyl test also reveals no triplet formation at high excitation energies. This is consistent with earlier data of Braun, Kato, and Lipsky.[99] In addition, the latter workers also examined triplet formation in condensed phase at 300°K where their results suggest that processes reducing radiative decay also reduce triplet formation. However, since variation in fluorescence and triplet formation in condensed phase occurs only when excitation initially pumps S_2 and S_3, those data do not comment directly on the vapor problem. A similar implication is offered by simultaneous examination of fluorescence and phosphorescence from benzene in rigid media at 77°K.[100]

On the other hand, photochemical activity is in marked contrast with triplet formation. Chemical relaxation in benzene *increases* with higher energy excitation, but the absence of even approximate primary quantum yields makes quantitative correlation of photochemical activity with other data impossible. No attempt will be made here to review the extensive literature on benzene photochemistry (see, e.g., Ref. 124). Suggestive evidence to link chemical activity with the fast nonradiative decay beginning with vibrational levels near 2500–3000 cm^{-1} above zero-point energy can be found in the paper of Kaplan and Wilzbach[126] describing their 2537 Å vapor phase experiments. Vapor yields of benzvalene formation and of the benzene-d_3 isomerization increase as excitation moves from 2537 to 2370 Å, but the sum of observed "photochemical yields" remain well below unity. The correlation is far from quantitative.

The preceding observations summarize what is known about the general problem of relaxation from levels above the thermal region of the $^1B_{2u}$ state. As excitation energy climbs higher than 2500 Å to initially populate levels with vibrational energy in and above the 2500–3000 cm^{-1} region, the pattern of relaxation changes. The lifetimes of those high levels is shortened sufficiently by nonradiative decay so that they cannot be intercepted by vibrational relaxation even at pressures of several hundred torr. Evidence for triplet formation in the vapor disappears and qualitative indications of chemical relaxation increase. Condensed phase data are consistent with these observations.

These data have important implications for the nature of relaxation from the thermal levels in the vapor. The high levels play an important role in condensed phase at 300°K where thermal activation to this fast "leak" from the electronic state is efficient, and presumably a similar situation is present in vapor experiments. In the vapor, however, the leak is severely limited by the lower rate of thermal activation. This small leak in the vapor plausibly accounts for the small but observable chemical relaxation from the thermal levels, and for the small but persistent pressure sensitivity of the thermal levels, and finally for the small temperature quenching[77] of fluorescence from the thermal levels. (Robinson has proposed another explanation of the collisional quenching of fluorescence.[135])

The most important implication of these data, however, is the statement they make about the nonradiative relaxation from the levels below 2500 Å. If the chemical activity in the thermally equilibrated experiments is limited to the high levels reached by thermal activation, then nonradiative relaxation from the levels below about 2500 cm^{-1} is a single process—intersystem crossing to the triplet state. Measurement of nonradiative decay rates from those levels could then provide an explicit picture of the dependence of intersystem crossing on the initial vibrational modes and energies. Unfortunately, proof of this proposition cannot be derived from the available data.

VII. RELAXATION FROM "ISOLATED" $^1B_{2u}$ MOLECULES AT LOW PRESSURES

The foregoing discussion has described facts about excited state relaxation that derive from conventional experiments with total pressures above about 10 torr. They reveal the presence of at least three electronic relaxation channels representing radiative decay, intersystem to a triplet state, and chemical relaxation. They also reveal an intriguing dependence of the relative rates of these processes on vibrational excitation. However, the fast vibrational relaxation in the gas phase at those pressures precludes a detailed study of the vibrational effect. It is difficult to classify the levels from which relaxation occurs more precisely than the headings "thermal levels" and "higher levels" indicate.

The obvious experiment is to retreat from high pressures and to attempt the more difficult task of working with low pressures where relaxation can occur from levels initially pumped in absorption. Then if we can use narrow-band excitation tuned to a single absorption band that terminates in a single vibronic level, a "vibrational spectrum" of excited state relaxation can be constructed. A detailed picture of the role of various excited modes and of vibrational energy in governing excited state relaxation will emerge.

Such data involving *single* vibronic levels do not presently exist for any polyatomic molecule except benzene. The first results from benzene experiments of this type were published in 1969.[64] Since then further and more quantitative results have come forth[75,84,97,136] and they are beginning to build a detailed picture of radiative and nonradiative decay in a large molecule. The published segments of those experiments will be discussed subsequently. However, all of these experiments were preceded by low-pressure studies using the 2536 Å Hg line for excitation of benzene, and they will be discussed first.

A. Relaxation from Levels Excited with the 2536 Å Hg Line

In 1932 Kistiakowsky and Nelles[59] showed that intense resonance fluorescence could be excited in low pressures of benzene with the 2536 Å Hg line. This revealed an opportunity to study "isolated molecule" relaxation, which was not used until 1964. Since that time, Kistiakowsky and co-workers as well as others have completed extensive studies of isolated molecule relaxation from the several levels populated by absorption of 2536 Å radiation. Although the experiments are now being superseded by those which explore relaxation from single vibronic levels, the 2536 Å experiments play an important role in answering fundamental questions about isolated molecule properties. They formed the first extensive body of kinetic data on large-molecule intramolecular electronic relaxation, and they still remain one of the most secure experimental descriptions of the intramolecular nature of intersystem crossing.

The low-pressure 2536 Å experiments were designed to explore the extent of intramolecular nonradiative transitions in a large isolated molecule. Indications of these processes already existed in the literature by 1964. In a widely ignored 1963 paper by Williams and Goldsmith[137] showed that $S_2 \leftarrow S_0$ absorption yielded $S_1 \rightarrow S_0$ emission in tetracene vapor at extremely low pressures. They argued that this was the direct consequence of intramolecular $S_2 \rightarrow S_1$ internal conversion in the isolated molecule. A similar conclusion derives from Pringsheim's 1938 study[138] of anthracene vapor emission as well as from Stockburger's 1962 spectra[93] from naphthalene vapor showing emission from S_1 after exciting S_2. Strickler and Watts[139] subsequently (1966) reconfirmed the naphthalene observation and reached conclusions similar to those of the tetracene workers.

The existence of radiationless transitions between electronic states within a large isolated molecule thus was demonstrated before 1964, but none of those studies bore on the issue of S–T transitions, nor did any explore this issue by measurements reflecting the kinetic aspects of the process. Kinetic information was obtained from benzene by measuring the fluorescence yield. The experiments simply sought to determine whether the fluorescence yield, which is reduced to 0.18 by intersystem crossing at high pressures, rose to unity at very low pressures where collisional perturbation of the excited singlet state becomes minimal.

The initial results[80] demonstrated that the yield did rise as total pressure was reduced from high values, and that it finally became constant at pressures below a few tenths of a torr. However, the yield rose by a factor of only 1.6, which puts it well below unity at low pressures. It is only about 0.29.

Douglas and Mathews[140] confirmed in a more accurate examination of the low-pressure region that the yields were truly constant from about 0.12 to about 0.005 torr. Anderson and Kistiakowsky[81] repeated the measurements over the entire range of high pressures to about 0.005 torr with results essentially similar to the earlier work (see Fig. 11).

The results of these studies were clear. At very low pressures (ca 0.01 torr) where the collision interval exceeds the $^1B_{2u}$ lifetime ($\sim 10^{-7}$ sec) by a factor of about 100, collisional influences on electronic relaxation disappear. Yet nonradiative decay still persists, and in fact about 70% of the decay from the isolated molecules uses that channel. Furthermore, observations of butene-2 isomerization[80,81] indicated that the triplet state is present under these conditions so that at least part of this isolated molecule relaxation is intersystem crossing. The thermal data described in preceding sections imply that the nonradiative channel may be entirely intersystem crossing, but the butene-2 method cannot confirm this. That method does not give quantitative results at very low pressures.[81]

Robinson[135] published a useful discussion of these experiments during the time that the foregoing checks on the initial work were being made. Those checks answer many of the questions raised in his article. Of prime concern was the question of whether the experiments had been extended to sufficiently low pressures to accurately observe isolated molecule behavior. This requires reduction of the *intermolecular* perturbations to a level equivalent or less than the average matrix element for intramolecular coupling of the initial and final zero-order states of the transition, and such a condition is obtained only when the average collisional interaction energies during the $^1B_{2u}$ lifetime do not exceed about 10^{-5} cm^{-1}. Pressures of about 10^{-3} torr or less are needed to achieve this.

To probe this question, fluorescence intensities were reexamined at pressures as low as 7×10^{-5} torr.[90] Still no hint of any collisional effects could be found. The fluorescence yield appeared constant from 7×10^{-5}

to 10^{-2} torr. It is evident that the behavior described in the first experiments truly describes the isolated molecule properties of levels near 2000 cm^{-1} and establishes the existence of intersystem crossing in isolated C_6H_6 molecules.

Similar results have been obtained with 2536 Å excitation of C_6D_6 vapor and of benzene-1,4,d_2 vapor.[81]. Sigal's suggestion[141] that triplet yields in C_6D_6 (measured by the butene probe) are pressure dependent below 0.1 torr is not borne out by the fluorescence data.[81] Severe problems are attendant with the butene probe at those low pressures[81,90]

It is interesting to note that calculations[90,135] place benzene on the borderline between the large molecule and the intermediate limit with respect to nonradiative transitions. From calculation alone, we would suspect the $^1B_{2u}$ state to be sensitive to collision-induced intersystem crossing, but all the data indicate otherwise.

This situation finds an interesting parallel in biacetyl. Biacetyl emits from both its singlet and triplet states, so that the initial and final state of the intersystem crossing both can be monitored. As a result, the role of environmental perturbation on this process is more carefully established than in any other molecule. The lifetime and quantum yield data show that the rate of $S_1 \rightarrow T$ intersystem crossing in biacetyl is essentially invariant as the system is taken from very low pressures (where the hard-sphere collision interval exceeds the S_1 lifetime by a factor of 10^3) to high pressures[142] and even into condensed phase.[142,143] Yet, as in the case of benzene, calculation fails to place biacetyl in the large molecule limit where crossing is truly an intramolecular phenomenon.[142] Intersystem crossing in biacetyl should be sensitive to collisional perturbation, but this is decidedly not the case. Neither biacetyl nor benzene seems to recognize the theoretical estimates of chemists.

Discussion of 2536 Å experiments ends with a note concerning the excited vibrational levels populated by that radiation. The perversity of nature demands that this convenient exciting line must coincide with a complex region of absorption in benzene, and this is borne out by inspection of the spectrum shown in Fig. 10. The rotational structure of at least five absorption bands overlaps the exciting line. Only one of these had been correctly assigned prior to 1971.

Very recently, a serious attempt has been made to identify the vibronic levels populated by 2536 Å excitation.[66–68] Five separate emitting states have been observed in the resonance fluorescence, and four of these have been identified. They are listed in Table VI. Very roughly, half of the resonance emission is from the levels $6^2 1^1$ with about 1965 cm^{-1} of vibrational energy. The remainder probably originates from levels within several hundred cm^{-1} of the $6^2 1^1$ levels.

TABLE VI
Some Excited Levels in $^1B_{2u}$ Benzene Seen in Resonance Emission Following Excitation of C_6H_6 Vapor with the 2536 Å Hg Resonance Line[66–68]

Vibrational origin of emission in the $^1B_{2u}$ state	Vibrational energy (cm^{-1}) above zero point
$2\nu_6(l_6 = 2) + \nu_1$	1,970
$2\nu_6(l_6 = 0) + \nu_1$	1,964
$2\nu_{11} + \nu_1$	1,953
$\nu_6 + 2\nu_{16} + \nu_{17}$	1,715
Unknown[a]	approx. 1,350–2,350

[a] Strong emission is seen from a level whose assignment is unknown. The energy of this level most probably lies between low and high limits set by absorption of 2536 Å light by $^1A_{1g}$ molecules in their zero-point levels or by $^1A_{1g}$ molecules in vibrational levels with about 1000 cm^{-1} of energy, respectively.

B. Relaxation from Single Vibronic Levels

Photochemists have long cherished the goal of mapping the course of excited state relaxation as a function of vibrational activity in the excited state. The most obvious experimental path is to use narrow-band excitation tuned to single absorption bands of gases at very low pressures. This technique, however, will be successful only with molecules possessing the following rather restrictive set of properties.

1. The molecule must have a discrete absorption spectrum so that "monochromatic" exciting light can truly pump single absorption bands not overlapped by others.
2. The absorption spectrum must be securely assigned to identify the terminating levels.
3. The excited state must have at least one relaxation channel that can be quantitatively monitored under isolated molecule relaxation conditions.
4. The excited state must have a sufficiently short lifetime so that isolated molecule relaxation conditions can be easily established in the laboratory.
5. Finally, and perhaps most important, the molecule should have theoretical accessibility by reason of symmetry or small size so that the data are *useful* in advancing theoretical descriptions of excited state relaxation.

Benzene possesses these characteristics to an unusual degree. Its spectrum is remarkably open for a molecule of its size and the SVL fluorescence

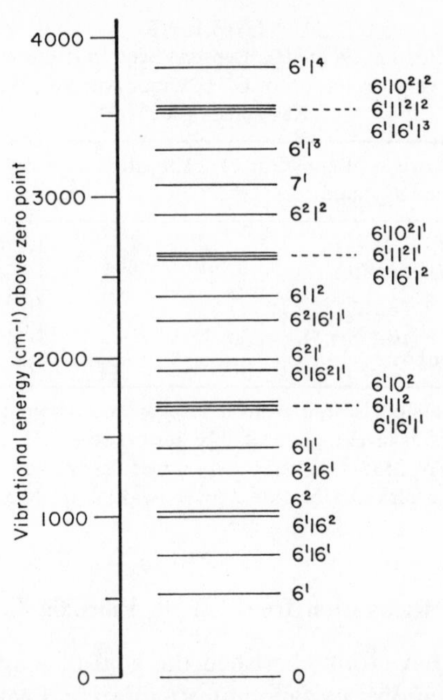

Fig. 14. The $^1B_{2u}$ vibronic levels accessible in C_6H_6 vapor for single vibronic level relaxation experiments with 20 cm^{-1} exciting band widths. See Section III for level notation.

spectra discussed in Section III demonstrate explicitly that single absorption bands can be satisfactorily pumped. They also confirm the assignments of the excited state reached in absorption. The fluorescence is a convenient experimental "handle" to monitor isolated molecule relaxation. The extensive spectroscopy reviewed in Section II indicates that many of the excited state constants necessary for calculation are available, and the high symmetry of benzene provides theoretical accessibility that few other systems of such large size can match. When all of these assets are viewed together with the large body of relaxation data already available from previous work, benzene seems an ideal choice for single vibronic level relaxation experiments.

The potential detail of these single-level studies derives from examination of the absorption spectrum. Figure 14 displays a compilation of discrete $^1B_{2u}$ vibronic levels below 4000 cm^{-1} that can be excited individually by pumping single absorption bands with "monochromatic" light having a bandpass of about 15 cm^{-1}. Sharper exciting sources would extend the list modestly. Not all of those levels in Fig. 14 can be reached with the *absolute* exclusion of other levels, but to a good approximation this is so.

The compilation in Fig. 14 has required a careful examination of the absorption spectrum. As discussed in Section III, some of the previous assignments were wrong, others were ambiguous, and in some regions the assignments were missing. Another problem is caused by overlapping absorption bands. An example concerns reaching levels in which successive quanta of the symmetric ring breathing mode ν_1 are excited. These levels can be reached by pumping the absorption bands $6_1^0 1_0^1$, $6_1^0 1_0^2$, etc., and Radle and Beck's tabulation[144] of band positions shows them well separated from strong neighboring transitions. However, when the spectrum is examined, the situation is seen not to be so simple. Figure 15 shows the B_1^0 band of Radle and Beck which is the absorption $6_1^0 1_0^1$. It is seen that this transition is overlapped by the rotational structure from the $6_0^1 16_1^1$

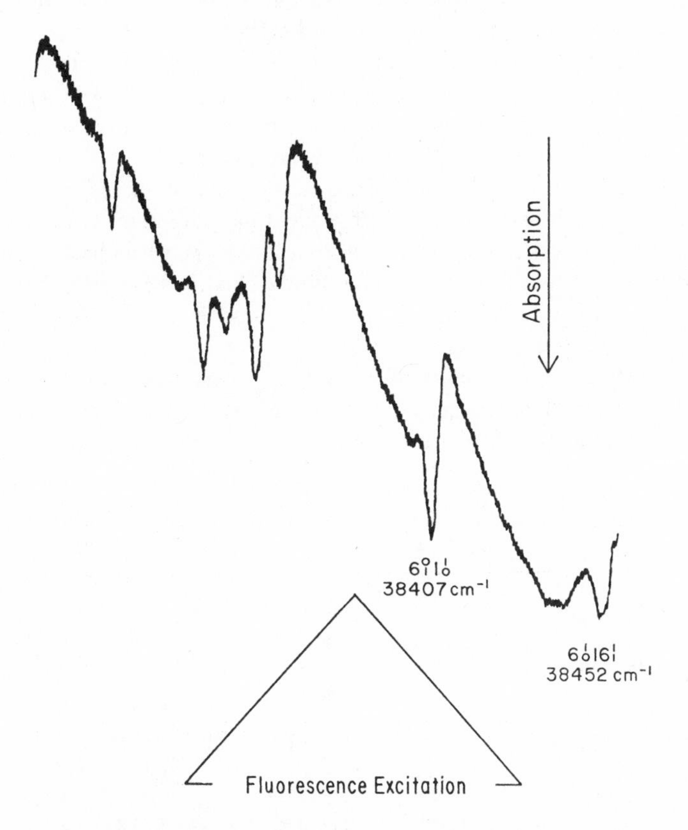

Fig. 15. The absorption spectrum of benzene near 38,400 cm^{-1}. Increasing absorption is down. The maximum assigned to $6_1^0 1_0^1$ is at 38,407 cm^{-1} and the higher energy of the pair assigned to $6_0^1 16_1^1$ is at 38,452.4 cm^{-1}. The rotational tail of the latter extends well underneath the $6_1^0 1_0^1$ absorption band.

transition, whose band maximum is over 40 cm^{-1} away. No source, however monochromatic, could pump the $6^0_1 1^1_0$ band without simultaneous and rather severe excitation of the other transition as well. The problem is confirmed in the resonance fluorescence spectrum shown in Fig. 16. The analysis shows prominent contributions from the excited state $6^1 16^1$ as well as from the 1^1 state. Similar overlapping precludes unique excitation of the overtones $2\nu_1'$, $3\nu_1'$, etc.

The resulting compilation in Fig. 14 shows a smooth and detailed accession of vibrational levels. Fundamentals or overtones or combinations of six modes spread over five symmetries and involving both in- and out-of-plane vibrations are included in the list.

Exploration of fluorescence from those levels is presently proceeding in a number of laboratories. The initial measurements of ϕ_f or τ_{obs} are presented and discussed in six publications.[64,75,84,97,136,145]

The first work[64] described the fluorescence spectra and gave the qualitative assessment of fluorescence yields shown in Fig. 13. A qualitative differentiation among vibronic levels appeared. Nonradiative decay appeared to completely and uniformly dominate relaxation from levels lying above 2500 cm^{-1}, whereas fluorescence spectra could be observed from all levels below this region. Subsequent measurements with improved equipment have extended this limit to the neighborhood of 3000 cm^{-1} but confirm the rapid decline in emission yields in the region 2000–3000 cm^{-1}. Nonradiative channels quickly become dominant as excitation climbs the vibrational ladder.

Quantitative measurements of ϕ_f and τ_{obs} from a few levels have now been published. The lifetimes come from the work of Gelbart et al.[75] and from Selinger and Ware[84] and they are compared in Table VII. The

TABLE VII

Relaxation Parameters of Some Single Vibronic Levels in the $^1B_{2u}(S_1)$ Electronic State of Isolated Benzene Molecules

Approximate exciting wave length (Å)	Vibronic level	Vibrational energy (cm^{-1})	τ_{obs} (nsec) Ware and Selinger[84]	Gelbart et al.[75]	ϕ_f[97]	k_r[b] (10^6 sec^{-1})	k_{nr}[b] (10^6 sec^{-1})
2,667	Zero point	0	90(164)[a]	103	0.22	2.1	7.6
2,589	ν_6	523	80(125)[a]	81	0.27	3.3	9.0
2,529	$\nu_6 + \nu_1$	1,446	68(84)[a]	74	0.25	3.4	10.1
2,471	$\nu_6 + 2\nu_1$	2,372	~50(~69)[a]	57	—	—	—

[a] Data in parentheses is for the indicated vibronic level of C_6D_6.
[b] Calculated from τ_{obs} of Gelbart et al.

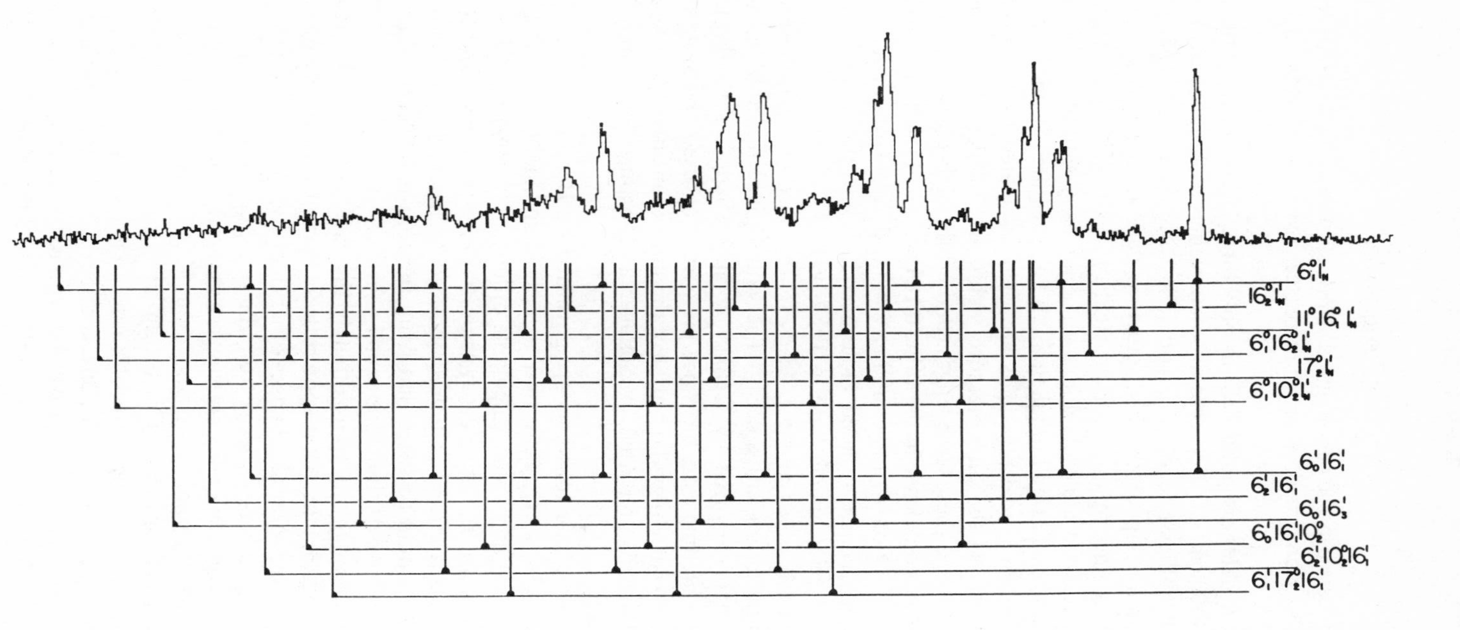

Fig. 16. Resonance fluorescence excited by absorption of the excitation shown by the triangular slit function of Fig. 15. The excitation half-bandpass is 45 cm^{-1} and the band-pass of the fluorescence spectrometer is 35 cm^{-1}. The assignment of structure is separated according to the two excited state origins 1^1 and $6^1 16^1$.

agreement between the sets is good. Both show a monotonic decrease in lifetime as vibrational energy increases from zero point to nearly 2400 cm^{-1}.

Quantum yields are published for only three levels, of which the highest is about 1500 cm^{-1}.[97] Accurate absolute values of fluorescence yields are notoriously difficult to obtain, and it is with these measurements that crucial relationship of the high-pressure data to the isolated molecule data first becomes apparent. Each yield in Table VII derives from comparison of isolated molecule fluorescence intensity with that from benzene with the same excitation but taken to "high" pressures by addition of inert gas. The isolated molecule yields are then calculated from Noyes' conclusion[95] that (1) the high-pressure yields are independent of the state initially excited at these wave lengths and (2) the high-pressure yields equal 0.18. We have confidence in the value 0.18 because of the numerous checks described in Section V, and confidence in (1) derives from direct measurements of the relative isolated molecule yields in other experiments.[66,98] The relative yields match those in Table VII.

Combination of ϕ_f and τ_{obs} leads to the absolute rate constants for radiative and nonradiative decay. The values given in Table VII use the set of lifetimes reported by Gelbart et al.[75] There is little discrepancy between the lifetime sets, but since Gelbart et al. used narrower exciting bandpass and since they did not have the nonexponential decay problems with the level $6^1 1^2$ reported by Selinger and Ware,[84] theirs may be the preferred set.

The radiative rate constants are governed by Herzberg-Teller vibronic mixing. The ${}^1B_{2u}$–${}^1A_{1g}$ transition is principally induced by $\Delta v_6 = \pm 1$ transitions, and as a result calculable differences are expected[84,136] in the rate constants for radiative decay from the zero-point level and the level 6^1. If the radiative decay is *entirely* induced by $\Delta v_6 = \pm 1$ transitions, the rate from 6^1 should be twice that from the zero-point level, whereas the rates from 6^1 and $6^1 1^1$ should be equal.

The data show this is approximately the case. However, the first ratio is only about 1.6, whereas it should be 2. Part of this discrepancy stems from the fact that vibronic changes other than $\Delta \nu_6 = \pm 1$ play a significant role in inducing radiative decay from the zero point level. The assignments of the SVL fluorescence spectra suggest that only about 65% of radiative decay from the zero-point level and about 95% from the level 6^1 is induced by $\Delta v_6 = \pm 1$ transitions. Correction of the rate constants for these crude estimates give the ratio of those components of radiative decay representing $\Delta \nu_6 = \pm 1$ transitions as about 2.3. A more careful examination of the spectra and data will be required to determine whether the discrepancy between 2.3 and 2.0 is due to failure of the theory or improper treatment of the data. It is clear, however, that a unique picture of vibronically induced radiative decay will ultimately derive from the completion of these experiments.

The nonradiative rate constants in Table VII present another magnitude of difficulty. We must first discover what channel or channels these rate constants describe. The data are useless until this question is answered, but the answer is hard to formulate. By definition, nonradiative transitions are hard to see, and those in benzene are no exception. There are no data that directly comment on this problem. The final state of the nonradiative channel has not been seen under the conditions used for the data in Table VII. The closest experiments are the 2536 Å isolated molecule exercises described in Section VIIA, and they show only that at least part of the relaxation from at least one state near 2000 cm^{-1} leads to triplet formation.

Once again, the high-pressure data—particularly that of Noyes and co-workers—plays the key role in the single vibronic level experiments. The data discussed in Section V suggest that the only nonradiative decay channel from the lower levels in the $^1B_{2u}$ state is intersystem crossing to the triplet state. They do not accurately define the domain of "lower" levels, but certainly the zero point and 6^1 level at 523 cm^{-1}, and probably levels somewhat beyond the $6^1 1^1$ level at 1446 cm^{-1} are included. Chemical relaxation cannot be excluded, but the sum of fluorescence and triplet yields from the thermal levels is near unity. Chemical relaxation must be small, and a plausible case can be built for its limitation to levels above 2500 cm^{-1} (Section VI).

It is thus on this tenuous basis that nonradiative relaxation from the three lower levels in Table VII has been equated with intersystem crossing. Although that assignment is in need of further experimental comment, there is nothing in that data or in any that are now coming forward in further studies to refute it.

The nonradiative decay rates reveal only a shallow dependence on vibronic excitation in the initial state, but the data in Table VII are too limited for very significant comment concerning sensitivity to modes and energies. Conceptual frameworks for discussion of vibronic effects on the rates of intersystem crossing have appeared in two papers,[75,145] and the accumulation of data on other levels in benzene will allow a test of these ideals.

The discussion here characterizes the beginning of a unique description of excited state relaxation in a large molecule. At the time of this writing, it is very much in the midst of construction. An extensive array of lifetimes and quantum yields from other vibronic levels will soon be published by Spears and Rice. A similar set of fluorescence yields and fluorescence spectra from many levels will be published by Parmenter, Schuh, and

Schuyler. Lifetime measurements of levels above 3000 cm^{-1} are being obtained by Callomon and co-workers which promise to display the sharp onset of a new nonradiative decay channel. Although all of the experiments are utterly dependent on the results of the more conventional high-pressure and condensed phase studies for interpretation, they extend into a new dimension of detail. The vibronic and even some rotational structure of relaxation channels is clearly visible. It is certain that these and similar studies in other systems will provide much stimulation for further theoretical developments in excited state relaxation.

Acknowledgments

I am grateful to my colleagues Drs. G. H. Atkinson, M. D. Schuh, and M. W. Schuyler for many helpful discussions. The National Science Foundation has provided the necessary financial assistance (Grant GP 22543).

References

1. G. N. Currie and D. A. Ramsay, *Can. J. Phys.*, **49,** 317 (1971) and references therein.
2. J. H. Callomon, T. M. Dunn, and I. M. Mills, *Phil. Trans. Roy. Soc.* (*London*), **259,** Ser. A, 499 (1966).
3. For a recent discussion and review see J. Jortner, S. A. Rice, and R. M. Hochstrasser, *Adv. Photochem.*, **7,** 149 (1969).
4. J. B. Birks, *Photophysics of Aromatic Molecules*, Wiley-Interscience, London, 1970.
5. S. P. McGlynn, T. Azumi, and M. Kinoshita, *Molecular Spectroscopy of the Triplet State*, Prentice-Hall, Englewood Cliffs, N.J., 1969, pp. 141–151.
6. M. Goeppert-Mayer and A. L. Sklar, *J. Chem. Phys.*, **6,** 645 (1938).
7. R. A. Parr, D. P. Craig, and I. G. Ross, *J. Chem. Phys.*, **18,** 1561 (1950).
8. R. Pariser and R. A. Parr, *J. Chem. Phys.*, **21,** 476, 769 (1953).
9. R. Pariser, *J. Chem. Phys.*, **24,** 250 (1956).
10. J. N. Murrell and K. L. McEwen, *J. Chem. Phys.*, **25,** 1143 (1956).
11. J. E. Bloor, J. Lee, and S. Gartside, *Proc. Chem. Soc.* (*London*), **1960,** 413.
12. J. W. Moskowitz and M. P. Barnett, *J. Chem. Phys.*, **39,** 1557 (1963).
13. W. E. Donath, *J. Chem. Phys.*, **40,** 77 (1964).
14. J. M. Schulman and J. W. Moskowitz, *J. Chem. Phys.*, **43,** 3287 (1965).
15. R. J. Buenker, J. L. Whitten, and J. D. Petke, *J. Chem. Phys.*, **49,** 2261 (1968).
16. G. Herzberg, *Electronic Spectra of Polyatomic Molecules*, Van Nostrand, Princeton, N. J., 1966, pp. 555–561, 665–666.
17. T. M. Dunn, *Studies on Chemical Structure and Reactivity*, J. H. Ridd, Ed., Methuen, London, 1966, p. 103.
18. H. Sponer, G. Nordheim, A. L. Sklar, and E. Teller, *J. Chem. Phys.*, **7,** 207 (1939).
19. C. K. Ingold (and in part H. G. Poole and A. P. Best), *J. Chem. Soc.* (*London*), **1948,** 406.
20. B. Honig, J. Jortner, and A. Szöke, *J. Chem. Phys.*, **46,** 2714 (1967).
21. P. R. Monson and W. M. McClain, *J. Chem. Phys.*, **53,** 29 (1970).
22. B. Katz, M. Brith, A. Ron, B. Sharf, and J. Jortner, *Chem. Phys. Lett.*, **2,** 189 (1968).
23. B. Katz, M. Brith, B. Sharf, and J. Jortner, *J. Chem. Phys.*, **52,** 88 (1970).
24. T. M. Dunn and C. K. Ingold, *Nature*, **176,** 65 (1955).

25. P. G. Wilkinson, *Can. J. Phys.*, **34,** 596 (1956).
26. R. Bensassen, R. Bonneau, J. Joussot-Dubien, and J. Faure, *Transitions Non Radiat. Mol., Réunion Soc. Chim. Phys., 20th,* 1969, p. 133 (Suppl. *J. Chim. Phys.* 1969–1970).
27. R. Cooper and J. K. Thomas, *J. Chem. Phys.*, **48,** 5097 (1968).
28. J. B. Birks, *Chem. Phys. Lett.*, **3,** 567 (1969).
29. D. R. Kearns, *J. Chem. Phys.*, **36,** 1608 (1962).
30. J. R. Platt, *J. Mol. Spectry.*, **9,** 288 (1962).
31. A. C. Albrecht, *J. Chem. Phys.*, **38,** 354 (1963).
32. D. F. Evans, *J. Chem. Soc.*, **1957,** 1351, 3885.
33. G. W. King and E. H. Pinnington, *J. Mol. Spectry.*, **15,** 394 (1965).
34. A. Grabowska, *J. Mol. Spectry.*, **20,** 96 (1966).
35. D. M. Burland G. Castro, and G. W. Robinson, *J. Chem. Phys.*, **52,** 4100 (1970).
36. S. D. Colson and E. R. Bernstein, *J. Chem. Phys.*, **43,** 2661 (1965).
37. H. Shull, *J. Chem. Phys.*, **17,** 295 (1949).
38. T. W. Ivanova and B. Y. Sveshnikov, *Opt. i Spektroskopiya*, **11,** 598 (1961); *Opt. Spectry. (USSR)*, **11,** 322 (1961).
39. S. Leach and R. Lopez-Delgado, *J. Chim. Phys.*, **61,** 1636 (1964).
40. G. F. Hatch, M. D. Erlitz, and G. C. Nieman, *J. Chem. Phys.*, **49,** 3723 (1968).
41. J. D. Spangler and N. G. Kilmer, *J. Chem. Phys.*, **48,** 698 (1968).
42. R. Pariser, *J. Chem. Phys.*, **24,** 250 (1956).
43. T. S. Godfrey and G. Porter, *Trans. Faraday Soc.*, **62,** 7 (1966).
44. R. Astier, A. Bokobza, and Y. H. Meyer, *Transitions Non Radiat. Mol., Réunion Soc. Chim. Phys., 20th,* 1969, p. 137 (Suppl. *J. Chim. Phys.* 1969–1970).
45. E. N. Lassettre, A. Skerbele, M. A. Dillen, and K. J. Ross, *J. Chem. Phys.*, **48,** 5066 (1968).
46. R. N. Compton, R. H. Huebner, P. W. Reinhardt, and L. G. Christophorou, *J. Chem. Phys.*, **48,** 901 (1968).
47. J. P. Doering, *J. Chem. Phys.*, **51,** 2866 (1969).
48. E. B. Wilson, Jr., *Phys. Rev.*, **45,** 706 (1934).
49. G. Herzberg, *Infrared and Raman Spectra*, Van Nostrand, Princeton, N.J., 1945, p. 118.
50. S. Brodersen and A. Langseth, *Kgl. danske Vidensk. Selsk., mat.-fys. Skrifter*, **1,** No. 1 (1956).
51. C. K. Ingold, W. R. Angus, C. R. Bailey, J. B. Hale, A. H. Leckie, L. G. Raisin, and J. W. Thompson, *J. Chem. Soc.*, *(London)*, **1936,** 912 ff.
52. C. K. Ingold, C. R. Bailey, A. P. Best, S. C. Carson, R. R. Gordon, J. B. Hale, N. Herzfeld, J. W. Hobden, H. G. Poole, I. H. P. Weldon, and C. L. Wilson, *J. Chem. Soc. (London)*, **1946,** 222 ff.
53. F. M. Garforth, C. K. Ingold, and H. G. Poole, *J. Chem. Soc. (London)*, **1948,** 491.
54. F. M. Garforth and C. K. Ingold, *J. Chem. Soc. (London)*, **1948,** 417.
55. G. H. Atkinson and C. S. Parmenter, forthcoming.
56. J. H. van der Waals, A. M. D. Berghuis, and M. S. de Groot, *Mol. Phys.*, **13,** 301 (1967).
57. A. L. Sklar, *J. Chem. Phys.*, **5,** 669 (1937).
58. G. Herzberg and E. Teller, *Z. physik. Chemie*, **B21,** 410 (1933).
59. G. B. Kistiakowsky and M. Nelles, *Phys. Rev.*, **41,** 595 (1932).
60. G. R. Cuthbertson and G. B. Kistiakowsky, *J. Chem. Phys.*, **4,** 9 (1936).
61. C. K. Ingold and C. L. Wilson, *J. Chem. Soc. (London)*, **1936,** 941.
62. D. P. Craig, *J. Chem. Soc. (London)*, **1950,** 2146.

63. W. L. Smith, *J. Phys.* (*Proc. Phys. Soc.*), **B1**, 89 (1968).
64. C. S. Parmenter and M. W. Schuyler, *Transitions Non Radiat. Mol.*, *Réunion Soc. Chim. Phys.*, *20th*, 1969, p. 92 (Suppl. *J. Chim. Phys.* 1969–1970).
65. C. S. Parmenter and M. W. Schuyler, *J. Chem. Phys.*, **52,** 5366 (1970).
66. M. W. Schuyler, Ph.D. Thesis, Indiana University, 1970.
67. G. H. Atkinson and C. S. Parmenter, *J. Phys. Chem.*, **75,** 1564 (1971).
68. G. H. Atkinson, C. S. Parmenter, and M. W. Schuyler, *J. Phys. Chem.*, **75,** 1572 (1971).
69. F. M. Garforth, C. K. Ingold, and H. G. Poole, *J. Chem. Soc.*, **1948,** 406.
70. J. D. Spangler and H. Sponer, *Spectrochim. Acta*, **19,** 169 (1963).
71. S. Leach, R. Lopez-Delgado, and L. Grajcar, *J. Chim. Phys.*, **63,** 194 (1966).
72. J. R. Henderson, *J. Chem. Phys.*, **41,** 580 (1964).
73. C. S. Parmenter and M. W. Schuyler, forthcoming.
74. G. H. Atkinson, Ph.D. Thesis, Indiana University, 1971.
75. W. Gelbart, K. G. Spears, K. F. Freed, J. Jortner, and S. A. Rice, *Chem. Phys. Lett.* **6,** 345 (1970).
76. H. Ishikawa and W. A. Noyes, Jr., *J. Chem. Phys.*, **37,** 583 (1962); *J. Am. Chem. Soc.*, **84,** 1502 (1962).
77. J. A. Poole, *J. Phys. Chem.*, **69,** 1343 (1965).
78. G. M. Brewer and E. K. C. Lee, *J. Chem. Phys.*, **51,** 3130 (1969).
79. M. Nishikawa and P. K. Ludwig, *J. Chem. Phys.*, **52,** 107 (1970).
80. G. B. Kistiakowsky and C. S. Parmenter, *J. Chem. Phys.*, **42,** 2942 (1965).
81. E. M. Anderson and G. B. Kistiakowsky, *J. Chem. Phys.*, **51,** 182 (1969); **48,** 4787 (1968).
82. L. Logan, I. Buduls, and I. G. Ross, *Molecular Luminescence*, E. C. Lim, Ed., Benjamin, New York, 1969, p. 53.
83. T. Chen and E. W. Schlag, *Molecular Luminescence*, E. C. Lim, Ed., Benjamin, New York, 1969, p. 381.
84. B. K. Selinger and W. R. Ware, *J. Chem. Phys.*, **53,** 3160 (1970); **52,** 5482 (1970).
85. E. C. Lim, *J. Chem. Phys.*, **36,** 3497 (1962).
86. J. W. Eastman, *J. Chem. Phys.*, **49,** 4617 (1968).
87. J. W. Eastman and S. J. Rehfeld, *J. Phys. Chem.*, **74,** 1438 (1970).
88. M. W. Windsor and W. R. Dawson, *Molec. Crystals*, **4,** 253 (1968).
89. C. A. Parker, *Anal. Chem.*, **34,** 502 (1962).
90. C. S. Parmenter and A. H. White, *J. Chem. Phys.*, **50,** 1631 (1969).
91. F. M. Garforth and C. K. Ingold, *J. Chem. Soc.* (*London*), **1948,** 427.
92. L. Logan, Ph.D. Thesis, University of Sydney, Australia, 1968.
93. M. Stockburger, *Z. Physik. Chem.* (*Frankfurt*), **35,** 179 (1962).
94. H. F. Kemper and M. Stockburger, *J. Chem. Phys.*, **53,** 268 (1970).
95. W. A. Noyes, Jr., and D. A. Harter, *J. Chem. Phys.*, **46,** 674 (1967).
96. W. A. Noyes, Jr., W. A. Mulac, and D. A. Harter, *J. Chem. Phys.*, **44,** 2100 (1966).
97. C. S. Parmenter and M. W. Schuyler, *Chem. Phys. Lett.*, **6,** 339 (1970).
98. K. G. Spears and S. A. Rice, forthcoming.
99. C. L. Braun, S. Kato, and S. Lipsky, *J. Chem. Phys.*, **39,** 1645 (1963).
100. M. D. Lumb, C. C. Braga, and L C. Pereira, *Trans. Faraday Soc.*, **65,** 1 (1969).
101. J. B. Birks, J. C. Conte, and G. Walker, *J. Phys.* (*Proc. Phys. Soc.*), **B1,** 934 (1968).
102. J. H. Callomon, private communication.
103. G. M. Almy and P. R. Gillette, *J. Chem. Phys.*, **11,** 188 (1943).
104. D. Phillips, *J. Phys. Chem.*, **71,** 1839 (1967).
105. C. S. Burton and W. A. Noyes, Jr., *J. Chem. Phys.*, **49,** 1705 (1968).

106. C. S. Burton and H. E. Hunziker, *J. Chem. Phys.*, **52,** 3302 (1970).
107. G. M. Brewer and E. K. C. Lee, *J. Chem. Phys.*, **51,** 3615 (1969).
108. R. B. Cundall and T. F. Palmer, *Trans. Faraday Soc.*, **56,** 1211 (1960).
109. C. S. Parmenter and B L Ring, *J. Chem. Phys.*, **46,** 1998 (1967).
110. D. R. Snelling, *Chem. Phys. Lett.*, **2,** 346 (1968).
111. E. K. C. Lee, G. A. Haninger, Jr., and H. O. Denschlag, *Ber. Bunsenges. Phys. Chem.*, **72,** 302 (1968).
112. S. Ho and W. A. Noyes, Jr., *J. Amer. Chem. Soc.*, **89,** 5091 (1967).
113. W. A. Noyes, Jr., and I. Unger, *Adv. Photo. Chem.*, **4,** 49 (1966).
114. R. B. Cundall and D. G. Milne, *J. Amer. Chem. Soc.*, **83,** 3902 (1961).
115. R. B. Cundall, F. J. Fletcher, and D. G. Milne, *Trans. Faraday Soc.*, **60,** 1146 (1964).
116. R. B. Cundall and A. S. Davies, *Trans. Faraday Soc.*, **62,** 1151 (1966).
117. S. Sato, K. Kikuchi, and M. Tanaka, *J. Chem. Phys.*, **39,** 239 (1963).
118. M. Tanaka, T. Terumi, and S. Sato, *Bull. Chem. Soc. Japan*, **38,** 1645 (1965).
119. E. K. C. Lee, H. O. Denschlag, and G. A. Haninger, Jr., *J. Chem. Phys.*, **48,** 4547 (1968); G. A. Haninger and E. K. C. Lee, *J. Phys. Chem.*, **71,** 3104 (1967).
120. W. A. Noyes, Jr., and D. A. Harter, *J. Amer. Chem. Soc.*, **91,** 7587 (1969).
121. W. A. Noyes, Jr., and D. A. Harter, *J. Phys. Chem.*, **75,** 2741 (1971).
122. W. A. Noyes, Jr., and C. S. Burton, *Ber. Bensenges. Phys. Chem.*, **72,** 146 (1968).
123. C. S. Burton and H. E. Hunziker, *Chem. Phys. Lett.*, **6,** 352 (1970).
124. D. Phillips, J. Lemaire, C. S. Burton, and W. A Noyes, Jr., *Adv. Photo. Chem.*, **5,** 329 (1968).
125. K. E. Wilzbach, A. L. Harkness, and L. Kaplan, *J. Amer. Chem. Soc.*, **90,** 1116 (1968).
126. L. Kaplan and K. E. Wilzbach, *J. Amer. Chem. Soc.*, **90,** 3291 (1968).
127. J. H. Callomon in Discussion of Reference 64.
128. E. C. Lim, *J. Chem. Phys.*, **36,** 3497 (1962).
129. J. W. Eastman, *J. Chem. Phys.*, **49,** 4617 (1968); *Z. Naturforsch.*, **A25,** 949 (1970).
130. C. A. Parker and C. G. Hatchard, *Analyst (London)*, **87,** 664 (1962).
131. J. W. Eastman and S. J. Rehfeld, *J. Phys. Chem.*, **74,** 1438 (1970).
132. J. R. Greenleaf and T. A. King, *Proc. Int. Conf. Luminescence*, Akademiai Kiado, Budapest, 1968, p. 212.
133. W. P. Helman, *J. Chem. Phys.*, **51,** 354 (1969).
134. D. M. Burland and G. W. Robinson, *J. Chem. Phys.*, **51,** 4548 (1969).
135. G. W. Robinson, *J. Chem. Phys.*, **47,** 1967 (1967).
136. W. R. Ware, B. K. Selinger, C. S. Parmenter, and M. W. Schuyler, *Chem. Phys. Lett.*, **6,** 342 (1970).
137. R. Williams and G. J. Goldsmith, *J. Chem. Phys.*, **39,** 2008 (1963).
138. P. Pringsheim, *Ann. Acad. Sci. Tech. Varsovie*, **5,** 29 (1938).
139. R. J. Watts and S. J. Strickler, *J. Chem. Phys.*, **44,** 2423 (1966).
140. A. E. Douglas and C. W. Mathews, *J. Chem. Phys.*, **48,** 4788 (1968).
141. P. Sigal, *J. Chem. Phys.*, **42,** 1953 (1965); **46,** 1043 (1967).
142. C. S. Parmenter and H. M. Poland, *J. Chem. Phys.*, **51,** 1551 (1969).
143. L. G. Anderson and C. S. Parmenter, *J. Chem. Phys.*, **52,** 466 (1970).
144. W. F. Radle and C. A. Beck, *J. Chem. Phys.*, **8,** 507 (1940).
145. W. Siebrand, *J. Chem. Phys.*, **54,** 363 (1971).

AUTHOR INDEX

When a number is followed by another in parentheses, the former identifies the page on which the author's work is discussed and the latter indicates the reference number by use of which the complete citations (appearing at the end of each chapter) may be discovered.

SUBJECT INDEX